Getting Started with SketchUp Pro

Embark on your 3D modeling adventure with expert tips, tricks, and best practices

David S. Sellers

<packt>

BIRMINGHAM—MUMBAI

Getting Started with SketchUp Pro

Group Product Manager: Pavan Ramchandani

Senior Editor: Divya Vijayan

Technical Editor: Simran Udasi

Copy Editor: Safis Editing

Project Coordinator: Sonam Pandey

Proofreader: Safis Editing

Indexer: Pratik Shirodkar

Production Designer: Alishon Mendonca

Marketing Coordinator: Marylou De Mello

First published: May 2023

Production reference: 1200423

Published by Packt Publishing Ltd.

Livery Place

35 Livery Street

Birmingham

B3 2PB, UK.

ISBN 978-1-78980-018-0

www.packtpub.com

To my wife and best friend, Alexis Sellers, thank you for supporting me through this long journey. I could not have continued to work on this project without your support and encouragement. And to everyone who never stopped believing that I could complete this project – here it is!

Contributors

About the author

David S. Sellers is a licensed architect and the owner and president of **Virtual Design & Construction Institute** (**VDCI**), a nationally accredited vocational school that provides training and education to AEC professionals. David works extensively with Autodesk, Adobe, and Trimble software, and has 15+ years of experience working with AutoCAD and Revit. David is an Autodesk Certified Instructor, Adobe Certified Instructor, and Certified 3D Warehouse Developer for SketchUp Pro.

I want to thank Packt for providing this opportunity to share SketchUp with a greater audience. I love teaching software, and this is my first attempt at teaching through a textbook. I hope that you find something helpful in this guide and that you find a passion for design software as I have. Thank you, Packt, for remaining patient and vigilant in this effort, and I hope this book was worth the wait!

About the reviewers

Rebecca Terpstra is an experienced interior designer, practicing both residential and commercial (landlord/tenant) design. Her passion as a designer lies in the technical details, using drafting software such as SketchUp to create textbook construction documents. Rebecca has been teaching collegiate-level beginner-to-advanced SketchUp and interior design courses since 2015. She has written effective, industry-based SketchUp curricula for architecture and interior design programs. She is now head of the CAD and interior design programs at Arapahoe Community College in Littleton, Colorado. You can find Rebecca presenting at SketchUp Basecamp or through LinkedIn Learning, teaching SketchUp basics and other SketchUp courses aimed at interior designers.

Alessandro Barracco is an aerospace engineer and architect (automatic BIM modeling from CAD drawings) and has a Ph.D. in the technology and management of aeronautical infrastructure – artificial intelligence. He is currently an assistant professor at the University of Palermo, the Kore University of Enna, and SUPINFO Paris, where he teaches informatics, mathematics, and Python programming. He is an IEREK scientific committee member (Egypt) and was a guest editor for the Parallelism in Architecture, Engineering & Computing Techniques conference at UEL (UK) in 2016. He is a member of the editorial board of several journals edited by John Wiley & Sons (US). He is a SketchUp Certified Specialist, the CEO of TECLAsoftware, and the author of several extensions for SketchUp. He wrote *SketchUp for Architecture*, a guide for design from concept to BIM models, published by Apogeo, in Italy.

Table of Contents

Part 2 – Views, Animations, and Materials

6

7

8

Materials   311

Part 3 – Advanced Modeling and Model Organization

9

Entity Info, Outliner, and Tags Dynamically Organize Your Models   387

10

Model Info and Preferences   429

11

Working with Components   469

12

Import, Export, 3D Warehouse, and Extensions   525

Preface

Welcome to *Getting Started with SketchUp Pro*, a comprehensive guide to learning SketchUp Pro for a range of 3D modeling applications. Whether you're a beginner or an experienced user, this book will provide you with the knowledge and skills you need to create high-quality 3D models with ease.

SketchUp Pro is a versatile and powerful 3D modeling software that is easy to learn and use, making it perfect for beginners and professionals alike. Whether you're an architect, interior designer, woodworker, or simply interested in 3D modeling, SketchUp Pro can help bring your ideas to life with its intuitive interface and robust set of tools.

In this book, we will guide you through every step of the SketchUp Pro workflow, starting with the basics of the user interface and working through the common toolbars. We will then cover Default Tray panels and advanced workflows, giving you a comprehensive understanding of SketchUp Pro's capabilities.

Our focus will be on a range of 3D modeling applications, including architecture, interior design, woodworking, landscape design, and industrial design, giving you the skills you need to create 3D models in a variety of settings. We'll provide you with practical examples and real-world scenarios to help you apply your new skills to your own projects.

By the end of this book, you'll have developed a strong understanding of SketchUp Pro and the confidence to tackle any 3D modeling project. Whether you're a student, a professional, or an enthusiast, we are confident that this book will serve as an invaluable resource for you. Let's dive into the fascinating realm of SketchUp Pro and unleash your creative potential!

Who this book is for

This book is for anyone interested in creating their own models in SketchUp Pro. SketchUp Pro is used professionally by architects, interior designers, landscape architects, woodworkers, hobbyists, industrial designers, cabinetmakers, and many more professions. However, SketchUp is not unapproachable or difficult to learn, and I would encourage anyone to try and make something for themselves. SketchUp is a versatile platform and can create drawings, renderings, animations, and 3D prints.

What this book covers

Chapter 1, Beginning with SketchUp Pro, takes a visual tour through the SketchUp Pro user interface, looking at the general structure, toolbars, and dialog boxes. In addition, this chapter will discuss how to use SketchUp tools and examines the rules that apply to all SketchUp geometry.

Chapter 2, Principal Tools, Axes, and Inferences, examines the Principal Tools used in almost every SketchUp workflow. We will also understand the importance of using Axes and Inferences while drawing and editing SketchUp geometry.

Chapter 3, Modeling with Groups and Components, discusses the principles of modeling in SketchUp when it comes to "sticky geometry." We will also discuss the importance of modeling using groups and/or components.

Chapter 4, Drawing Tools – We Begin Modeling!, begins with modeling! More importantly, we will start to understand how to use the drawing tools to make sure we can effectively create basic geometry.

Chapter 5, Editing Tools – Making Big Changes!, uses the Editing Tools to make changes to basic geometry. We will talk about how we can use the Editing Tools to make changes to our models and how to create new Edges and Faces using existing geometry.

Chapter 6, Camera Options, examines 3D camera options in SketchUp Pro. We will discuss the Perspective, Two-Point Projection, and Parallel Projection views as we look at example geometry. We will also analyze the available Camera Tools to introduce different ways to place the SketchUp camera and navigate the model.

Chapter 7, View Options, examines the various options to change the view in SketchUp Pro. We will explore Styles, including the Edge, Face, Component Styles, Shadows, and Fog, including how this can be set by a Geo-location. We will also discuss Sections that allow for dynamic views of a SketchUp model. Finally, we will review how to save these view options into snapshots, known as Scenes, which can be animated!

Chapter 8, Materials, looks at Materials in SketchUp Pro. We will discuss how to use the Paint Bucket Tool and the Materials Panel to select and apply Materials to Objects. We will look at creating Materials as colors and as image imports, and we will finish the chapter by looking at advanced texture Material editing techniques.

Chapter 9, Entity Info, Outliner, and Tags Dynamically Organize Your Models, examines three Default Tray panels: Entity Info, Outliner, and Tags. We will discuss how the Entity Info panel can be used to view and edit geometry attributes. Then, we will discuss the Outliner panel and start to see our models in a new way. We will also look at the Tags panel (formerly called the Layers panel), which can be used to assign visibility options to sets of geometry.

Chapter 10, Model Info and Preferences, looks at two advanced dialog boxes in SketchUp Pro: Model Info and Preferences. We will discuss how to utilize the Model Info dialog box to update settings for the current file, including Objects, Statistics, and Units. We will also discuss utilizing the Preferences dialog box to change overall SketchUp settings, including Accessibility, Drawing, and General.

Chapter 11, Working with Components, explores Components and what makes them special in SketchUp Pro. Components allow us to keep our files more organized, easily use Objects in multiple files, and automatically align geometry in our models. Component attributes can be set and edited to achieve many different functions in SketchUp Pro. We will wrap up this chapter by briefly discussing Dynamic and Live Components.

Chapter 12, Import and Export, 3D Warehouse, and Extensions, talks about the different types of files that SketchUp Pro can import and the different options for exporting information from our SketchUp models, including images, 2D and 3D files, and animations. We will also talk about the 3D Warehouse for downloading and uploading files and sending our SketchUp files to LayOut. We will finish by looking at Extensions in the Extension Warehouse and Extension Manager.

To get the most out of this book

This book is a hands-on exploration of SketchUp Pro, so it is recommended to have a version of SketchUp Pro for Windows **operating system** (**OS**) installed on your computer. This book includes features from the latest version of SketchUp Pro, but older versions of SketchUp Pro will be compatible with this book. Using a full keyboard and mouse with left-click, right-click, and a scroll wheel is highly recommended.

Software/hardware covered in the book	Operating system requirements
SketchUp Pro 2023	Windows

The SketchUp for Web and SketchUp Make user interfaces are different from SketchUp Pro. Additionally, multiple workflows are discussed in this book that are exclusive to SketchUp Pro, including DWG import.

Download the color images

We also provide a PDF file that has color images of the screenshots and diagrams used in this book. You can download it here: `https://packt.link/CHJYk`.

Conventions used

There are a number of text conventions used throughout this book.

`Code in text`: Indicates code words in text, database table names, folder names, filenames, file extensions, pathnames, dummy URLs, user input, and Twitter handles. Here is an example: "If working in a feet and inches file, you can also type mm (millimeters), `cm` (centimeters), m (meters), and `yd` (yards)."

Bold: Indicates a new term, an important word, or words that you see onscreen. For instance, words in menus or dialog boxes appear in **bold**. Here is an example: "Axes can also be hidden by going to the **View** menu dropdown and choosing **Axes**."

> **Tips or important notes**
> Appear like this.

Get in touch

Feedback from our readers is always welcome.

General feedback: If you have questions about any aspect of this book, email us at customercare@ packtpub.com and mention the book title in the subject of your message.

Errata: Although we have taken every care to ensure the accuracy of our content, mistakes do happen. If you have found a mistake in this book, we would be grateful if you would report this to us. Please visit www.packtpub.com/support/errata and fill in the form.

Piracy: If you come across any illegal copies of our works in any form on the internet, we would be grateful if you would provide us with the location address or website name. Please contact us at copyright@packt.com with a link to the material.

If you are interested in becoming an author: If there is a topic that you have expertise in and you are interested in either writing or contributing to a book, please visit authors.packtpub.com.

Share Your Thoughts

Once you've read *Getting Started with SketchUp Pro*, we'd love to hear your thoughts! Scan the QR code below to go straight to the Amazon review page for this book and share your feedback.

https://packt.link/r/1789800188

Your review is important to us and the tech community and will help us make sure we're delivering excellent quality content.

Download a free PDF copy of this book

Thanks for purchasing this book!

Do you like to read on the go but are unable to carry your print books everywhere?

Is your eBook purchase not compatible with the device of your choice?

Don't worry, now with every Packt book you get a DRM-free PDF version of that book at no cost.

Read anywhere, any place, on any device. Search, copy, and paste code from your favorite technical books directly into your application.

The perks don't stop there, you can get exclusive access to discounts, newsletters, and great free content in your inbox daily

Follow these simple steps to get the benefits:

1. Scan the QR code or visit the link below

https://packt.link/free-ebook/9781789800180

2. Submit your proof of purchase
3. That's it! We'll send your free PDF and other benefits to your email directly

Part 1 –
User Interface and
Beginning Modeling!

The learning objectives in *Part 1* of this book are to help you understand the basic functionalities of SketchUp Pro and to get you familiar with navigating the user interface. We will begin by reviewing the user interface and Principal tools, including new tools introduced in 2023. Then, we will expand on the "rules" of SketchUp Pro, including how to model with Groups and Components. *Part 1* also includes an introduction to all of the tools on the Drawing and Editing toolbars, including the Line, Rectangle, Arc, Move, Push/Pull, and Flip tools.

This part has the following chapters:

- *Chapter 1, Beginning with SketchUp Pro*
- *Chapter 2, Principal Tools, Axes, and Inferences*
- *Chapter 3, Modeling with Groups and Components*
- *Chapter 4, Drawing Tools – We Begin Modeling!*
- *Chapter 5, Editing Tools – Making Big Changes!*

1
Beginning with SketchUp Pro

Welcome to SketchUp Pro! I am excited to work through this introductory book with you.

In this book, we will examine the tools, workflows, and best practices to get started with using SketchUp Pro. SketchUp Pro is powerful software for anyone interested in 3D modeling and design, 3D printing, and even visualization and rendering! We will start with the basics and slowly work our way through to showcase some advanced professional workflows at the end of the book.

In this chapter, we will take a visual tour through the SketchUp Pro User Interface, looking at the general structure, toolbars, and dialog boxes. Then, we will discuss navigating through a 3D SketchUp model using some standard SketchUp tools but, more importantly, the keyboard and mouse. Finally, we will look at some of the underlying rules that apply to all SketchUp geometry. We will start with a quick discussion about the available SketchUp versions you may have seen online.

The following topics are covered:

- SketchUp Versions
- SketchUp Pro User Interface
- Understanding toolbars and the default tray
- Beginning to navigate through SketchUp
- SketchUp geometry basics and rules

Let's get started!

SketchUp Versions

SketchUp Pro is a desktop software and is the traditional format for utilizing SketchUp. Recently, SketchUp has released a new way to access SketchUp through a web browser app, known as SketchUp for Web. While the newer version of SketchUp will be more accessible and does not require any installation time, SketchUp Web does not support Extensions or desktop integrations. Because of this, SketchUp Pro is the preferred software for working professionals.

SketchUp for Web has been included in many pricing models outlined on the SketchUp website. This includes SketchUp Go, SketchUp Shop, and SketchUp for Schools. SketchUp Studio has also been introduced, which includes SketchUp Pro and SketchUp for Web as well as a collection of design and analytics tools that go beyond the scope of this book.

A full comparison of the different SketchUp options can be found on the SketchUp website.

You can see the interfaces for SketchUp Pro and SketchUp for Web in the following figures. We will discuss the parts of the SketchUp Pro User Interface later in this chapter.

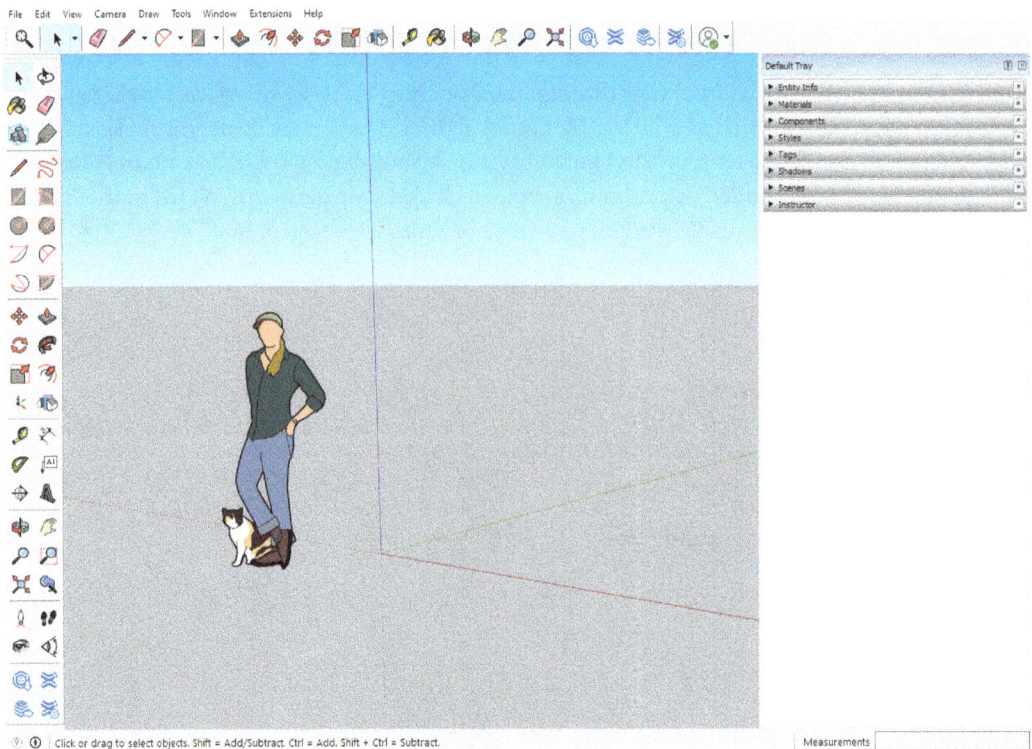

Figure 1.1: SketchUp Pro User Interface

You can see that SketchUp for Web needs to be run in a web browser and in this case, it is Microsoft Edge:

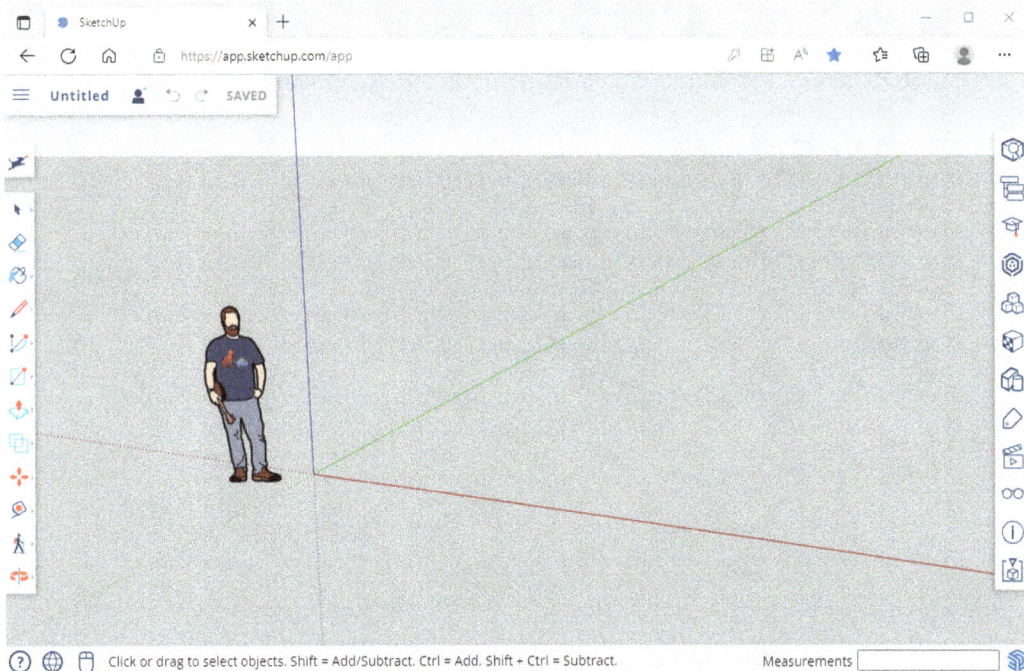

Figure 1.2: SketchUp for Web User Interface

Exploring the SketchUp Pro User Interface

SketchUp Pro is available for both the Windows OS and macOS environments. This book will focus on using SketchUp Pro in the Windows OS environment using a PC keyboard (with *Shift*, *Ctrl*, and *Alt* keys) and a 3-button mouse.

SketchUp provides quick reference guides for both Windows and macOS on their website. I would suggest downloading the Quick Reference Card for your preferred software. This book will exclusively focus on the SketchUp Pro interface toolbars and buttons, and keyboard and mouse controls for the Windows OS environment.

> **Note**
>
> It will be possible to use this book while working in SketchUp Pro in a macOS environment, but you will notice differences in the User Interface and keyboard and mouse shortcuts!

SketchUp Pro Quick Reference Card | Windows   SketchUp 2022

Large Tool Set

Select (Spacebar)	Make Component	
Paint Bucket (B)	Eraser (E)	
Line (L)	Freehand	
Rectangle (R)	Rotated Rectangle	
Circle (C)	Polygon	
Arc	2 Point Arc (A)	
3 Point Arc	Pie	
Move (M)	Push/Pull (P)	
Rotate (Q)	Follow Me	
Scale (S)	Offset (F)	
Tape Measure (T)	Dimensions	
Protractor	Text	
Axes	3D Text	

Orbit (O)   Pan (H)
Zoom (Z)   Zoom Window
Zoom Extents   Previous
Position Camera   Walk
Look Around   Section Plane

Solid Tools
Outer Shell   Intersect (Pro)
Union (Pro)   Subtract (Pro)
Trim (Pro)   Split (Pro)

Sandbox (Terrain)
From Contours   From Scratch
Smoove   Stamp
Drape   Add Detail
Flip Edge

Standard Views
Iso
Front
Back
Top
Right
Left

Style
X-Ray
Wireframe
Shaded
Monochrome
Back Edges
Hidden Line
Shaded with Textures

Dynamic Components
Interact
Component Options
Component Attributes

Location
Add Location
Toggle Terrain

Warehouse
3D Warehouse
Share Component
Send to LayOut (Pro)
Extension Warehouse
Share Model
Classifier (Pro)

Middle Button (Wheel)

Scroll	Zoom
Click-Drag	Orbit
Shift+Click-Drag	Pan
Double-Click	re-center view

Tool	Operation	Instructions
2 Point Arc (A)	Bulge	specify bulge amount by typing a number and Enter
	Radius	specify radius by typing a number, the R key, and Enter
	Segments	specify number of segments by typing a number, the S key, and Enter
3 Point Arc	Alt +	use Option '+' or Option '-' to change the number of segments
	Arrows	lock direction; up = blue, right = red, left = green, and down = parallel/perpendicular
Circle (C)	Shift	lock current inferences
	Radius	specify radius by typing a number and Enter
	Segments	specify number of segments by typing a number, the S key, and Enter
Eraser (E)	Ctrl	soften/smooth (use on edges to make adjacent faces appear curved)
	Shift	hide
	Alt	unsoften/unsmooth
Follow Me	Alt	use face perimeter as extrusion path
	Expert Tip!	first Select path, then choose the Follow Me tool, then click on the face to extrude

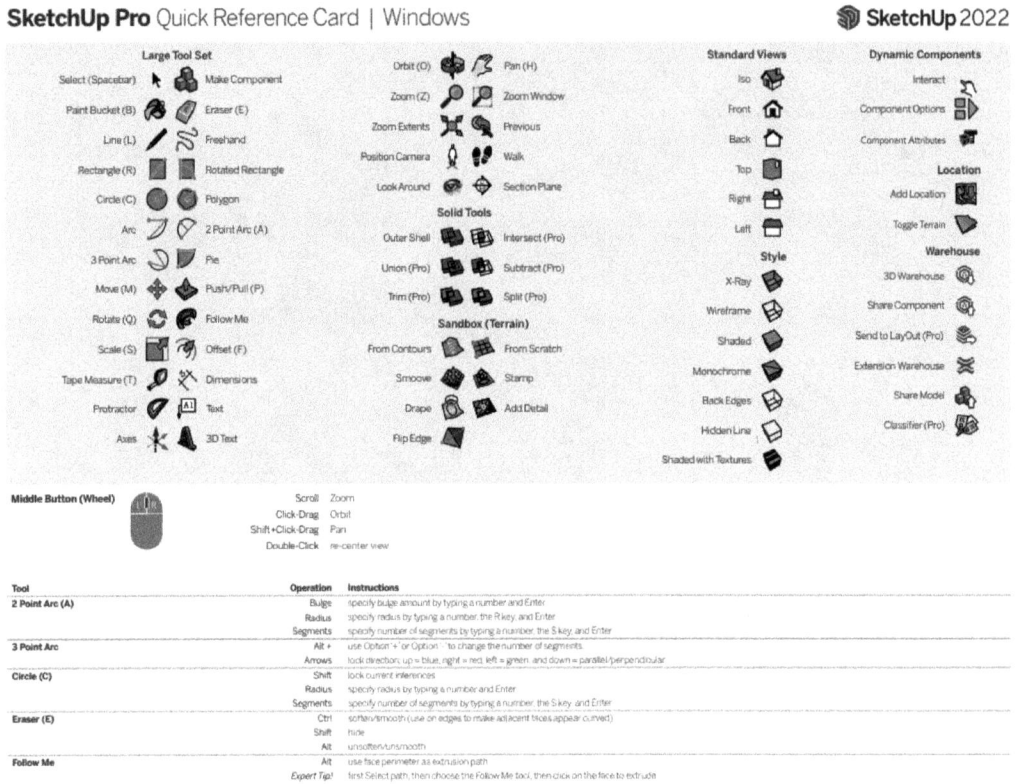

Figure 1.3: SketchUp Pro Quick Reference Card for Windows OS

> **Note**
> You can find the enlarged image here – `https://download.sketchup.com/HC-QRC2022-en-SU-win-v01.pdf`.

The SketchUp User Interface is composed of six main elements:

1. Drawing Area
2. Title Bar
3. Menu Bar
4. Toolbars (the Getting Started Toolbar)
5. Status Bar (Prompts & Measurements)
6. Default Tray (containing Default Panels)

Each of these elements is identified in the following screenshot:

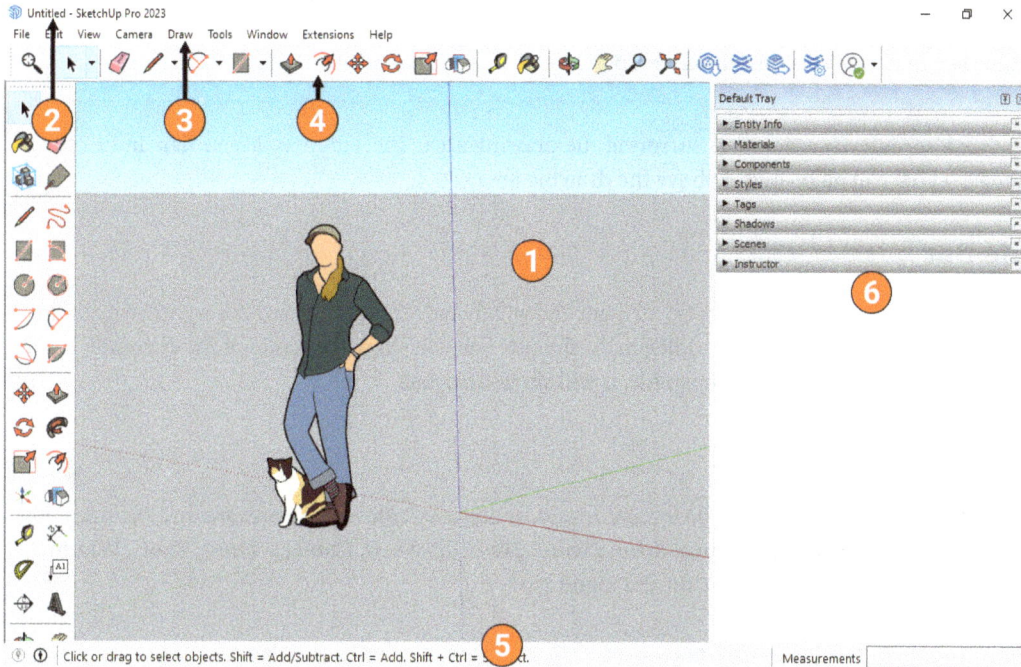

Figure 1.4: SketchUp Pro User Interface with Labeled Elements

Let's look at each of them in detail.

Drawing Area

The largest part of the User Interface is the drawing area, which is the large modeling section in the middle of the screen. This is where all SketchUp geometry is created and edited! The drawing area is always a fully 3D environment, meaning that you can draw and move geometry in any direction, and you can move the camera to view geometry at any distance or angle.

The drawing area will contain the drawing axes (red, green, and blue lines) and typically will show a model of a person to get a sense of scale. All tools are used by interacting with the drawing area. We will talk more about templates and using the drawing area in *Chapter 3, Modeling with Groups and Components*.

> **Note**
> The drawing area traditionally had a green ground and blue sky in older versions of the Simple SketchUp Pro template. This template has been updated to have a blue-gray ground and bright blue sky in newer versions of SketchUp.

All other User Interface elements surround the drawing area. The Title Bar, Menu Bar, and Getting Started Toolbar can all be found above the drawing area.

Title Bar

The title bar is a standard Windows OS element and will include standard Windows controls, including minimize, maximize, and close. Additionally, the title bar will show the name of the currently open file. If you have not yet saved an open file, it will show **Untitled**.

Menu Bar

The menu bar is another standard element found in many Windows OS software and includes the majority of all SketchUp tools in drop-down menus. **File**, **Edit**, **View**, **Camera**, **Draw**, **Tools**, **Window**, and **Help** are the main menu titles for the menu bar:

Figure 1.5: Title Bar (Top), Menu Bar (Middle), and Getting Started Toolbar (Bottom)

Getting Started Toolbar and Other Toolbars

The **Getting Started** toolbar is meant to provide a visual and linear guide for creating 3D models in SketchUp Pro. The Getting Started toolbar is the only toolbar that is displayed when opening SketchUp Pro for the first time. All toolbars in SketchUp Pro follow the same visual design, with colorful icons that represent the tools.

While the **Getting Started** toolbar contains an excellent guide for working linearly through a SketchUp project, it does not contain all tools that are in SketchUp Pro. Additional toolbars can be displayed by right-clicking on the gray space directly to the right of the getting started toolbar and selecting additional toolbars. Additionally, toolbar visibility can be toggled by doing the following:

1. Accessing the Menu Bar
2. Selecting **View | Toolbars**

This workflow will open the Toolbar dialog box:

Figure 1.6: Toolbar Right-Click Contextual Menu and Toolbar Toggle Options

More toolbars will be introduced alongside professional workflows as we move through this book. It is recommended that the getting started toolbar remains visible and in the original position. However, all toolbars can be hidden or displayed, can be moved to dock at the sides of the drawing area, or can float in front of the drawing area.

To move a toolbar, do the following:

1. Click and hold on the line of small dots on the left or top of the toolbar.
2. Drag the toolbar to the preferred location.

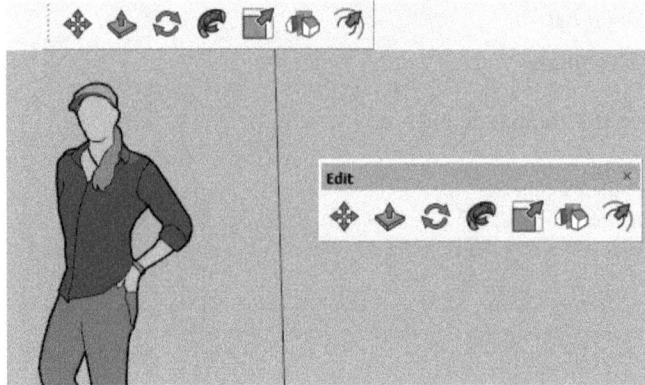

Figure 1.7: "Docked" Toolbar with Small Dots (Left) and "Floating" Toolbar (Right)

Turn on Toolbars and try moving them around for yourself! SketchUp Pro users will often have different sets of Toolbars open – you will need to find the right Toolbars and locations for your workflows.

Status Bar

The Status Bar can be found at the bottom of the screen. The Status Bar contains the Geo Location button, the Attribution button, Prompts, Measurement Label, and Measurements Value. These elements cannot be moved to other locations in the User Interface but can have their display toggled by right-clicking on the Status Bar.

> **Note**
> The Status Bar may appear to be frozen at times. This may be because SketchUp Pro is performing an autosave of the SketchUp Model. Try to back out of the tool and redo the workflow.

For now, we will focus on the prompts (or tooltips) and the measurements value. Both the prompts and measurements will constantly update as you work through a SketchUp Model. It is essential to know how to read and use both elements.

Prompts

The prompts will clarify what to do next while using a tool. Even before a tool is selected, the prompt will give a brief description of the tool's function. Additionally, the prompt will provide options that expand on the primary function of the tool, typically accessed by using the *Shift*, *Ctrl*, and *Alt* keys on the keyboard.

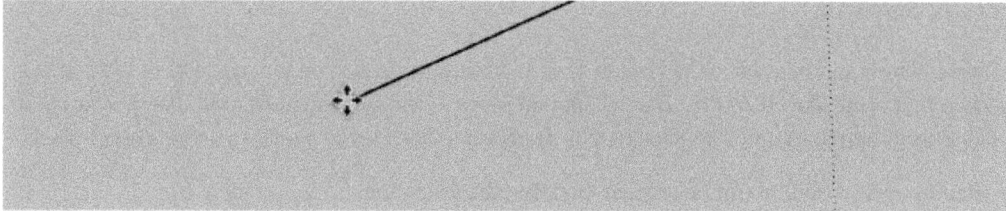

Pick two points to move. Ctrl = toggle Copy, hold Shift = lock inference, Alt = toggle Auto-fold.

Figure 1.8: SketchUp Pro Status Bar Prompt for the Move Tool

Measurements

The measurements label and value (measurements box collectively) should be used when creating precise geometry. SketchUp geometry can occasionally be created without specific dimensions, but typically professional projects require precise dimensions.

> **Note**
>
> SketchUp Pro is incredibly precise, down to 15 decimal places. For reference, writing 1/64 as a decimal only requires 6 decimal places (0.015625).

The measurements value will behave similarly to the prompts in that it will update to show different values depending on the type of tool being used. The value will be a distance (dimensional) value when drawing, offsetting, or extruding with the Push/Pull tool. Similar to showing a Distance value, the measurements box value can also be an angle, like when using the Rotate or Protractor tool. Additionally, the measurements box value could represent a whole number like when using the Copy tool to create an array.

We will use the measurements box often during this book and so it would be good practice to recognize what the measurements label is prompting during different workflows:

Figure 1.9: SketchUp Pro Status Bar with Measurements Box Showing a Typed Distance (Dimensional) Value

> **Note**
> SketchUp Pro will recognize Imperial and Metric units in the same SketchUp model. If working in a feet and inches file, you can also type mm (millimeters), cm (centimeters), m (meters), and yd (yards).

Default Panels

The Default Panels are collected in the Default Tray, and these are docked on the right side of the drawing area when first using SketchUp Pro. The default panels are essentially the most used dialog boxes that are tied to tools and workflows, so SketchUp is designed to keep these panels in a convenient place.

The default panels' visibility can be toggled by doing the following:

1. Accessing the menu bar

2. Selecting **Window | Default Tray**

3. Selecting the appropriate panels:

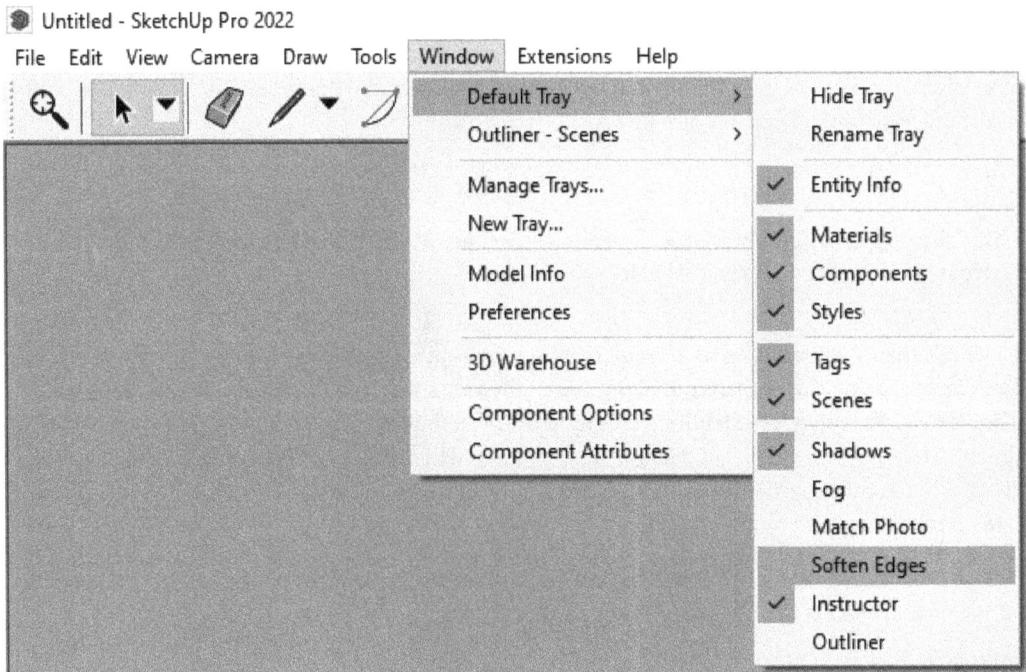

Figure 1.10: Menu showing Visibility Options for the Default Tray

The **Manage Trays…** option in the **Window** menu will also open a more comprehensive dialog box for managing the **Default Tray**:

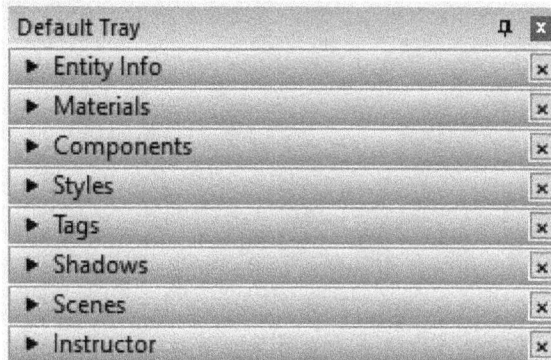

Figure 1.11: SketchUp Pro Default Tray

The default tray is located on the right of the screen to start, but this tray can be located on any side of the drawing area or can float above the drawing area.

To move the default tray, do the following:

1. Click and hold on the blue section at the top of the tray.
2. Drag to the preferred location. Alternatively, drag on top of the placement guides that will appear.

The panels in the default tray can be minimized or expanded by clicking on the title text or on the black arrow to the left of the title text. Panels can also be hidden by hitting the small gray "X" button, but they must be unhidden (shown) by accessing the menu as discussed at the beginning of this section.

> **Note**
> The Default Tray can also be "minimized" by unpinning the Tray. Trays can be unpinned by clicking the Pin icon in the top blue bar, which will "minimize" the bar to the side of the screen. When the Tray is hovered over with the mouse, the Tray will appear. This is helpful for smaller screens where the screen can get crowded.

Default tray panels will be explored as we move through this book. However, it would be a good idea to view the Instructor panel while going through this book, as the Instructor panel provides helpful animations and expanded tooltips for the SketchUp tools:

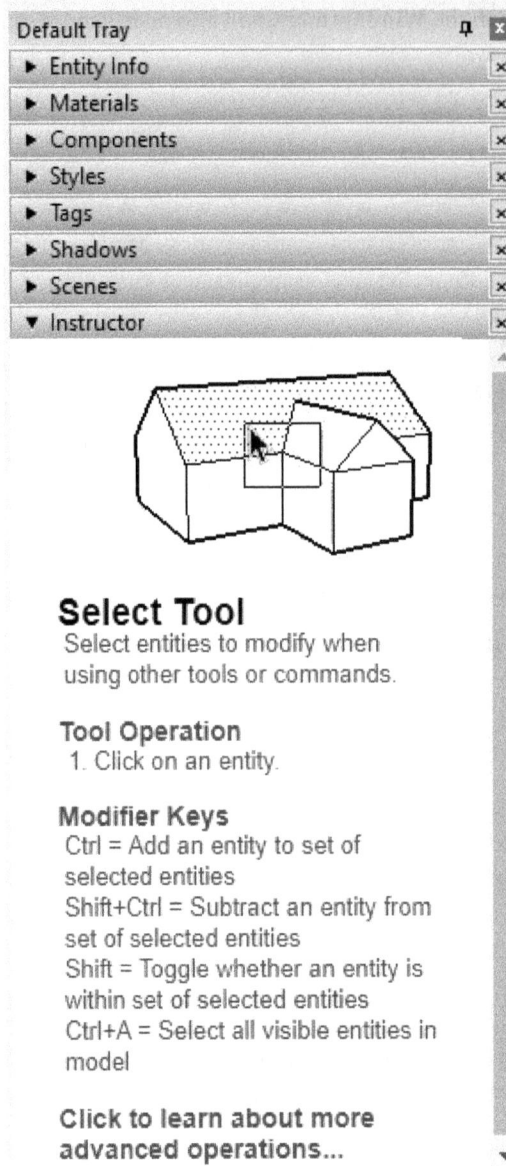

Figure 1.12: Instructor Panel Included in the Default Tray

Note

Additional User Interface elements, such as dialog boxes and context menus, will be explored later in this book. As you move forward with your skills in SketchUp Pro, this Tray can be hidden as screen space is highly valuable when modeling.

The SketchUp Pro User Interface may seem complicated at first! Please take some time to explore, move toolbars around, and begin to get comfortable knowing where things are. You could even start clicking on tools to read the Prompts and see the Measurements Box update. Don't worry, we'll discuss all of the basic tools as we move through the next few chapters.

The first set of tools that we must become familiar with is the Navigation tool set. SketchUp is a 3D modeling program, and we first need to learn how to navigate through our 3D space!

Navigating in SketchUp Pro

Navigation is required in all steps of a SketchUp workflow, from drawing and editing to visualizing the model. The primary navigation tools can be found on the Getting Started Toolbar, and these are the Orbit, Pan, and Zoom tools. While we will discuss how to use these tools, we will also discuss the relevant keyboard and mouse shortcuts that can be used to activate these tools. Using the keyboard and mouse shortcuts will drastically increase modeling productivity for all tools, but especially for the navigation tools.

We will also explore some other ways of navigating through SketchUp, which includes introducing the Views Toolbar, Zoom Extents, Zoom Window, and Zoom Previous. While these options are used far less than Orbit, Pan, and Zoom, they all have unique use cases that you could find helpful as you navigate your SketchUp models.

> **Note**
>
> The Orbit, Pan, and Zoom tools all use the scroll wheel on a 3-button mouse. Please make sure that you are using a 3-button mouse while using SketchUp Pro. When the tools are described, "scroll" means to roll the scroll wheel, and "press" means to click down and hold the scroll wheel to use it like a button.

Figure 1.13: Orbit, Pan, Zoom, and Zoom Extents on the Getting Started Toolbar

Orbit

The Orbit tool is the most essential tool for working effectively in SketchUp. SketchUp requires that the drawing area camera be placed at specific angles to create and edit geometry, and orbiting the camera is a quick way to align your view with the desired modeling result.

The Orbit tool button looks like a vertical black line being "orbited" by red and green arrows:

Figure 1.14: Orbit Tool Button

The Orbit tool can be activated by doing the following:

1.   Clicking the orbit button

2.   Pressing the left mouse button

3.   Moving the mouse

Or, at any time (including while using any other tool), the Orbit tool can be activated by doing the following:

1.   Pressing the scroll wheel

2.   Moving the mouse

Pressing and dragging using the middle mouse wheel is the most common and effective way of using the Orbit tool!

Figure 1.15: Orbit Tool Reference Guide for 3-Button Mouse

> **Tip**
> Try moving the camera with the Orbit tool! It will take some practice. Orbit will always rotate the camera around the middle of the screen, so it is best to practice using Orbit alongside the Pan and Zoom tools.

Pan

The Pan tool is a perfect companion to the Orbit tool. While the orbit tool rotates the camera, the Pan tool moves the camera laterally from side to side or up and down. This means that you will always face the same direction when using the Pan tool while the camera slides around the 3D environment.

The Pan tool button is an open hand:

Figure 1.16: Pan Tool Button

Just like the Orbit tool, the Pan tool can be activated by doing the following:

1. Clicking the Pan button
2. Pressing the left mouse button
3. Moving the mouse

Or, at any time (including while using any other tool), the Pan tool can be activated by doing the following:

1. Pressing the scroll wheel
2. Pressing *Shift* at the same time
3. Moving the mouse

Just make sure that you hold down *Shift* at the same time as the mouse scroll wheel while you move the mouse to pan!

Figure 1.17: Pan Tool Reference Guide for 3-Button Mouse

Tip

The Pan tool and the Orbit tool work the same way with the 3-button mouse; just add *Shift* to activate the Pan tool! You can even try holding the scroll wheel and alternating between pressing *Shift* and releasing *Shift* to switch back and forth between Orbit and Pan. Note, however, that this method is not commonly used in professional workflows.

Zoom

The Zoom tool is the third primary navigation tool alongside the Orbit and Pan tools. While Pan and Orbit move the camera side to side, up and down, and around objects in the drawing area, the Zoom tool moves the camera in and out. The Zoom tool is not moving the objects – it is sliding the camera!

The Zoom tool button is a magnifying glass. There are other zoom tools, and these also have a magnifying glass, but the Zoom tool is the tool with only the magnifying glass:

Figure 1.18: Zoom Tool Button

The Zoom tool is easy to use! The Zoom tool can be activated by doing the following:

1. Clicking on the button
2. Pressing the left mouse button
3. Move the mouse up and down

Or, at any time (including while using any other tool), the Pan tool can be activated by scrolling the mouse scroll wheel:

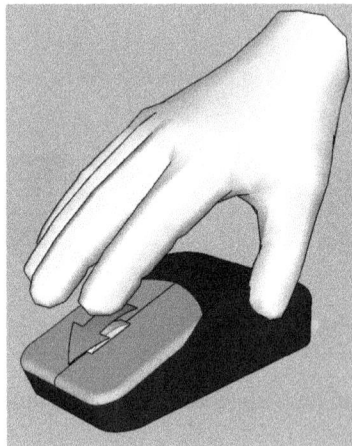

Figure 1.19: Zoom Tool Reference Guide for 3-Button Mouse

> **Tip**
> Scrolling the wheel "up" will zoom in and scrolling the wheel "down" will zoom out. Pointing the mouse button at Geometry on the screen will make SketchUp zoom quickly, and pointing the mouse at empty space will make SketchUp zoom slowly. Try this for yourself!

Field of View

The camera in SketchUp Pro has a default Field of View value of **35.00**. When activating the Zoom tool, you will see the Measurements Box update to have a Field of View value.

The Field of View can be changed by using the Field of View tool in the Menu Bar under the camera. In SketchUp Pro 2023 a Field of View button was added to the Large Tool Set Toolbar:

Figure 1.20 Field of View Button on the Large Tool Set Toolbar

The Field of View can be adjusted by pressing and holding the Left mouse button and dragging up or down to change the field of view.

Alternatively, the Zoom tool can activate the Field of View tool. Do the following:

1. Pressing and holding *Shift*
2. Pressing and holding the left mouse button
3. Moving the mouse

Field of View can be changed at any time while you are modeling, but it is not recommended to do so unless it is necessary for the project.

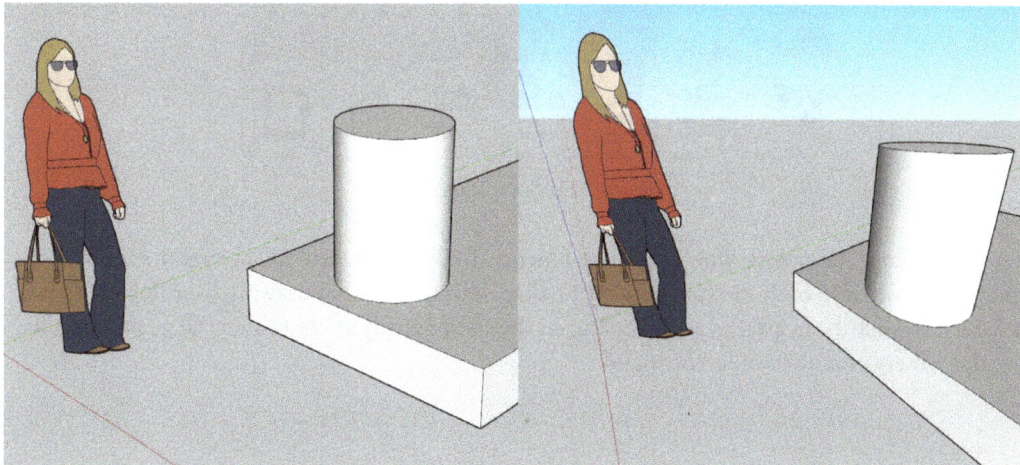

Figure 1.21: Standard Field of View of 35.00 (Left) and Extreme Field of View of 100.00 (Right)

> **Note**
> SketchUp workflows will often involve getting the right camera angle for a tool or process. Using Orbit, Pan, and Zoom in concert will allow you to set up your camera view for each workflow. This takes patience and practice!

Zoom Extents

Zoom Extents is a fast way to automatically adjust Zoom and Pan to fill the drawing area with all visible geometry. This means that you will see your whole model very quickly! The camera view will not be orbited when using Zoom Extents.

The Zoom Extents tool button looks like a magnifying glass with three red arrows:

Figure 1.22: Zoom Extents Button

Zoom Extents has a slightly more complicated keyboard shortcut, *Ctrl + Shift + E*, so it is suggested to use the toolbar button to activate this zoom option.

Views Toolbar

The **Views** toolbar includes six tools, Iso (Isometric), Top, Front, Right, Back, Left, and Bottom:

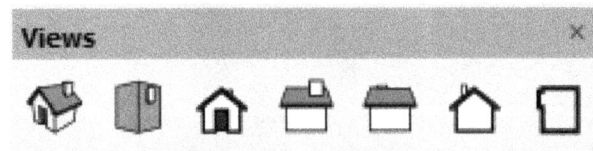

Figure 1.23: Views Toolbar

These tools change the camera view like Zoom Extents. The updated view will change the zoom level slightly, but it will also orbit the camera view to align with the original drawing coordinates. These standard views may be helpful when modeling, as they will quickly zoom to show all geometry, and move the camera around to different angles:

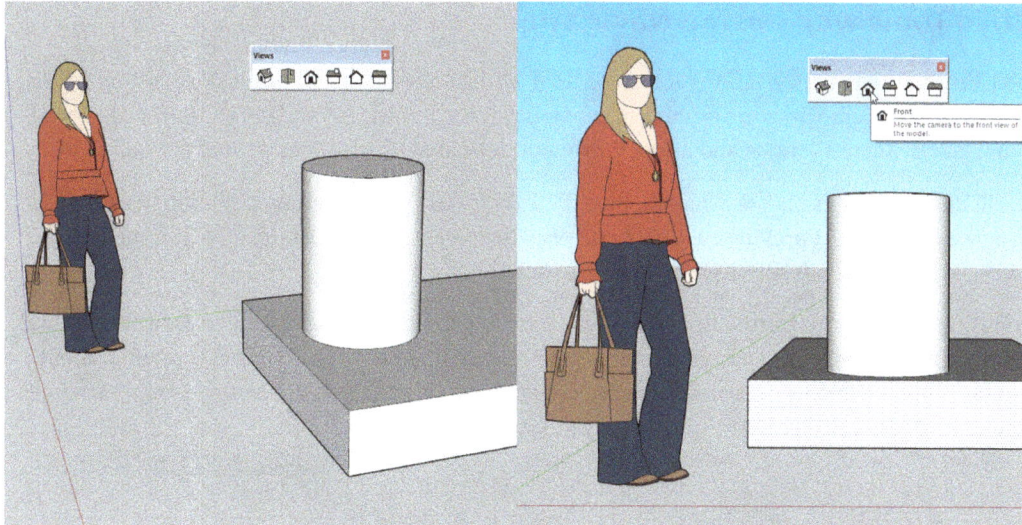

Figure 1.24: Model Shown at a Random Angle (Left) and Model at Front View (Right)

> **Note**
>
> Zoom Extents and Views are not the same tool, but they are very similar! You may never need to use the standard Views while modeling, but it is suggested to try to use Views and Zoom Extents. We will discuss more views and camera options in *Chapter 5*.

Other Zoom Tools

Two additional Zoom tools can be found on the Large Tool Set. These are Zoom Window and Zoom Previous Tools. Zoom Window will allow you to drag a window to zoom into a specific section of your model. Zoom Previous can be clicked to move back through your recent camera views. This can be very helpful when setting up Scenes.

Figure 1.25: Zoom Window (Top) and Zoom Previous (Bottom)

Active Tools while using Navigation

SketchUp allows for all navigation tools to be used while in the middle of using any other tool. This includes when you are drawing or editing geometry in the model! All six tools, Orbit, Pan, Zoom, Zoom Extents, Zoom Window, and Zoom Previous, can be used without interrupting the editing tools.

We will look at a few examples using the Push Pull Tool. You do not need to follow along with these examples; they are just provided for context. We will talk about the Push Pull Tool in more detail *Chapter 4, Drawing Tools: We Begin Modeling!*.

In this first example, we start with a circle drawn on a box. After activating the Push Pull Tool, we want to orbit the camera angle to see the geometry from a different angle. Using the 3-button mouse shortcut, the orbit tool is activated, and the camera angle is changed. When the 3-button mouse shortcut is released, SketchUp resumes the Push Pull tool.

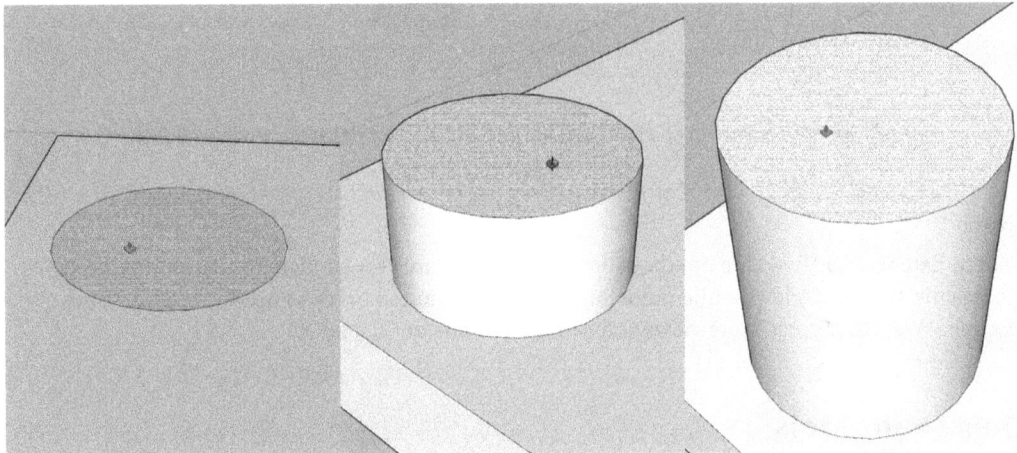

Figure 1.26: Starting the Push Pull Tool (Left), Orbit (Middle), Pan, and Zoom (Right)

> **Note**
> In the example, the Push Pull tool was not cancelled while the view was changed with the navigation tools! You can even use the toolbar buttons, but you must click the same creation or editing tool to resume the workflow without canceling the tool.

In this second example, we are using Push Pull on the circle again. After activating Push Pull on a new camera angle, the Pan tool is selected from the toolbar. When the camera has been panned into the location that we want, Push Pull can be reactivated from the toolbar and the workflow will be resumed.

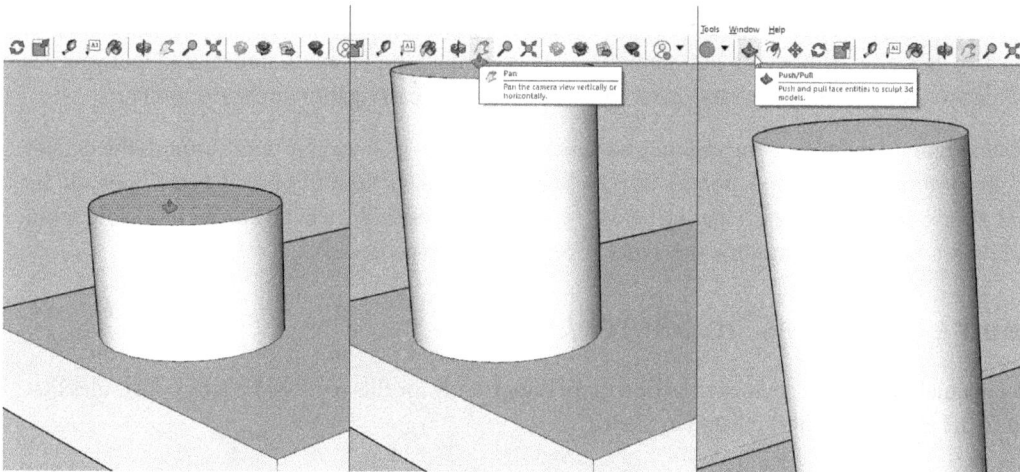

Figure 1.27: Push Pull, then Pan, then Push Pull, which resumes the previous Push Pull workflow

In this third example, we will see how a workflow can be canceled accidentally. Again, we are beginning with the Push Pull tool, then selecting the Pan tool from the toolbar. But in this case, after the pan is completed, the Select tool is chosen instead of Push Pull. Even though it appears as though the Push Pull tool created a cylinder of geometry, the workflow has not completed. When a different tool was chosen, the workflow was canceled, so no work was completed.

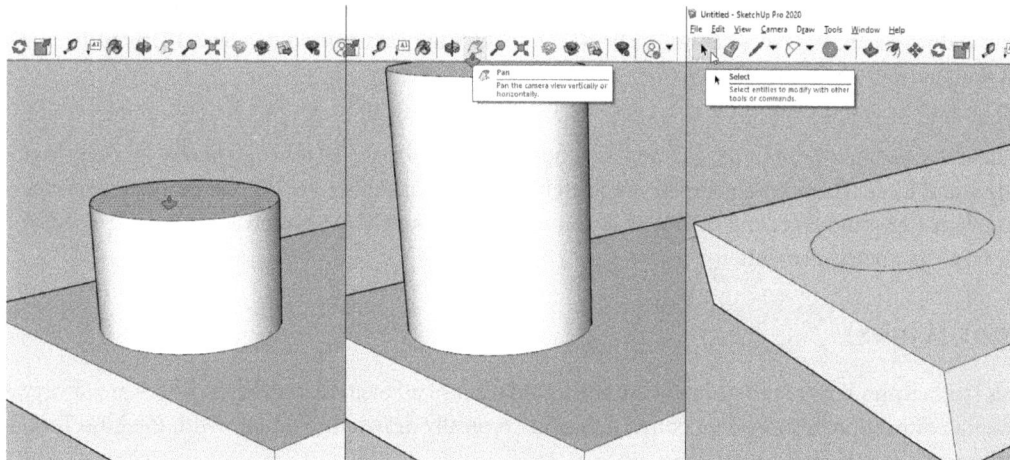

Figure 1.28: Push Pull, then Pan, then Select, which has canceled the workflow

It is so very helpful to understand how to use the 3-button mouse shortcuts while navigating in SketchUp. But even without a 3-button mouse, it is even more important to understand that you can continue working while navigating, even when clicking on the navigation tools in a toolbar.

I would highly suggest that you practice navigating in SketchUp, even if it is just around the default person component when you create a new file! Navigation takes time and practice, and it would be good to have some practice navigating before you begin to model. We will talk about modeling soon, but first, we want to look at some rules that govern SketchUp geometry.

Geometry "Rules" in SketchUp

It is undeniable – SketchUp does rule! But, more importantly for this section, SketchUp has rules that govern the geometry that is created in SketchUp.

The Geometry rules are as follows:

- All edges are straight lines between two points
- All faces are bounded (surrounded) by edges
- All faces are two-dimensional (flat)

> **Note**
> Even though faces are two-dimensional, they do not have to align with any of the model axes, but it does mean that faces cannot be bent! If a surface appears curved in a SketchUp model, it is actually made of many flat faces.

All SketchUp geometry can be categorized as either an edge or a face. All other geometry is comprised of edges and faces. There is no geometry in SketchUp that breaks this rule! We will evaluate the rules that govern edges and faces and then look at some common SketchUp geometry that is comprised of these basic geometry elements.

Edges (Lines)

Edges (sometimes referred to as lines) are the foundation of all SketchUp models. Edges are always a straight connection between two points. Edges are typically drawn individually with the Line Tool, but all creation tools make new edges:

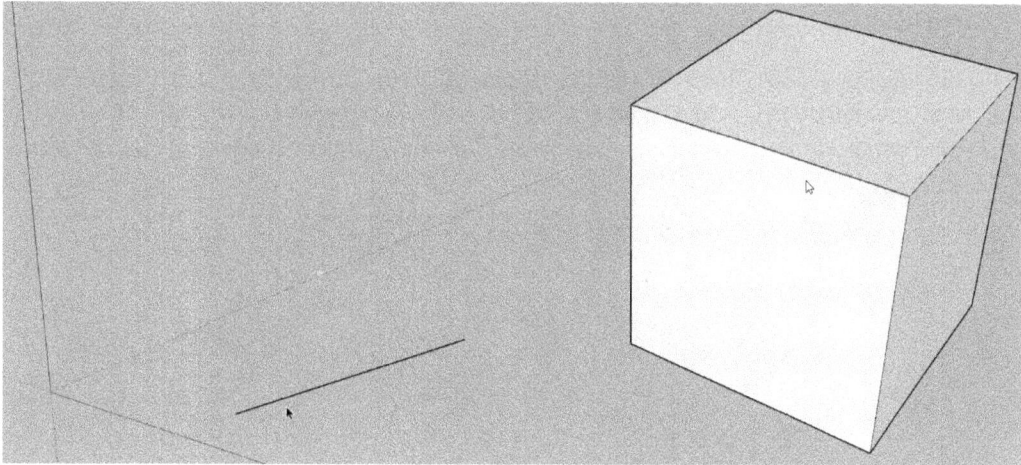

Figure 1.29: A Single Edge (Left) and an Edge on a Cube (Right)

Faces

Faces are two-dimensional surfaces that are bounded (surrounded) by multiple edges. There is no limit to the number of edges that can surround a face, but the edges must all be in the same plane. Edges cannot be bent or warped; they are always two-dimensional or flat. It is important to remember that faces cannot exist without their bounding edges, but edges can exist without faces. We will expand on this concept in later chapters.

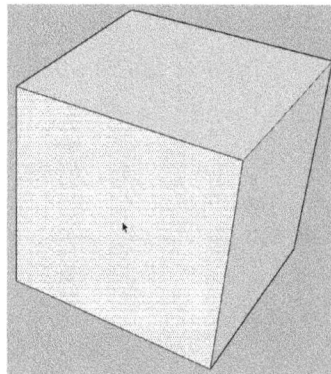

Figure 1.30: Face on a Cube

If all edges are straight, and all faces are flat, then how can we create complex and "real-world" models in SketchUp? What about the parks, plants, buildings, and interiors that we see online? How is that geometry created? SketchUp has built-in tools that allow us to get a head start on our modeling so that we do not have to draw each edge or face one by one.

Rectangles

Rectangles are familiar shapes that have a square or rectangular face surrounded by four edges. This is one of the most common shapes to start with in SketchUp. The Rectangle tool and Rotated Rectangle tool are used to quickly create rectangles, but it is good to know that any shape can be drawn with the Line tool!

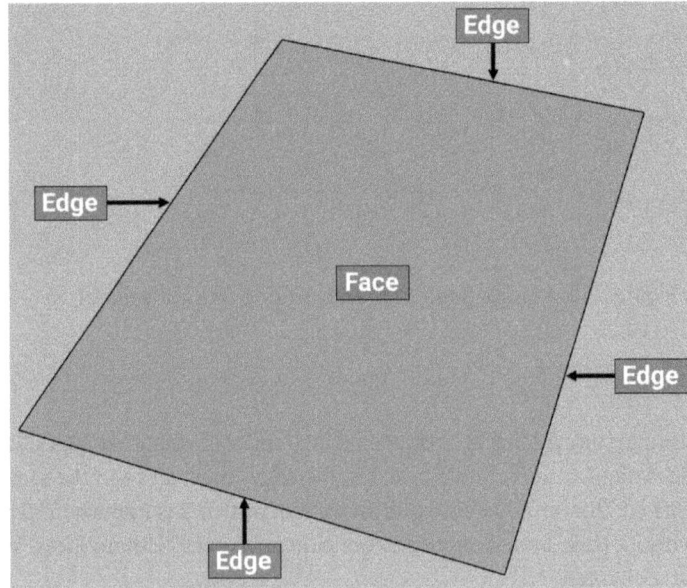

Figure 1.31: Rectangle with Faces and Edges

Polygons

As mentioned before, there is no limit on the number of edges surrounding a face. In some cases, the lengths of the edges and the angles connecting them are equal, which creates regular polygons. These shapes are also very common in designs, and SketchUp helps us model these with the Polygon tool.

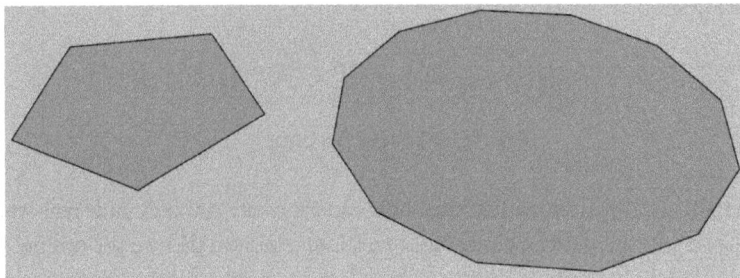

Figure 1.32: Pentagon and 12-gon (Dodecagon)

Polygons in SketchUp typically refer to regular polygons because that is what the Polygon tool creates. Any loop of edges that creates a face is technically a polygon, but we will not use the word polygon to refer to these objects in this book. We will use the term "shape" for any irregular, closed loop of edges surrounding a face:

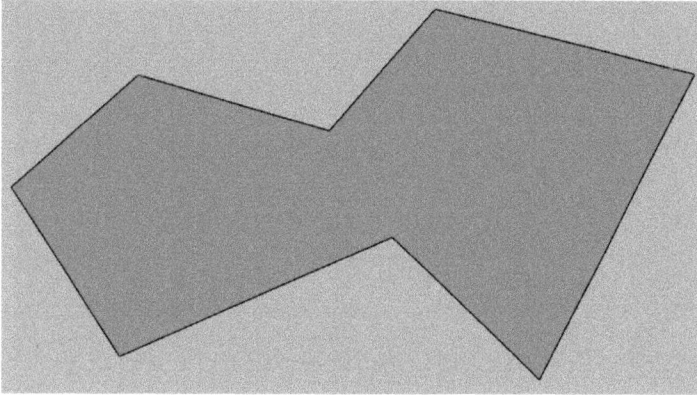

Figure 1.33: An 8-sided Shape

Curves (Welded Edges)

Curves in SketchUp are not just arcs or squiggles; Curve is a blanket term that includes any collection of edges, connecting into a continuous path, that are Welded together. This could include any squiggle or path, two-dimensional or three-dimensional, and includes arcs and the edges of circles and polygons. Essentially, this is a collection of lines that are stuck together.

Arcs and circles seem to break the rule that all edges in SketchUp are straight. Of course, we know that arcs and circles have curved edges! To make sure that the first rule of SketchUp is not broken, arcs and circles have a series of short, straight lines that collectively look like a smooth curve. We will talk more about arcs and circles when we discuss the creation tools.

> **Note**
>
> By default, the Freehand, Arcs, Circle, Polygon, and Pie Tools all create curves. The Line and Rectangle Tools do not create curves because their edges are not welded together.

This is a curve that is comprised of 12 segments in a standard Arc. Remember, this is a curve because the edges are connected:

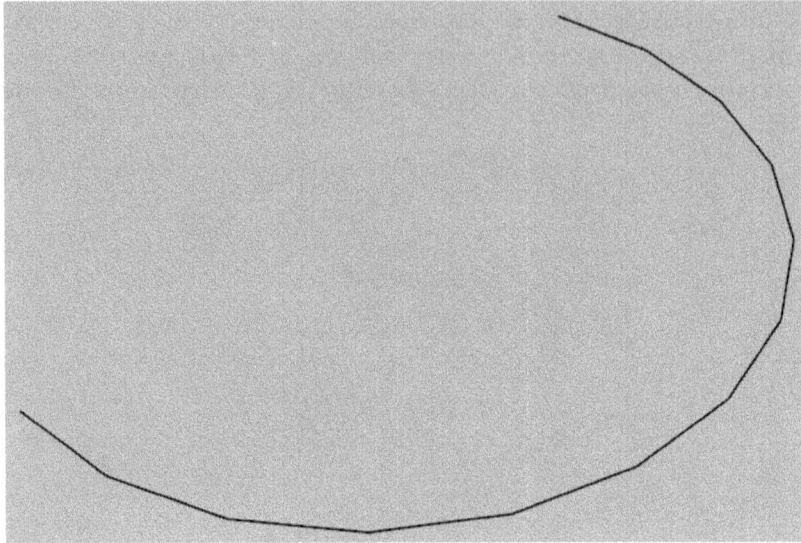

Figure 1.34: Curve – Arc

This is a "closed" curve because it creates a loop. This was created using the polygon tool:

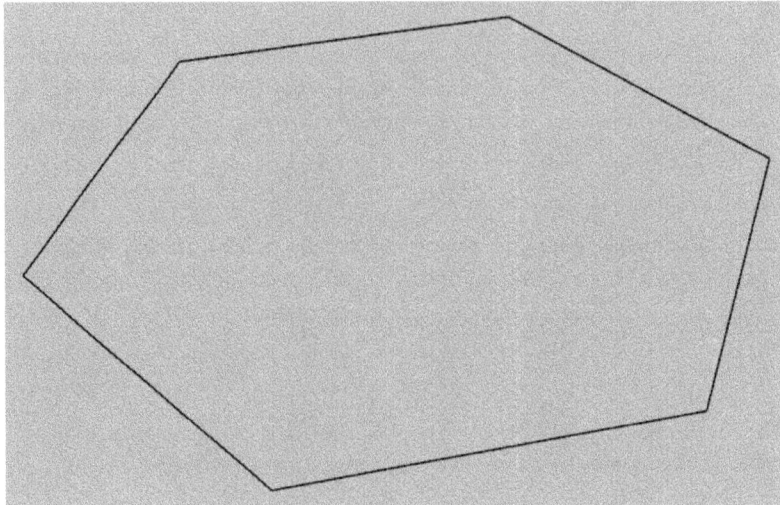

Figure 1.35: Curve – Polygon (Hexagon)

This is a curve that was created manually by drawing using the Pencil tool and using Weld Edges. We will discuss this process later in the book.

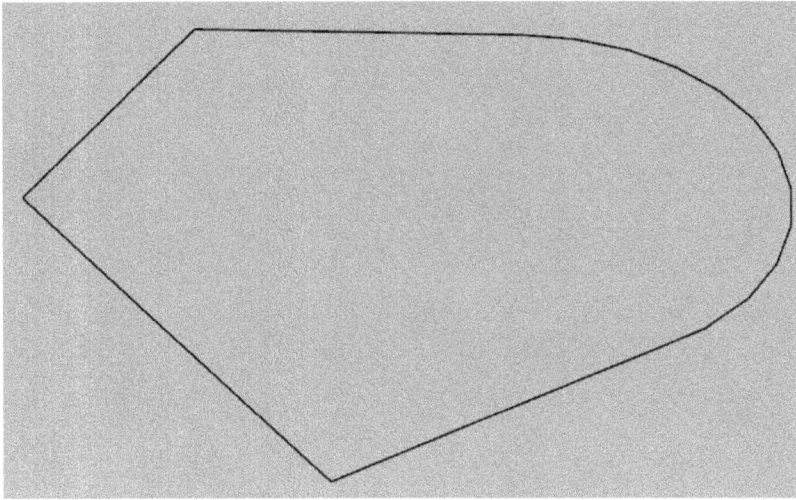

Figure 1.36: Curve – Welded Edges

Summary

In this chapter, we quickly compared SketchUp Pro and SketchUp For Web, explored the User Interface and associated toolbars and dialog boxes, learned how to navigate through 3D space in SketchUp models, and looked at the underlying rules for SketchUp geometry. We will build on all of these as we move through the rest of the book.

In the next chapter, we will look at some of the most used Tools – the Primary tools – as well as Inferences. These tools and Inferences will be essential to every SketchUp model you work on, so let's get started!

2

Principal Tools, Axes, and Inferences

In this chapter, we will examine two of the Principal tools that are used in almost every SketchUp workflow! The Select and Eraser tools can do more than you might think at a first glance! We will also understand the importance of using Axes and Inferences while drawing and editing SketchUp Geometry.

The following topics are covered in this chapter:

- The Principal Tools

  A. Select

  B. Lasso Select

  C. Eraser

- Axes

- Snaps (Point Inferences)

- Inferences

Working with the Principal Tools

If you have skipped ahead to this chapter, I understand! We always want to jump right into any new design software or tool and begin to see our designs realized! However, I highly suggest that you review the previous chapter before you begin modeling in SketchUp Pro. It is very, very important to understand the basics before beginning your SketchUp journey.

We will continue with a couple of other basics before we look at the Drawing Toolbar and the Drawing Tools. Five of the most essential and most common tools have been collected in the Principal Toolbar. These five tools are Select, Lasso Select, Paint Bucket, Eraser, and Make Component. We will cover Make Component and Paint Bucket extensively in later chapters, so for now, we will focus on the Select Tool, Lasso Select Tool, and the Eraser Tool.

The Principal Toolbar typically does not need to be individually opened and shown on screen. Instead, it is common to have the Large Tool Set Toolbar open. This toolbar is typically docked to the left-hand side of the screen. Remember that toolbars can be shown by right-clicking on the gray space directly to the right of the Getting Started toolbar and selecting additional toolbars. Also, toolbars can be moved and docked by clicking and holding the line of small dots on the left or at the top of the toolbar and dragging the toolbar to your preferred location. You can see the Large Tool Set Toolbar docked to the left-hand side of the drawing area in the following figure:

Figure 2.1: Large Tool Set Toolbar docked to the left of the Drawing Area

You can see that the four Principal Tools can be found at the top of the Large Tool Bar:

Figure 2.2: Select (Top Left), Lasso Select (Top Right), Paint Bucket (Middle
Left), Eraser (Middle Right), and Make Component (Bottom Left)

> **Note**
> The Tag Tool is included in the top section of the Large Tool Set Toolbar but is not one of the
> five Principal Tools. The Tag Tool will be reviewed in *Chapter 9, Entity Info, Outliner, and Tags
> Dynamically Organize Your Models*.

We will take a deep dive into the Make Component and Paint Bucket Tools in later chapters. So, for
now, we want to focus on the Select Tool, Lasso Select Tool, and the Eraser Tool.

The Select Tool

The Select Tool should be the most frequently used tool in any project! The Select Tool is suggested to
be the "home" or "default" tool that is used after every workflow. This includes after long workflows,
but also after using any Drawing or Edit tool – unless there is a clear chain of Tools that you are using.
There is nothing more frustrating than accidentally reactivating a Drawing or Editing Tool after
finishing a workflow. Returning to the Select Tool is a great habit to get into!

Figure 2.3: The Select Tool

The Select Tool can be found on the Getting Started Toolbar, the Large Tool Set Toolbar, and the
Principal Toolbar. The Select Tool can be activated as follows:

- Clicking on any of the Select Tool Buttons
- Hitting the *Spacebar* on the keyboard

Hitting the *Spacebar* is the easiest and most convenient way to activate the Select Tool!

> **Note**
>
> Unlike some other modeling software, the *Esc* (escape) key does not activate a "home" or "default" state. Instead, we should use the *Spacebar* to activate the Select Tool.

The Select Tool can be used in multiple ways to select objects and geometry. These ways include left-clicking, dragging windows, using keyboard shortcuts, and using modifier keys.

Left-Clicking

The most common use of the Select Tool is done with the left-click button on a three-button mouse. Using a single-, double-, or triple-click with the left button will result in different results.

A single click will select the object or geometry that is directly under the tip of the mouse cursor. Using a single click is a great way to make the exact selection you are attempting to highlight!

A double click will either do the following:

- For Objects: Open the contextual edit for the Object. This will "drill" one layer into the Object.
- For Edges: Select the Edge that is under the cursor, and then select all connected Faces.
- For Faces: Select the Face that is under the cursor, and then select all connected Edges.

> **Note**
>
> For Edges that are not connected to any Faces, there is no difference between a single click or double-click.

We will go into detail about Objects in SketchUp Pro in the next chapter. Groups and Components are known as Objects in SketchUp. Please see that chapter for more information about editing in the contextual edit for Objects.

When selecting standard Geometry, a double-click is a fast and easy way to select geometry and the geometry that is immediately connected. This will only select the first connection from edges to faces or vice versa. This will not select edges that are connected endpoint to endpoint, or a large collection of faces. Please see *Modifier Keys* later in this section for more information about selecting custom selection sets.

You can see that when the edge is double-clicked that the two connected faces are also selected. Similarly, the edges that bound the rectangular face are selected when double-clicking the rectangular face.

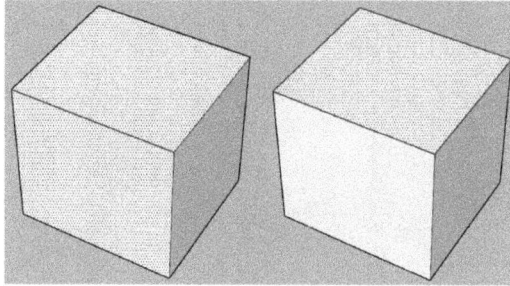

Figure 2.4: Double-Clicked Edge (Left) and Double-Clicked Rectangular Face (Right)

If you would like to select all connected Geometry, you can use a triple-click. Triple-clicking on any geometry will not just select the immediately connected geometry but also any geometry that is connected to that geometry in a chain! This can get out of hand very quickly if not using Groups and Components to separate your Geometry! We will learn much more about Sticky Geometry and separating Geometry using Groups and Components in the next chapter.

We can see a triple-click in three stages for the following Geometry. The first click selects the face, the second click selects the face and the bounding edges, and the third click selects all connected geometry. The cube to the side is not selected because it is a Group:

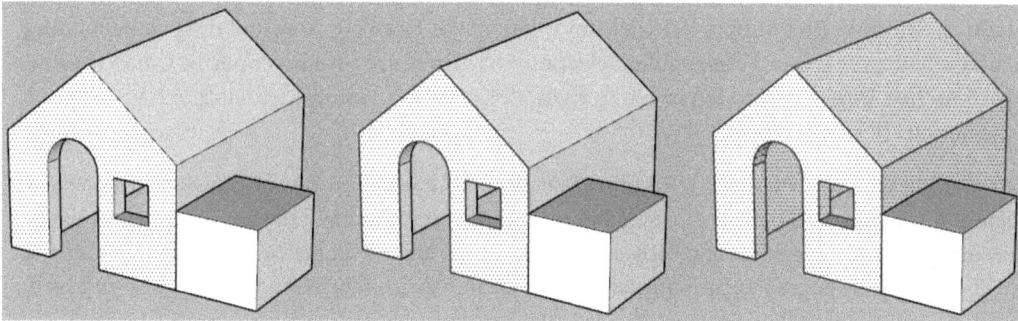

Figure 2.5: Geometry selected by single- (Left), double- (Middle), and triple-clicking (Right)

> **Note**
>
> A triple-click will select geometry even if it is connected with endpoint-to-endpoint edges or intersecting edges, but a triple-click will not select intersecting faces if the bounding edges of those faces never intersect!

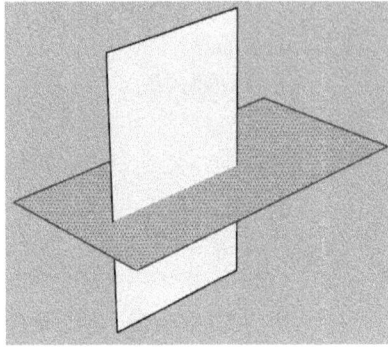

Figure 2.6: Intersecting faces are not selected because their edges do not connect or intersect

You should practice single-, double-, and triple-clicking Geometry as you work through the rest of this chapter. You will notice that you have to click quickly to activate the multiple-click options: this takes time and practice to get right!

Selection Boxes

Creating Selection Boxes in SketchUp is an excellent way to select multiple Objects or Geometry at once!

To create a Selection Box, simply left-click anywhere in the Drawing Area and drag while holding the mouse button to create a rectangular Selection Box. There are two directions in which you can drag a Selection Box – right to left or left to right. Both are done using a left-click and holding the mouse button.

A left-to-right window will make a Window Selection. This Selection Box is represented by a solid outline and will only select objects that are fully encased in the Window Selection. You can see in the following figure that some Edges and Faces are partially inside of the Window Selection, and they are not highlighted. Only the Edges and Faces that are fully inside the left-to-right Window Selection are selected:

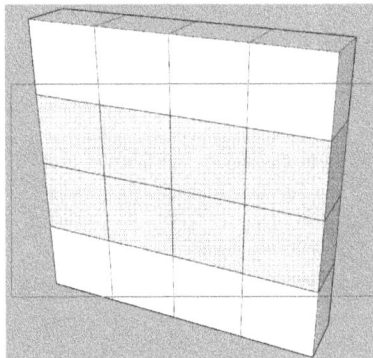

Figure 2.7: Window Selection using the Select Tool

A right-to-left window will make a Crossing Selection. This Selection Box is represented by a dashed outline and will select all objects that cross the window or are fully enclosed. You can see in the following figure that all Edges and Faces that are crossed or that are inside the Crossing Selection are selected:

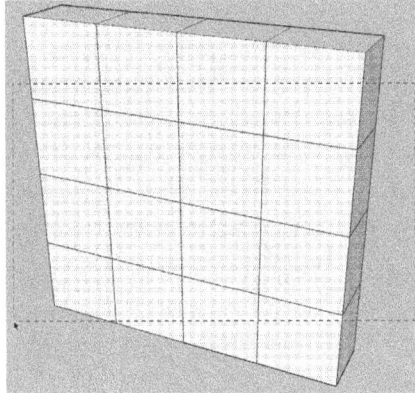

Figure 2.8: Crossing Selection using the Select Tool

Be careful! When using Selection Boxes with the Selection Tool, you will select anything that is in the window. This includes Geometry behind other geometry. You can see in the following figure that it looks like one Face and the surrounding edges have been selected, but when the model is orbited, it is clear that some Edges and Faces from the back of this model were also selected.

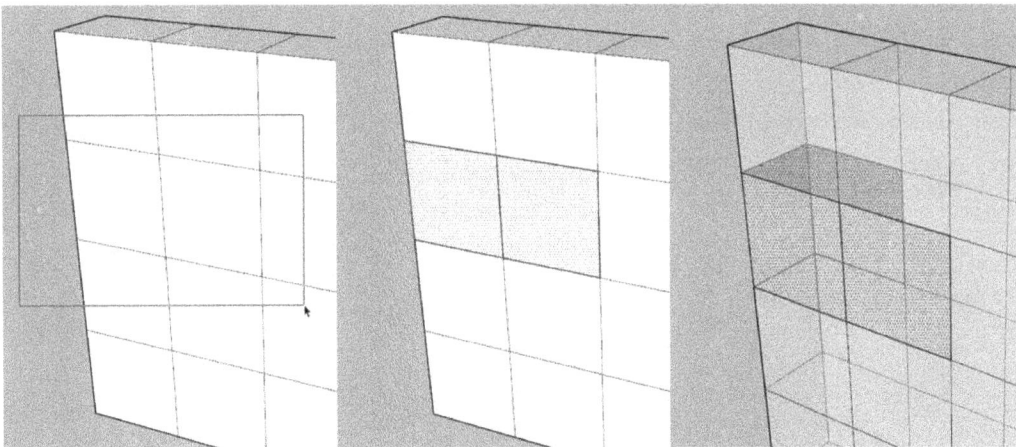

Figure 2.9: Window Selection (Left), Intentionally Selected Geometry
(Middle), and Unintentionally Selected Geometry (Right)

Some more practice will be very helpful with Selection Boxes as well! It is beneficial to understand which type of Selection Box is created with the left-to-right motion (Window Selection) and right-to-left motion (Crossing Selection).

Keyboard Shortcuts

There are a couple of keyboard shortcuts that may be helpful when working with the Select Tool. These shortcuts are as follows:

- Select All: *Ctrl + A*

- Invert Selection: *Ctrl + Shift + I*

Both of these shortcuts will also be found under the Edit dropdown in the Menu Bar and in the right-click menu when using the Select Tool. Select All will select all the geometry or Objects in the active Drawing Area, and Invert Selection will deselect the current selection and select everything else.

If and when these keyboard shortcuts are helpful will depend on your SketchUp model and workflow. If you understand these shortcuts, they can help you make fast and accurate selections.

> **Note**
> Remember, the keyboard shortcut for activating the Select Tool is the *Spacebar*!

Modifier Keys

The Select tool has two modifier keys that change how objects are added or removed from the selection set. These keys are *Ctrl* and *Shift*. There are three possible combinations of these keys, and each combination changes how objects are added and removed.

> **Note**
> Any combination of modifier keys can be used for clicking or creating Selection Boxes.

Shift

Holding *Shift* is the most common use of modifier keys for the Select Tool. Holding *Shift* adds a plus and a minus sign next to the cursor:

Figure 2.10: Cursor with the Shift modifier key

Now, when the Select Tool is used, the previous selection set is maintained, and either of the following happens to the selected objects:

- If the geometry is currently in the selection set, it is removed from the selection set (deselected)

- If the geometry is not currently selected, it is added to the selection set (selected)

To put it more simply, you can select more than one object with different selection workflows or remove objects from your current selection with more precision!

Ctrl

Holding *Ctrl* activates only one of the options that is activated while using *Shift*. *Ctrl* will show a plus sign next to the cursor and will only be able to add geometry to the selection set:

Figure 2.11: Cursor with the Ctrl modifier key

Ctrl + Shift

Holding *Ctrl* and *Shift* at the same time will activate the other option, which is the opposite of the *Ctrl* key. *Ctrl* and *Shift* will show a minus sign next to the cursor and will only remove geometry from the selection set:

Figure 2.12: Cursor with Ctrl and Shift modifier keys

This option is often activated when Geometry is accidentally added to the selection when using the *Ctrl* modifier key. By also holding the *Shift* key momentarily, the accidentally selected Geometry can be quickly removed from the selection, and the *Shift* key can then be released to continue adding Geometry with the *Ctrl* modifier key.

> **Remember**
> When there are no modifier keys pressed, the current selection set is deselected when you click or drag a new Window Selection or Crossing Selection.

The Lasso Select Tool

The Lasso Select Tool was added to SketchUp in 2022. The Lasso Select Tool works identically to the Select Tool except for the Selection Box is not a box, but a freehand "lasso." The Lasso Select Tool is represented by a cursor like the Select Tool, but an additional "lasso" has been drawn around the cursor.

Figure 2.13: The Lasso Select Tool

The Lasso Select Tool can be found on the Large Tool Set next to the Select Tool, but on the Getting Started Toolbar and the Principal Toolbar, the Select Tool and Lasso Select Tool are nested in a dropdown. The Lasso Select Tool can be accessed by clicking on the dropdown arrow next to the Select Tool and choosing the Lasso Select Tool. The Lasso Select Tool does not have a keyboard shortcut.

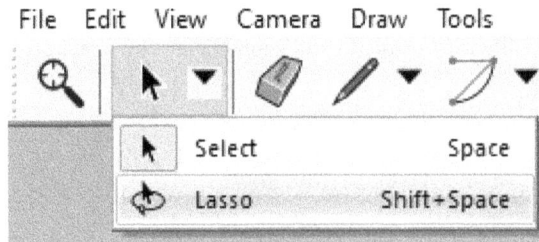

Figure 2.14: The Lasso Select Tool and Select Tool Nested in the Getting Started Toolbar

When the Select Tool is clicked and held, a rectangular Selection Box will appear. Moving this to the right will create a Window Selection and moving it to the left will create a Crossing Selection. The same workflow can be used to generate a Window Lasso or a Crossing Lasso – it depends on which direction the lasso is drawn. To create a Window Lasso using the Lasso Select Tool, the lasso must be drawn in a clockwise direction. Window Lassos can be recognized because the straight line connecting the starting and ending points will be solid like a Window Selection Box.

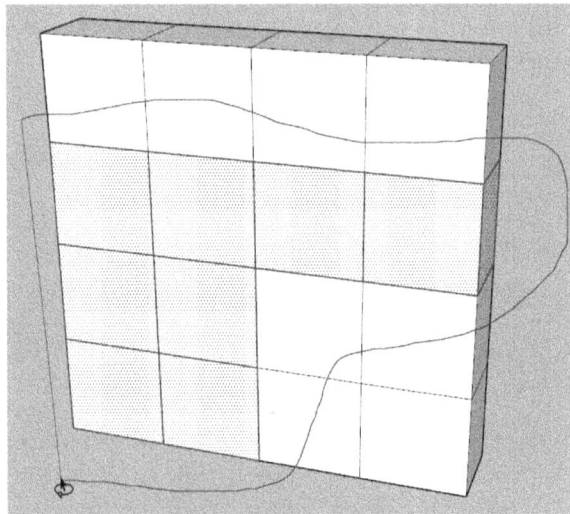

Figure 2.15: Window Lasso Drawn Clockwise

To create a Crossing Lasso using the Lasso Select Tool, the lasso must be drawn in a counterclockwise direction. Crossing Lassos can be recognized because the straight line connecting the starting and ending points will be dashed like a Crossing Selection Box.

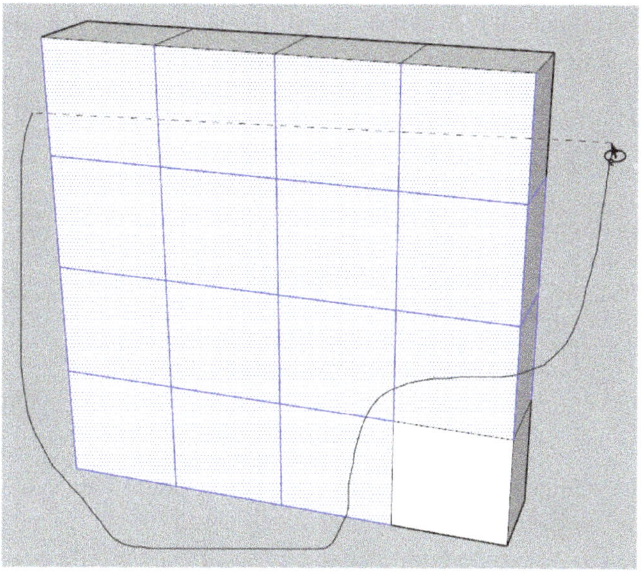

Figure 2.16: Crossing Lasso Drawn Counterclockwise

> **Note**
>
> The lasso can be drawn using any shape – it does not need to be roughly circular. The Lasso Select Tool is a great option when selecting many small Objects, specific corners of Geometry, or when working inside of a complex model.

The Lasso Select Tool can use the same Modifier Keys as the Select Tool and can be used to single-click on Geometry, right-click, and double- and triple-click to select Geometry and activate Groups and Components.

The Eraser Tool

The Eraser Tool might not be the second most-used tool in any project, but it might be the second-most important! The Eraser Tool is the primary way that geometry or objects are erased (or deleted) from a project. The Eraser Tool is represented by a pink block eraser:

Figure 2.17: The Eraser Tool

The Eraser Tool can also be found on the Getting Started Toolbar, the Large Tool Set Toolbar, and the Principal Toolbar. The Eraser Tool can be activated as follows:

- Clicking any of the Eraser Tool Buttons
- Hitting *E* on the keyboard

It is not as important to memorize the keyboard shortcut for the Eraser Tool as it is for the Select Tool (*Spacebar*), but it is helpful to become familiar with all keyboard shortcuts. You can always look up the Quick Reference Guides that SketchUp produces for SketchUp Pro that outline all keyboard shortcuts.

As already mentioned, the Eraser Tool erases (removes or deletes) geometry. This is done by either doing one of the following:

- Clicking on an Edge of the Geometry or Object
- Clicking anywhere in the Drawing Area and dragging the Eraser icon across an Edge of some Geometry or an Object

It may seem like the Eraser Tool and the *Delete* key perform the same functions. But this is not the case. Some differences include these:

Eraser Tool	Delete Key
Can remove Edges but not Faces	Can remove any selected Geometry or Object, including Faces
Does not remove the current selection set when activated	Removes the current selection set when pressed
Can be used in a dragging workflow to remove multiple objects at once	Only affects the selection set, which may involve complex Select Tool workflows that include the use of Modifier Keys
Can hide and soften Edges	Can only delete Geometry

> **Note**
>
> The Eraser Tool cannot erase Faces by itself! However, when a bounding Edge is erased, the Face will be removed because there always needs to be a full ring of Edges surrounding a Face.

It will be up to you to decide between using the Eraser Tool and the *Delete* key while working through your projects. We will look at a few examples of when one workflow might be better than the other.

Example 1 – removing a Face

In this example, a Face should be removed so that it can later be connected to another part of the model. The Face cannot be erased using the Eraser Tool by itself. Also, if one of the bounding Edges is erased, it will break other adjoining Faces, which is not a good workflow for removing the face:

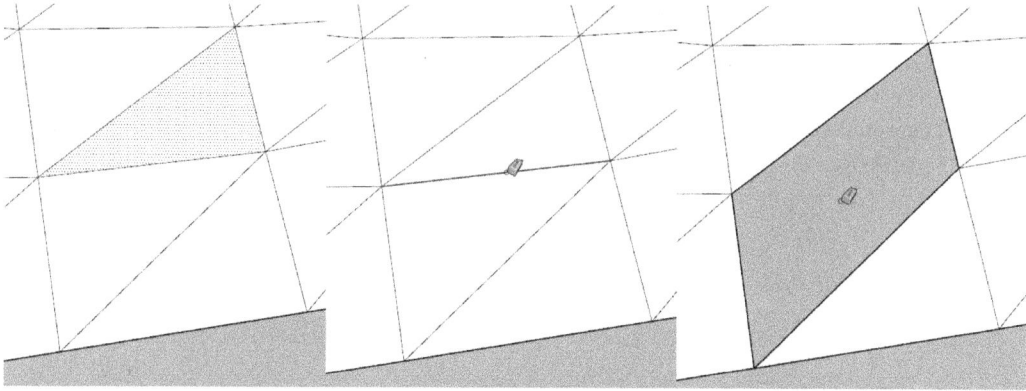

Figure 2.18: Geometry with highlighted Face (Left) Erasing the
bottom Edge (Middle) Erased Adjoining Face (Right)

Instead, the *Delete* key produces the right result with fewer steps. The Face is selected with a single click using the Select Tool and the *Delete* key is pressed to remove the face:

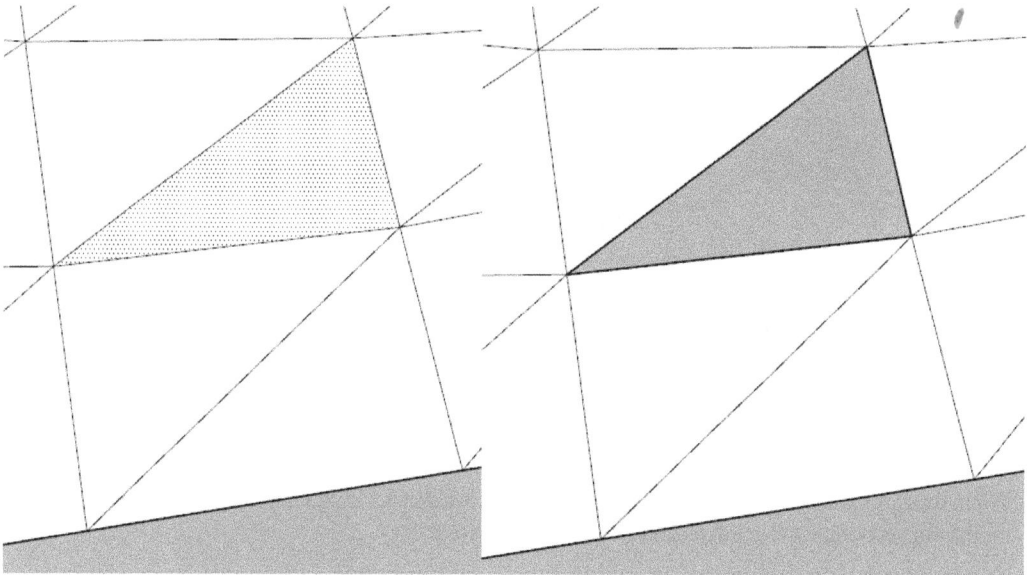

Figure 2.19: Geometry with Highlighted Face (Left) removed after hitting the Delete key (Right)

Example 2 – removing Multiple Edges and Faces

This example will focus on splitting two sections of a mesh surface where one of the rows of Faces needs to be removed. The mesh surface in this example is a complex model, but this workflow can be applied in many ways, including any selection set of multiple Objects or Edges and Faces. We will discuss creating complex Geometry in a later chapter.

In the example, the blue faces need to be removed. If the *Delete* key were to be used, the selection would have to be made with a series of clicks, dragged windows, and Modifier Keys with the Select Tool. Then, the *Delete* key would need to be pressed to remove the Edges and Faces. This would take some time and could involve over 20 clicks with the Select Tool before the right selection set can be established:

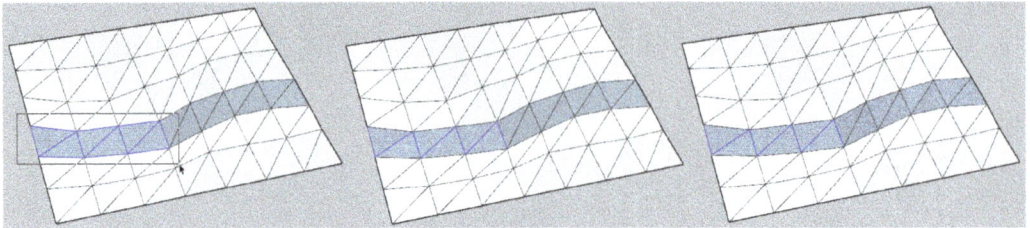

Figure 2.20: The Selection Set is started by dragging windows (Left) and
completed by picking with Modifier Keys (Middle and Right)

Alternatively, the Eraser tool could remove these Edges and Faces in one action! The Eraser Tool can be dragged across the Edges from left to right or right to left across the screen, catching the edges and removing them. Once the Edges have been erased, the Faces are also removed because they have lost some of their bounding Edges:

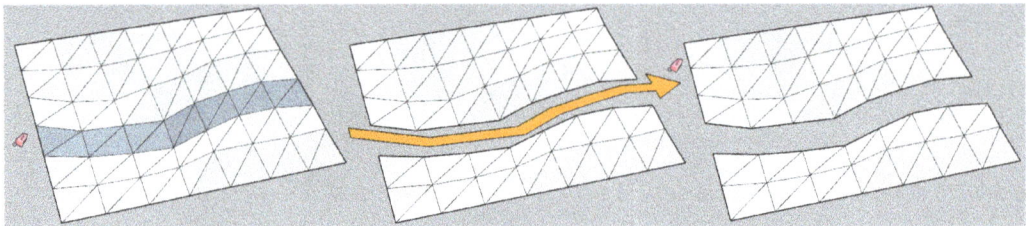

Figure 2.21: Eraser Tool pressed and held (Right) dragging across the
screen (Middle) and Geometry is removed (Right)

> **Note**
> When dragging the Eraser Tool, make sure that you move slowly! If you move too fast, SketchUp might not recognize what Edges you are dragging across!

Example 3 – Modifier Keys to Hide and Smooth/Soften Edges

A more advanced workflow for the Eraser Tool includes the Modifier keys, specifically *Shift* and *Ctrl*. *Shift* will Hide Edges and *Ctrl* will Smooth/Soften Edges. Smoothing or Softening edges is very similar to Hiding edges, except for the Geometry appears to have a smooth finish instead of a sharp or faceted finish. In the first figure, we can see the same Toposurface from the previous example. This Toposurface has all edges Shown. To get a more finished appearance, we can use the Eraser tool with *Shift* held down to hide the Edges:

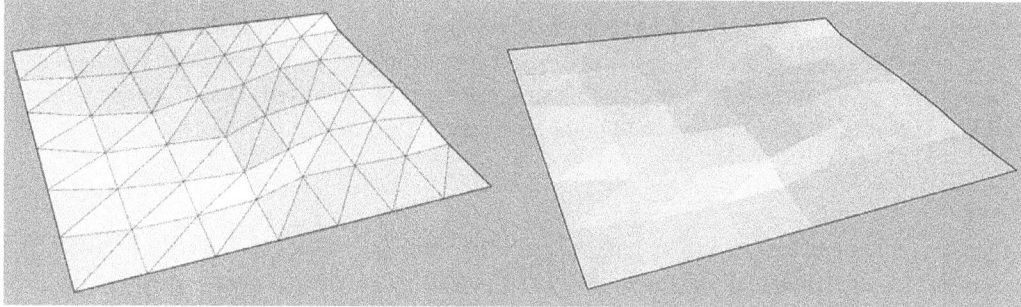

Figure 2.22: Toposurface with Shown Edges (Left} and Hidden Edges (Right)

> **Note**
> Edges can also be hidden by selecting them and choosing Hide from the Edit Menu dropdown or the right-click context menu. However, using the Eraser Tool and the *Shift* Modifier key can be an easier solution!

Now, you can see the difference when the *Ctrl* key is used instead for the Edges, which produces a Smooth or Softened appearance across the Toposurface. This type of finish can also be applied using the Soften Edges Panel in the Default Tray. However, the Eraser Tool and the *Ctrl* Modifier key will allow for greater control with smaller selection sets on a larger object.

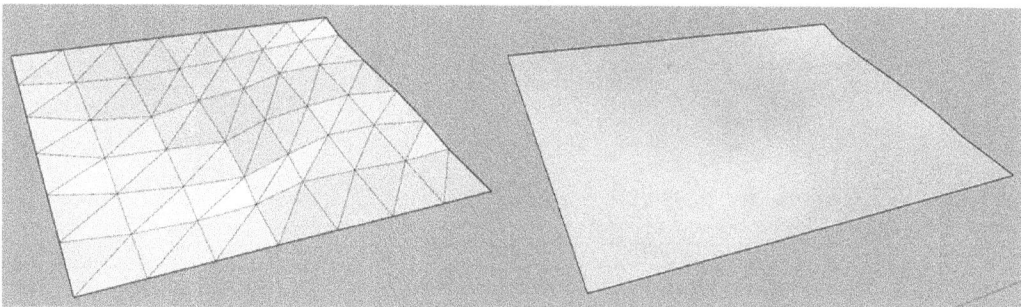

Figure 2.23: Toposurface with Shown Edges (Left) and Smooth/Softened Edges (Right)

The Principal tools are used in every SketchUp workflow. It is essential to understand how to effectively use the Select and Eraser tools while drawing and editing your SketchUp models. Three other drawing and editing tools must be understood before we can begin modeling, and those are Axes, Snaps, and Inferences.

Axes – orienting in our Model

All versions of SketchUp organize the Drawing Area in Three-Dimensional space using an $x/y/z$ coordinate system. This system, also known as the 3D Cartesian coordinate system, keeps track of all SketchUp geometry by assigning each endpoint to a coordinate. Of course, endpoints are at the ends of Edges, and Edges bound Faces, so endpoints are involved in all SketchUp Geometry! These points are assigned an x-, y-, and z-coordinate so that they can be found relative to 0, 0, 0 (or the model Origin), as well as the other endpoints and other geometry in the model.

> **Note**
>
> There is some complicated math happening every time an Edge or Face is created, especially when the Edit Tools are used! It is a good thing that SketchUp Pro takes care of this math for us!

SketchUp helps us understand this 3D environment by including the Axes in the Drawing Area:

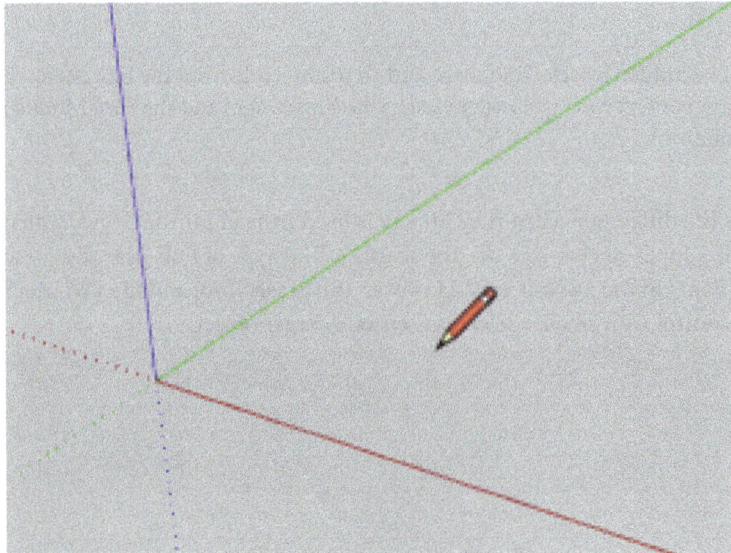

Figure 2.24: Image of Axes

These Axes are color-coded so that we can quickly identify the cardinal directions in our coordinate system. These are outlined in the following table:

Coordinate System	Axis Line Type and Color	Image Orientation
Positive x	Solid Red	Moving Right
Negative x	Dashed Red	Moving Left
Positive y	Solid Green	Moving Forward
Negative y	Dashed Green	Moving Backward
Positive z	Solid Blue	Moving Up
Negative z	Dashed Blue	Moving Down

Note

The image orientation listed here will only align this way for this image or similar views. As you orbit around your model, you will see the red, green, and blue Axes changing how they are viewed on the screen. They are not moving with the Orbit tool – remember that you are moving the camera!

Where the Axes intersect is the Origin point. This is 0X, 0Y, and 0Z in the coordinate system. You can see this highlighted in the following figure:

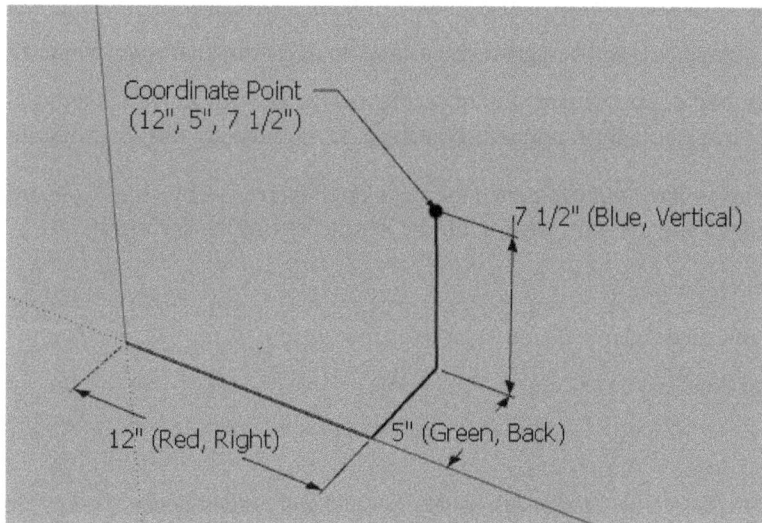

Figure 2.25: Coordinate Points relative to the Origin

The coordinates are organized using the project Units, which are defined by the project template. In this book, we are going to use inches in the Feet and Inches Template. As Geometry coordinates move further away from the Origin, the units will increase! Of course, make sure you check whether the Geometry is in the Positive direction or the Negative direction – the number of Units will increase positively or negatively depending on where the Geometry is relative to the Origin.

The Axes provide a great visual reference to orient our modeling around the middle of the 3D world, which is the SketchUp Drawing Area, but the Axes provide more assistance when we create and edit our SketchUp models!

Axes help to orient drawing and editing movements in SketchUp by aligning the cursor in six directions when the angle appears to be similar on the screen. This is a default setting in all versions of SketchUp.

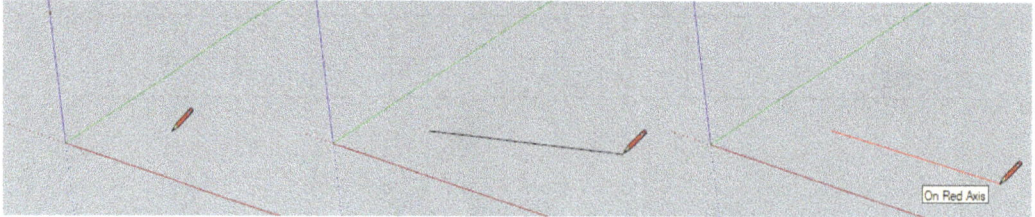

Figure 2.26: The Pencil Tool started Drawing (Left) with no Axis alignment
(Middle) and aligning with the positive Red Axis (Right)

You will see this in action as we begin to draw and edit Geometry in SketchUp. It is important to keep an eye on this for multiple reasons:

- Most basic Geometry in SketchUp is aligned with one of the major Axes.

- Aligning with an Axis will be a great visual check to make sure that your Geometry is aligned.

- Watch out! Make sure that you do not accidentally align with an Axis you did not mean to align with. It is especially important to watch out for accidentally aligning with the Blue Axis.

Axes can be aligned by attempting to match the angle on the screen or by using keyboard shortcuts. To activate and lock a specific Axis, you can hit the Arrow Keys on the keyboard:

- **Left Arrow**: Green Axis

- **Right Arrow**: Red Axis

- **Up Arrow**: Blue Axis

> **Note**
>
> *Down Arrow* locks in other Inferences, usually parallel and perpendicular. This is covered in the *Snaps and Inferences* section.

Additionally, each Axis or Inference direction can be locked in by holding *Shift* while briefly showing the aligned Axis. Then, while *Shift* is held, the cursor can be moved to any point of the Drawing Area and the Axis will be locked in as if one of the Arrow Key keyboard shortcuts was activated.

We will practice aligning to these Axes in later chapters while drawing and editing our SketchUp Geometry.

> **Note**
> Do you not like the Axes' colors? Any of the Axes or Inference colors can be changed in **Preferences | Accessibility**. Feel free to choose whatever color you like, although the Axes will still be called Red, Green, and Blue in SketchUp.

The Axes can be controlled by right-clicking on the Axes in the drawing area. The Axes options are as follows:

- **Place**
- **Move**
- **Reset**
- **Align View**
- **Hide**

Place

Activating the **Place** option is the same as hitting the Axes tool on the Large Tool Bar:

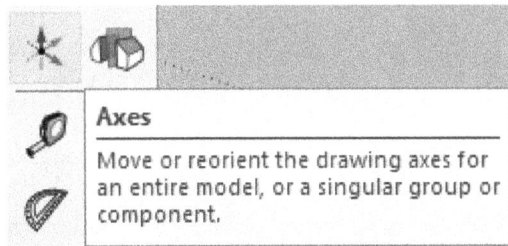

Figure 2.27: The Axes Button

Axes can be moved and placed anywhere in the Drawing Area. This can be extremely helpful if drawing many Edges that are at a different angle to existing Geometry. Be careful! If even slightly off, askew Axes might mess up your SketchUp Geometry! You can reset your Axes at any time – see the *Reset* section.

To place the Axes, you must do the following:

1. Choose a new Origin. This will not reset the coordinates of the model, but it will act as a visual guide for your Axes. Click anywhere in the Drawing Area to place the Axes' visual origin.

2. Choose a new Red direction by clicking in the Drawing Area. By default, this will remain in the existing Red/Green plane (flat), but it can be overridden by choosing another Geometry.

3. Choose a new Green direction by clicking in the Drawing Area. Once the visual origin, Red direction, and Green direction have been chosen, the Axes are placed. The Blue direction is inferred by the locations chosen for the Red and Green direction.

> **Note**
>
> Remember, the SketchUp Axes can be changed! This means that the associated coordinates might be different – especially once the Axes move. You might want to double-check by resetting the Axes to make sure that they are at the original 0, 0, 0 point – especially if you or someone else might have moved them in the past!

Move

Move is very similar to **Place** but is done with a dialog box. In this dialog box, there are options to apply **Move** and **Rotate** operations on the Axes:

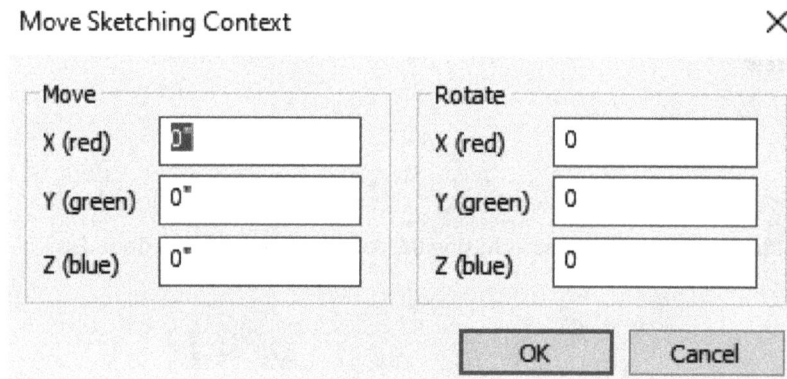

Figure 2.28: Move dialog box – Move Sketching Context

The Axes can be moved in the Red, Green, and Blue directions with positive or negative unit increments. Additionally, the Axes can be rotated in each direction. Moving the Axes in this way is very precise, but it may be better to use **Place** for the Axes when using Geometry that is already in the model.

Reset

Reset will simply move the Axes back into the original model orientation and location. This may help to reorganize the model if it gets confusing, or it could be useful after changing the Axes for a specific task. Either way, resetting the Axes will always return them to their original state!

> **Note**
>
> If you do not have to, do not use **Place** or **Move** for the Axes before drawing your Geometry! It may be difficult to re-align with the angles that you chose, but by using **Reset**, you can always reset to the original location and directions!

Align View

Align View works similarly to the Standard Views toolbar. However, it will align to look toward the Axes' origin along the Axis that was clicked. This is extremely helpful when attempting to look in a specific direction. This is similar to the right-click | **Align View** option when working with Faces, but it can be done regardless of what Geometry or Objects are visible in the Drawing Area:

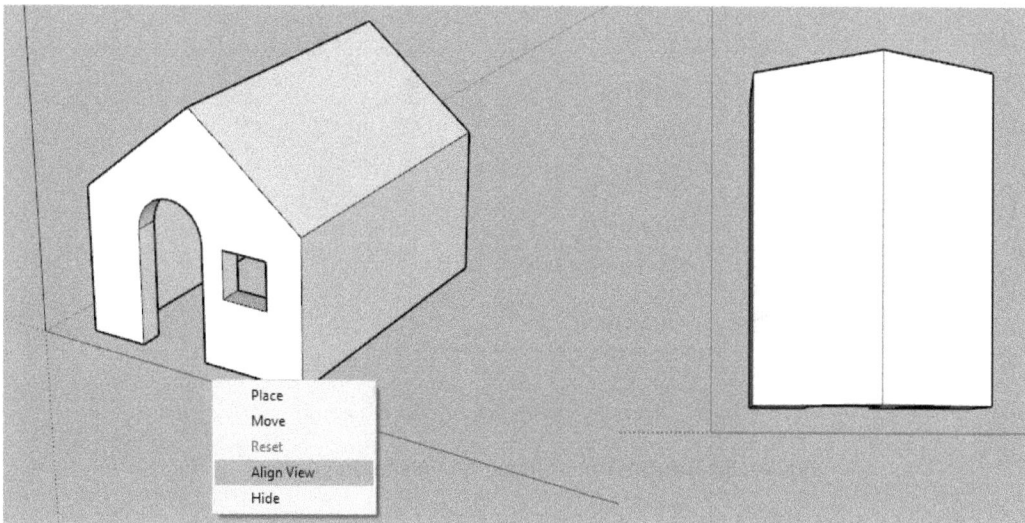

Figure 2.29: Right-Click on Axis (Left) and Align View (Right)

Hide

Axes can be Hidden like Geometry and Objects. The Axes inferences will continue to work even while the Axes are hidden. Who knows, the Axes might be ruining your screenshot!

Axes can also be hidden by going to the **View** menu dropdown and choosing **Axes**. Axes can be shown using the same workflow:

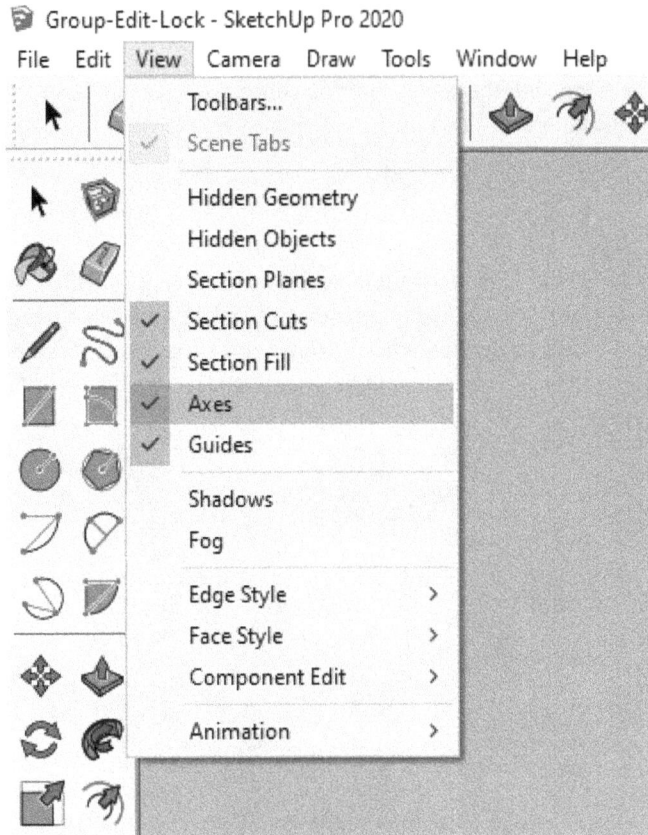

Figure 2.30: The Axes option in the View Menu Dropdown

Axes are extremely helpful to use when drawing accurate and organized models. Orienting to the Red, Green, and Blue Axes will happen over and over in every workflow and SketchUp model! Axes organize our model and assist with creation and editing by guiding the design. A couple of other built-in tools that work similarly to Axes are Snaps and Inferences.

Understanding Snaps and Inferences

Inferences are always used in SketchUp! Inferences are used in drawing and editing and are essential to many workflows. Snaps are coordinate-based Inferences in SketchUp that are selected for **Create** or **Edit Tool** workflows.

Snaps

Snaps may also be referred to as Point Inferences. This is because they are the list of Inferences that are single points in your SketchUp model:

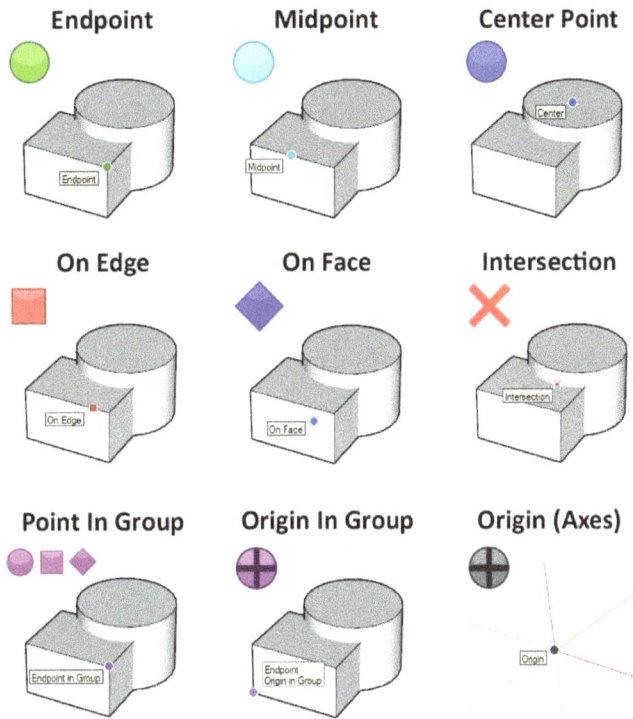

Figure 2.31: Snaps (Point Inferences)

Endpoint

This is represented by a green circle at the endpoint of any Edge.

Midpoint

This is represented by a light blue (Cyan) circle at the midpoint of any Edge. This point is exactly halfway between the two endpoints of the edge.

Center Point

This is represented by a dark blue circle at the center point of a Circle, Arc, or regular Polygon. This point may not appear when hovering around the center of an object. The Edge of the object should be hovered over first to show the center point and then the center point can be more easily selected.

On Edge

You will see a red square along any Edge. This is distinctly not the Endpoint or Midpoint of an Edge and will typically be a random point along the Edge. This is not preferred when attempting to find exact measurements along an Edge.

On Face

This is represented by a dark blue diamond on any Face. Like **On Edge**, this will be anywhere on the Face, and will typically be at a random location. This is also not preferred when attempting to find exact measurements on the Face, especially without first invoking the Linear Inferences (see the *Inferences* section).

Intersection

You will see a red cross (X) where two Faces or Edges intersect. This Snap may be activated when the following applies:

- Two Faces intersect without generating an Edge along that point

- The Geometries are in separate groups and two Edges intersect

- An Edge punctures a Face without intersecting with any of the bounding Edges

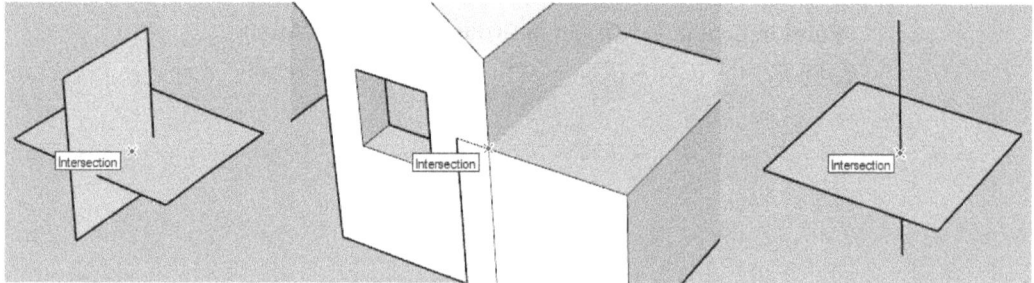

Figure 2.32: Two Intersecting Faces (Left) Groups with Intersecting
Edges (Middle) and an Edge puncturing a Face (Right)

Point in Group/Component Instance

These are Magenta/Purple Snaps that represent any of the previously listed Snaps where the Snap is referencing something in an Object. The Object name will be listed (if it is a named Component) or the Snap will be identified as **Point in Group**. This Snap helps to identify Objects even if they are not selected. We will discuss other Object properties (for Groups and Components) in the next chapter.

Origin

This is represented as a dark gray circle with a vertical cross that is either of the following:

- At the origin of the red, green, and blue Axes.

- At the Origin of an Object. This will be represented in magenta/purple.

Inferences

Inferences take Snaps one step further and allow for complex modeling and editing workflows. These Inferences are all activated by pausing in specific locations while performing different workflows until a Cyan or Magenta dashed or solid highlight appears. These Inferences may be considered too high-level for this stage in your SketchUp journey, but please review them and remember them as you see these Inferences appear while you model.

Linear Inferences

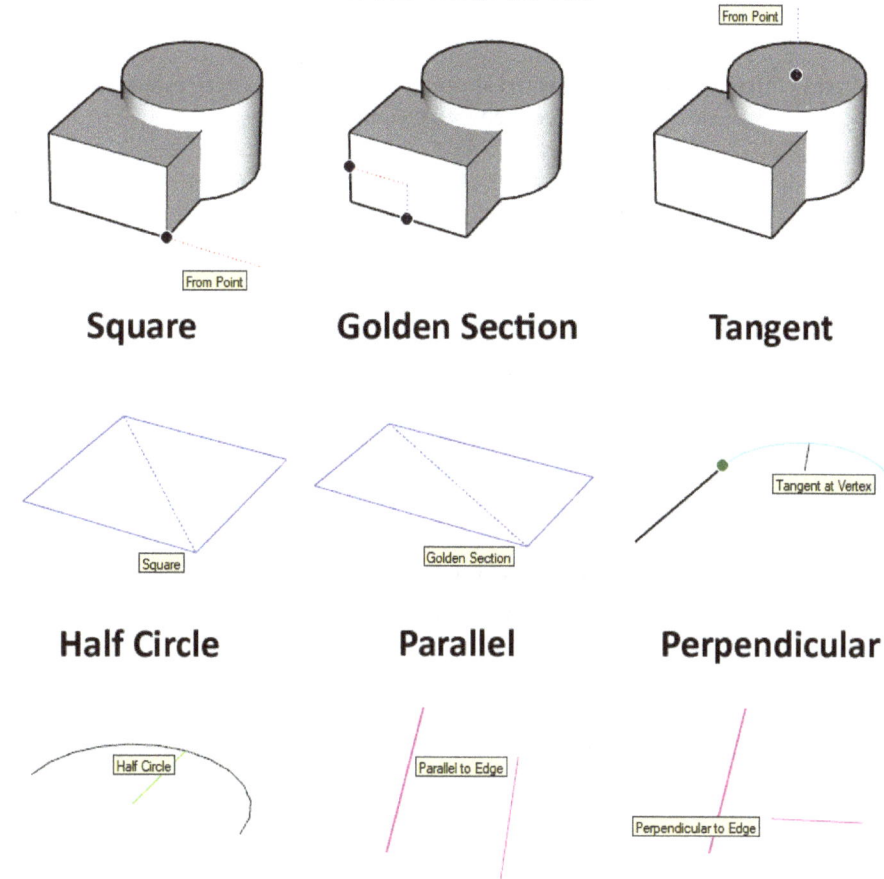

| Square | Golden Section | Tangent |

| Half Circle | Parallel | Perpendicular |

Figure 2.33: SketchUp Inferences

Linear Inferences

Linear Inferences allow for extensions from existing Geometry even without creating Guides or other temporary Geometries! These imaginary Inference lines always reference a Snap (Point Inference). Linear Inferences are an advanced way to model without using too many Guides!

To activate a Linear Inference, you should do the following:

1. Hover over a Snap for at least 1 second while showing the Snap Icon.
2. Move the cursor away from the Snap in the direction of the (hopeful) Linear Inference. This dotted or dashed line will now be able to be referenced while the current Tool is active.

SketchUp will show linear inferences along the Red, Green, and Blue Axes, and sometimes it will recognize a Magenta perpendicular or parallel Inference as well.

> **Hint**
>
> Multiple Linear Inferences can be activated at once! Use multiple Linear Inferences to find specific locations on a Face or even spaces between Geometries!

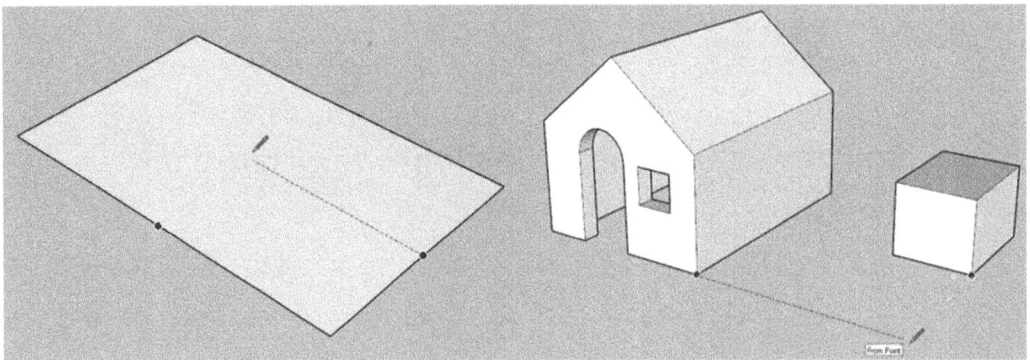

Figure 2.34: Linear Inferences show the middle of a Face (Left) and an
imaginary point is found between two Groups (Right)

Square

This will produce a dotted or dashed line inside of the Rectangle preview while using the Rectangle tool. The square will have four equal Edges bounding the Face.

Golden Section

This will produce a dotted or dashed line inside of the Rectangle preview while using the Rectangle tool. The Golden Section, or the Divine Proportion, is a mathematically defined and aesthetically pleasing ratio that is often used in design projects.

Tangent

There will be a Cyan highlight when using the Arc Tool in this case. It can detect tangency to an Edge or the endpoint of an Edge.

Half Circle

There is no highlight when using an Arc Tool in this case: it can snap to a half-circle shape. This will be tricky to see – there is no highlighting color, only the **Half Circle** text next to the cursor.

Parallel

There is a Magenta highlight when using the Line Tool in this case: it can create an Inference that is parallel to an Edge that is not aligned with an Axis.

Perpendicular

There is a Magenta highlight when using the Line Tool in this case: it can create an Inference that is perpendicular to an Edge that is not aligned with an Axis.

Summary

In this chapter, we discussed the Select Tool and the Eraser Tool. These tools are used in every SketchUp project! We also reviewed Snaps and Inferences. I hope you have begun to play with these Tools and Inferences – it will be great to get a handle on these elements before you begin to model!

At this point, you are probably excited to begin modeling, but before we get to that, we want to look at Groups and Components and how they must be used to create professional SketchUp models in SketchUp Pro.

In the next chapter, we will examine Groups and Components in SketchUp and introduce the concept of Sticky Geometry.

3
Modeling with Groups and Components

In this chapter, we will discuss principles of modeling in SketchUp when it comes to "sticky geometry." Sticky geometry is why we need to have Groups and Components, among other reasons. We will also discuss the importance of modeling using Groups and/or Components.

The following topics are covered:

- Sticky Geometry

- Cleaning up models with Groups and Components

- Understanding when to use Groups and Components

Working with "Sticky Geometry"

In this section, we will talk about "sticky geometry." What is sticky geometry? Why should we be concerned about it? And, most importantly, what does this have to do with Groups and Components in our SketchUp models? Let's spend a few sections discussing these questions.

What is Sticky Geometry?

All versions of SketchUp follow the same basic rules that we covered in the last chapter. Those are as follows:

- All edges are straight lines between two points

- All faces are bounded (surrounded) by edges

- All faces are two-dimensional (flat)

There is another rule that we need to add to the list, which has to do with the way geometry interacts when edges and faces intersect:

- Geometry in SketchUp will interact with the geometry that it touches in either of the following cases:

  - Both geometry objects are in the same group

  - Both geometry objects are in the same component

  - Neither geometry object is in a group or component (in the top level of the model)

> **Note**
>
> This includes geometry that is hidden and geometry that is on different Tags (previously Layers).

So, what does this mean? This could be shown in several different ways.

> **Note**
>
> You do not need to follow along with any of the sticky geometry examples. These are just for you to review for now! As a reminder, we will be covering all editing tools in later chapters!

In this first example, we will start with a single Edge:

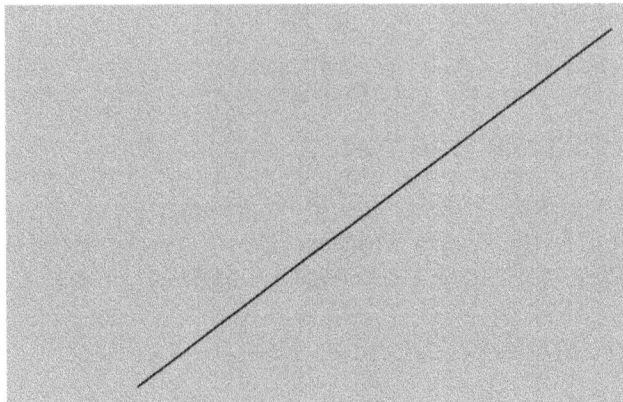

Figure 3.1: Single Edge

Now, when a Rectangle is drawn that touches this line, it is visually attached:

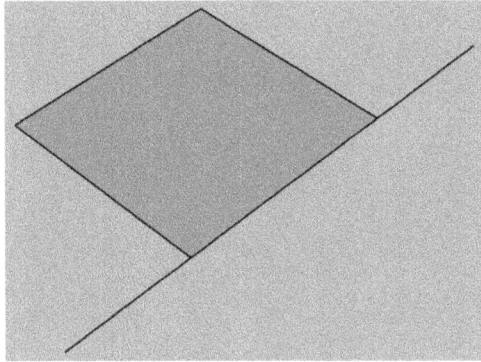

Figure 3.2: Rectangle drawn to align with existing Edge

Because both elements are on the same "level" (meaning that they are both at the top level of the model, not in a group or component), they will interact. To begin, the rectangle has split the edge into three separate line segments.

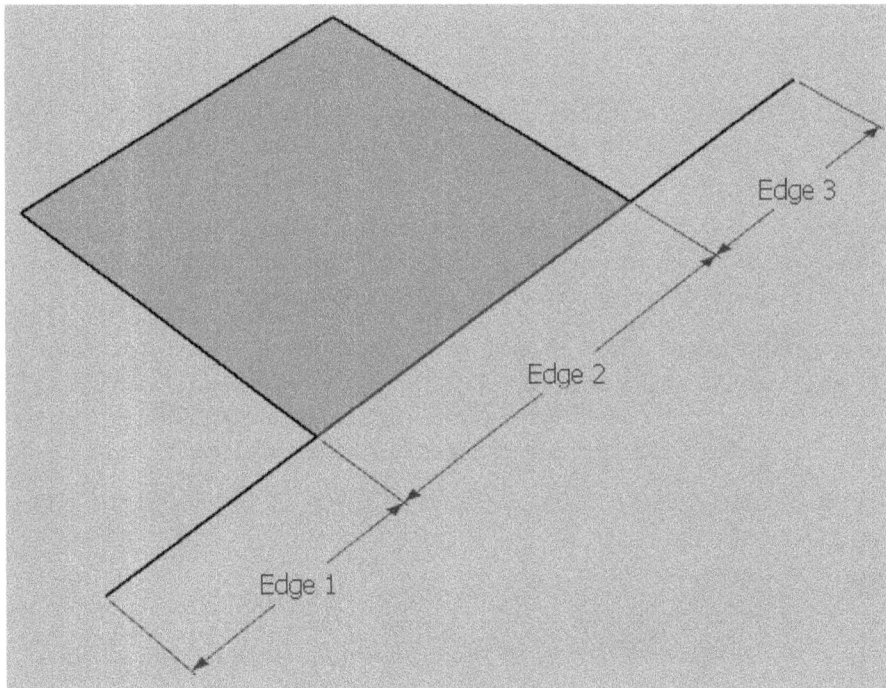

Figure 3.3: The original edge has been split into three segments

This will happen with colinear geometry (lines that are on top of each other) and with intersecting geometry (lines that cross).

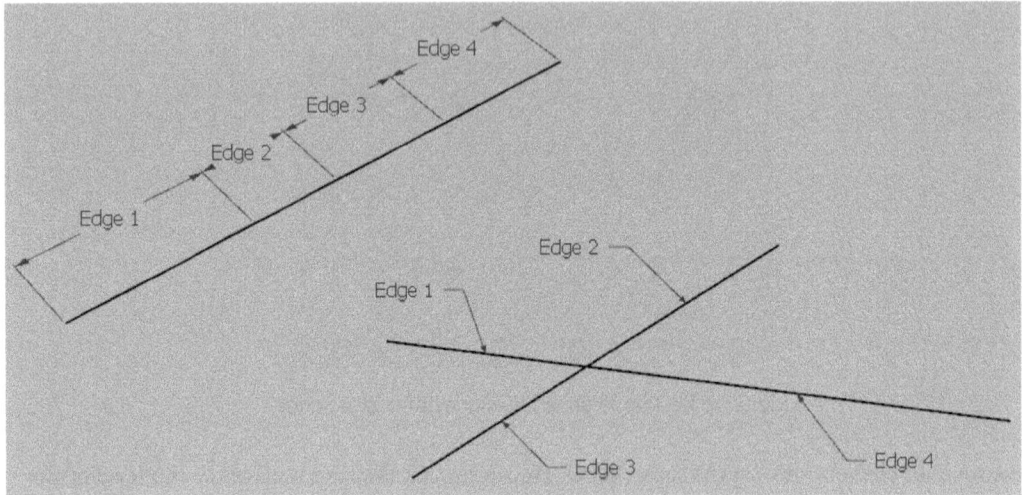

Figure 3.4: Colinear Edges (Left) and intersecting Edges (Right)

> **Note**
>
> All these interactions happen in three dimensions. This does not just happen with "flat" geometry! Any vertical or diagonal edges, shapes, or 3D geometry will also interact when they intersect.

This interaction can be very useful. And, a lot of the time, this is intentional. But the interaction is not a static process. Once geometry begins to interact with the geometry around it, it does not want to stop interacting. This is what we mean when we say "sticky geometry."

Geometry does not stick and move in its originally drawn shape, but only the parts that are connected. We think of edges as solid lines, but SketchUp "sees" that edge as two endpoints *connected* by a line. So, if one of the endpoints moves, the edge may change length, angle, or both. This could occur when connected objects are moved, scaled, rotated, or any of the edit tools are used.

In the following examples, we will look at how an edge reacts with sticky geometry. This will be the starting image for the next three examples:

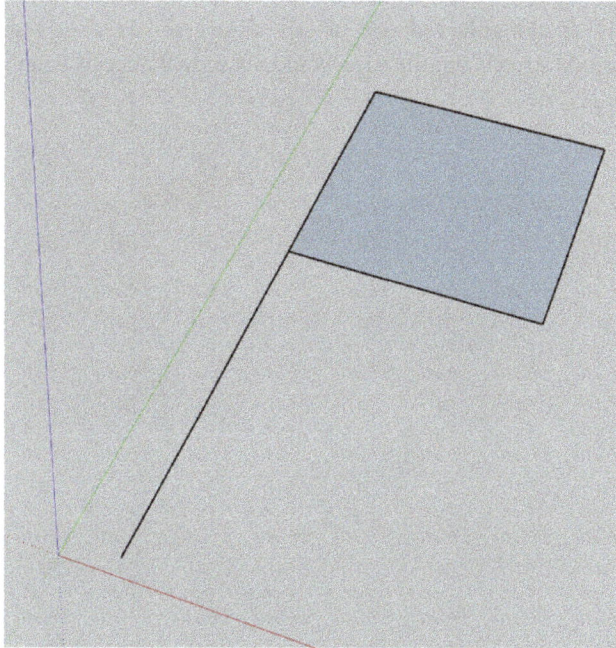

Figure 3.5: An Edge that is connected to a Rectangle

When the rectangle is moved along the green axis, the edge will stick to the rectangle with one endpoint, and the line will stretch to change its length.

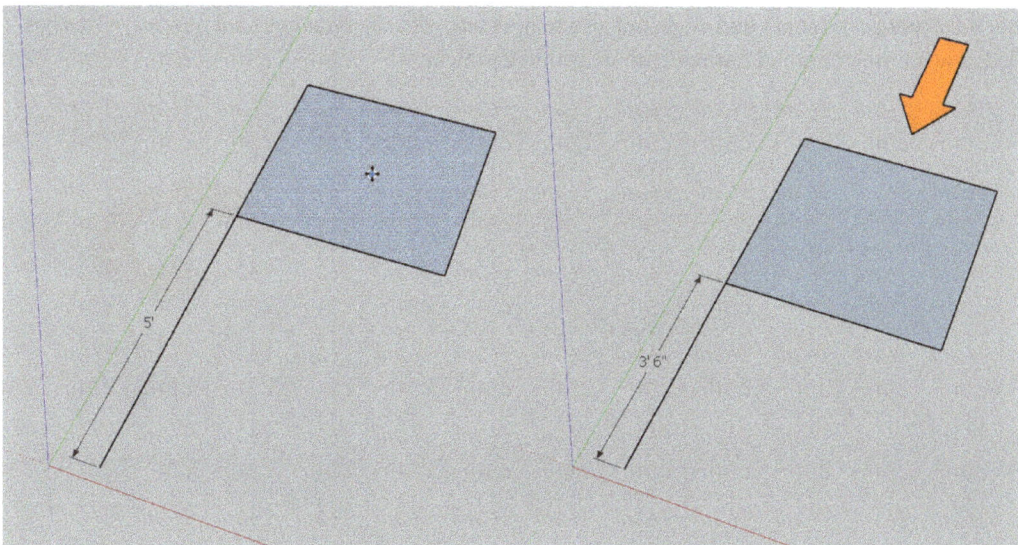

Figure 3.6: The original geometry (Left), the rectangle is moved, and the Edge changes its length (Right)

If the rectangle were moved along the red axis instead of the green axis, the edge endpoint would still stick and the edge would stretch, but the edge would not only change its length but also its angle.

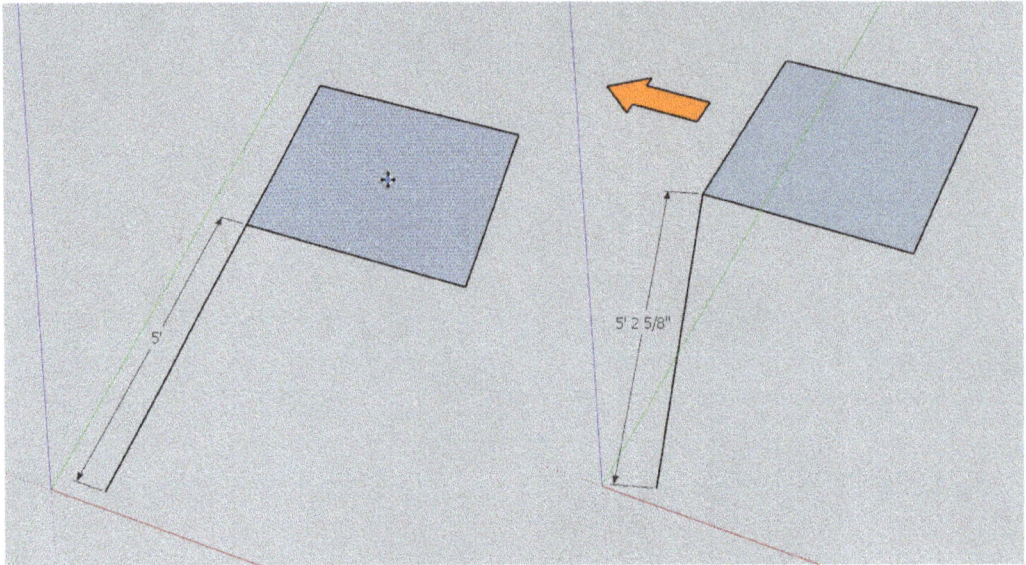

Figure 3.7: The original geometry (Left), the rectangle is moved,
and the Edge changes length and angle (Right)

This geometry could change due to any of the edit tools, and this will change the edge in many ways due to sticky geometry. In this last example, the rectangle has been rotated. This has caused the edge to not only change its length and angle but also to intersect with the rectangle in a new way! This has added new intersections and has made more geometry sticky.

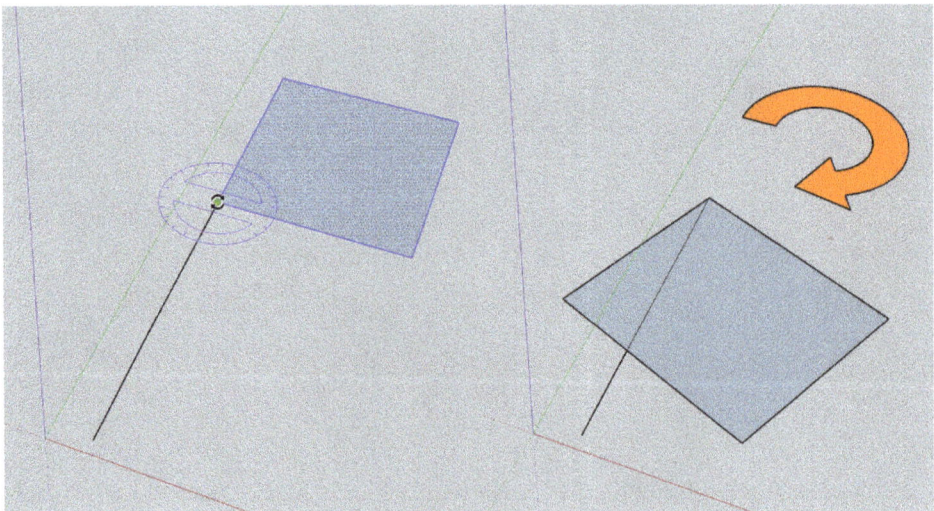

Figure 3.8: The original geometry (Left), the rectangle is rotated (Right)

You can see that where we began with five edges (the line has one and the rectangle has four), we now have seven! The line has been split into two edges, and one of the sides of the rectangle has been split as well. We began with one face (inside the rectangle), and that has also been split.

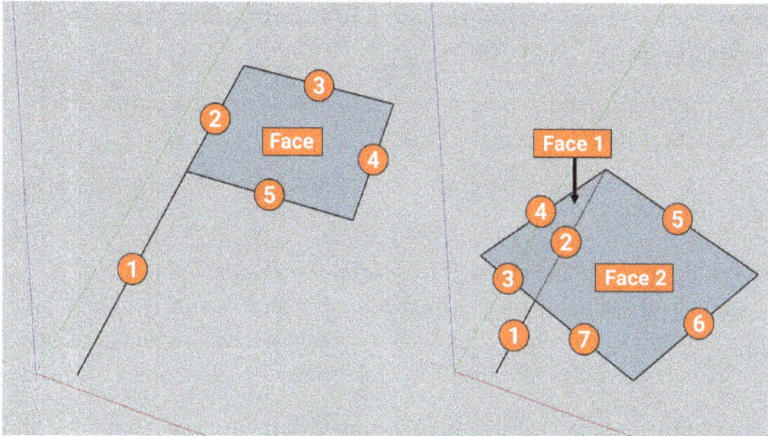

Figure 3.9: Original five edges and one face (Left) and updated seven edges and two faces (Right)

Sticky geometry can be frustrating for these reasons, but it can also be helpful. When and how should we avoid sticky geometry? And when should we use sticky geometry, and does it really make our modeling workflows more efficient? We will spend some time examining these questions.

Avoiding Sticky Geometry

How can we avoid Sticky Geometry? Separate the Geometry by creating a Group or Component! We will examine Groups and Components in depth later in this chapter, but first, we want to talk about why they are so important!

Geometry should **always** be grouped or made into a Component while modeling. You do not have to group the geometry right away, but always before you move on to another part of the model! We will discuss how to create Groups and Components later in this chapter. Also, we will talk about the differences between the two, and when to use one over the other.

But how do Groups and Components stop sticky geometry? From a general perspective, they do not.

Remember our rule:

- Geometry in SketchUp will interact with geometry that it touches in either of the following cases:

    A.   Both geometry objects are in the same group

    B.   Both geometry objects are in the same component

    C.   Neither geometry object is in a group or component (in the top level of the model)

Even within groups and components, sticky geometry can still cause an issue! This is where we may need to use **nested** groups or components. We will talk about modeling best practices – including nested groups and components – at the end of this chapter.

The final takeaway: If two things are not supposed to stick together, make sure that you first separate them using groups and components.

There are occasions where we can use sticky geometry to our advantage. It is purposefully programmed into SketchUp Pro, so there must be some use cases!

Using Sticky Geometry

Sticky geometry is often used in SketchUp in high-level workflows. Typically, sticky geometry is confusing and frustrating for beginning SketchUp Pro users. However, we will quickly look at some more advanced workflows to understand where sticky geometry can be used! As with previous examples in this chapter, these are only for context. Please do not feel the need to follow along with these examples.

SketchUp may refer to this process as **Stretching Geometry** or **Autofold**. Stretching and Autofold may automatically be applied when using editing tools, but in some cases, it may need to be forced by using a toggle key. If it is an option, it will appear in the Prompts bar.

> Example
> Autofold is an option when using the move tool. It can be toggled by using the *Alt* key.

Example #1

In this first example, we have a simple hole in a rectangle. This might represent a hole for a window in a wall:

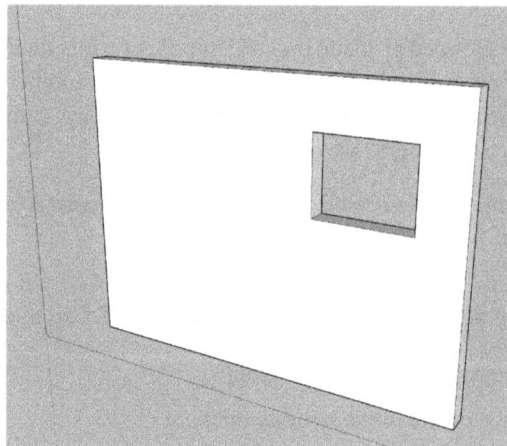

Figure 3.10: Rectangular hole in a vertical box

We will soon learn that Push Pull can extrude single faces in *Chapter 4, Drawing Tools: We Begin Modeling!*. But sticky geometry can also be used for this purpose. By selecting the face inside the hole and using the Move tool, the hole can be expanded or stretched.

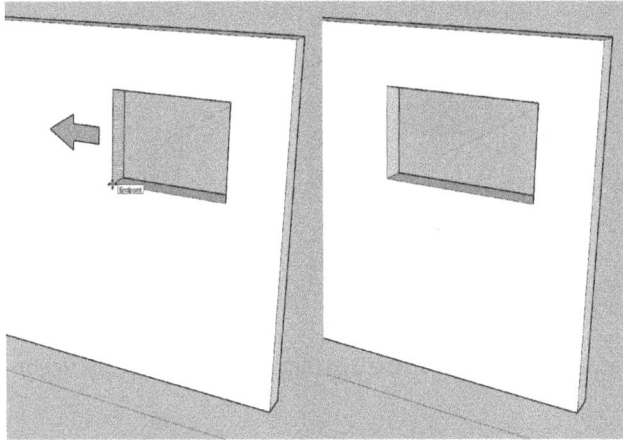

Figure 3.11: The face is selected (Left) and moved to expand the hole (Right)

This can also be used for more than one face. If the entire hole needed to be moved to the side and down along the face of the big box, it would require four separate Push Pull workflows. Instead, by selecting all the geometry that is connected to the hole (four faces and surrounding edges) and using the move tool, the entire hole can be moved.

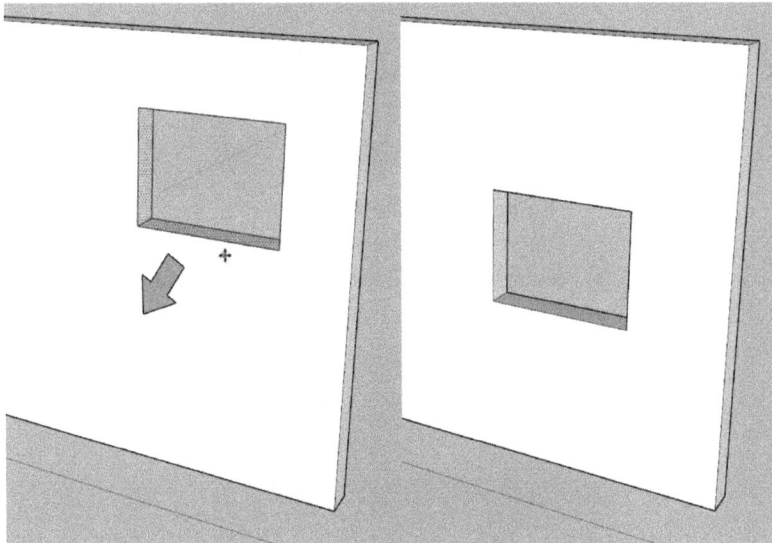

Figure 3.12: The hole geometry is selected (Left) and moved to
the side and down using the Move tool (Right)

It may appear, at a first glance, that not much happened in these two workflows. But the faces on the box were changed without us even thinking about it! You can see that the original shape looks much different than the updated shape:

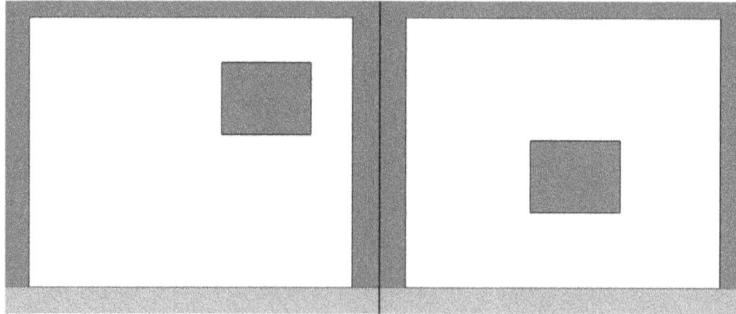

Figure 3.13: Original face shape (Left) and updated face shape (Right)

Example #2

This example will focus on "folding" faces. This is an interesting workflow that can be used for a variety of applications. The simplest form is to create a tent shape. First, start with a simple rectangle. This rectangle has four edges and one face:

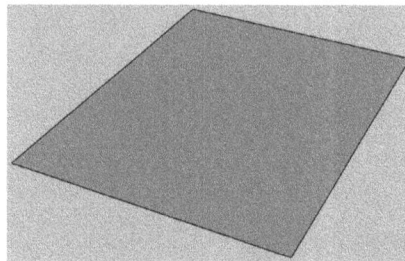

Figure 3.14: Simple Rectangle

Then, an edge is drawn to split the rectangle in two. This creates two faces and six edges in the original rectangle. The added edge makes the total geometry two faces and seven edges:

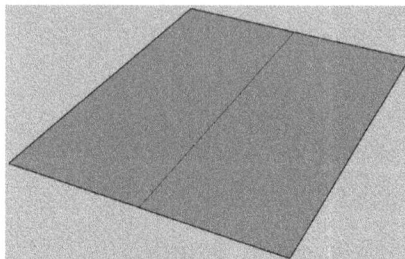

Figure 3.15: Rectangle split into two faces and seven edges

The new edge can be moved up in 3D space to make a tent shape:

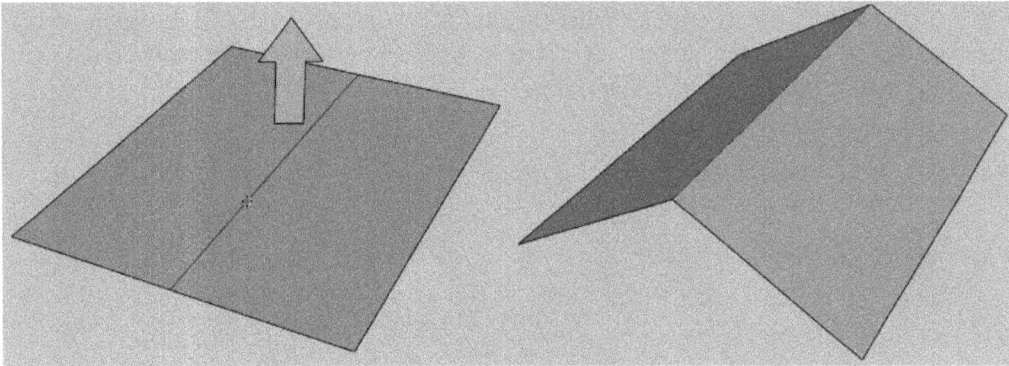

Figure 3.16: Line moved with the Move tool (Left) and the resulting tent shape (Right)

This new geometry has two faces and six edges. The edges on the left and right sides of the rectangle were not changed at all, and the middle dividing edge was not changed except that it was moved vertically in the space. The big changes were that the original edges at the top and bottom of the shape were stretched, which caused their lengths to increase and changed the angle. Because of this change, the faces that were inside the edges appeared to rotate "up" to create the tent shape:

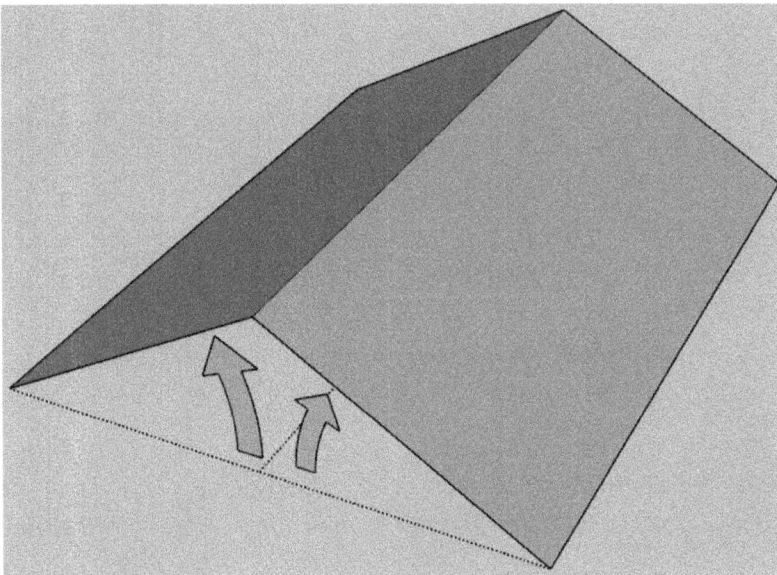

Figure 3.17: Edges stretched and edges appear to rotate

This was a very clean approach to "folding" faces. But this approach can also be used to fold faces in several ways.

Example #3

For this last example, we will continue to look at folding faces. However, it will be important to understand that using different editing tools will result in a slightly different geometry. You can see, in the figure, that we will begin with a simple rectangle:

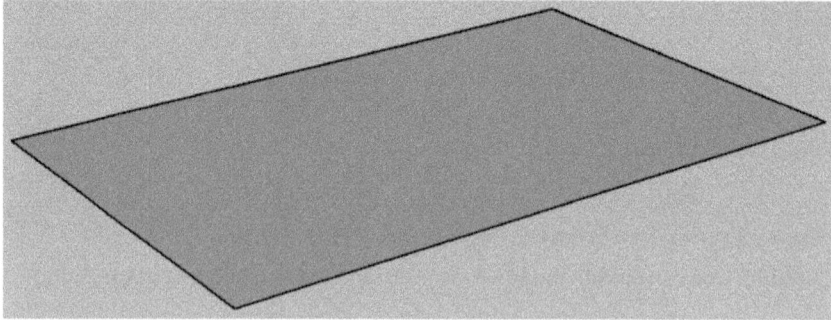

Figure 3.18: Simple rectangle

We have added an edge that intersects with two connected edges, creating a new triangular face at one of the corners. Again, this has resulted in two faces and seven edges:

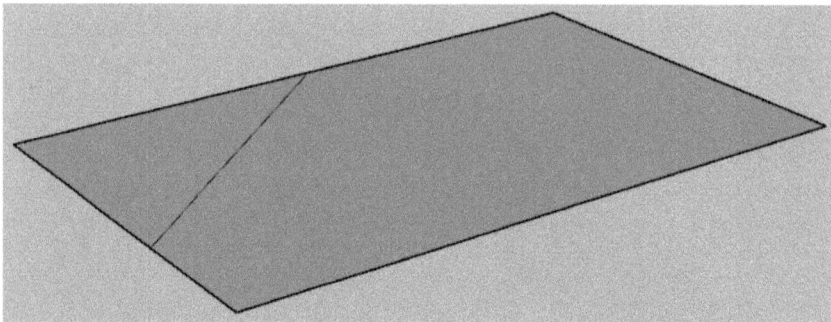

Figure 3.19: Rectangle split to create two faces and seven edges

There are two ways to "fold" the isolated corner faces up in the vertical direction:

- Move the endpoint up in the vertical space
- Rotate the triangular corner face

Moving the endpoint is like the workflow that we looked at when creating the tent in the previous example. This appears to rotate the faces, but it only stretches and lengthens them. We can see how this triangle has been stretched and lengthened by moving the endpoint:

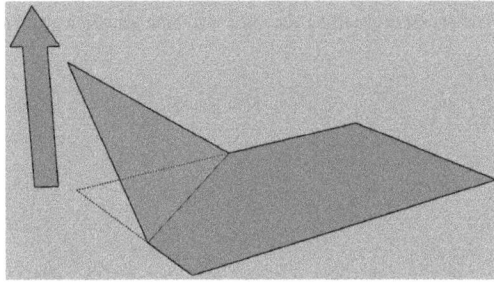

Figure 3.20: Endpoint moved in the vertical direction to stretch the triangular face and connected edges

Notice how the original footprint of the triangle is the same size when viewed from above:

Figure 3.21: Original rectangle outline and folded triangle viewed from above

The second way to "fold" the triangle is more like how we would fold a piece of paper in real life. This does not change the size of the triangle or connected edges, but will instead only change the angle of the edges and face.

> **Note**
> This is a more advanced workflow and will include the use of additional geometry. We will discuss how to use the Rotate tool in *Chapter 4, Drawing Tools: We Begin Modeling!*.

We will start with the same rectangle with split faces to create a triangular face in one corner:

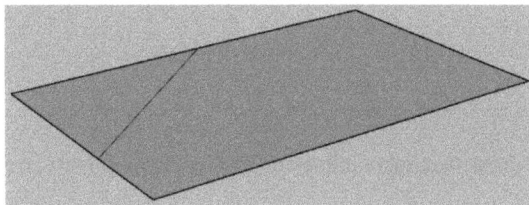

Figure 3.22: Rectangle split to create two faces and seven edges

Instead of moving the endpoint to stretch the face, we will use the Rotate tool to rotate the face and connected edges:

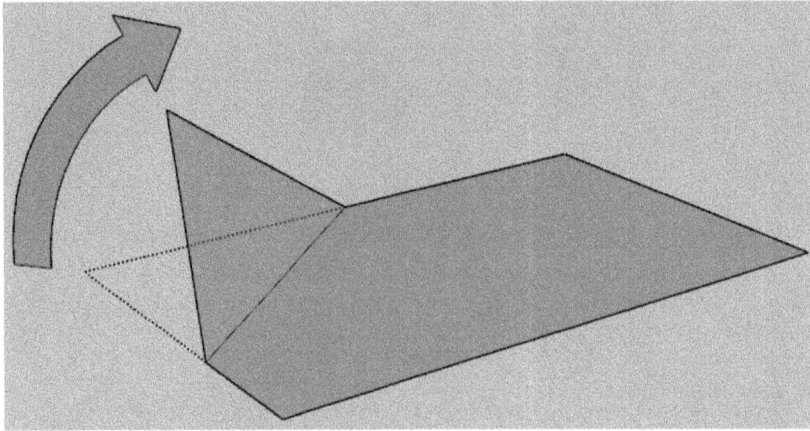

Figure 3.23: Rotating the rectangle around the edge connected to the other face

Note

Again, this is an advanced workflow. This will fully depend on the angle of the Rotate tool and the basepoint to get these results.

Notice how, this time, the original footprint is outside of the folded rectangle when viewed from above:

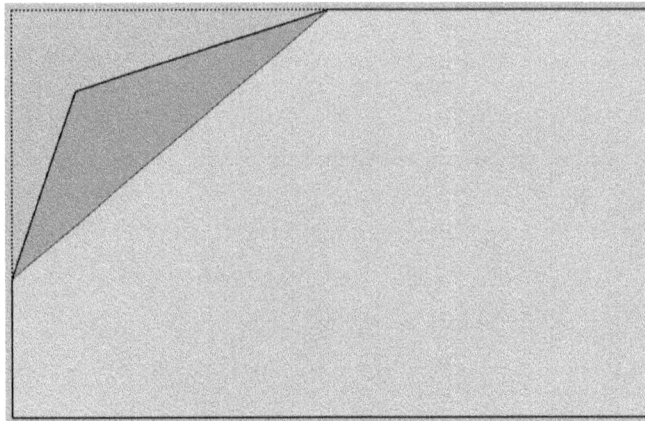

Figure 3.24: Original rectangle outline and folded triangle viewed from above

The difference between these options can most easily be seen when looking at the shapes from the side:

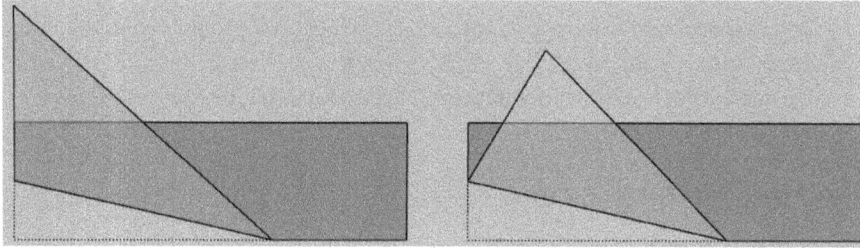

Figure 3.25: Rectangle with moved endpoint and Autofold (Left), and rectangle with rotated face (Right)

Note

Because Autofold is a feature in the Move Tool, we will review some Autofold examples in *Chapter 5, Editing Tools – Making Big Changes!*.

Sticky geometry and Autofold are great tools to have in your back pocket when modeling. But we still need to focus on the basics! Sticky geometry and Autofold are advanced-level workflows. To begin, we need to learn more about separating our geometry so that it does not stick together or fold when we do not want it to. As we mentioned earlier, this is achieved through **Objects**.

Model with Groups and Components (Objects)

Groups and Components are known as Objects in SketchUp. We will spend some time discussing these Objects, as you should use them in every single project you work on in SketchUp Pro. Understanding how to create, edit, and work with Groups and Components is an essential skill in SketchUp!

Objects in SketchUp

SketchUp Pro defines Groups and Components as Objects. In fact, they are the only two Objects in the software! So, what is an Object?

SketchUp defines four essential ideas that help to define Objects and their behavior:

- Objects hold other entities (geometry)
- Objects can be nested within other objects
- Objects do not stick to other entities outside their own object
- Objects can be locked

Source: `help.sketchup.com`

We will quickly examine each of these defining statements.

It is difficult to define groups and components without using the word "group." In the simplest terms, SketchUp organizes (groups) and isolates geometry when you create a group or component. This is what SketchUp defines as "holding other entities." We have defined those entities so far as basic geometry, or edges and faces.

Objects Hold Other Entities

Groups and components both hold entities in the same way. When a group or component is created, the held/collected geometry acts as a single object. Creating Objects is a simple process. After you have selected the geometry and/or any objects you would like to group or add to a new component, you can do the following:

- Right-click and choose **Make Component...**

- Right-click and choose **Make Group**

  Choose either of these options from the edit **Menu** Bar dropdown

- Pick the **Create Component** tool from the **Large Tool Set** Toolbar

You will notice the ellipsis after **Make Component** in both the right-click contextual menu and the Edit drop-down menu. Choosing this option will open the **Create Component** dialog box, which we will discuss in *Chapter 11, Working with Components*. Creating a group will happen without any additional dialog boxes or inputs required.

When you create a group or component, you will notice that it immediately acts differently when you click on it with the Select tool. Non-grouped geometry can have its individual edges and faces selected, while grouped geometry will appear as a bounding box when selected. This is the same for components. We can see this in the following figure:

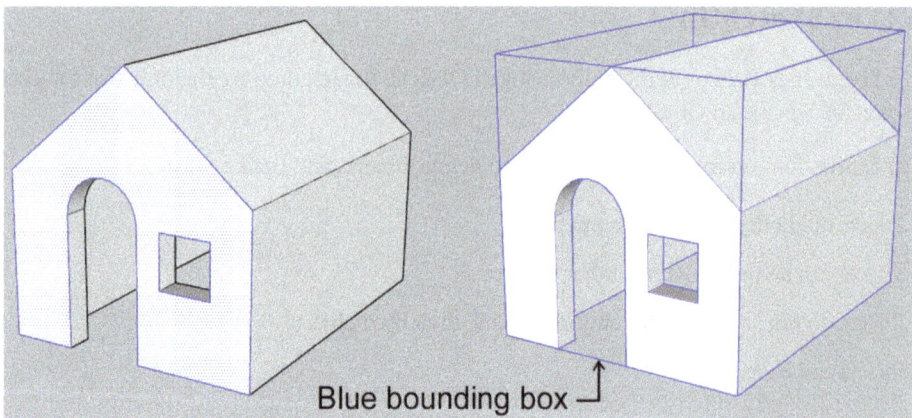

Figure 3.26: Non-grouped geometry and selected face (Left), and grouped geometry with bounding box (Right)

> **Note**
>
> The bounding box is calculated by showing the outermost geometry in the Axes directions. The bounding box will quickly tell you if you have selected the correct geometry to create the object.

Once an object is created, it is not locked forever! The geometry inside the object can be edited, but first, the contextual edit for the group or component must be opened. This can be achieved in two ways:

- Using the **Select Tool**, double-click on the object
- Right-Click on the object and choose **Edit Group/Edit Component**

> **Note**
>
> The Select tool must be used to edit the object using a double click, but not to activate the right-click context menu.

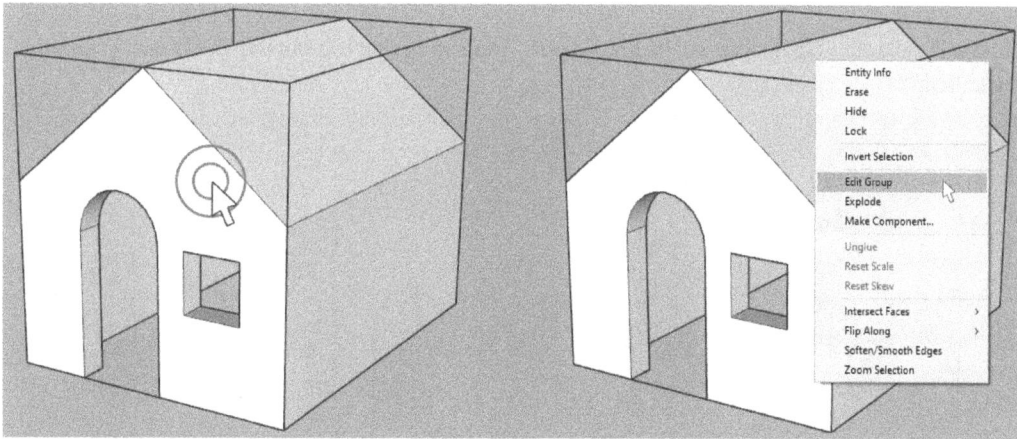

Figure 3.27: Double-click using the Select tool (Left), and using the right-click context menu (Right)

> **Note**
>
> In either case, the object will first be selected to show the bounding box as if it were only clicked once with the Select tool.

Then, after the double-click, or after **Edit Group** or **Edit Component** has been selected, a new box will appear. This is represented by dotted lines and will seem to "float" outside of the object's bounding box. This will confirm that you have opened the Object and can now edit the Object's contents.

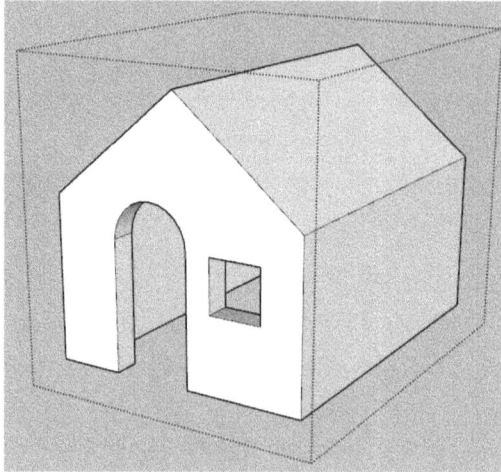

Figure 3.28: Dotted box showing object is open and can be edited

Additionally, all other Geometry will be grayed out – indicating that it is outside the Object currently being edited.

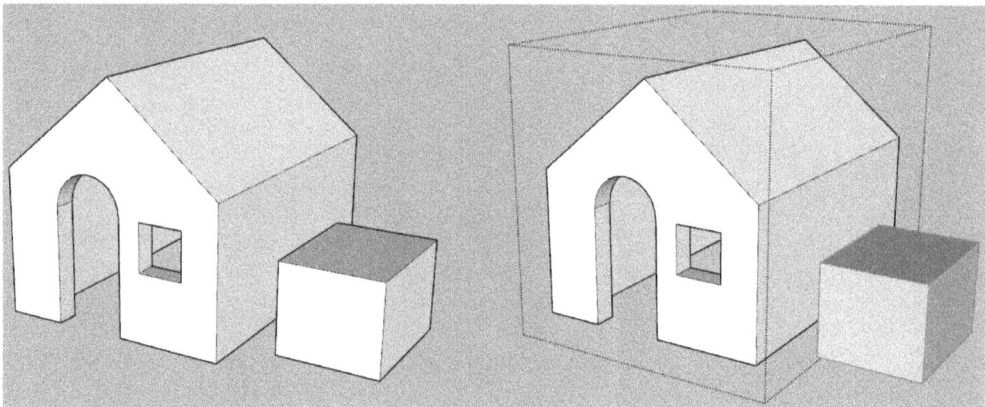

Figure 3.29: Object and Box are Selectable (Left), and Object is Opened and Box is Grayed Out (Right)

The object's contextual edit can be closed similarly to how it was opened, by doing either of the following:

- Using the select tool, click anywhere in the drawing area outside of the context box
- Right-Click and choose **Close Group/Close Component**
- Choose **Edit | Close Group/Component** in the Menu Bar

Note

We will cover advanced workflows for editing components in *Chapters 9* and *10*.

Objects can also be exploded, which removes the group or component but leaves the geometry behind.

> **Note**
> Exploding objects should only be done when you are sure that you do not need the object any longer. Objects should not be exploded instead of editing the object from within the context box.

To explode an object, simply Right-Click on the object and choose **Explode**.

Objects Can Be Nested Within Other Objects

We have also mentioned that groups and components can be nested within each other.

Nesting simply means putting an object inside another object. There is no limit to the number of groups or components that can be contained in each other. Nesting is an essential workflow when creating complex models in SketchUp Pro.

Nesting does not require special tools or workflows. All it takes is to create a new object while selecting at least one other object and/or other geometry.

We will discuss workflows for nesting objects in much more depth in *Chapter 9*.

Objects Do Not Stick to Other Entities Outside Their Own Object

Of course, we have spent much of this chapter discussing sticky geometry. We should now understand how important it is to make sure that our geometry does not stick to other geometry!

Objects Can Be Locked

It is important to understand the difference between locking an object and how objects automatically separate geometry in a SketchUp model. Even though object geometry is separated from the rest of the model, the object itself can still be edited using the edit tools, such as Move, Rotate, and Scale. Also, the geometry inside of the object can be accessed and edited.

When an object should not be edited in any way, including editing the grouped geometry, it can be Locked. An object can be locked by right-clicking the object and choosing **Lock**. A locked object can be Unlocked using the same workflow.

> **Note**
> Locking objects should not be confused with hiding objects. We will discuss Hiding objects and other geometry later in this book.

Locked geometry will be immediately recognizable, as it will have a red bounding box when selected instead of a blue bounding box.

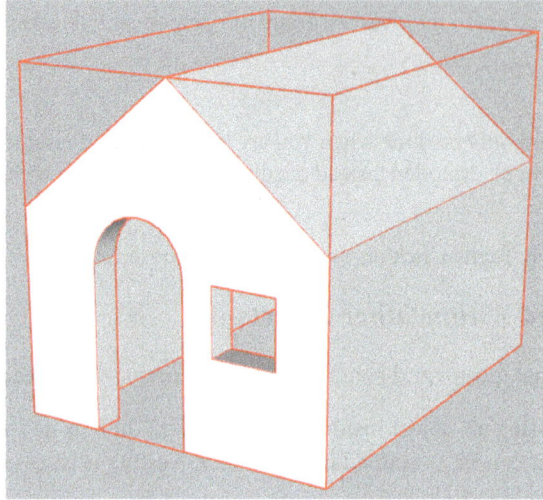

Figure 3. 30: Red bounding box around a locked object

Now that we know what Objects are, how and when should we use Groups and Components? And, how do we know when to use one over the other?

Deciding When to Use Groups versus Components

When modeling, it can sometimes be confusing to decide between using a group versus a component. They are both objects, and both fully align with the object rules we discussed earlier. However, components are much more complex, and we will discuss these complexities in *Chapters 9* and *10*.

So, when do we use a group and when do we use a component? There are a few suggested rules that will make your model and your workflow more efficient. There is no "penalty" for using one over another, but you might get frustrated with yourself! Some general guidelines are listed in the following table:

Use a Group When:	Use a Component When:
The object will only be used once.	The object will be used more than once, especially if it will be copied/repeated.
The object is a custom design for a specific use.	The object is a general placeholder.
The object is a very complex, singular entity that defines a large part of the model. Example: Connected walls in a building model.	The object is a small, reusable object that is used in many parts of the model. Example: A door, window, or piece of furniture.

> **Note**
>
> One thing to keep in mind: Many copies of Groups will increase a file size much faster than many instances of Components – if something will be used many, many times, it should be a Component!

To state this another way, any time something will be repeated/copied, use a component instead of a group. Any time something has a custom design or is very complicated, use a group instead of a component.

Summary

This chapter might not make much sense right now, and that is okay! You will start to get a feel for the differences between using Groups and Components. We will be practicing with Groups and Components as we move forward with the rest of this book! It is important to remember that if your Geometry sticks together, you might need to create a Group or Component.

To practice with Groups and Components, we first need to know how to create Geometry. It is finally time to begin modeling! We will start to look at the Drawing tools in the next chapter.

4
Drawing Tools – We Begin Modeling!

In this chapter, we will begin modeling! More importantly, we will start to understand how to use the Drawing tools to make sure we can effectively create basic geometry. We will also look at how faces may heal themselves while drawing and discuss some simple workflow organizations for creating 2D Geometry in SketchUp Pro.

The following topics are covered in this chapter:

- Modeling with the Drawing Tools:

    Line, Freehand, Rectangle, Rotated Rectangle, Circle, Polygon, Arc, 2 & 3 Point Arc, Pie

- Creating 3D, Solid, and Watertight Geometry
- Understanding how Faces heal

Modeling with the Drawing Tools

We learned about the different types of SketchUp Geometry in *Chapter 1, Introduction to SketchUp Pro*, in the *Geometry "Rules"* section. Those rules are set out here:

- All edges are straight lines between two points
- All faces are bounded (surrounded) by edges
- All faces are two-dimensional (flat)

Using these rules, we can create straight Edges and flat Faces. We also discussed that we could use Edges and Faces to create Rectangles, Polygons, and Curves (Welded Edges). The Drawing tools are the tools we use to create all of this Geometry!

The Drawing tools can all be found on the Getting Started toolbar, the Large Tool Set toolbar, and the Drawing toolbar. The Getting Started toolbar utilizes dropdowns to show the nested tools, which create miniature collections of the Drawing tools:

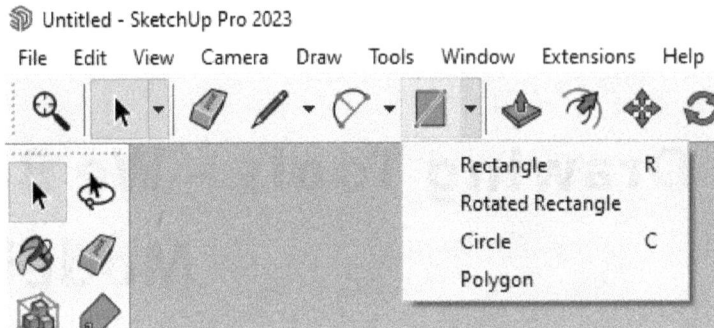

Figure 4.1 – Getting Started Toolbar and Drawing Tool collections

These collections are the Line tools (including the Line and Freehand tools), the Arc tools (including Arc, 2 Point Arc, 3 Point Arc, and Pie), and the Polygon tools (including Rectangle, Rotated Rectangle, Circle, and Polygon). These collections can help us organize how we think about each of these tools. The Large Tool Set toolbar organizes these tools into two vertical rows, which almost creates a series of pairs, as shown in the following screenshot:

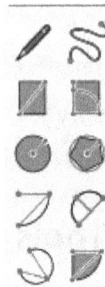

Figure 4.2 – Large Tool Set Toolbar and Drawing Tool "Pairs"

In this case, the collections are further broken into pairs. Here, we can clearly see that Line and Freehand, Rectangle and Rotated Rectangle, Circle and Polygon, Arc and 2 Point Arc, and 3 Point Arc and Pie are paired across from one another.

> **Note**
> I would suggest that Arc and Pie are a better pair. Look at that later in this chapter!

You do not need to have the Drawing toolbar open if the Getting Started toolbar or the Large Tool Set toolbar is open. The Drawing tools should always be accessible on your screen!

Where Can You Draw Geometry?

When most SketchUp projects are started, the first bit of Geometry is often drawn on the ground plane (which is the Red-Green plane). But in SketchUp, you can draw Geometry pretty much anywhere, as outlined here:

- On the ground plane

- On a vertical plane

- On existing faces

- Separate from existing geometry (aligned to an axes plane)

- Inferenced from existing geometry

How can we set up which area to draw on? How does SketchUp "know" that you want to draw a vertical Rectangle or a circle on an angled Face? This all comes down to the camera angle!

Typically, our camera in SketchUp is looking mainly from the top down onto the ground plane. This makes sense—this is how we view the real world! But if you want to draw on a different vertical plane (such as the Red-Blue or Red-Green Axis planes) or on a vertical or angled Face, the best solution is to "look at" the Face!

This does not have to be perfect; you could orbit the camera to roughly look in the right direction. But you can always "look at" a face by using the **Align View** option. To align the camera with a Face, proceed as follows:

1. Select the Face.

2. Right-click and choose **Align View**.

You can see this in the following screenshot where the Face is selected (Left in the following figure), we Right-Click and choose **Align View** (Middle), and the camera orients to the Face (Right):

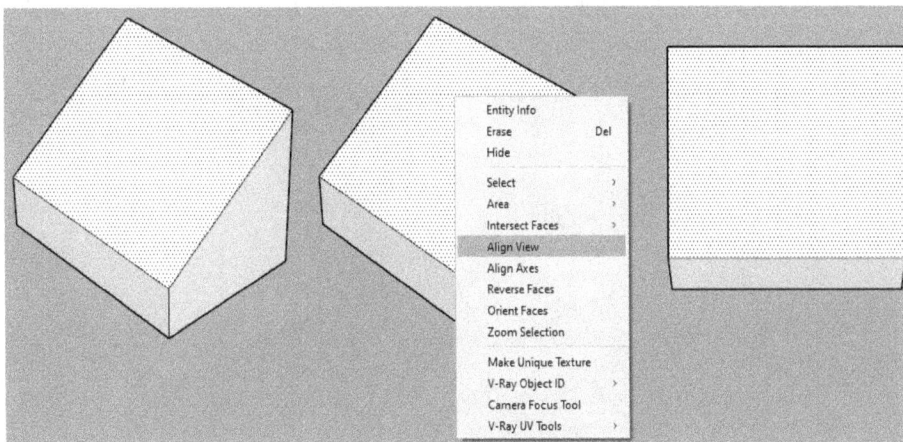

Figure 4.3 – Align View

Once the camera angle is set up, you can start drawing on any surface or plane! When a drawing is started, it makes sense to begin on one of the Axes planes, usually the ground plane:

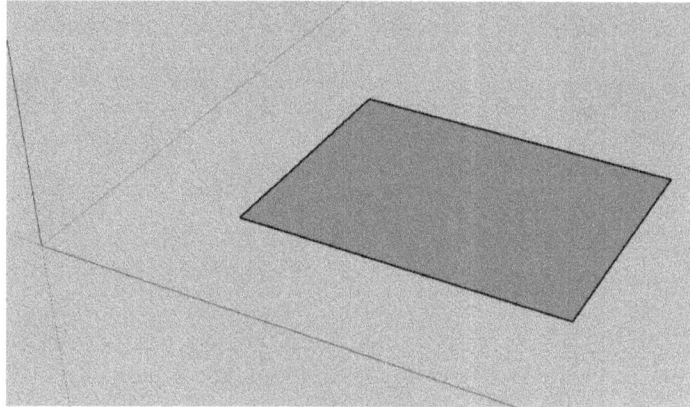

Figure 4.4 – Rectangle drawn on the ground plane

Then, after some more Geometry has been added to the model, new Geometry can be created using any of the existing shapes, or even with additional Inferences. You do not have to draw directly on a Face to match the angle—you might only want to use it as an Inference.

To use a Face as an Inference plane, proceed as follows:

1. Start a Drawing tool.

2. Hover over the Face and hold *Shift*.

3. Draw the Geometry once you can see the Magenta Inference.

You can see this workflow in the following screenshot where we have the Active Drawing tool (Left), the cursor aligns to the Face (Middle), and Geometry is drawn using a Magenta Inference (Right):

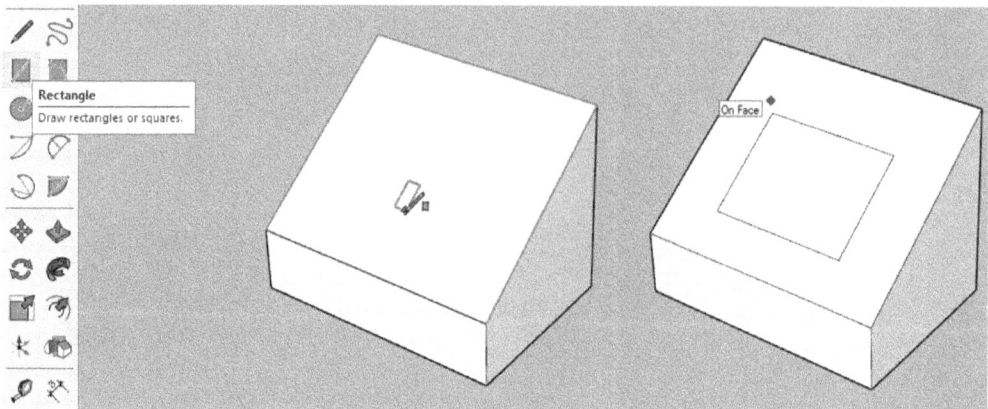

Figure 4.5 – Using a Face as an Inference plane

> **Note**
>
> If you ever start drawing on the wrong Face or Inference, you can hit the *ESC* key to stop the current workflow. The tool will remain active, so you can give it another shot!

Now that we know where to draw, let's learn about the tools used to create geometry from the Drawing toolbar.

Line Tool

The Line tool is a great place to start any SketchUp model! The Line tool is used to create straight Edges (also called line entities). The Line tool is represented by a pencil:

Figure 4.6 – The Line Tool

The Line tool can be activated by doing the following:

- Clicking on any of the Line tool buttons
- Hitting *L* on the keyboard

The Line tool can create an Edge by clicking anywhere in the Drawing Area. This is achieved by doing the following:

1. Activating the Line tool
2. Clicking to start to draw an Edge
3. Moving the cursor along an axis, or in any direction, and clicking to finish the Edge

> **Note**
>
> The Line tool will continue to draw Edges from the last point! The Line tool will only stop if the new Edge connects to another Edge, which creates a Face.

We can see how to draw a simple Edge in the following screenshot:

Figure 4.7 – Starting the Edge and moving the cursor (Left); Clicking to complete the Edge (Right)

You can also see that when we draw a shape with the Line tool, we are still in the tool to draw back-to-back Edges. Then, once the shape is completed, a Face is created and the Line tool ends the drawing:

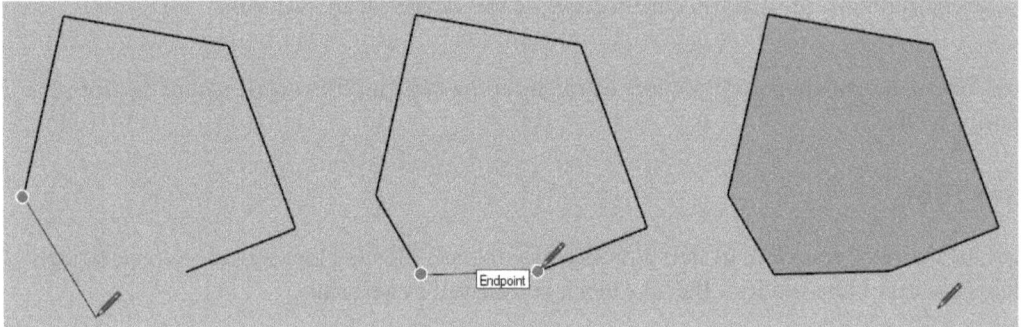

Figure 4.8 – Drawing a shape (Left), closing the shape (Middle), and the Line Tool ends the drawing (Right)

> **Note**
>
> This is a hint at the process of Auto-Heal or how Faces are created by only drawing Edges. This will be further explained later in this chapter.

The Line tool is simple in concept, but it can be used in many, many workflows! Practice using the Line tool to create complex Geometry, but do not use it exclusively! SketchUp features many other Drawing tools that can help us with some Geometry.

Freehand Tool

The Freehand tool is paired with the Line tool. This is clear on the Large Tool Set toolbar and when looking at the Cursor for the Freehand tool, which is represented by a pencil and a freehand scribble:

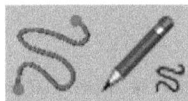

Figure 4.9 – Freehand Tool Button (Left) and Freehand Tool Cursor (Right)

The Freehand tool can be activated by clicking on any of the Freehand tool buttons. There is no keyboard shortcut for the Freehand tool. The Freehand tool creates a freehand curve (Welded Edges) by clicking the left mouse button and dragging on the screen, like so:

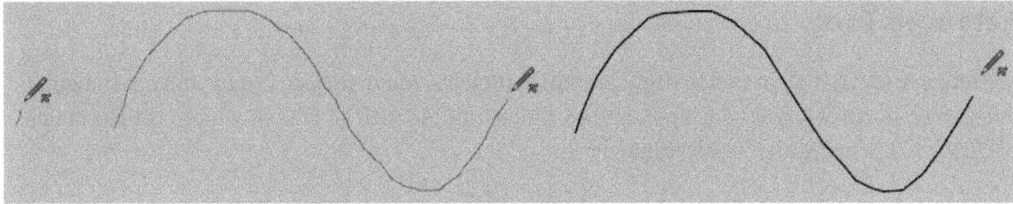

Figure 4.10 – Clicking and holding (Left); Moving the mouse around the
screen (Middle); Releasing to end the Freehand Curve (Right)

The Freehand tool will only end when the mouse is released. It does not automatically end when it intersects other Geometry—including itself! This means that the Freehand tool can snap to inferences, but it can also intersect Geometry at random locations to create Faces and split Edges:

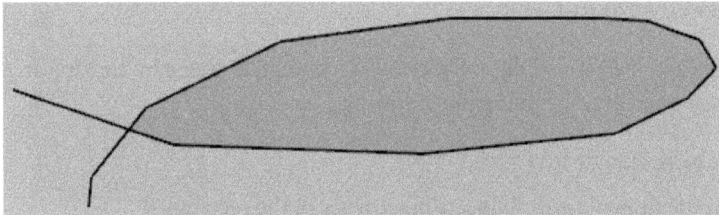

Figure 4.11 – Face created by a Freehand shape that created a loop

The Freehand tool will always create a Curve of Welded Edges. The number of Edges in the Curve is defined by how long the mouse button was held. The two Curves in the next screenshot follow roughly the same path, but the Curve on the left was created in roughly 1 second, creating 10 Edge Segments, and the Curve on the right was created in roughly 5 seconds, creating 29 Edge Segments:

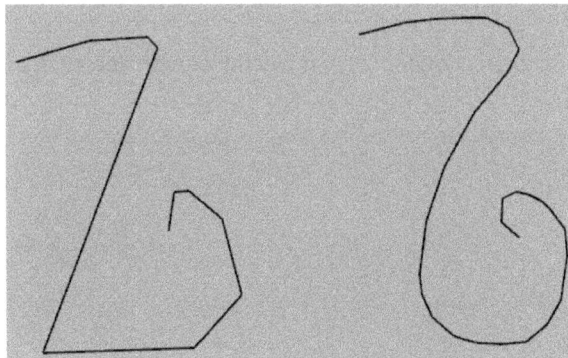

Figure 4.12 – 1 second, 10 Edge Segments (Left); 5 seconds, 29 Edge Segments (Right)

Additionally, any small mouse movements might be captured by the Freehand tool and could add to the Edge Segment count.

Rectangle Tool

The Rectangle tool is a great tool to begin SketchUp projects. Many projects begin with a rectangular shape and are further refined during the project workflow. As with all Drawing tools, the Rectangle tool cursor is a pencil with a small rectangle:

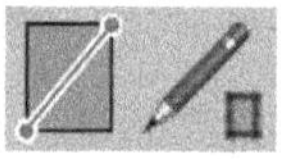

Figure 4.13 – Rectangle Tool Button (Left) and Rectangle Tool Cursor (Right)

The Rectangle tool can be activated by doing the following:

- Clicking on any of the Rectangle tool buttons
- Hitting *R* on the keyboard

The Rectangle tool can create four Edges that create a rectangular shape by clicking anywhere in the Drawing Area. This is achieved by doing the following:

1. Activating the Rectangle tool
2. Clicking on the drawing area to locate one corner of the rectangle
3. Moving the cursor in any direction and clicking to set the opposite corner of the rectangle

> **Note**
> The sides of the rectangle will automatically align with the Red, Green (or Blue for vertical) Axes. To create any other rectangle, you can use the Rotated Rectangle tool, or manually rotate the rectangle after it is created.

For example, click to start the rectangle (Left), move the cursor, and click to set at the opposite corner (Right):

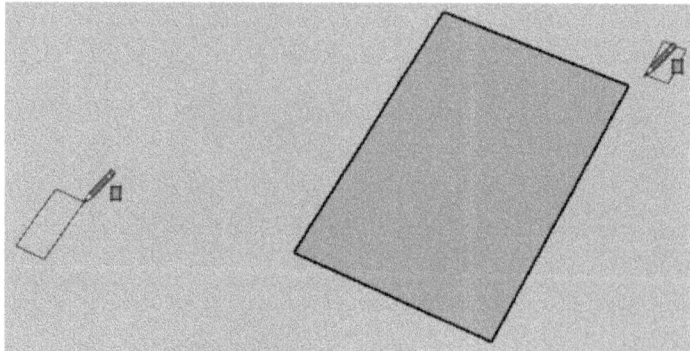

Figure 4.14 – Creating a rectangle

You can also create rectangles with a desired size. You can achieve this by doing the following:

1. Activating the Rectangle tool

2. Clicking on the drawing area to locate one corner of the rectangle

3. Typing in two numbers separated by a comma (these represent the length and width)

4. Hitting *Enter* on the keyboard

> **Note**
>
> If there are no specified units (inches, meters, and so on), then SketchUp will use the drawing units. You can type in any recognized distance unit.

In the following example, you can see a rectangle of 9 inches by 2 feet. The Length value is 9 inches (along the Red Axis) and the Width value is 2 feet (along the Green Axis). This is represented in the Measurements box with a typed value of (9",2'):

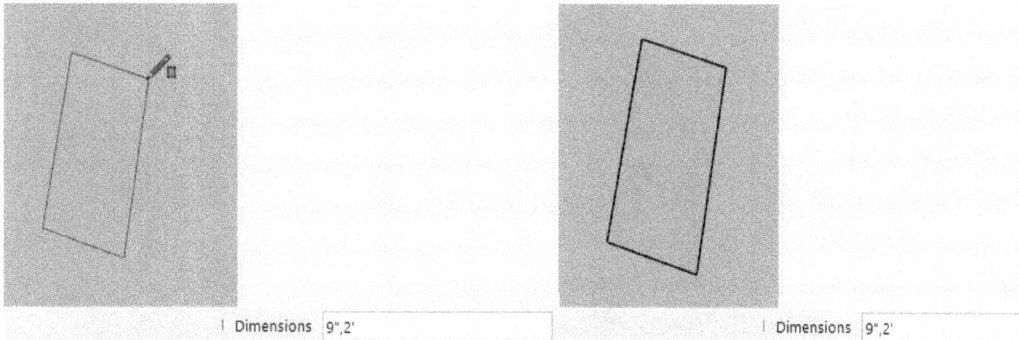

Figure 4.15 – Rectangle started (Left); Measurement typed into
Measurements Box (Middle); Finished Rectangle (Right)

> **Note**
>
> You do not need to click in the Measurements box before you type a value. Once a Drawing tool has been started (by clicking on the screen to start drawing), you can begin typing. Also, you can update the size of a rectangle or other Geometry even after it has been drawn if no other tool has been activated.

The Rectangle and Rotate Rectangle tools can activate two specific Inferences: Square and Golden Section. These are discussed in detail in *Chapter 2, Principal Tools, Axes, and Inferences*.

Rotated Rectangle Tool

The Rotated Rectangle tool can be useful when you need to draw a rectangle whose face is at an angle to the default Red, Green, or Blue Axes or to other Geometry. Instead of clicking two opposite corners to create a rectangle, the Rotated Rectangle tool uses two clicks to create one of the sides of the Rectangle (this is the length), and then a third click to establish the angle and width of the rectangle. This can be achieved by doing the following:

1. Activating the Rotated Rectangle tool

2. Clicking on the drawing area to locate one corner of the rectangle

3. Clicking again, this time to establish one Edge of the rectangle

4. After a rotate gizmo appears, clicking a third time to establish the other end of the rectangle

We can see this in the following example:

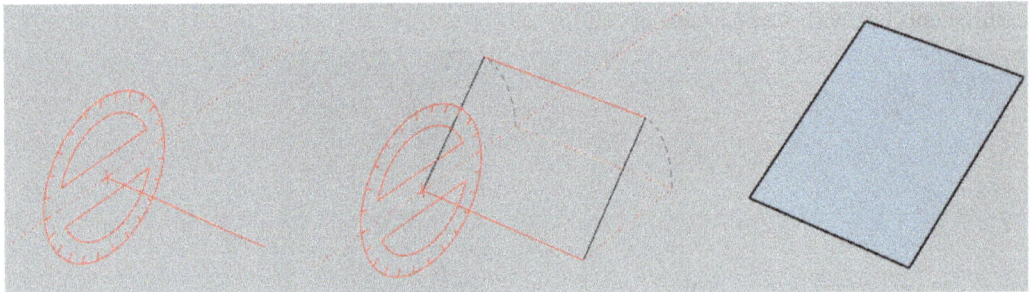

Figure 4.16 – Two clicks establish one Edge (Left), the gizmo shows the angle
and width (Middle), and the rectangle is completed (Right)

The Rotated Rectangle tool can also create rectangles with a desired size. You can achieve this by doing the following:

1. Activating the Rotated Rectangle tool

2. Clicking on the drawing area to locate one corner of the rectangle

3. Typing in a distance and hitting *Enter* after showing the desired direction (this represents length)

4. After a rotate gizmo appears, typing in two numbers separated by a comma (these represent the angle (in degrees) and width)

5. Hitting *Enter* on the keyboard

> **Note**
>
> Any combination of entering Measurements and clicking can be used to create a rectangle using the Rotated Rectangle tool.

In the following example, we will again show a rectangle of 9 inches by 2 feet, but this time it will be rotated off the ground plane (Red-Green Axis plane) by 30 degrees. The Length value is 9 inches (along the Red Axis), the angle is 30 degrees, and the width value is 2 feet (toward the positive Green Axis direction). This is represented in the Measurements box with a typed value of 9″ then 2′ , 30:

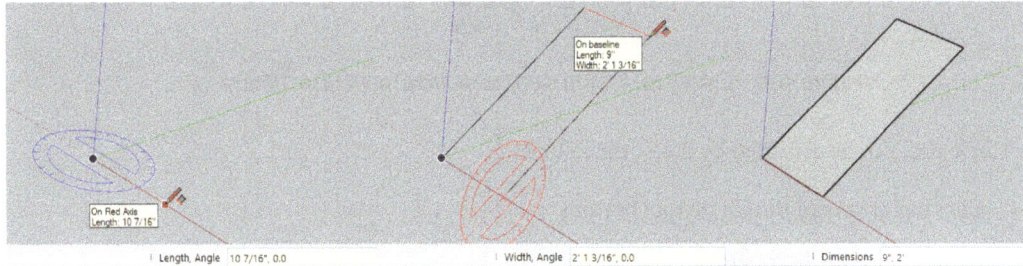

Figure 4.17 – Rectangle started with Length Measurement (Left); Angle and Width Measurements (Middle); Rectangle (Right)

> **Note**
> The Rotated Rectangle tool can find Parallel to Edge Inferences when completing the angle and width portion of the tool workflow.

Circle Tool

Circles and Arcs in SketchUp are drawn with a series of Line Segments that appear to create a smooth curve. This will be essential to understand before drawing any curved shapes with the Circle or Arc Drawing tools. After activating any of the Circle or Arc tools, the Measurements Box will prompt for a **Sides** value.

Before beginning to draw the Geometry, you can type in a number and hit *Enter* on the keyboard to set the number of sides for that Geometry. The number of sides is mainly a visual change, but it will impact the performance of the model if too many sides are chosen. It will depend on you to decide how many Sides each circle or arc should have. In most cases, unless the shape is a large feature in the project, it is sufficient to have 8 to 24 Sides, like so:

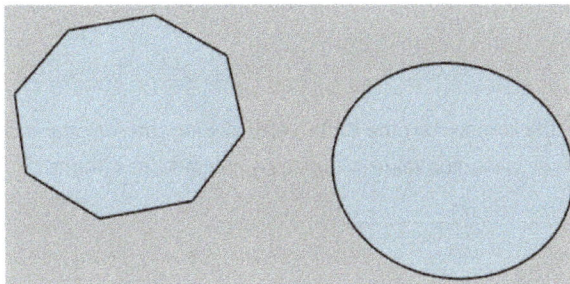

Figure 4.18 – A circle with 8 sides (Left) and a circle with 64 sides (Right)

The Circle tool is the only way to create circles in SketchUp. As with all Drawing tools, the Circle tool cursor is represented by a pencil with a small circle:

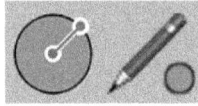

Figure 4.19 – Circle Tool Button (Left) and Circle Tool Cursor (Right)

The Circle tool can be activated by doing the following:

- Clicking on any of the Circle tool buttons
- Hitting *C* on the keyboard

The Circle Tool creates a circle using three values: the number of Sides, a center point, and a radius value. This is achieved by doing the following:

1. Activating the Circle tool
2. Ensuring that the number of sides is correct, typing the value if it should be changed, and hitting *Enter*
3. Clicking to set a center point
4. Clicking to set a second point to set the radius distance value

You can see the result in the following screenshot:

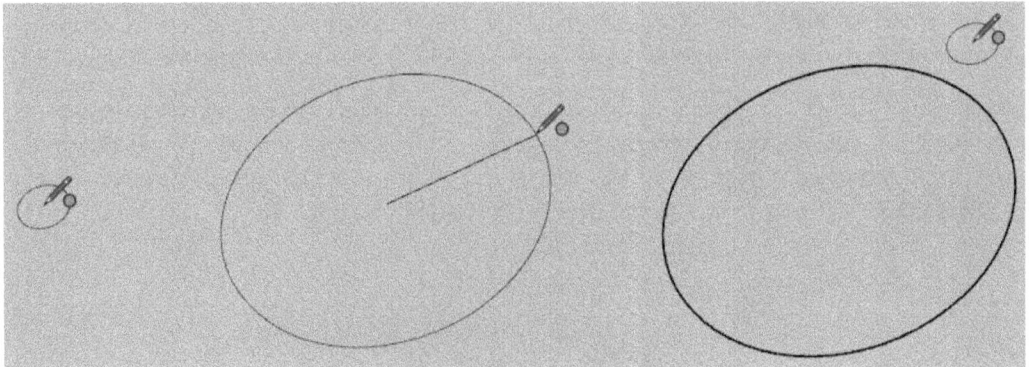

Figure 4.20 – Clicking to start the circle (Left); Moving the cursor and clicking to set the radius value (Middle); completed circle (Right)

> **Note**
> Circles can be drawn on any surface or plane, just as with the Rectangle tools!

You can also create circles with a desired size. You can achieve this by doing the following:

1. Activating the Circle tool
2. Ensuring that the number of sides is correct, typing the value if it should be changed, and hitting *Enter*
3. Clicking to set a center point
4. Showing a direction and typing in a radius value
5. Hitting *Enter* on the keyboard

If we wanted to draw a circle with 20 Sides and a radius of 10', we would start by activating the circle tool. Then, we would update the sides by typing 2 0 into the Measurements Box and then hitting *Enter*:

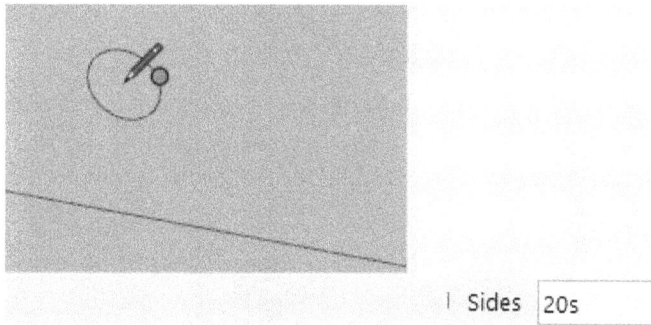

Figure 4.21 – Circle Tool activated (Left); Sides updated in Measurements Box (Right)

Once the number of Sides has been locked in, we can set the center point and type in our radius value:

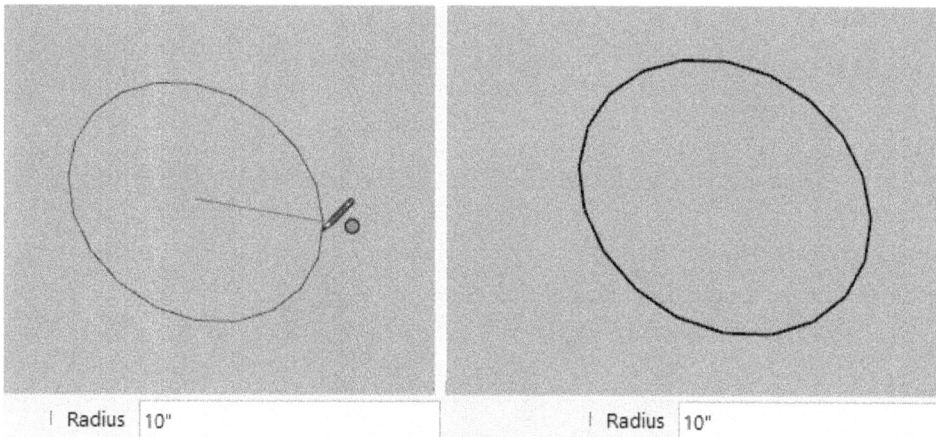

Figure 4.22 – Clicking to set center point (Left); Radius value entered using Measurements Box (Middle); Finished Circle (Right)

> **Note**
>
> The Entity Info dialog box can be used to update the number of Sides or the radius value even after a circle has been drawn. Make sure to select the outer edge of the circle (not the face) to update the information in the Entity Info dialog box.

Polygon Tool

The Polygon tool is paired with the Circle tool. Instead of Line Segments acting as a smooth curve, the Polygon tool creates distinct straight Edges that create a regular polygon. We can see an example pentagon (five sides) on the Polygon tool button and Cursor, which is represented by a pencil and a polygon:

Figure 4.23 – Polygon Tool Button (Left) and Polygon Tool Cursor (Right)

The Polygon tool can be activated by clicking on any of the Polygon tool buttons. There is no keyboard shortcut for the Polygon tool.

The Polygon tool creates a polygon using the same three values as the Circle tool: the number of Sides, a center point, and a radius value. This is achieved by doing the following:

1. Activating the Polygon tool

2. Ensuring that the number of sides is correct, typing the value if it should be changed, and hitting *Enter*

3. Clicking to set a center point

4. Clicking to set a second point to set the radius distance value

> **Note**
>
> There are two additional options: **Inscribed** and **Circumscribed**. We will look at these later in this section.

The result of this can be seen in the following screenshot:

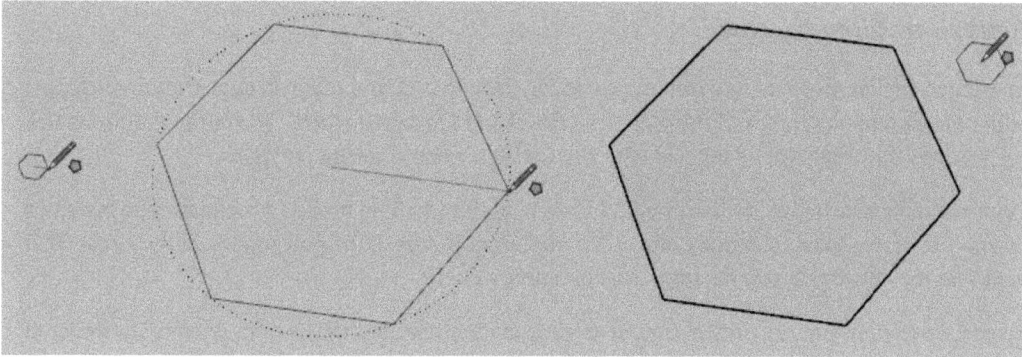

Figure 4.24 – Clicking to start the polygon (Left); Moving the cursor and
clicking to set the radius value (Middle); completed polygon (Right)

You can also create polygons with a desired size. You can achieve this by doing the following:

1. Activating the Polygon tool
2. Ensuring that the number of sides is correct, typing the value if it should be changed, and hitting *Enter*
3. Clicking to set a center point
4. Showing a direction and typing in a radius value
5. Hitting *Enter* on the keyboard

If we wanted to draw a polygon with six Sides and a radius of 5', we would start by activating the Polygon tool. Then, we would update the sides by typing "6" into the Measurements Box and then hitting *Enter*:

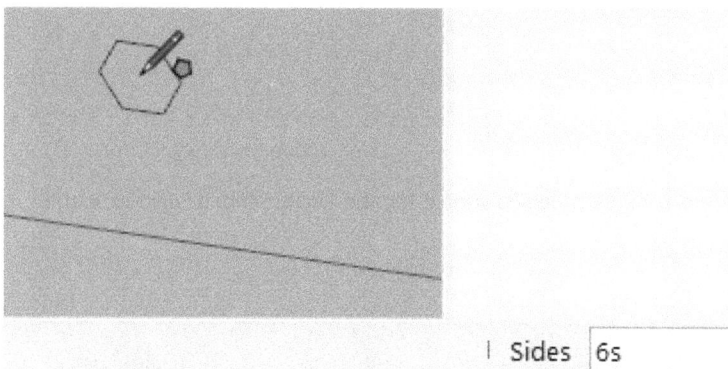

| Sides 6s

Figure 4.25 – Polygon Tool activated (Left) Sides updated in Measurements Box (Right)

Once the number of Sides has been locked in, we can set the center point and type in our radius value.

Inscribed or Circumscribed?

You will notice that there are two other options for the Polygon tool that will appear in the Prompts and the Measurements Box, and these are **Inscribed** and **Circumscribed**. One of these options will be active, and the other will be toggled with the *Ctrl* key when drawing polygons.

What is the difference between these options? Does it matter when we model? Absolutely, yes! Inscribed polygons are drawn in an imaginary circle, and all the endpoints of the polygon touch the circle. That means that the entire polygon fits inside the imaginary circle.

Circumscribed polygons are drawn around an imaginary circle, and all the midpoints of the polygon touch the circle. The Edges of the polygon are tangent to the imaginary circle. That means that the polygon is slightly larger than the imaginary circle.

In both cases, a Radius value is used. For Inscribed polygons, this will be the distance from the center point to the endpoints of the Edges of the polygon. For Circumscribed polygons, this will be the distance from the center point to the midpoints of the Edges of the polygon. We can see this in the following screenshot:

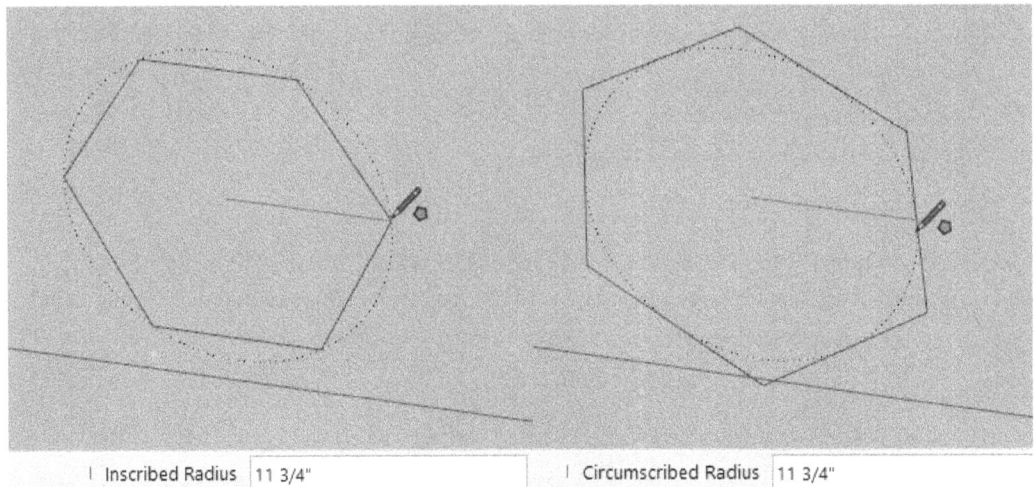

Figure 4.26 – Inscribed Polygon (Left) and Circumscribed Polygon (Right)

> **Note**
> These polygons are drawn so that the widest point is along the Red Axis. Inscribed and Circumscribed polygons might need to be drawn along different Axes to ensure that the Geometry is in the desired orientation. The Inscribed Polygon was drawn with the radius along the Red Axis, and the Circumscribed polygon was drawn with the radius along the Green Axis.

Arc Tool

We discussed earlier that Arcs in SketchUp are drawn with a series of Line Segments that appear to create a smooth curve. After activating any of the Arc Tools, the Measurements Box will prompt for a "Sides" value. Before beginning to draw the Geometry, you can type in a number and hit *Enter* on the keyboard to set the number of sides for that Geometry. The number of sides is mainly a visual change, but it will impact the performance of the model if too many sides are chosen. As with all Drawing tools, the Arc Tool cursor is represented by a pencil with an arc:

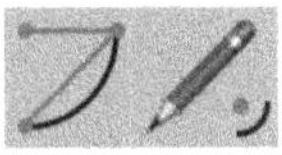

Figure 4.27 – Arc Tool Button (Left) and Arc Tool Cursor (Right)

The Arc Tool can be activated by clicking on any of the Arc tool buttons. The *A* keyboard shortcut will activate the 2 Point Arc tool, not the regular Arc tool. The Arc tool creates an arc using the same three values as the Circle tool: the number of Sides, a center point, and a radius value. Additionally, a radial value is required to represent the length of the Arc, and this is represented by an angle value. This is achieved by doing the following:

1. Activating the Arc tool

2. Ensuring that the number of sides is correct, typing the value if it should be changed, and hitting *Enter*

3. Clicking to set a center point

4. Clicking to set a second point to set the radius distance value

5. Clicking to set the arc length (represented by an angle around an imaginary circle)

This will give you the following output:

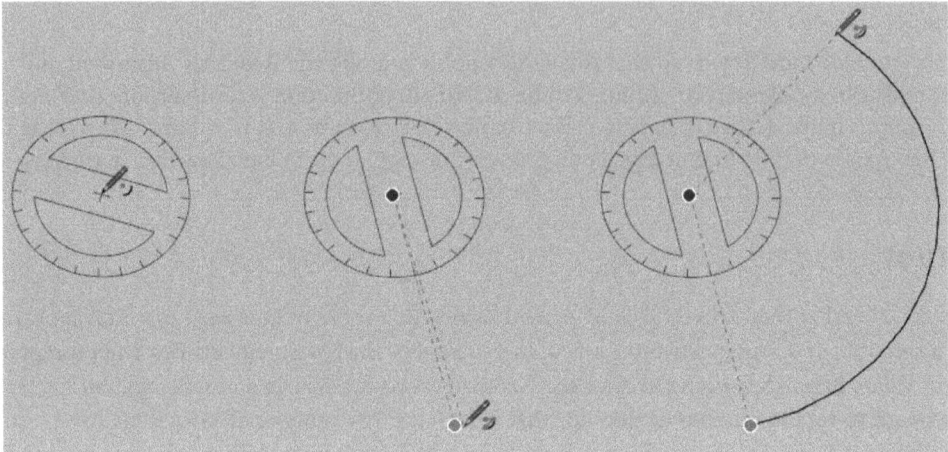

Figure 4.28 – Clicking to start the arc (Left); Moving the cursor and clicking to set the radius value (Middle); angle value and completed arc (Right)

You can also create arcs with a desired size. You can achieve this by doing the following:

1. Activating the Arc tool
2. Ensuring that the number of sides is correct, typing the value if it should be changed, and hitting *Enter*
3. Clicking to set a center point
4. Showing a direction and typing in a radius value to set the first point of the arc
5. Hitting *Enter* on the keyboard
6. Showing a second direction and typing an angle value to set the arc length
7. Hitting *Enter* on the keyboard

Any combination of entering measurements and clicking can be used to create an arc using any of the Arc tools.

The Arc tool should be used if the center point is clear. Many times, the center point is not clear when modeling, and one of the other arc tools might be more appropriate. It may be helpful to try the different Arc tools to find the right workflow!

2 Point Arc Tool

The 2 Point Arc tool does not require a center point to be set at the beginning of the workflow. Instead, the 2 Point Arc tool first establishes the start and end of the arc segment, then a "Bulge" distance. This Bulge distance is represented by the distance from the point exactly in between the start and end points and the middle of the arc:

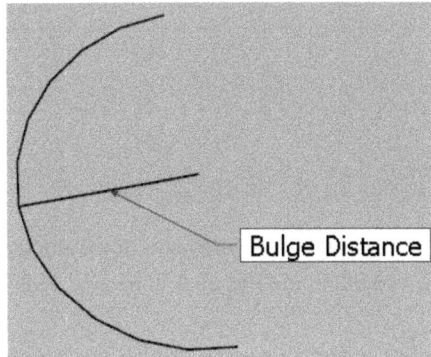

Figure 4.29 – Bulge Distance in a 2 Point Arc

> **Note**
>
> The Bulge in a Half Circle is equal to the radius!

The 2 Point Arc tool is a fast way to connect existing Geometry with an arc. As with all Drawing tools, the 2 Point Arc tool cursor is represented by a pencil with an Arc. The Cursors for the Arc tools are similar, with different red dots that represent the first step in each workflow:

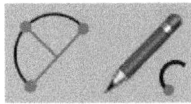

Figure 4.30 – 2 Point Arc Tool Button (Left) and 2 Point Arc Tool Cursor (Right)

The 2 Point Arc tool can be activated by doing the following:

- Clicking on any of the 2 Point Arc tool buttons
- Hitting *A* on the keyboard

2 Point arcs can be drawn by doing the following:

1. Activating the 2 Point Arc tool
2. Clicking to set a start point
3. Clicking to set an end point
4. Moving the mouse to show the Bulge direction and clicking to set the Bulge distance

The Bulge distance is commonly set by typing in a distance in the Measurements Box or finding a common Inference such as a Tangent or Half Circle:

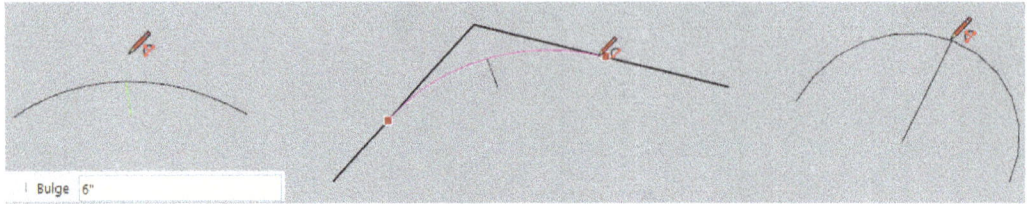

Figure 4.31 – Bulge set by Measurements Box (Left), Tangent
Inference (Middle), and Half Circle Inference (Right)

The 2 Point Arc is commonly used to create tangencies for filleting corners. In the following example, there is a box that should have a rounded corner. We will start by aligning the camera to look at the side of the box with the Orbit tool. We can then activate the 2 Point Arc tool and find a point on the top Edge of the box:

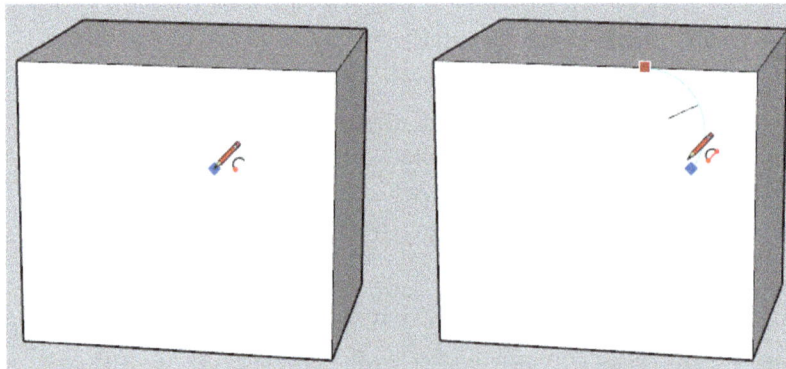

Figure 4.32 – Box with 2 Point Arc Tool active (Left); First point of arc selected (Right)

Then, when we move the mouse along the right Edge of the Face, we will see the **Cyan** Tangency line switch to a Magenta line; this infers that there is a double Tangency—at both the top and right Edges:

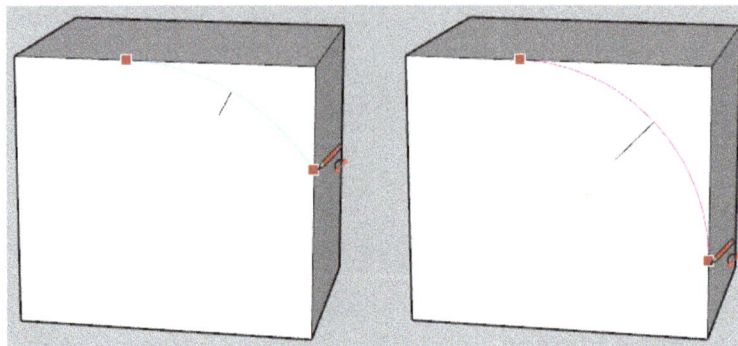

Figure 4.33 – Cyan showing single Tangency (Left); Magenta showing double Tangency (Right)

Then, using a double-click, the arc can be locked in at the double tangency. If a single click is used, a Bulge distance will be required. Because we are looking for the double tangency, we do not need to be concerned with the Bulge distance—a double-click is a more appropriate choice:

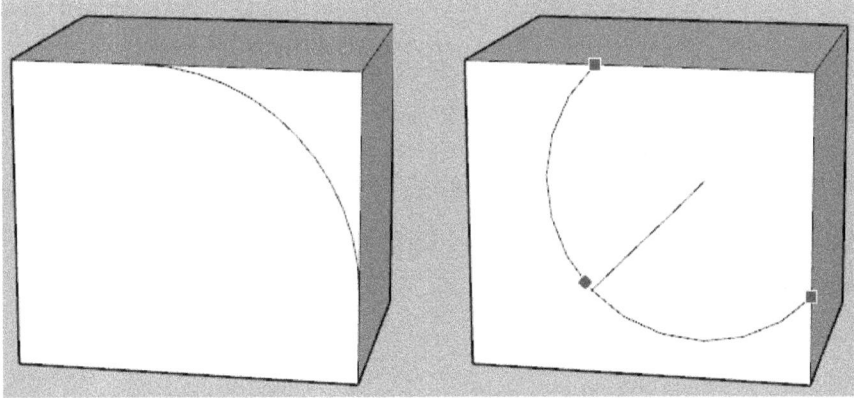

Figure 4.34 – Double-clicking to lock in the double Tangency (Left)
or single-clicking to activate the Bulge option (Right)

If this action is performed on a 2D Face that is not connected to other Geometry, the 2 Point Arc will Fillet the Face. Filleting the Face will remove the hard corner of the Face and replace it with a smooth, double-tangent curve:

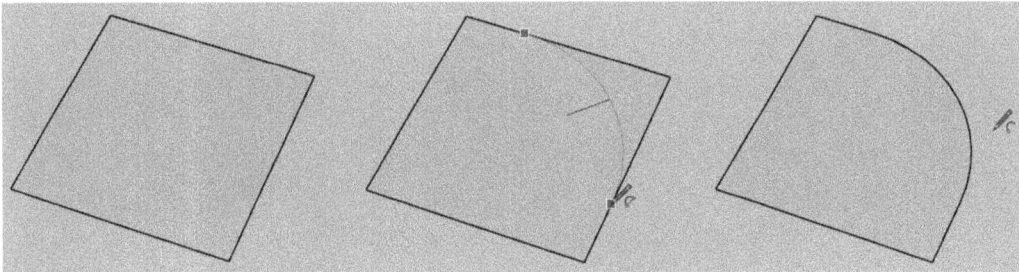

Figure 4.35 – Rectangular Face (Left); Double Tangency locked in
with a Double-Click (Middle); Filleted corner (Right)

The 2 Point Arc tool is the most common tool for creating Fillets. The 3 Point Arc tool can also be used, but it can be more challenging. Let's take a look at the 3 Point Arc tool next.

3 Point Arc Tool

As with all Drawing tools, the 3 Point Arc tool cursor is represented by a pencil with an Arc. The Cursors for the Arc tools are similar, with different red dots that represent the first step in each workflow:

Figure 4.36 – 3 Point Arc Tool Button (Left) and 3 Point Arc Tool Cursor (Right)

As with the 2 Point Arc tool, the 3 Point Arc tool does not require a center point to be set at the beginning of the workflow. Instead, the 3 Point Arc tool chooses the starting, second, and end points of the arc segment. The 3 Point Arc tool can be activated by doing the following:

1. Activating the 3 Point Arc tool by clicking on any of the 3 Point Arc Buttons
2. Clicking to set a start point
3. Clicking to set a second point
4. Clicking to set an end point

> **Note**
> There is only one possible arc that can be created with three points when two are set as the starting and end points!

You can see the three-step process for drawing a 3 Point Arc here:

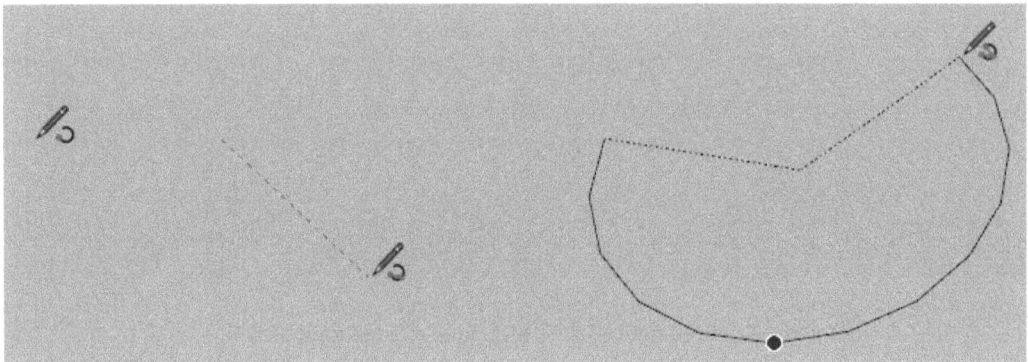

Figure 4.37 – First (Start) Point (Left), Second Point (Middle), and Third (End) Point (Right)

The 3 Point Arc is a great way to see the shape of an arc while drawing and before choosing the end point. When the first two points are set, the third point will drastically change the shape of the arc as you move the cursor around the screen:

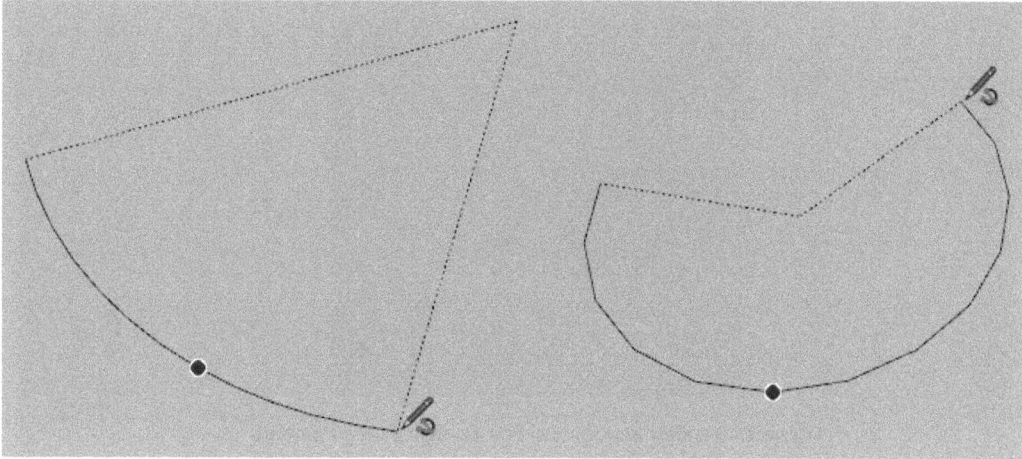

Figure 4.38 – Two points set (Left); mouse moved to show a
different shape with the same two points (Right)

If the third (end) point is closer to the second point, the arc will have a short radius and a greater angle. If the third (end) point is farther away from the second point, the arc will have a large radius and will represent less of the total angle:

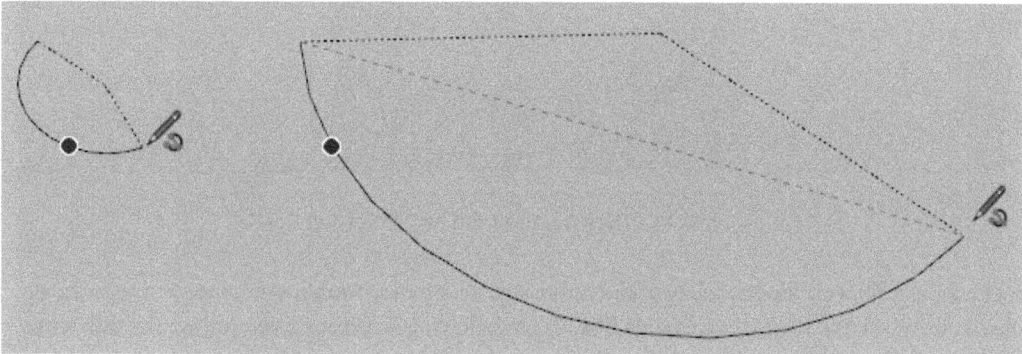

Figure 4.39 – Third point close (Left) and third point far (Right)

When a 3 Point arc has been drawn, the overall length of the arc can be changed using the Angle prompt in the Measurements Box. In this example, we can see that a 3 Point arc is being drawn, with the start and second points already clicked:

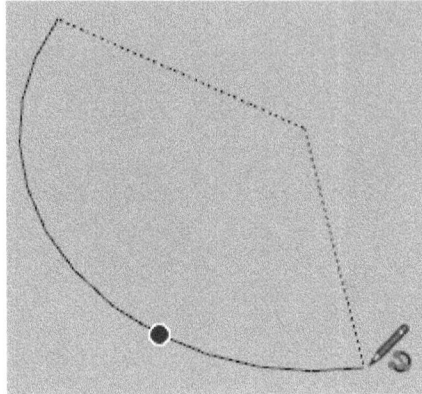

Figure 4.40 – Start and second points clicked for 3 Point arc

Then, a third (end) point must be established. It is important to set the third point with the Radius point in mind because we can change the overall length after!

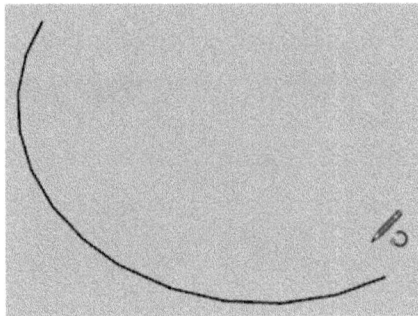

Figure 4.41 – Third point clicked for 3 Point arc

Now, once the 3 Point arc has been drawn, and before any other tool or workflow becomes active, an Angle value can be typed into the Measurements Box. In the following screenshot, you can see the difference between an angle of 20 degrees, 180 degrees (half circle), 225 degrees, and 360 degrees (full circle):

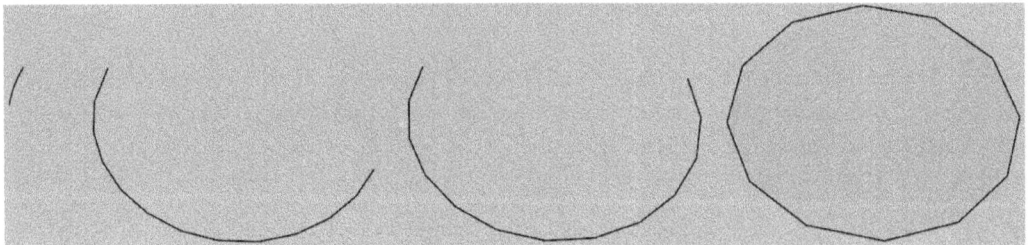

Figure 4.42 – Arc updated with 20 degrees (Left), 180 degrees (Left Middle), 225 degrees (Right Middle), and 360 degrees (Right)

The Arc tools create Edges unless the Arc creates a complete circle or connects to existing Geometry. The next tool also creates an Arc Edge, but we will see how it also creates a Face.

Pie Tool

The Pie tool is essentially the same as the Arc tool with a key difference—the Pie tool creates a Face! A Face is created by adding two lines from the center point to the start point and end points, creating a pie shape! As with all Drawing tools, the Arc tool cursor is represented by a pencil with a pie:

Figure 4.43 – Pie Tool Button (Left) and Pie Tool Cursor (Right)

The Pie tool can be activated by clicking on any of the Pie tool buttons. The Pie tool creates an arc using the same three values as the Arc tool: the number of Sides, a center point, and a radius value. This is achieved by doing the following:

1. Activating the Pie tool
2. Ensuring that the number of sides is correct, typing the value if it should be changed, and hitting *Enter*
3. Clicking to set a center point
4. Clicking to set a second point to set the radius distance value
5. Clicking to set the arc length (represented by an angle around an imaginary circle):

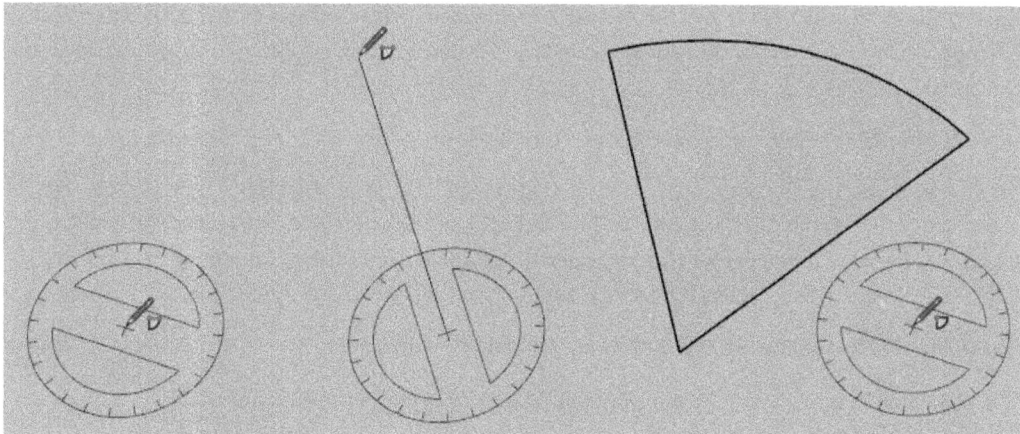

Figure 4.44 – Clicking to start the pie (Left); Moving the cursor and clicking to set the radius value (Middle); angle value and completed pie (Right)

As with arcs, you can also create pies with a desired size. You can achieve this by doing the following:

1. Activating the Pie tool

2. Ensuring that the number of sides is correct, typing the value if it should be changed, and hitting *Enter*

3. Clicking to set a center point

4. Showing a direction and typing in a radius value to set the first point of the arc

5. Hitting *Enter* on the keyboard

6. Showing a second direction and typing an angle value to set the arc length

7. Hitting *Enter* on the keyboard

The Pie tool is helpful for jumping immediately into the Edit tools when turning our 2D shapes into 3D Geometry. But what makes a SketchUp model 3D? Are these shapes already in 3D Geometry? Do we need to have Faces for 3D Geometry? Can 3D Geometry be made only using the Line tool and other Drawing tools?

Creating 3D, Solid, and Watertight Geometry

These are all great questions! A key distinction in SketchUp is the difference between 3D (three-dimensional) and Solid or Watertight Geometry. We can explain the differences by answering the previous questions.

Q: What makes a SketchUp model 3D?

A: All SketchUp models exist in the 3D Drawing Area. Also, all Geometry has *X*, *Y*, and *Z* coordinates. So, all Geometry exists in 3D space. But we would not call an Edge or single Face 3D Geometry—it is the collection of many Edges and Faces that sit on the Red, Green, and Blue Axes that create true 3D Geometry.

Q: Are 2D Shapes 3D, Solid, or Watertight Geometry?

A: 2D Shapes are technically in the 3D Drawing Area in SketchUp. And, even if the *Z* (Blue Axis) value is 0, it is still counted! But 2D Geometry cannot be Solid or Watertight—because it is flat! Solid and Watertight Geometry requires that there is space inside the object—space that could be completely filled ("Solid") or space that could hold water without letting any leak out through holes ("Watertight").

Q: Do we need to have faces for 3D, Solid, or Watertight Geometry?

A: Yes, absolutely! Faces are crucial for Solid and Watertight Geometry, and almost all 3D models in SketchUp need Faces to show the design.

Q: Can 3D Geometry be made only using the Line Tool and other Drawing Tools?

A: Yes, but it is tricky and time-consuming. It is always suggested to use the best tools and workflows to create Geometry, and that should be a combination of the Drawing and Edit tools. However, it is possible and sometimes easier to use the Drawing tools to create small-scale 3D Geometry.

This might all be a little overwhelming, but do not worry! You do not need to have a perfect model from start to finish when modeling! There is always the Undo button, but additionally, SketchUp is built to help you heal your model! Specifically, SketchUp will help heal Faces to ensure that your model returns to a Solid and Watertight state.

> **Note**
> SketchUp has added a new way to view watertight models in SketchUp Pro 2023. The Solid Inspector Extension can be viewed as an Overlay in SketchUp Pro to provide real-time feedback on "holes" in a SketchUp Model. We will discuss Extensions and Overlay-compatible Extensions in *Chapter 12, Import, Export, 3D Warehouse, and Extensions.*

Understanding how Faces Heal

SketchUp will help you heal your model! Faces in SketchUp can be created automatically, even when using tools that do not seem to generate Faces. The Line tool can even be used to help heal your SketchUp models when Faces seem to be missing! Let's explore how some of these tools can be used to generate Faces.

Creating Faces by Drawing Edges

Faces are generated whenever a flat, closed shape is created. This is automatically done whenever using the Rectangle, Rotated Rectangle, Circle, Polygon, and Pie tools. Additionally, Faces can be created when using the Push/Pull and Follow Me tools, but this is a slightly different process. We will discuss the Edit tools in the next chapter.

It is important to see that there is no "3D Face Tool," "Mesh Tool," or any other tool that creates independent Faces in SketchUp Pro. There may be plugins or extensions in the SketchUp universe, but these tools are not included in SketchUp Pro. This is because SketchUp will only allow Faces to be created when a frame of Edges is in place, and that can be done with the Drawing tools at any time. As long as SketchUp creates Faces automatically, the Edges are all we need to draw!

An easy example to try on your own is to draw a shape with the Line tool. After activating the Line tool, draw any shape on the Drawing Area and connect the Edges in a loop:

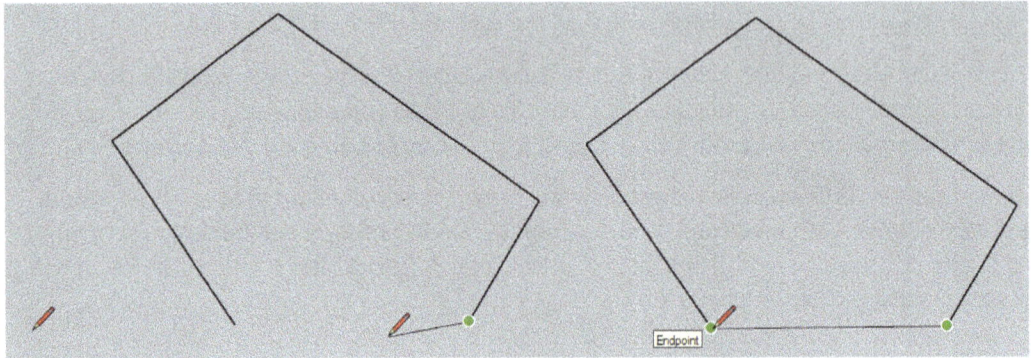

Figure 4.45 – Line Tool Active (Left); Drawing Edges (Middle); Closing the shape in a loop (Right)

Once the Line tool connects to an Edge, the tool will stop the Line tool from drawing more Edges, because SketchUp believes that you were intending to draw a loop. Also, a Face should appear!

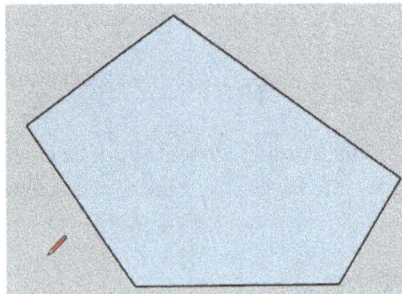

Figure 4.46 – Face appears after closing the loop with the Line Tool

If a Face did not appear, it may be because one of the lines was drawn in the Blue Axis direction. Orbit the model to see your shape from the side! Faces can only generate as flat Geometry, so any change in the vertical direction will not allow a Face to generate!

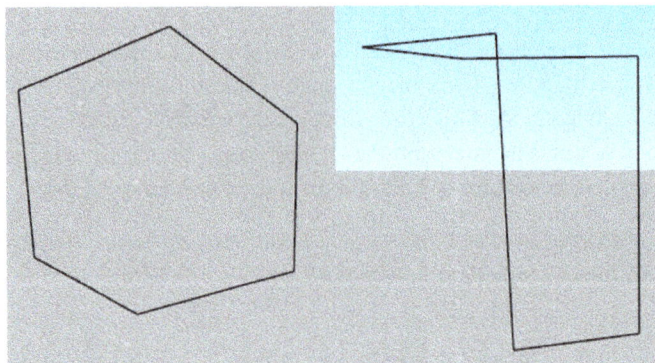

Figure 4.47 – Closed loop shape (Left); Orbited shape to show vertical lines (Right)

The Line tool is not the only tool that can create Faces. All Drawing tools can be used to create additional Faces. Next are some examples of creating Faces with other Drawing tools:

- 2 Point Arc tool:

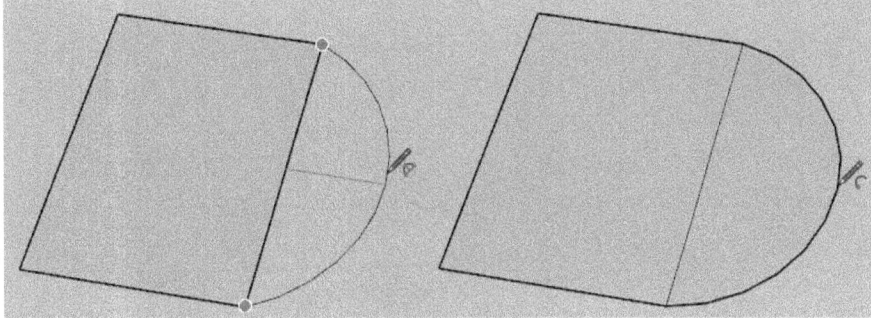

Figure 4.48 – The 2 Point Arc Tool is used to add a semicircular Face to the side of a rectangle

- The Rectangle tool:

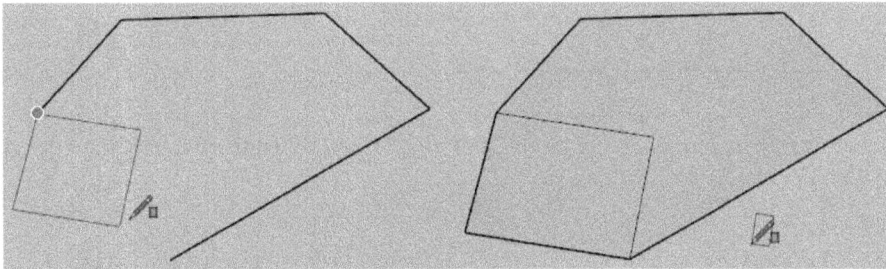

Figure 4.49 – The Rectangle Tool is used to complete an open loop of Edges

- The Freehand tool:

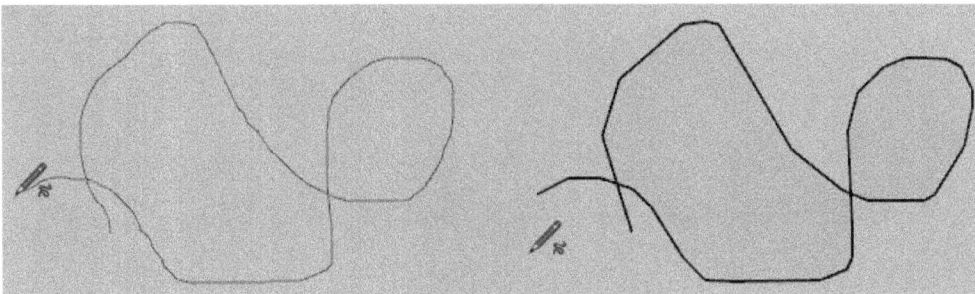

Figure 4.50 – The Freehand Tool crosses itself and creates Faces in the closed sections

Faces can be created any time a tool creates a complete loop of Edges. In the next section, we will look at what happens when Faces are broken, and how we can fix those broken Faces!

Healing "Broken" Faces

At some point in your modeling, you may accidentally delete a Face. We can use the Drawing tools to heal these Faces and bring them back! Let's see this in action. In this example, we will start with a rectangle. Then, using the Eraser tool, we will erase one of the Edges:

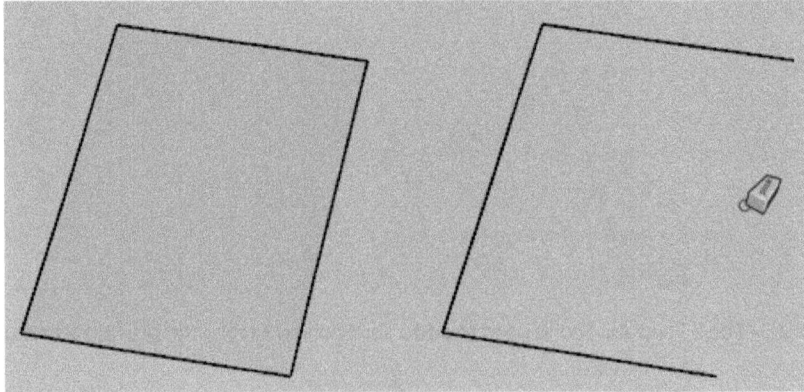

Figure 4.51 – Rectangle (Left) and Eraser Tool to erase an Edge (Right)

> **Note**
> Remember that Faces cannot exist without bounding Edges. When the frame of the Edges is broken, the Face can no longer exist.

At this point, it may seem like the best way to fix this situation is to redraw the rectangle using the Rectangle tool. After all, the Rectangle tool creates a Face every time! But we can heal this Face by only using the Line tool! With the Line Tool, the Edge can be redrawn by clicking on the broken Edges at their endpoints:

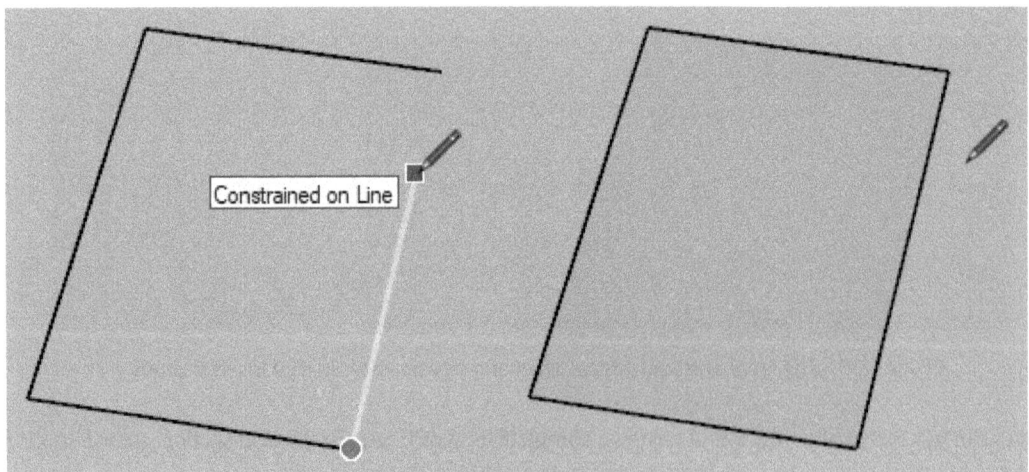

Figure 4.52 – Line Tool used to redraw line (Left) and healed Face (Right)

This will redraw the line, and when the frame is redrawn, the Face will reappear. This will also end the Line tool from drawing more Edges because SketchUp believes that it has healed the model to an appropriate amount. Of course, you can click again to start new Edges while the Line tool is still active.

An Edge might be connected to multiple Faces that might need to heal. Remember—a Face can only exist in a single plane as flat Geometry. Faces cannot be bent to wrap around Geometry. In the following example, we can see that there is a hole in the mesh surface:

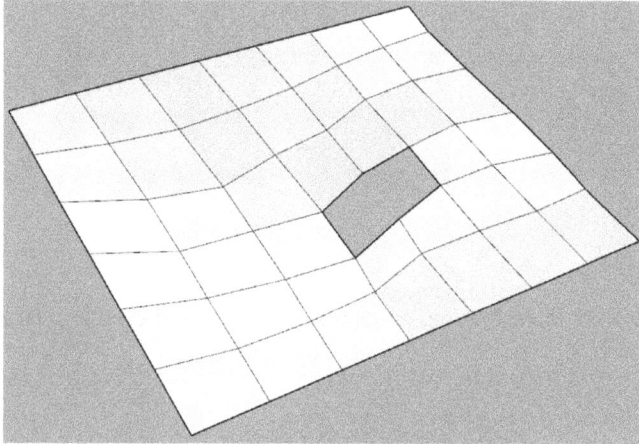

Figure 4.53 – Mesh surface with a hole

This Geometry cannot be healed by adding a single line down the middle. It would appear to create two rectangles, but these Edges are slightly off from flat relative to one another:

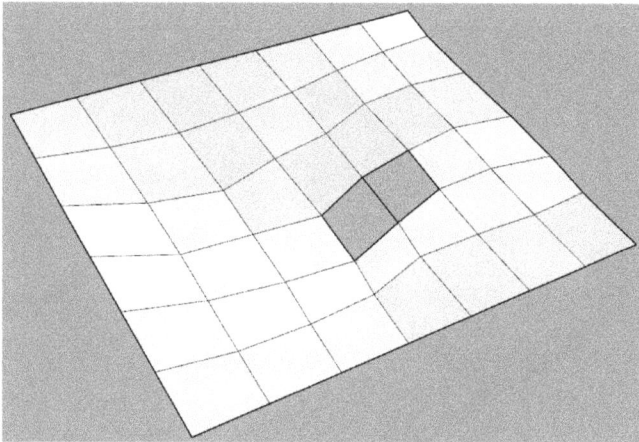

Figure 4.54 – Single Edge will not heal Faces

In this case, additional Edges must be drawn with the Line tool to create triangles. Triangles are always flat because they only have three Edges and three endpoints. By adding enough lines to create triangles, you can always ensure that the Faces in your model will heal:

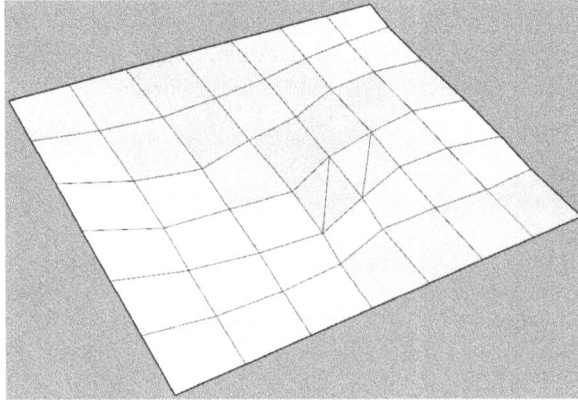

Figure 4.55 – Multiple Edges create triangles that heal the mesh surface

> **Note**
> Remember to use the Eraser tool to hide or smooth/soften extra Edges that might need to be in the model for Faces to generate, but that should not be seen!

In this last example, we will see a complex face that is broken. This model is a small building model, and we can see that the top of the wall Geometry has been removed:

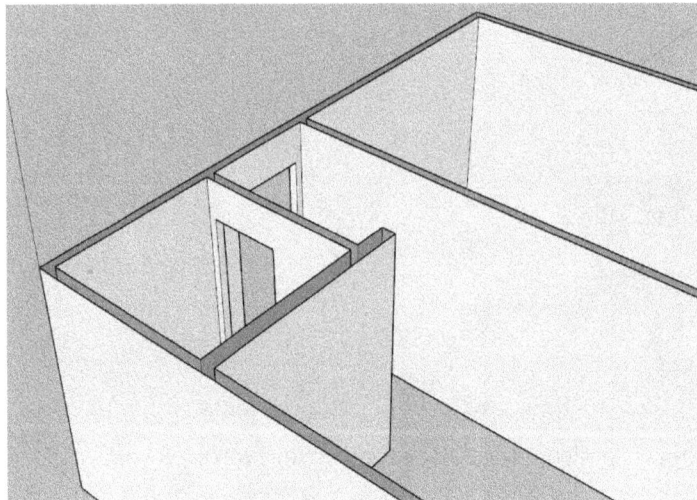

Figure 4.56 – Wall with top face removed

In this case, all the bounding Edges in the model seem to be in the correct place. The Face was most likely removed by using the *Delete* key, and the Eraser tool was not used. Brand-new Edges do not need to be created to complete the frame around a Face. In some cases, the Face can be healed by drawing an Edge on top of another Edge. For this example, we can use the Line tool and trace any of the Edges on the top of the wall, and we will see the top Face heal:

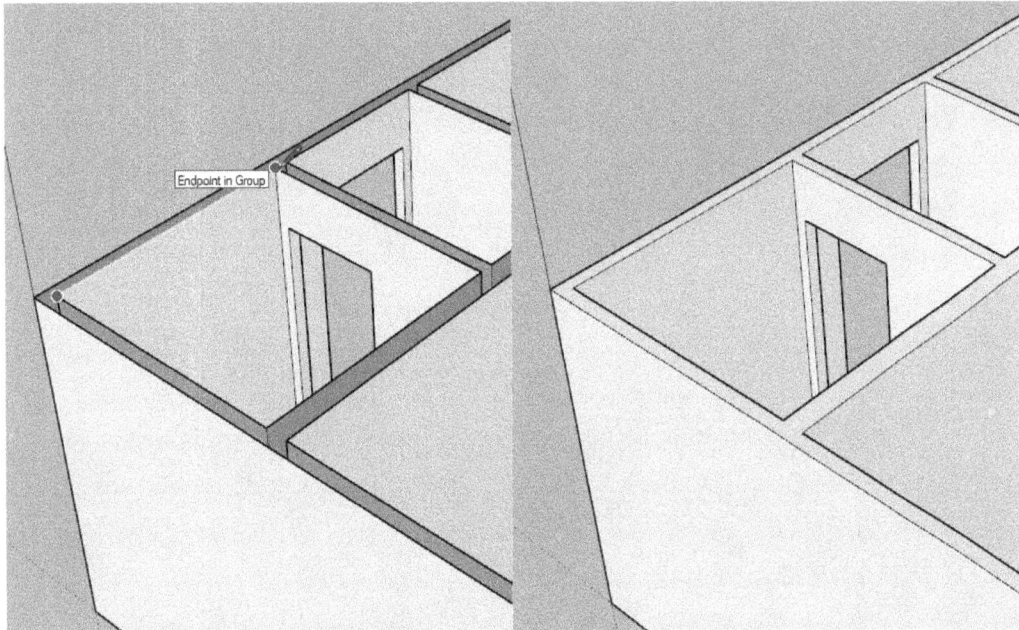

Figure 4.57 – Tracing one of the Edges (Left) and the Face heals (Right)

Healing Faces using the Drawing tools might be a necessary part of your modeling workflow, but healing Faces is not the preferred way to create in SketchUp. We'll look at how we can create Faces and Edges in the next chapter.

Summary

In this chapter, we discussed the Drawing tools and their fundamental uses in SketchUp Pro modeling workflows. The Drawing tools are a necessary beginning to any SketchUp modeling project and will be used throughout every workflow. However, to move to the next level in modeling workflows, we must understand how to edit the Geometry we have created! In the next chapter, we will discuss the Edit tools and how we use these tools to create new Faces and Edges in exciting ways!

5
Editing Tools – Making Big Changes!

In this chapter, we will use the Editing tools to make changes to basic Geometry! We will talk about how we can use the Editing tools to make changes to our Models and how to create new Edges and Faces using existing Geometry. While examining the Editing tools, we will also discuss how this impacts our Faces and Edges in the model, by looking at how and when to create and edit Geometry, how our edits might impact Face orientation, and how to use Guides to assist in our Editing workflows.

The following topics are covered:

- Changing Geometry with the Editing Tools
- Preselecting Geometry—Workflow Order of Operations
- Move, Rotate, Scale, Push/Pull, Follow Me, Offset, and Flip Tools
- Editing Existing Geometry and Creating New Geometry
- Face Orientation
- Tape Measure and Protractor Guides

Changing Geometry with the Editing Tools

We have just learned about how to create 2D Geometry using the Draw tools, and now, we want to move into 3D space. What can we do with our Edges and Faces? The Edit tools are the answer! We are ready to jump in and learn about editing our Geometry— not only to extrude shapes into 3D with Push/Pull, or create sweeps and revolves with Follow Me, but also to move, copy, rotate, and scale our Geometry! However, before we jump right in, we need to consider an element of the SketchUp workflow that we have not needed to use until now—the selection of existing Geometry.

We have picked on snaps and aligned with inferences, but we have not needed to select existing Geometry before or during a creation workflow. All of the Edit tools have this in common: at some point, they use existing Geometry to edit the model, and this requires that we click or pick on something!

Clicking with the mouse is the preferred way to pick our Geometry in an Edit workflow, and we have seen selection options (including single-, double- and triple-clicking) in *Chapter 2, Principal Tools, Axes, and Inferences*. We know how we need to select Geometry, but now we need to consider when to select our Geometry. Geometry can either be selected as soon as the Edit tool is active, or it can be selected before, which is referred to as preselecting. We can see a regular selection in the following screenshot:

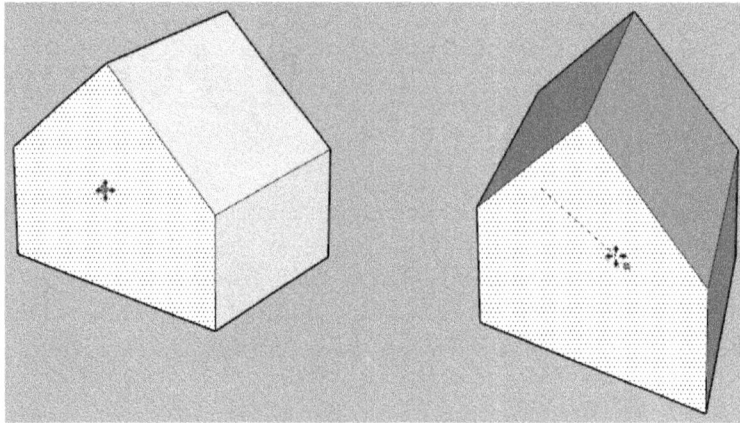

Figure 5.1: Move Tool Activated First (Left) and Object Picked Second (Right)

And preselection in the following screenshot:

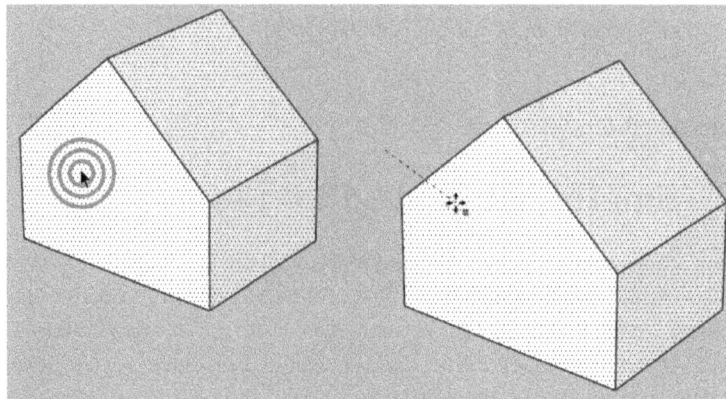

Figure 5.2: Object Preselected First with Select Tool (Left) and Move Tool Activated Second (Right)

When should we preselect? Let's learn more about the order of operations in the next section.

Preselecting Geometry – Workflow Order of Operations

So, when do we want to preselect our Geometry? We can preselect in any Edit workflow, and some modelers prefer to preselect every time they use an Edit workflow. When using an Edit tool that can select more than just a single piece of Geometry, such as Move, Scale, or Rotate, preselecting is the only way to select multiple Edges, Faces, or Objects. When an Edit tool is activated, you cannot drag a selection window or use a modifier key to add other Geometry to the selection set—the first element that is clicked will be the active selection in the Edit workflow. In the following screenshot, we can see that while using the Rotate tool, any Face that is clicked will choose only the Face and activate the next step in the Rotate workflow:

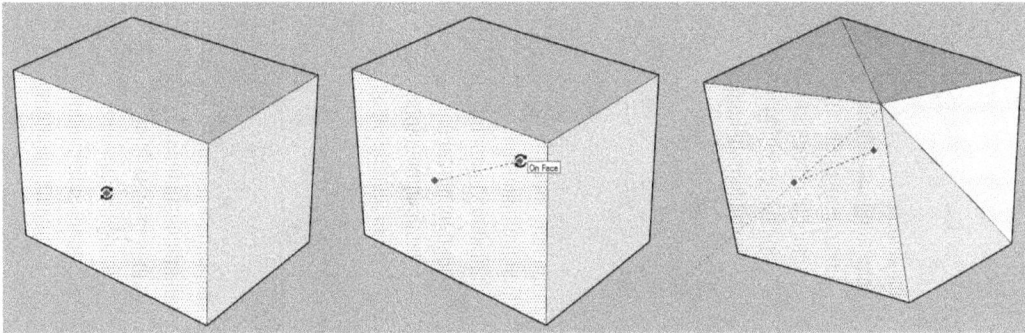

Figure 5.3: Rotate Tool Activated (Left); Face is Selected and Basepoint
Set (Middle); Rotate Activates Autofold (Right)

In the following screenshot, we can see the box selected with a triple-click, and then the Rotate workflow affects all selected Geometry:

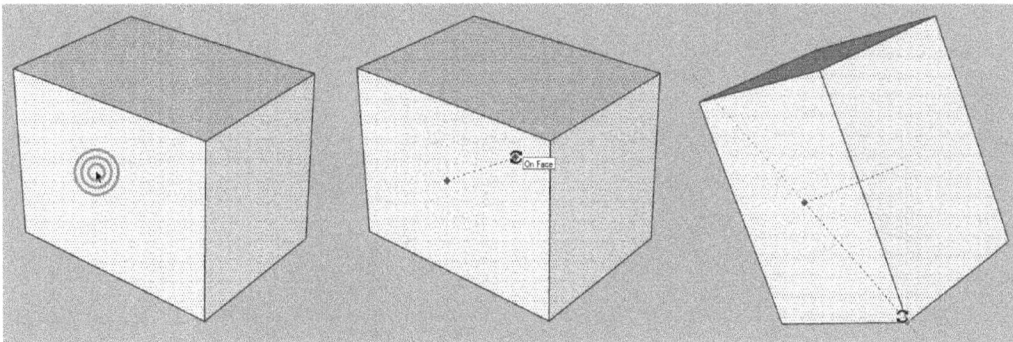

Figure 5.4: Box Preselected with Triple-Click (Left); Rotate Tool
Activated (Middle); All Geometry Rotated (Right)

> **Note**
>
> At any time when an Edit tool is active, the *Esc* key can be pressed to deselect it! This can be very helpful if the wrong Geometry is preselected before activating an Edit tool!

When an Edit tool is activated without preselecting the Geometry, it can be difficult to choose the right selection and the workflow basepoint at the same time. This may be impossible when trying to snap to an Edge for a workflow, but the selection set should be a Face. In the next screenshot, you can see that the lower corner should be the basepoint for the Move tool, but the Face is the Geometry trying to be moved. Without preselecting the Face, the endpoint snap cannot be selected at the same time as the Face:

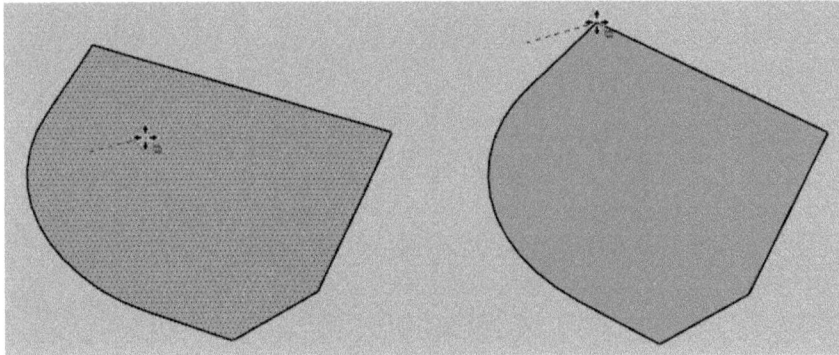

Figure 5.5: Face in Selection (Left); Endpoint in Selection (Right)

With preselection, it frees up the second part of the workflow to be entirely focused on setting the basepoint. This is especially helpful in **Move**, **Copy**, and **Rotate** workflows. In the following screenshot, we can see a Move workflow, where the Group has been preselected, the basepoint is chosen on a different point of the model entirely, and the destination point has been calculated to move the Group to the perfect location:

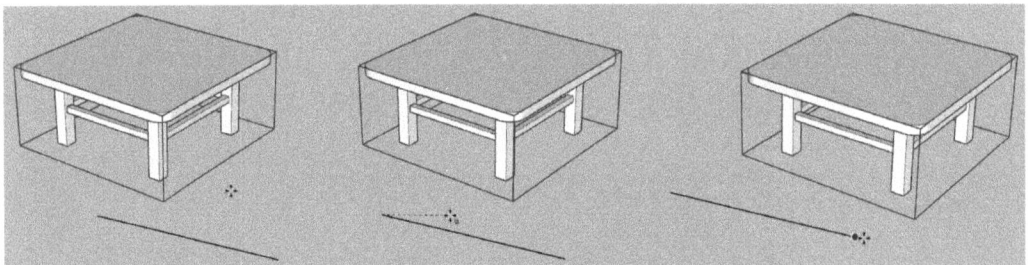

Figure 5.6: Group Preselected (Left); Basepoint Picked Away from
Selected Group (Middle); Move Complete (Right)

Preselecting is potentially less valuable when working in Push/Pull and Offset, as these tools are limited to one Face at a time. However, if the Face is occluded or if the Face and the destination point are not easily seen in the same view, it could be a good idea to preselect.

> **Note**
>
> Be careful! Sometimes, you might accidentally have a Face or other Geometry preselected before you start using an Edit tool, and you might not know it! You can get locked into using that Face, and when you attempt to choose a different Face, you could accidentally edit some Geometry without knowing it! If you are not preselecting, it could be good practice to activate the Select tool and click on empty space before activating an Edit tool, or to hit *Esc* right after activating the Edit tool to clear the selection.

This is different with the Follow Me tool, as preselecting with the Follow Me tool helps to quickly complete the workflow. Anything preselected with the Follow Me tool will be chosen as the path, and this can be a Face for a full loop or a series of Edges. We will talk more about the Follow Me tool later in this chapter, but remember that the preselection for Follow Me will always be the path—never the profile!

In your own workflows, you can always preselect, never preselect, or use a mixture. My best suggestions are outlined here:

- **Move, Rotate, Scale, Edit -> Copy & Paste**: Always preselect, especially with ungrouped Geometry.

- **Push/Pull**: Preselection is not necessary, unless in very specific circumstances. You will know when you get there!

- **Follow Me**: Preselection can be very helpful if you know your desired path. The preselection will always go the full distance of each Edge, so preselection might not be desired when using Follow Me to follow a path for a specific distance.

- **Offset**: As with Push/Pull, preselection is not necessary.

Now that we better understand preselection, we can learn more about each Edit tool. Let's start by taking a look at Move/Copy!

The Move Tool

The Move tool might be the easiest Edit tool to conceptualize—it moves Geometry through our 3D modeling environment! The Move tool has a couple of very important modifier keys as well—*Ctrl* to activate **Copy** (and **Array**) and *Alt* to activate **Autofold**. The Move tool is represented by four arrows pointing left, right, up, and down:

Figure 5.7: Move Tool Button

> **Note**
>
> Remember—even though we are changing our Geometry in 3D space, a specific axis/direction can always be locked using the arrow keys (to find Red, Green, and Blue cardinal directions) or by holding *Shift* to lock in the current inference.

The Move tool can be activated by doing the following:

- Clicking on any of the Move tool buttons
- Hitting *M* on the keyboard

The Move tool can move Geometry using two methods: Point-to-Point and Distance and Direction.

Geometry can be moved using the Point-to-Point method by doing the following:

1. Activating the Move tool.
2. Selecting the object to be moved (if not preselected) with a click. This will also set the move basepoint (first point). If the object is preselected, the first point can be clicked anywhere on the screen.
3. Clicking anywhere to set the destination point (second point). SketchUp will calculate the distance between these two points automatically and will move the selected geometry according to the two selected points.

We can see how the Point-to-Point method can be used in the following example. In this example, there is a small box that we want to move to sit on top of the larger box:

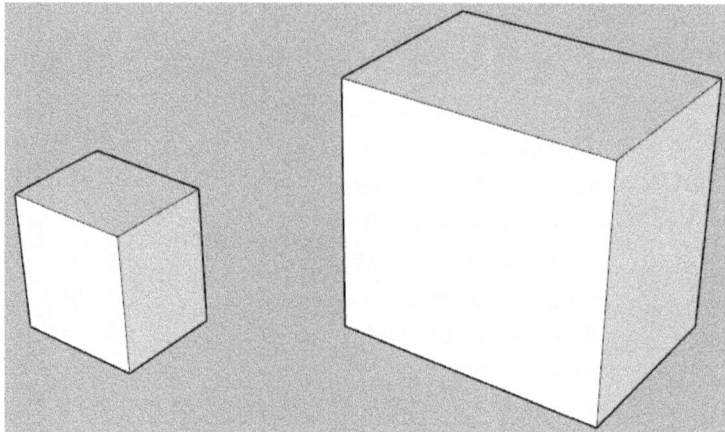

Figure 5.8: The Small Box (Left) will be Moved on Top of the Larger Box (Right)

In this case, we should visualize which points will eventually be in the same place in 3D space, or coincident:

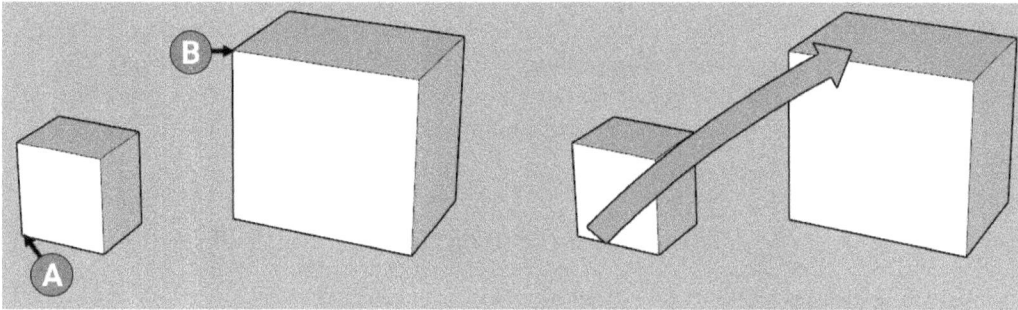

Figure 5.9: The Basepoint (A) and Destination Point (B) are Identified (Left); Visualization of Move (Right)

Once we have identified the two points, we can use them as the basepoint and destination point for the move:

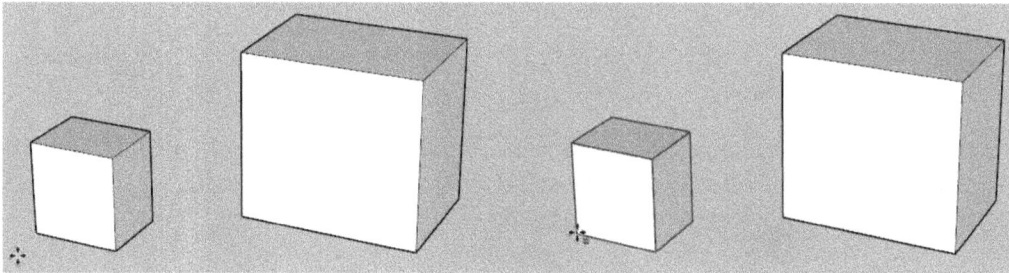

Figure 5.10: The Move Tool is Activated (Left) and the Small Box is Selected by Picking the Basepoint (Right)

The box can then be moved!

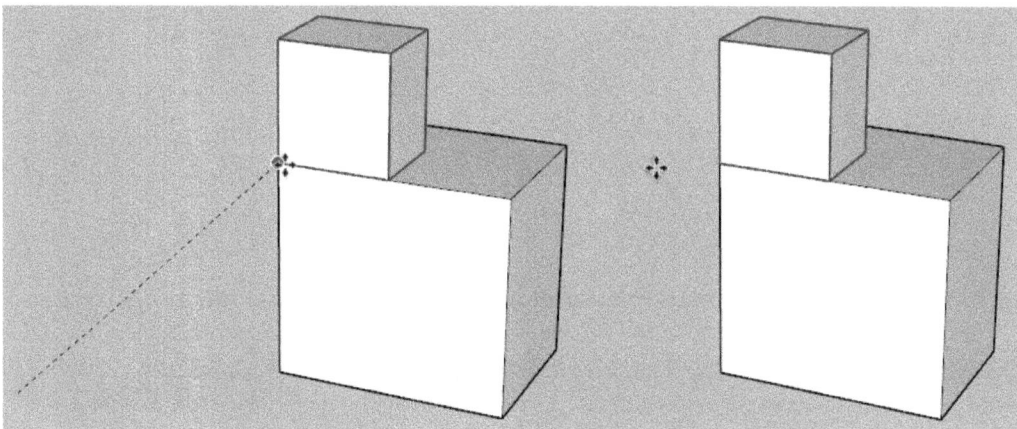

Figure 5.11: The Small Box is Moved (Left) and Placed by Picking the Destination Point (Right)

> **Note**
>
> Use Point-to-Point moves when you can clearly visualize the move in your head, or when you can say a sentence out loud to help narrate your workflow—for example, "The Small Box needs to Move on top of the Larger Box." One object is moving to snap to another object—this is Point-to-Point!

Geometry can be moved using the Distance and Direction method by doing the following:

1. Activating the Move tool.

2. Selecting the object to be moved (if not preselected) with a click. This will also set the move basepoint (first point).

3. Moving the cursor to show the appropriate direction, typically along an inference. The inference can be locked in if desired.

4. Typing in a distance into the Measurements Box and hitting *Enter*.

5. SketchUp will use the current shown direction and will move the Geometry in that direction to the specified distance.

We can see how to use a Distance and Direction move in the following example. In this example, the small box should be moved away from the wall at a distance of 10 inches. In this case, "away" is perpendicular to the wall, and we can see that is in the red-axis direction:

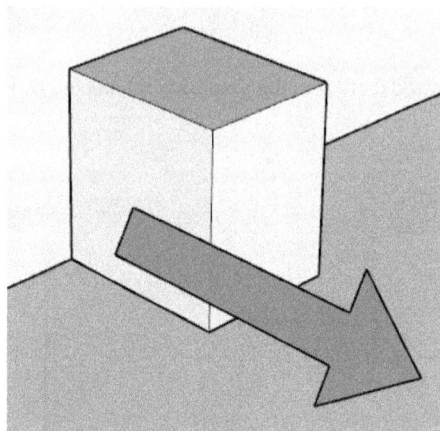

Figure 5.12: The Small Box will be Moved in the Red Direction Away from the Wall at a Distance of 10 Inches

In this example, we do not need to set a specific basepoint, because we are using a Distance and Direction workflow. As long as we can show the desired Move direction in our camera view, the Distance and Direction move is possible. However, we might need to preselect if we are working with Ungrouped Geometry—especially when working with Sticky Geometry. In this example, the box is a Group:

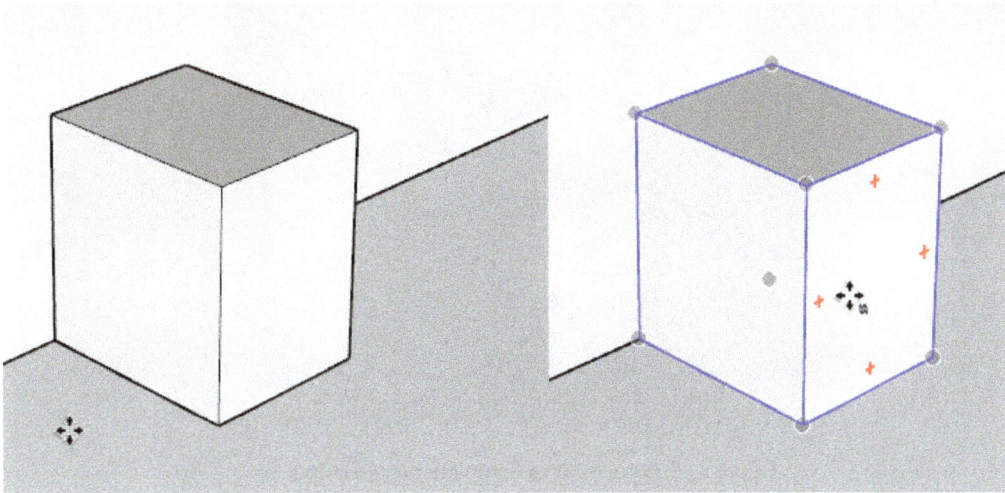

Figure 5.13: The Move Tool is Activated (Left) and the Small Box
Group is Selected by Picking any Basepoint (Right)

Once the basepoint is chosen on the box, we can show the red direction (lock in the red direction by hitting the right arrow on the keyboard), type in 10", and hit *Enter*:

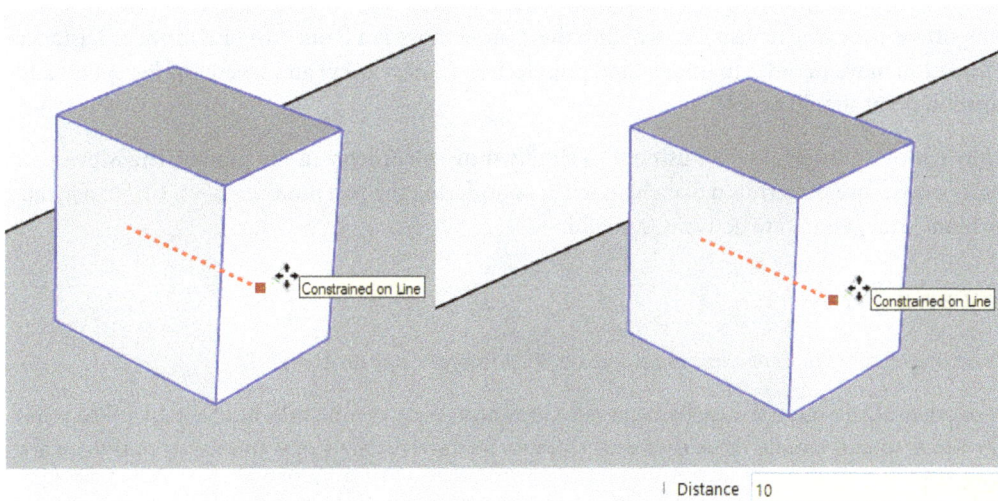

Figure 5.14: The Red Direction is Shown and Locked In (Left) and
10" is Typed into the Measurements Box (Right)

Using precise distances typed into the Measurements Box gives us the correct Move distance of 10":

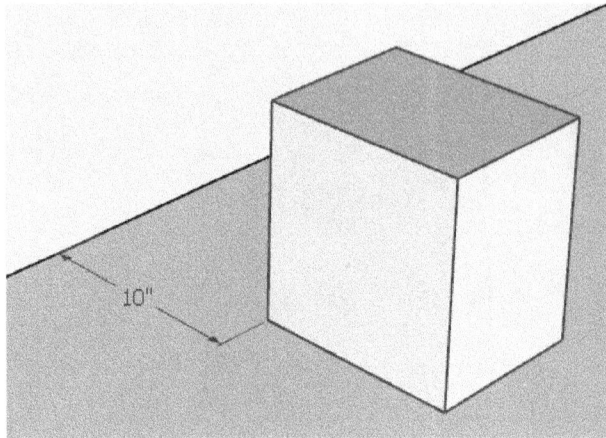

Figure 5.15: Enter is Hit to Complete the Move Tool

Note

Use Distance and Direction moves when the sentence you say in your head that describes the move includes a distance and a direction, such as this: I need to move the box **away** (in the Red Direction) at a Distance of **10 inches**. Or: the Geometry should move **up 5 feet**.

It is important to be able to visualize whether the desired move is a Point-to-Point move or Distance and Direction move in order to understand preselection (if necessary) and where the basepoint and destination point should be set.

The Move tool is powerful when using it to simply move Geometry in the model! The Move tool becomes even more powerful in SketchUp when considering the two modifier keys: *Ctrl* to activate Copy (and Array) and *Alt* to activate Autofold.

Copy and Array

There are no separate tools or commands in SketchUp Pro for Copy or Array!

A lot of other 3D design software has separate Copy and Array commands, but SketchUp Pro relies on the Move tool to handle these useful workflows. So, how do the Copy and Array options work in SketchUp?

Copy is essentially a Move tool, but it leaves a Copy of the Geometry in the original location too! It is easy to see this while creating a copy. In this example, we will start by selecting the Group with the Move tool active:

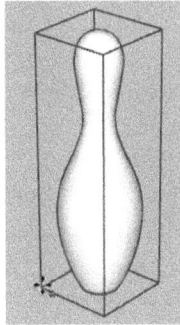

Figure 5.16: The Move Tool is Active, and the Basepoint has been Set on the Group

As soon as *Ctrl* is hit on the keyboard, a copy of the original Geometry will appear in the original location. Hitting *Ctrl* again will toggle the Copy modifier, and the copy will disappear:

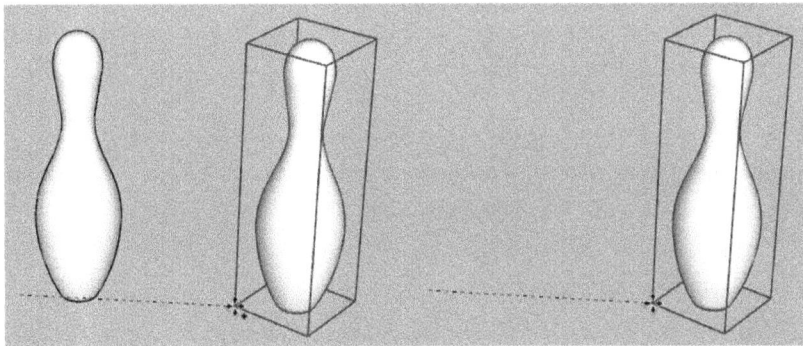

Figure 5.17: Hitting Ctrl will Toggle a Copy in the Original Location to Appear (Left) and Disappear (Right)

The Move can then be completed by selecting a destination point. The Moved Geometry will remain selected, and the Copy in the original location will be unselected:

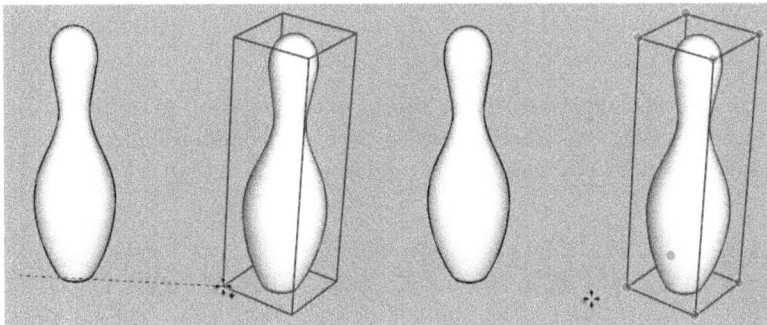

Figure 5.18: Copy Toggled On (Left) and the Move is Completed by Clicking to Set Destination Point (Right)

> **Note**
>
> Copies are an essential part of modeling in SketchUp. Copies can also be made by bringing in multiple instances of a Component. Take a look at the *Components* section of this book to learn more! When the Copy option is toggled and a copy is made, it will toggle off after the first Move/Copy workflow is completed. The Move tool will remain active, but you will have to hit *Ctrl* again to make additional copies. That is, of course, unless you use Array!
>
> Array in SketchUp is not exactly an Array in the traditional sense, but rather is multiple copies that go along a straight line. The Rotate tool also has this Array (or multiple copies) option, but it happens around a circular shape based on the center point of the rotation.
>
> Polar Arrays are performed with the Rotate tool. We will discuss this in the next section.

Arrays or multiple copies require one more step after the Copy workflow, which is adding a multiplier value or a divisor value. The multiplier value ("x" and a number) adds extra copies that continue past the first copy, and the divisor value ("/" and a number) adds multiple copies equally spaced between the original object and the copy. Multiple copies (or a one-directional Array) can be created by doing the following:

1. Activating the Move tool.
2. Selecting the object to be moved (if not preselected) with a click. This will also set the move basepoint (first point). If the object is preselected, the first point can be clicked anywhere on the screen.
3. Hitting *Ctrl* to activate the Copy option.
4. Moving the object using Point-to-Point or Distance and Direction workflows (see the previous instructions in the *Move Tool* section).
5. After the object has been copied, and both copies are on the screen, typing in either a whole-number multiplier value (x#) or a divisor value (/#) into the Measurements Box and hitting *Enter*.
6. If desired and before activating any other tool, the multiplier value or divisor value can be changed by typing in new values.

> **Note**
>
> There is no way in the default SketchUp Pro Toolset to create a 2D or 3D grid Array with one command—instead, create multiple copies using the Move Tool in one direction, then copy those in another direction. You can even repeat the step a third time to create a 3D grid of copied objects!

Let's take a look at a few examples of Array/Multiple Copies. In these examples, we will start with the same object—a small cylinder—and we will begin by Copying the cylinder 12" along the Red Axis:

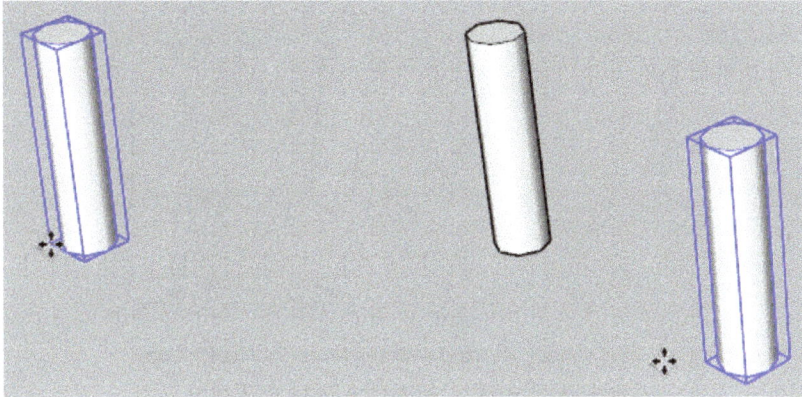

Figure 5.19: Cylinder selected with the Move Tool (Left); Copy
Toggled and Placed 12" along the Red Axis (Right)

Next, before anything is clicked on the screen or another tool is activated, we can type in a multiplier or divisor value. For this example, we can type x3 into the Measurements Box and hit *Enter*. When x3 is typed, two additional Copies of the object appear (this creates three total copies). These will be placed 12" from the previous copy each time, with the final object being 3" from the original object:

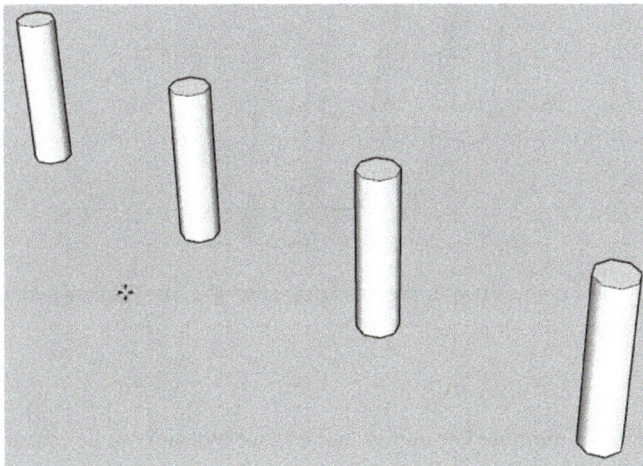

Figure 5.20: x3 Multiplier Creates Two Additional Copies Past the First Copy

This is an excellent workflow for creating objects when you know the distance between two objects. This could be useful when spacing out chairs and desks in classroom design or placing many objects side by side, such as books on a bookshelf.

In this next example, we will use the same cylinder and setup, creating one copy at 12" along the Red Axis:

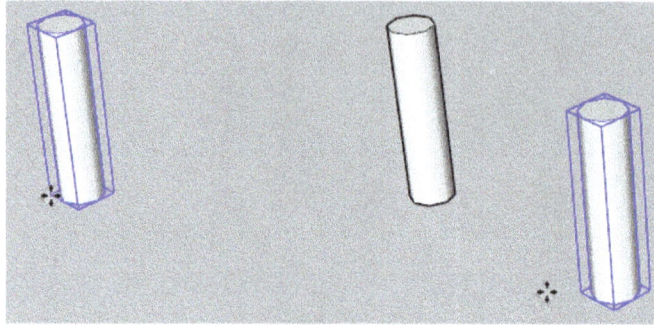

Figure 5.21: Cylinder selected with the Move Tool (Left); Copy
Toggled and Placed 12″ along the Red Axis (Right)

This time, however, instead of typing in a multiplier value, we will type a divisor value, /3. When /3 is typed, two additional copies are added, but this time, they are added between the original object and the copied object. The new copies are spaced evenly apart using the original copy distance as the maximum value. So, these new copies will be spaced 12/3 inches, or 4 inches apart:

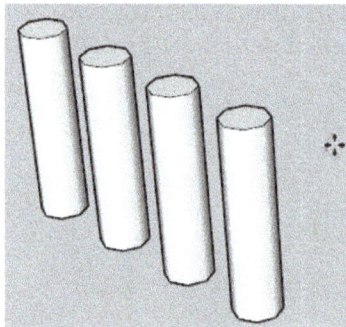

Figure 5.22: /3 Divisor Creates Two Additional Copies Between the Original and First Copy

Autofold

Autofold is the second major modifier key option in the Move tool and can be activated by hitting the *Alt* key. Autofold allows the Move tool to "break the rules" during a Move workflow, and SketchUp will add creased and divided faces to keep the Geometry connected. This is the "fold" in Autofold—faces are creased and folded into new angles from their original plane. Autofold allows you to move a point, edge, or face in any direction, automatically creating new edges and faces.

This process might be easier to visualize than to read about. Let's take a look at a few examples. In this first example, we have a box with an Offset rectangle on the top face. This rectangle is on top of the face of the box—they are in the same plane. When the Move tool is activated, the rectangular face is allowed to "slide" around the top of the box—it is locked into that plane:

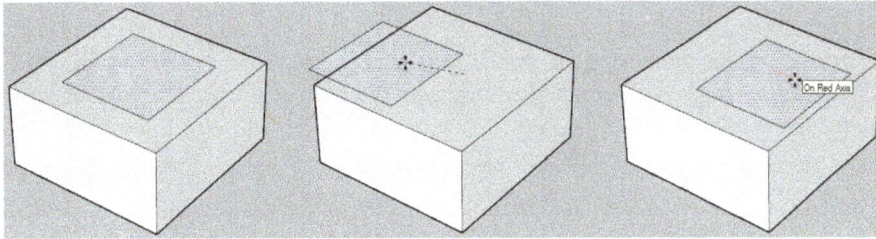

Figure 5.23: Box with Rectangular Face Selected (Left); Move Tool
Activated and Rectangle Can "Slide" (Middle and Right)

Autofold breaks the rule of being locked into "sliding" on top of the box—the rectangular face can move in any direction! When the *Alt* key is pressed to toggle Autofold, we can see that the rectangle can move in any direction, and in this example, we will move up along the Blue Axis. When the rectangle moves up, the square donut face—the leftover face geometry that made the top of the box—is forced to automatically fold to keep the sides of the box connected to our moving rectangle! This creates chamfered edges around the top of the box, splitting one face into four faces by adding four edges that connect from endpoint to endpoint of our rectangles:

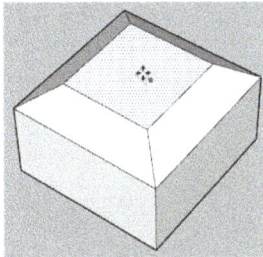

Figure 5.24: Autofold Toggled On and Rectangle Moved Up along Blue Axis

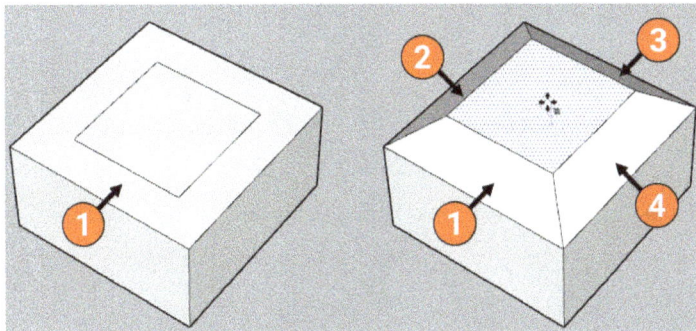

Figure 5.25: The Square Donut Before Autofold as One Face (Left)
and the Four Created Edges and Split Faces (Right)

> **Note**
>
> Locking in an Axis Direction using the Arrow Keys can produce similar results to Autofold—especially when moving in a perpendicular direction. But remember that Autofold allows the move to happen in any direction!

Autofold is different from the Move we used in the last chapter, as that did not create new Edges during the command. Autofold is distinct in that it creates Edges and splits faces—typically one face that is all on a singular plane—into multiple chamfered or folded faces. Let's look at one more example. In this example, one corner of a box will be pushed to the center of the top face:

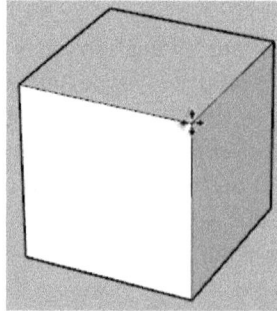

Figure 5.26: The Move Tool is Activated, and the Corner is Selected

This will create a folded appearance:

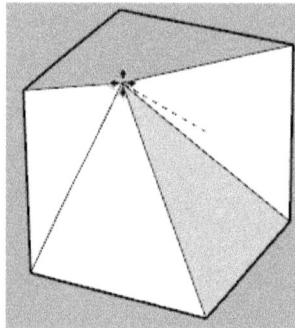

Figure 5.27: The Corner Point is Moved to the Middle of the Top Face

> **Note**
>
> When moving points, Autofold does not need to be turned on—it will work this way automatically!

In this case, two new edges were created that folded the vertical faces of the box. These two edges were created according to SketchUp's Autofold rules, and these rules might not always be what we want! In this case, we might prefer the edges to have been created to fold the rectangle in the other direction, but Autofold chose differently. If we know which direction the fold "should" go in order to match our design, we can always draw those edges in before the Move workflow. This defeats the purpose of Autofold—we set it up for SketchUp! But if it allows our design to match our design intention, then we do not always need to rely on Autofold:

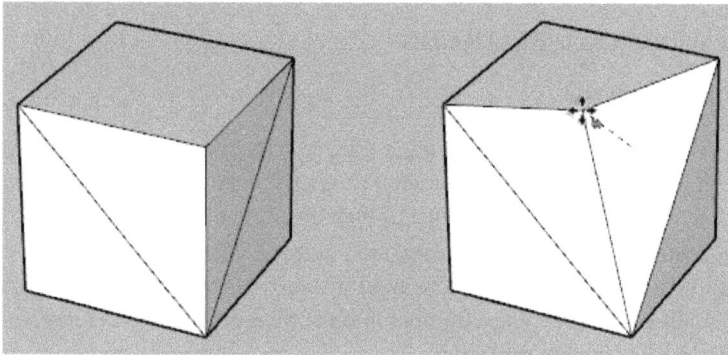

Figure 5.28: Edges are Drawn Before the Move Workflow (Left);
Move is Completed and Edges are Folded (Right)

> **Note**
> The Sandbox tools and other plugins and extensions have tools and Workflows for flipping edges, and you can even erase and redraw edges with Auto-Heal! There are many ways to reach the desired outcome, but if you know that outcome from the start, it can be best practice to add a step before the Move instead of after the Move.

All of the tools in this chapter will take practice to understand intuitively—and Move is no exception. This is especially true when considering the modifier key options of Copy and Autofold! Practice and play around in SketchUp with Move, Copy, and Autofold before you begin working on your own projects!

The Rotate Tool

The Rotate Tool adds one more variable to the Move tool workflow—the Rotate tool rotates objects around a specified center point. The Rotate tool does not just do internal Object Rotation. Objects can be rotated around any point in a drawing! The Rotate tool uses the *Ctrl* key as the Copy modifier key, and this works the same way as Move for copies and multiple copies. The Rotate tool is represented by two curved arrows turning around each other:

Figure 5.29: Rotate Tool Button

The Rotate tool can be activated by doing the following:

- Clicking on any of the Rotate tool buttons
- Hitting Q on the keyboard

The Rotate tool selects the Geometry at the same time as selecting the central rotation point and rotation plane. It is highly recommended to preselect geometry when starting out with the Rotate tool. Once the Geometry is selected, three points must be picked to define the rotation angle. The first point (center point) will define the angle vertex, the second point will define the start of the angle, and the third point will define the angular distance. The angle distance can also be set by selecting the vertex and starting angle point and then typing a number into the Measurements Box representing degrees. Objects can be rotated by doing the following:

1. (*Optional but recommended*) Preselecting Geometry.
2. Activating the Rectangle tool.
3. Hovering over the desired plane and clicking to set the center point, then using *Shift* to lock a reference plane or the Arrow Keys to align with the default Axis references.
4. Clicking anywhere to set the first reference direction, picking specific Geometry if defining the angle with a second click, or clicking anywhere if typing a degree value.
5. Clicking to set the second reference angle or typing a degree value into the Measurements Box and hitting *Enter*. SketchUp will rotate the Geometry around the center point using the specified angle.

> **Note**
> When choosing the first point for our rotation, it is important to also align our rotation plane—we can see this on the Gizmo. We can use the Arrow Keys and *Shift* to lock in our reference planes.

The Rotate tool can be used very powerfully either when rotating Objects relative to other drawing Geometry or as an internal rotation (rotating an Object by itself). Most Rotate copies are achieved by rotating around other objects—let's first take a look at an example of rotating around other Geometry.

In this example, we have a dial on a vertical surface. The dial has one tick mark on the side of the connected face, but we want to rotate it to a different position. If we wanted to use an internal rotation for this workflow, we would also need to move it to the new location, so it makes sense to use an external rotation for this workflow. We would need to first preselect the rectangular group, then activate the Rotate tool. Then, we could select the center of our dial as the center for the rotation by either seeing the Green Gizmo or—as in this case—holding *Shift* to activate the Magenta reference Gizmo:

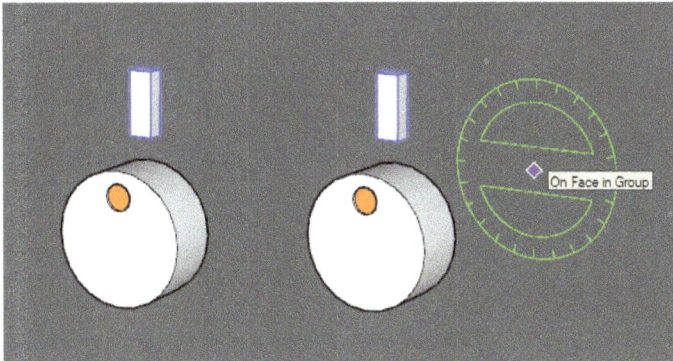

Figure 5.30: The Dial and Tick with the Tick Preselected (Left); Rotate
Tool Activated and Green Gizmo Shown (Right)

Then, we can select the center of the circle as the center point of our rotation. Once it is selected, we can choose the midpoint of the rectangle as the first point for the angle:

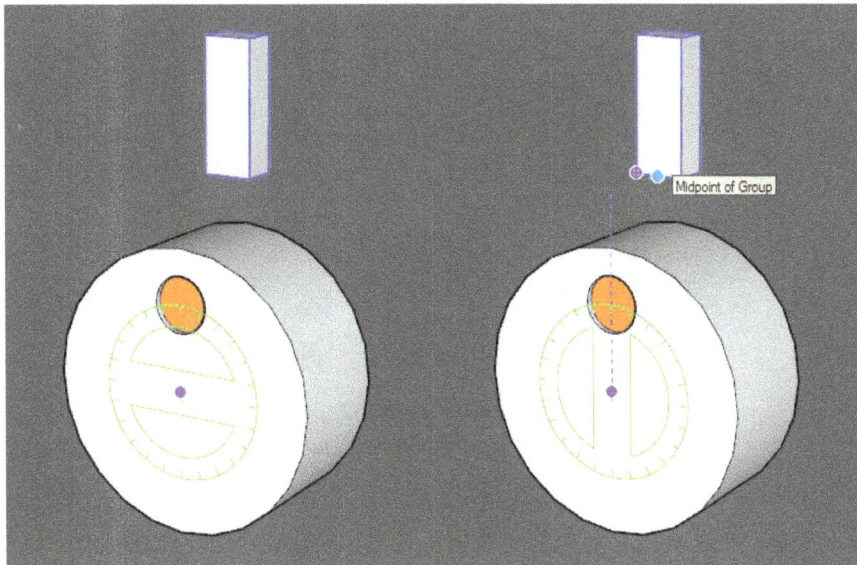

Figure 5.31: The Center Point is Chosen (Left) and the First Angle Point is Set as the Midpoint (Right)

Now, we can choose a destination point. In this case, we can show the rotation going Clockwise and type **30** into the Measurements Box. When we hit *Enter*, we see the tick mark has been rotated around the dial:

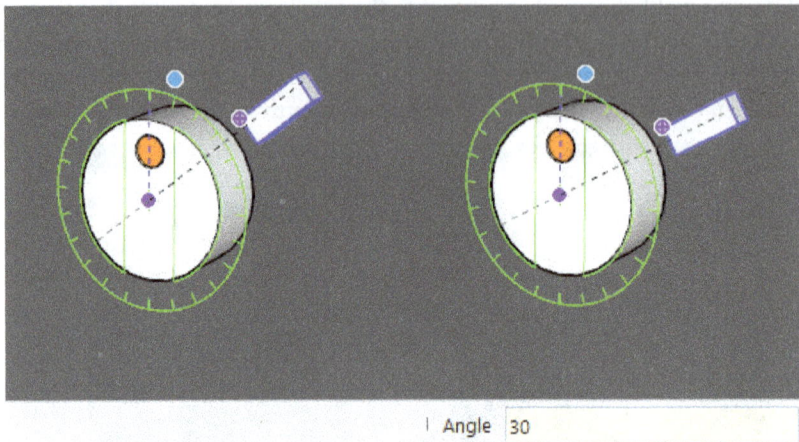

| Angle 30

Figure 5.32: Rotation Shown Clockwise (Left) and a Degree Angle of 30 is Entered (Right)

If we'd wanted to make a copy of the tick mark, the workflow would have been exactly the same, but we would have toggled Copy by hitting *Ctrl* before we typed in our degree angle of **30**. Then, we could also type in a multiplier value—in this case, **x8**—and see a radial Array or multiple radial copies being created!

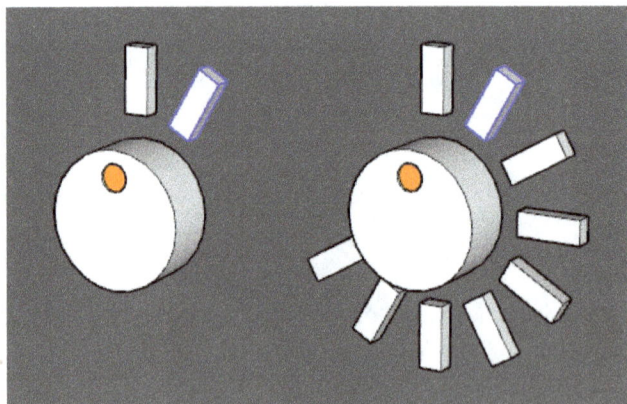

Figure 5.33: Copy Created at 30-Degree Angle (Left); 8 Total
Copies Created using x8 Multiplier Value (Right)

> **Note**
>
> Radial Copies are typically created using this workflow. We would rarely want an internal rotation to create copies—that just puts Geometry inside itself!

Internal rotations are achieved the same way, except that we can choose a center point on the geometry itself. In the previous example, if we wanted to rotate the dial, we could follow the same steps, but by preselecting or selecting the dial, we would rotate it instead!

There is one more internal rotation tool built into SketchUp Pro, and it can appear when using the Move tool on Groups or Components. This is slightly counterintuitive as it is for internal rotation, not moving, but SketchUp has added this as a bonus tool so that Move and Rotate workflows might be a little easier. Let's take a look at the dial and ticks that we looked at in the previous examples. If we wanted to internally rotate the dial on its own, we'd first need to ensure that it is a Group or Component. We can do this by preselecting it, and we see a blue bounding box appear:

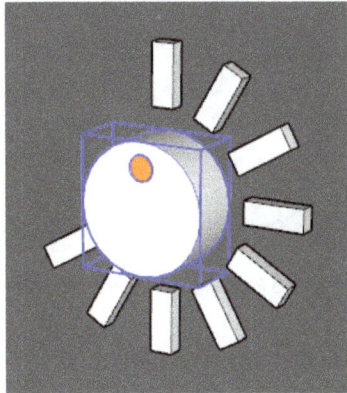

Figure 5.34: Dial Preselected with Blue Bounding Box Shown

Now, when we activate the Move tool and move the mouse over the bounding box, we see four red plus (+) signs appear on the side of the bounding box. These plus (+) signs are at the middle of the bounding box on the top, bottom, and sides of each face:

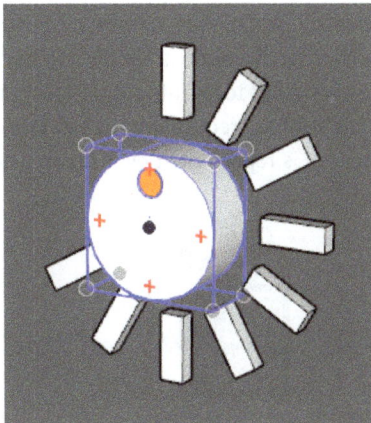

Figure 5.35: Bounding Box Red Plus (+) Signs

When we hover over one of the plus (+) signs, we can see that a Rotate Gizmo automatically appears and that the initial angle is previewed along the line between the center of the Object and the plus (+) sign. By clicking on a plus (+) sign, we are automatically in an internal rotation around the exact center of the group! We can then type an angle or click to rotate the Object. In this example, the dial can be snapped to any of the tick marks by selecting a midpoint:

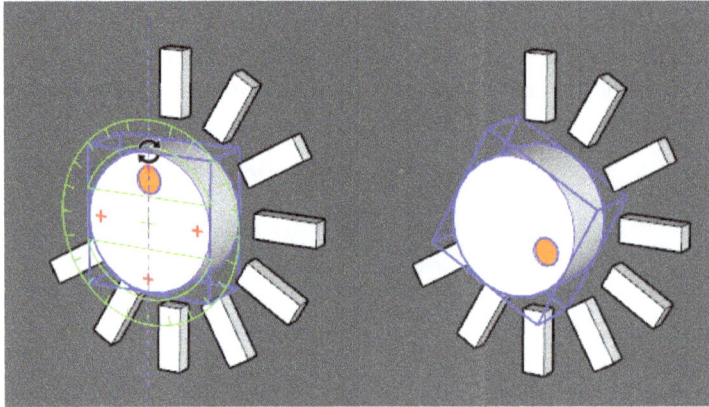

Figure 5.36: Rotate Activated Using the Top Plus (+) Sign (Left); Rotation Snapped to Tick Mark (Right)

Using the Move and Rotate tools together can create some very efficient and powerful workflows either by using Move and the built-in rotate grips or by using both the Move tool and Rotate tool. Let's continue our exploration of the Edit tools by looking next at the Scale tool.

The Scale Tool

The Scale tool is another familiar tool from other software—it changes the scale, or size, of an object. The Scale tool in SketchUp is driven primarily by the Object or Geometry bounding box and can change the scale of the three axes independently. That means that it can stretch something to be skinnier in one direction and taller in another—the Geometry does not have to keep the same proportions. There are two modifier keys for the Scale tool: *Shift* for scaling uniformly, and *Ctrl* for scaling about the center. The modifier keys for the Scale tool are required to be held throughout the workflow, unlike other tools' modifier keys. The Scale tool is represented by a square with an arrow showing how the square could increase:

Figure 5.37: Scale Tool Button

The Scale tool can be activated by doing the following:

- Clicking on any of the Scale tool buttons
- Hitting *S* on the keyboard

The Scale tool can be used on Groups, Components, or other Geometry. Even if the Geometry is not in a Group or Component, the Scale tool will still create a bounding box around the selected Geometry. The bounding box must represent at least two Axes directions, so individual Edges that follow an Axis will not be able to be scaled with the Scale tool. Geometry that is flat relative to the Axes will create a bounding rectangle rather than a bounding box. Once the Scale tool is active, green grips will appear on the bounding box at each corner point, the midpoint of each Edge, and the center point of each Face. These grips can be used to stretch and scale the Geometry along those axes. Additionally, once a grip has been activated by clicking and dragging, a positive or negative value can be typed into the Measurements Box. A value of 1 is no change, and a value of 2 is twice as big, whereas a value of 0.5 is half as big. Negative values are the same but will invert the Geometry so that it is flipped over the opposite grip or center. Geometry can be scaled by doing the following:

1. Activating the Scale tool.
2. Selecting the object to be scaled (if not preselected) with a click. A Scale bounding box will appear.
3. Hovering over a grip point and reading the tooltip popup, then activating any modifier keys to achieve the desired scale location and type.
4. Clicking and dragging the grip to change the scale of the Geometry.
5. *Optional*: Typing in a positive or negative scale value.

> **Note**
> Remember to hold the modifier keys in the Scale tool—they need to be held throughout the entire workflow in order to be utilized!

The best way to think through using the scale tool is to think about the following questions:

- Where to Scale from?
- Which Scale proportions should be used?
- How much Scale should be applied (Scale Amount and flipped)?

Where to scale from?

Choosing where to scale from is important. If some geometry needs to stay in the same place, that is where we scale from! So, if something should stay on the surface of a table and should scale to be larger, we would choose to scale from the opposite point. The grabbed grip will always move, and the opposite grip will always stay in place. In this example, we have a bowl on a surface, and we want the bowl to be larger. By grabbing one of the top grips—in this case, the center of the top face—when we scale the bowl to be taller, the bottom of the bowl stays on the table:

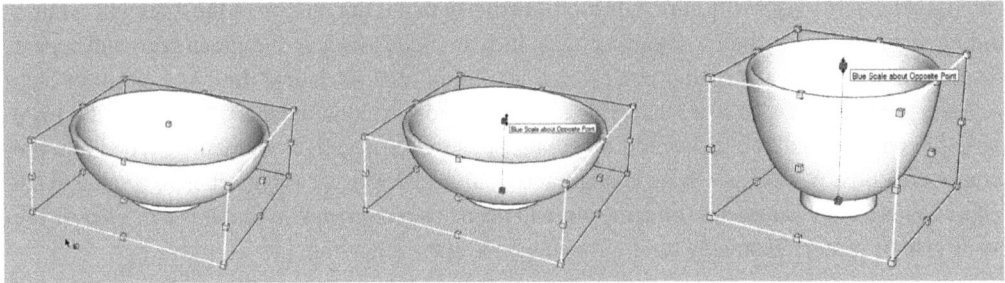

Figure 5.38: Bowl Selected with Scale Tool Active (Left); Center Grip of
Top Face Grabbed (Middle); Bowl Scaled Up (Right)

If we had an object sitting in a corner of a room and we wanted it to stay in the corner, we would choose the opposite corner grip point. In this example, we want to scale down a bookshelf. The bookshelf is in the corner of the room, and we want the height to remain the same. So, we can choose the right-Edge midpoint if we want the bookshelf to scale in width and depth:

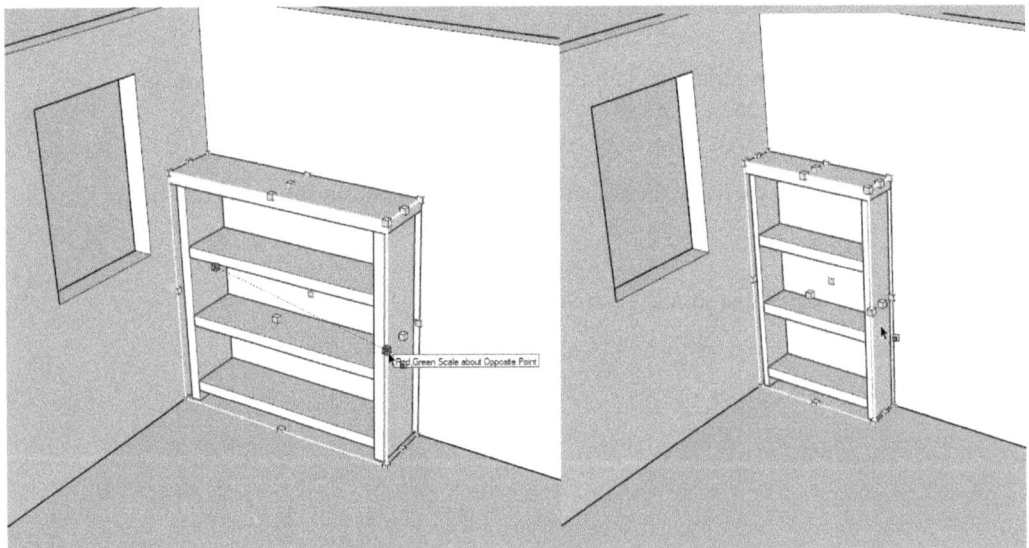

Figure 5.39: Right-Edge Midpoint is Chosen (Left) to Scale the Width and Depth (Right)

Or, we can choose the right-Face center point if we only want to scale the width:

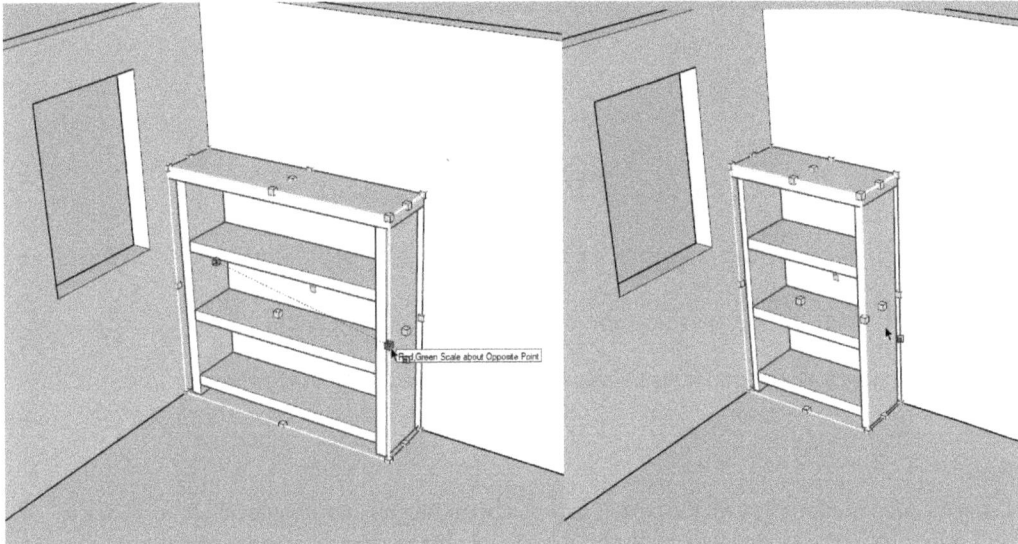

Figure 5.40: Right-Face Center Point is Chosen (Left) to Scale only the Width (Right)

> **Note**
> This is not the best way to change the dimensions of Groups and Components as it changes the size of the Object parts. Even though we did not change the height or depth of the shelves, the vertical members on the sides of the bookshelf got much, much skinnier. This might be good for a quick fix, but it is better to use Group Edit, Push/Pull, and Move to appropriately change the dimensions of this Group.

The final scale location can be the center of the Geometry! The center could be used when working with unconnected Geometry, or when Objects are floating in space. To scale from the center, you can use the *Ctrl* modifier key, but remember to hold it the whole time.

Which Scale Proportions?

The only two proportions we need to consider when using the Scale tool are keeping the original proportions along the Red, Green, and Blue Axes or breaking the original proportions. The tooltips will explain whether proportions are being kept or not, by saying: "Uniform Scale" or "Color(s) Scale about Opposite Point." When "Uniform Scale" is shown, the object will retain its original proportions. This is something that we can easily rationalize—the object is simply growing or shrinking but not changing shape. Uniform Scale will automatically be applied when scaling from a corner of the bounding box, and the *Shift* modifier key will need to be used to lock in Uniform Scale on all other grip points:

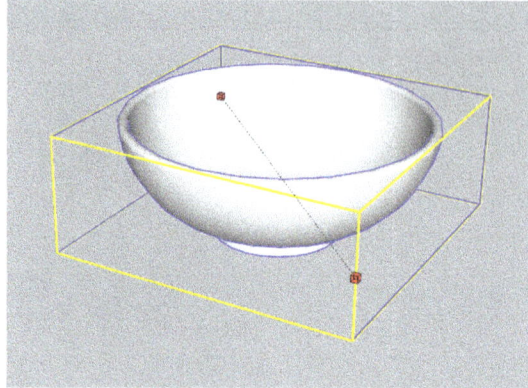

Figure 5.41: Uniform Scale Using a Midpoint Grip and the Shift Modifier Key

Note

Using the *Shift* modifier key on the corner grip of a bounding box will toggle off Uniform Scale. This might be tricky to navigate, as all three Axes are able to change. In this case, it might be easier to scale non-uniformly in multiple steps!

Non-uniform scaling is where our SketchUp Geometry can start to get wonky. When scaling along only one Axis, it might be simple—for a box, the result might look like we have only moved a face or used the Push/Pull tool. However, scaling more complex Geometry can result in obviously stretched or squeezed Geometry, so be careful when scaling along only one or two Axes. This example shows a modest approach to a single-Axis scale—the round bowl is turned into an oval shape by only scaling along the Red Axis:

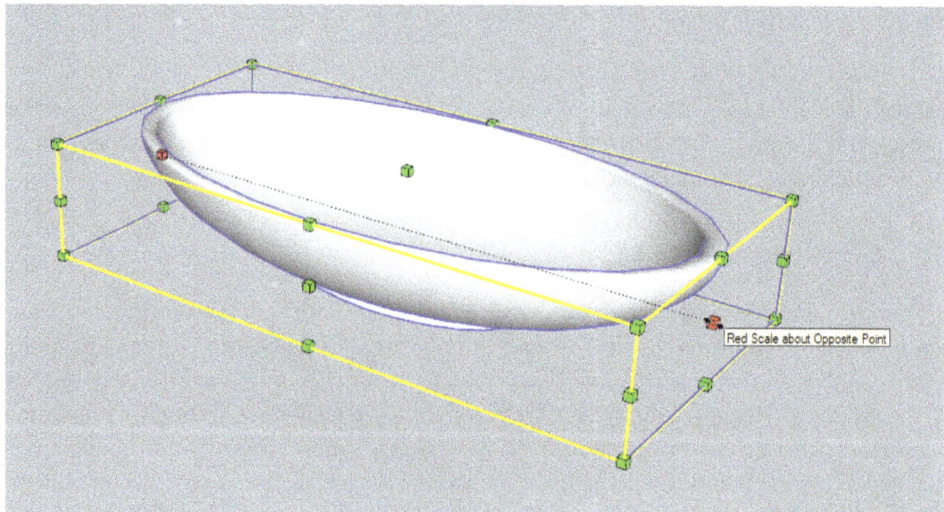

Figure 5.42: The Bowl is Scaled along the Red Axis Non-Uniformly

How Much Scale Should Be Applied? (Scale Amount and Flipped?)

Scaling Geometry to the correct size can be very tricky in SketchUp, especially when scaling along two Axes. Sometimes the Scale amount is defined by other Geometry in the model, but in most cases, it is suggested to type Scale values into the Measurements Box. In this first example, we can see a perfect use case for scaling to other Geometry when we do not know the exact scale value. We want to fill this landscaping box with sand, and we have a block of sand created from an earlier project. But the distances don't match up, and we need the sand to fill the box. In this case, we can place the sand Component in one corner of the box:

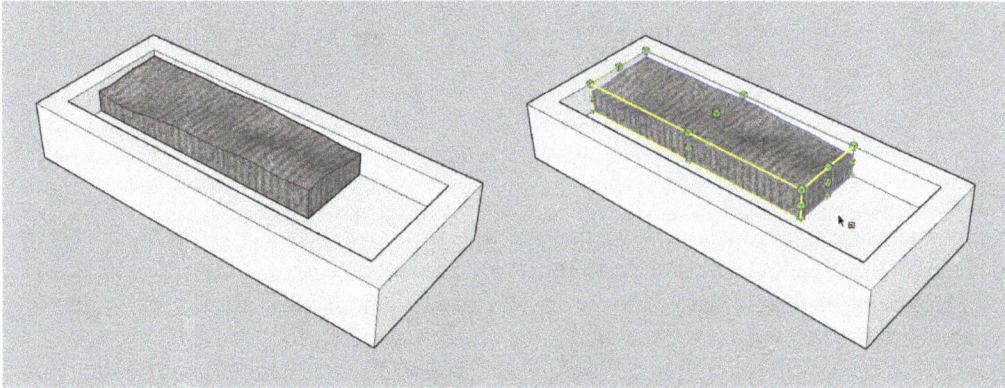

Figure 5.43: Sand in the Landscaping Box (Left) and Selected with the Scale Tool (Right)

Then, Scale using the Midpoint Grip on the opposite Edge:

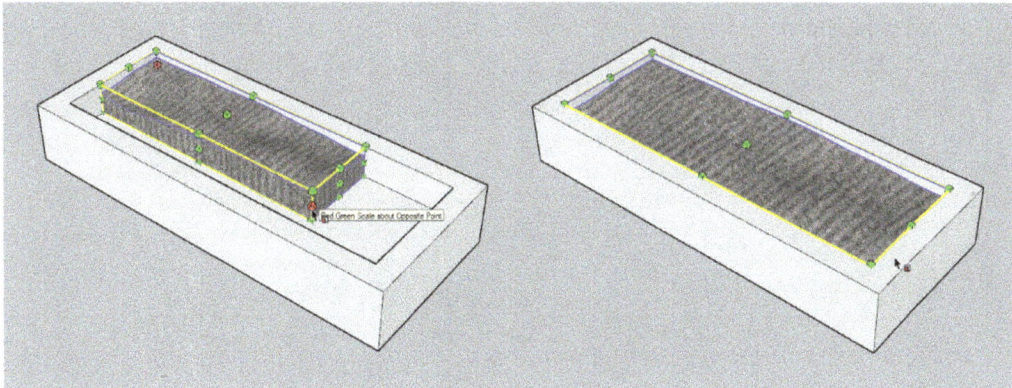

Figure 5.44: Opposite Edge Midpoint Grip Chosen (Left) and Scaled
using the Landscaping Box Endpoint Snap (Right)

This will be a Red,Green Scale about Opposite Point and will allow us to snap to the opposite corner of the box:

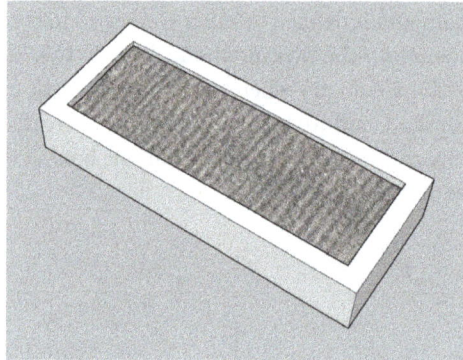

Figure 5.45: Finished Landscaping Box

However, most of the time, we have either an exact value in mind to scale our Geometry or at least a good idea of the magnitude of the scale. In either case, the best practice is to use a scale multiplier in the Measurements Box. When a scale workflow is activated by clicking on one of the grips, the Measurements Box will update the prompt to show which Axes will be included and in what order. When more than one Axis is used, the values should be separated by a comma. Remember—the base scale is 1, so greater than 1 will increase the scale and less than 1 will decrease the scale.

In this example, we want the bowl to get two times taller and nothing else, so we choose the top Face center grip and drag. Then, without activating any tool or clicking anywhere else, we can type 2 and click *Enter*. The bowl was only scaled using the Blue Axis based on our grip selection, so only one value needed to be added:

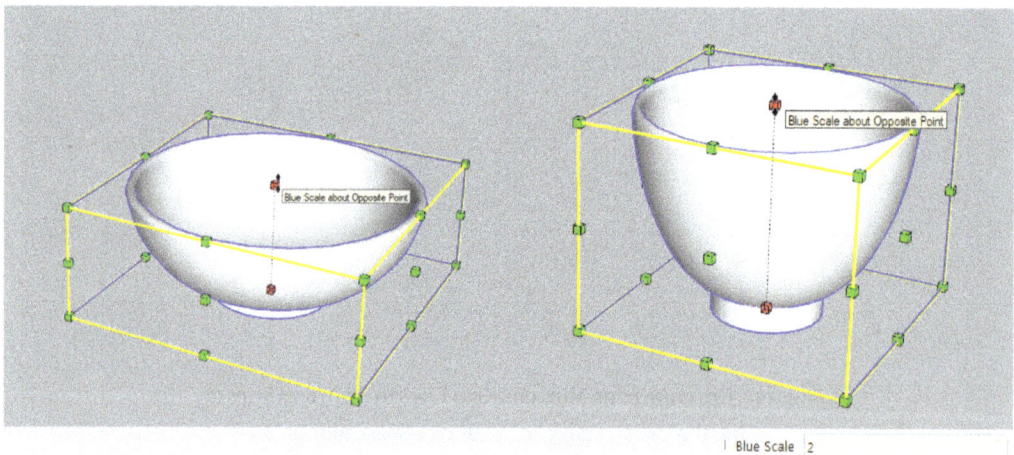

Figure 5.46: Bowl Selected with Scale and the Top Center Grip is
Activated (Left); Scaled 2x Along Blue Axis (Right)

In this next example, we want to scale the coffee Table to be 2 times longer and 75% as deep so that we can grab the opposite edge midpoint grip:

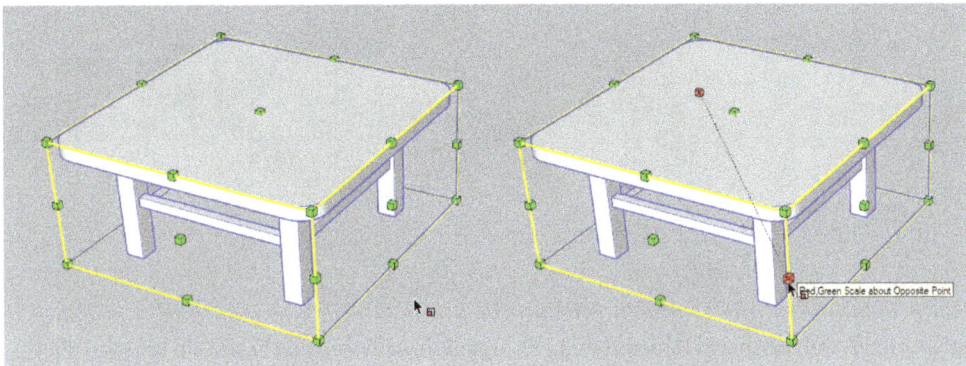

Figure 5.47: Coffee Table Selected with Scale (Left) and Opposite Edge Midpoint Grip Active (Right)

Then, we do not know which value will be first—Red or Green—so we can approximate the size of the scale using the Grip. The Measurements Box value updates to something close to what we want, but not exactly the right values. But we can see that the Red Axis is the length as it grew larger and close to our desired length of 2, and the Green Axis is the depth as it grew shorter. So, we can type "2,0.75" into the Measurements Box and hit *Enter* to set the exact values:

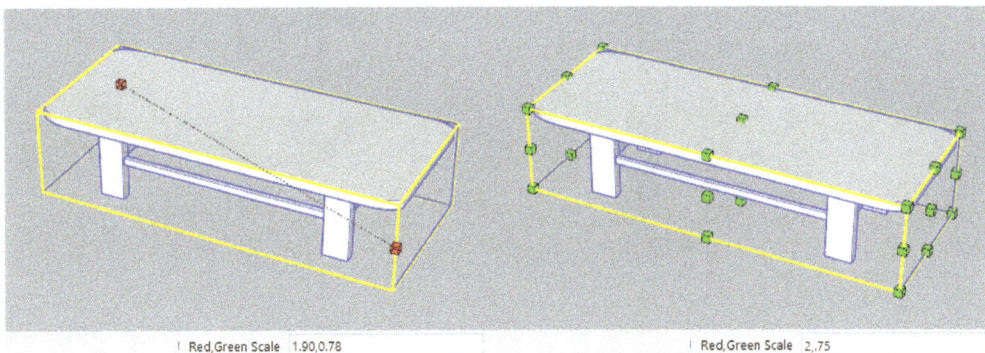

Figure 5.48: Hand-Scaled Estimated Size and Values (Left); Typed-In Exact Size and Values (Right)

We can see how this changes the Geometry inside the coffee table Group, and this is a good reminder that non-uniform Scaling not be the best workflow for changing the size of Groups or Components.

Objects in SketchUp can also be scaled to exact dimensions by using the Scale tool. In this next example, we will see a block that has not been drawn in the correct dimensions. However, without knowing the exact dimensions that have been drawn, it could be difficult to resize this model without multiple Move or Push/Pull workflows. In this case, it is easier to scale the model using the Scale tool:

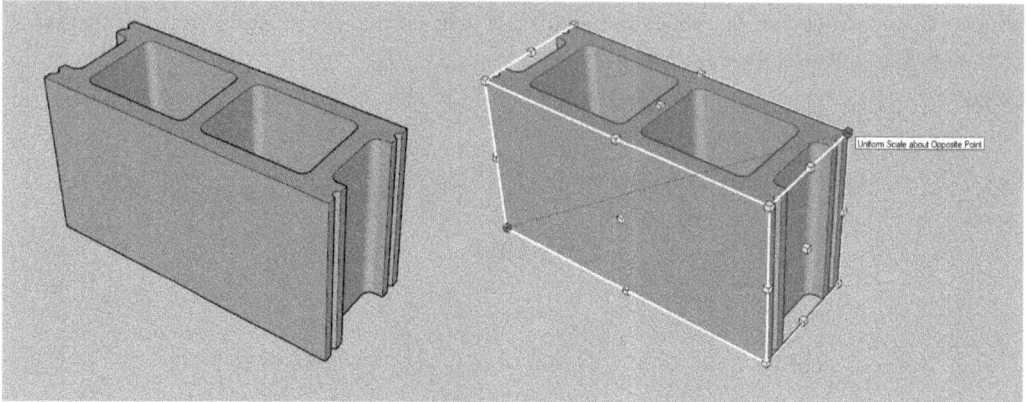

Figure 5.49: Block with Unknown Dimensions (Left); Scale Tool Activated and Scale Grip Selected (Right)

Once the Scale tool has been activated, the exact dimensions can be typed into the Measurements Box. In this example, the nominal values are typed as 1'4", 0'8", and 0'8":

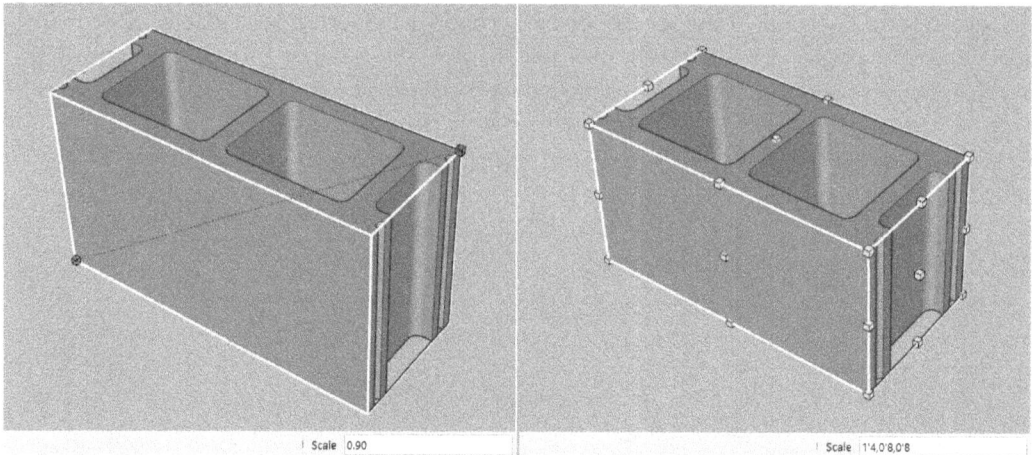

Figure 5.50: Scale Tool Activated (Left); Exact Values Typed in Measurements Box (Right)

> **Note**
> Keep track of your model Units! This model is set to Feet as the primary unit, so all measurements must start with feet! That is why we have used the leading 0' amounts when we only need inches.

Remember that SketchUp is very precise! We could also type in the exact values for this block, although it is more challenging to complete this process in one step. The exact values need to be typed out with tick marks and quotation marks for feet and inches, and each value must be separated with a comma. In this example, the exact values are typed as 1'3 5/8", 0'7 5/8", and 0'7 5/8":

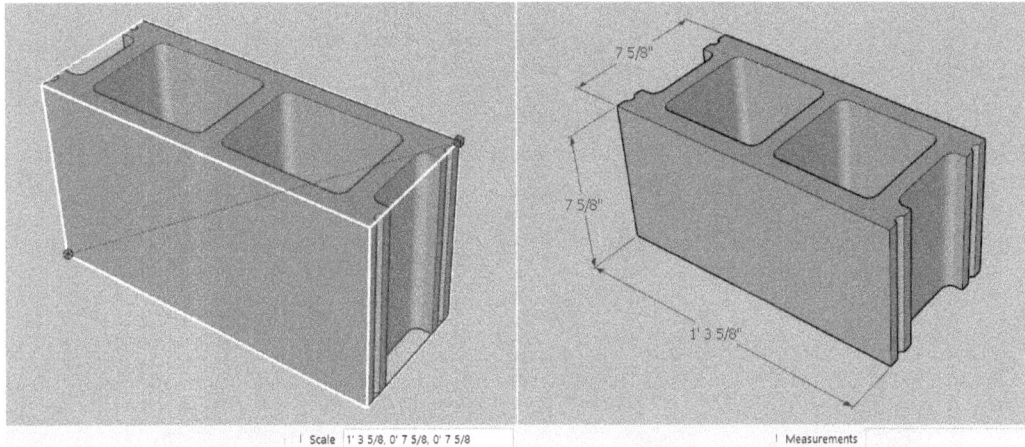

Figure 5.51: Exact Values Typed into Measurements Box (Left); Finished Size (Right)

The Tape Measure tool can also be used to scale a model. The Tape Measure tool will scale all Geometry in the active Group, Component, or Model to a desired size. The Tape Measure tool is a great option when scaling in Imported Geometry at the beginning of a design workflow, such as when bringing in 2D drawings. The Tape Measure tool can scale Geometry by doing the following:

1. Activating the Tape Measure tool. Ensure that the Toggle Guides option is off.

2. Clicking to set a Basepoint and clicking again to set the measurement distance.

3. Typing a new value into the Measurements Box that the specified distance should match.

In this example, we can see that a floorplan has been brought into SketchUp but that it is at the wrong scale and none of the dimensions is showing the expected values:

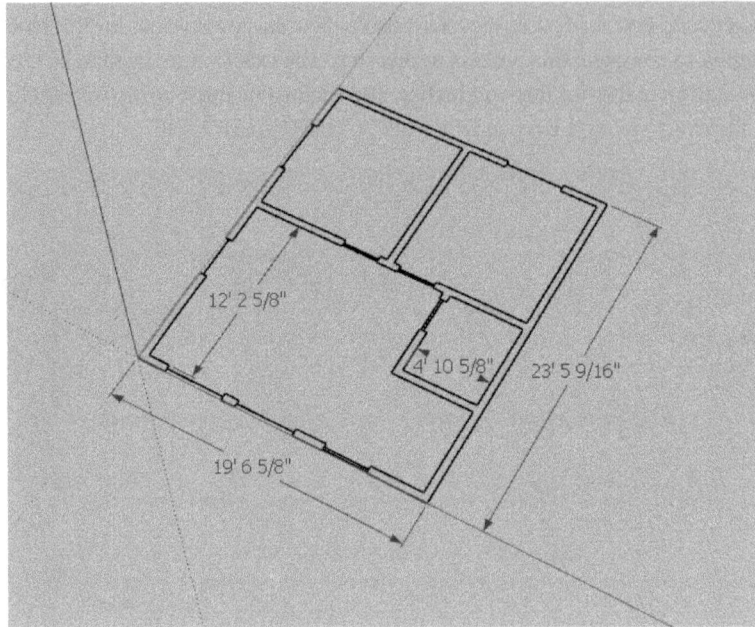

Figure 5.52: Floorplan in SketchUp at Incorrect Scale

We want to scale the entire drawing, so we can activate the Tape Measure tool and ensure that Guides are toggled off. Then, the origin point can be clicked to set the first point for the Scale workflow:

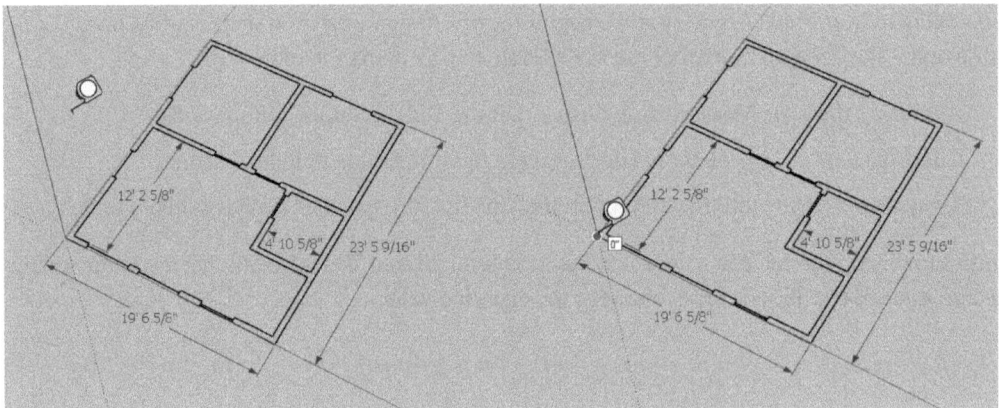

Figure 5.53: Tape Measure Activated with Guides Toggled Off
(Left); Tape Measure First Point Selected (Right)

With the first point selected, the second point can be clicked to set the distance on the model. In this case, the adjusted distance should be 20'0", so 20' is typed into the Measurements Box. When a measurement is entered, SketchUp will prompt with a warning box asking if you want to resize the model:

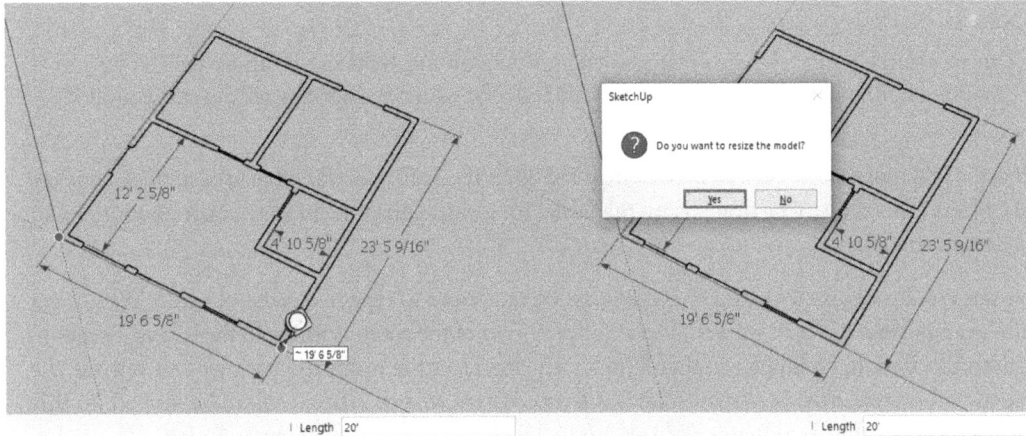

Figure 5.54: Second Point Clicked and Measurement Added (Left); Warning Box (Right)

Once the measurement is entered, the entire model will scale, and we can see the dimensions update to reflect the correct dimensions!

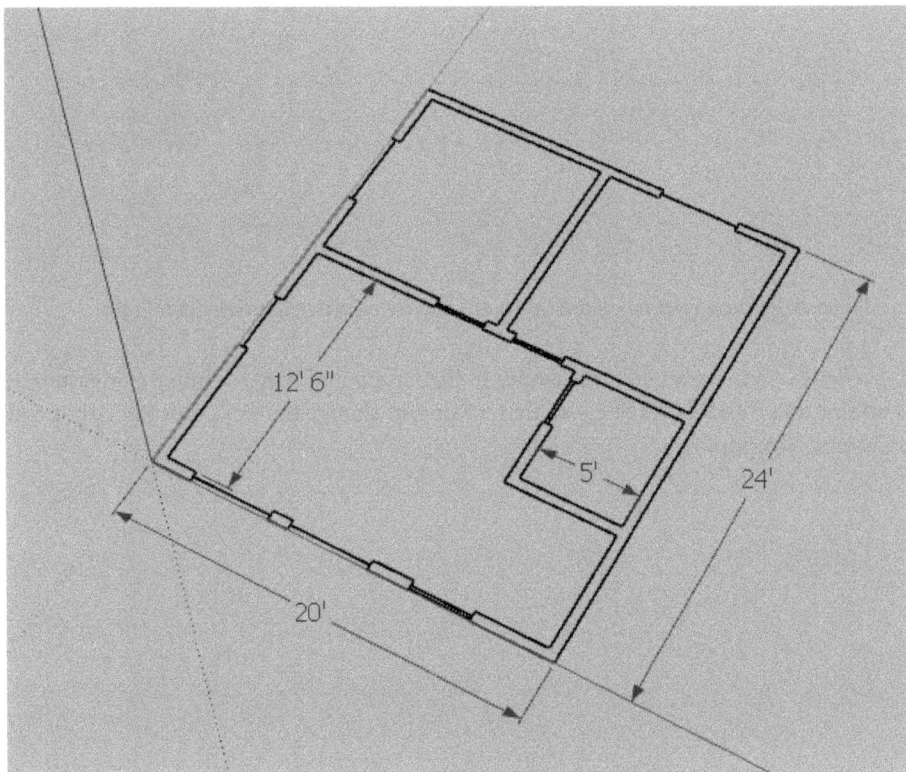

Figure 5.55: Resized Model with Correct Dimensions

> **Note**
>
> This workflow will scale the entire model unless you are working in an active Group or Component. It might be better to use the Scale tool if you do not want to scale the entire model!

Before the Flip tool was introduced in SketchUp Pro 2023, the Scale tool was often used to flip Geometry and Objects. This can still be done using the Scale Tool, although you can learn more about the Flip tool later in this chapter.

If we want to flip Geometry using the Scale tool, we only need to type in a negative value! We can see in this example that the desk we want to put in the corner of the room is facing in the wrong direction. We need this desk to be flipped so that it fits nicely and faces the room. To start, we can activate the Scale tool, then select the grip that isolates the Axis where we need the Group to be flipped. In this case, the desk only needs to be flipped Left/Right, so we can use the Face Center Grip:

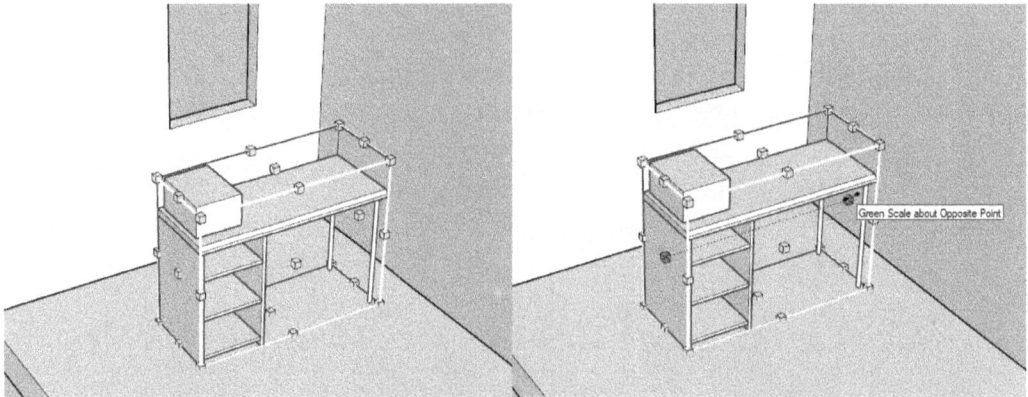

Figure 5.56: Desk Selected with Scale (Left) and Face Center Grip Selected (Right)

Then, we can hold *Ctrl* to scale about the center of the Group and type -1 into the Measurements Box. This retains the overall Scale of 1—so that we do not change the size—but the Group will be flipped along that direction:

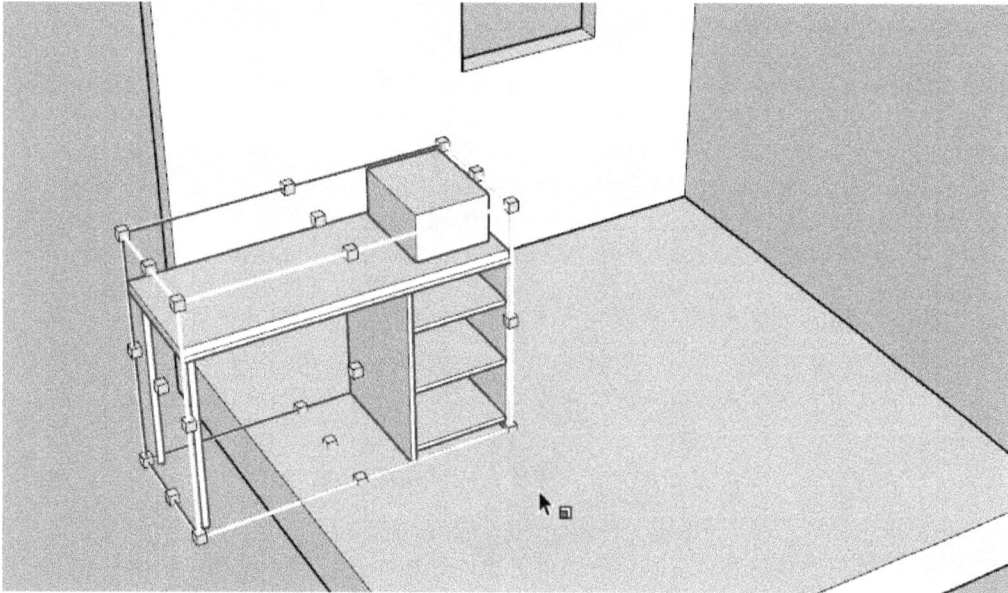

Figure 5.57: -1 Value in Measurements Box (Left); Desk is Flipped (Right)

When done, the desk can be moved into place:

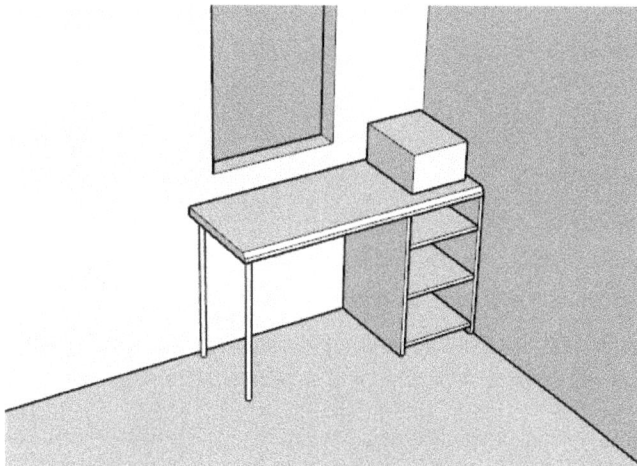

Figure 5.58: Desk Moved into Corner

Move, Rotate, and Scale are all common tools in 3D modeling software. In the next section, we will look at a unique tool that has helped to differentiate SketchUp Pro—the Push/Pull tool.

The Push/Pull Tool

The Push/Pull tool is SketchUp Pro's version of Extrude, but it comes with some extra-powerful features! Push/Pull can do two different things—it can create or remove Geometry, as the name Push/Pull suggests. Push/Pull creates a 3D shape from a 2D Face in the perpendicular direction—it cannot go in any direction, unlike the Move/Autofold tool. Push/Pull can extrude new Geometry using any Face, and it can also cut holes into any watertight Geometry. Also, if two faces of watertight Geometry are in the same orientation, Push/Pull can cut a hole through the entire object. Push/Pull has one modifier key—*Ctrl*—to create new starting faces. The Push/Pull tool is represented by an arrow pointing up from a flat box:

Figure 5.59: Push/Pull Tool Button

The Push/Pull tool can be activated by doing the following:

- Clicking on any of the Push/Pull tool buttons
- Hitting *P* on the keyboard

The Push/Pull tool can change and create Geometry using the same workflow. It just depends on whether you go "in/through" or "away from" other Geometry. Geometry can be changed and created by doing the following:

1. Optionally preselecting one or more Faces. Only one Face can be selected at a time once the Push/Pull tool is activated.

2. Activating the Push/Pull tool.

3. Selecting a Face to be moved (if not preselected) with a click. The starting plane is the plane of the Face, so no other basepoint needs to be set.

4. Clicking anywhere to set the destination point (second point). SketchUp will either extrude the face to create a new 3D shape or will cut away Geometry from the connected watertight Geometry.

We can see Push/Pull in action in the following examples. We will start with creating extruded Geometry. In this example, we can see a Face that we want to pull into 3D. We can click on the Face and move the mouse up and click again or type a distance. In this case, the distance is 6":

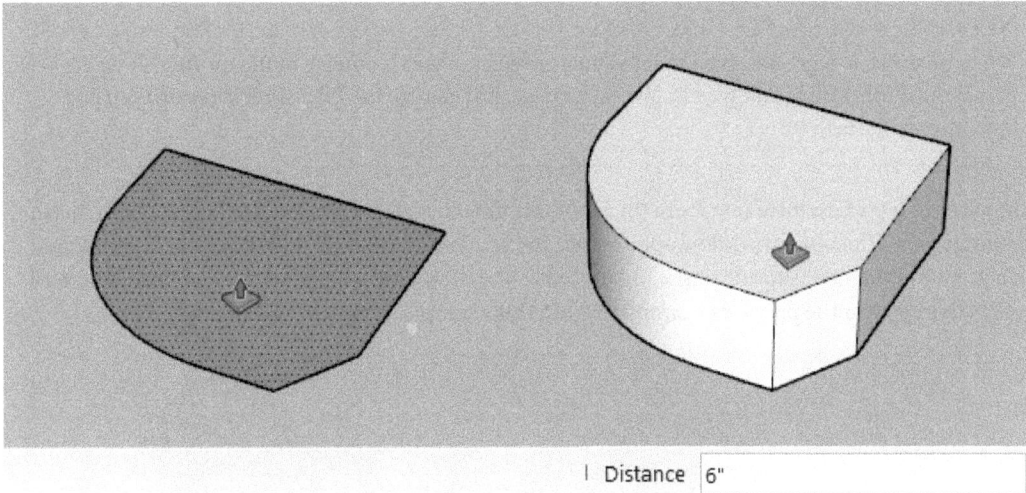

| Distance   6"

Figure 5.60: Face is Selected with Push/Pull Tool Active (Left); 6" is
Typed into Measurements Box and Enter is Hit (Right)

Push/Pull can also be snapped to existing Geometry in the model. Because Push/Pull will always create or cut away Geometry in the perpendicular direction, there is no need to lock in an inference or direction. That makes it very easy to select Geometry anywhere in the model without being concerned that the Push/Pull workflow will create the Geometry in the wrong spot. In this example, both posts should be the same height. The top face of the first post is selected with Push/Pull, and then the second post can be picked to set the same height:

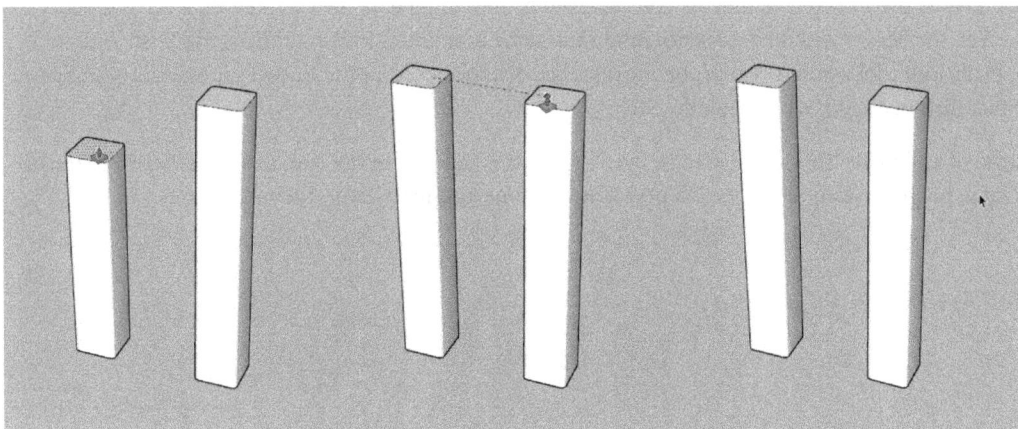

Figure 5.61: First Post Selected with Push/Pull (Left); Second Post
is Clicked (Middle); Posts Same Height (Right)

> **Note**
>
> When possible, it is great to type exact values or select other Geometry in the model. The great part about Push/Pull is that we can always go back and change the Push/Pull value again using the same workflow!

Let's look at that post example one more time, but this time the taller post needs to be shrunk to match the shorter post. The same workflow applies, but the face of the taller post is clicked first with Push/Pull active, and then the shorter post can be used as the destination height data. The Push/Pull tool created Geometry in the previous example, and in this example, it removes Geometry:

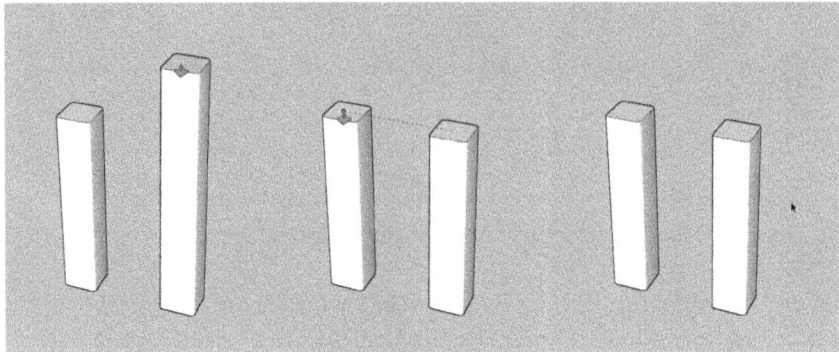

Figure 5.62: Taller Post Selected with Push/Pull (Left); Shorter Post
is Clicked (Middle); Posts Same Height (Right)

The previous two examples could also be achieved by moving the top faces in the Blue Axis direction. However, the Move workflow does not retain the same amount of information as the Push/Pull tool. The Push/Pull tool can remember the most recent distance used in the Push/Pull workflow and can use that distance again with a double-click.

Instead of two posts, this time we have five. One post is taller than the rest, but the other four are all the same height. We could click each post one at a time and match it to the taller post:

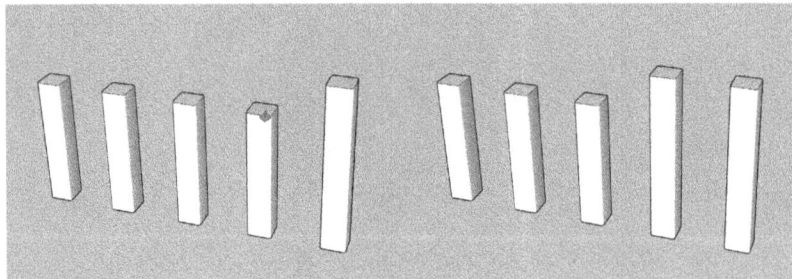

Figure 5.63: Five Posts and Shorter Post Selected with Push/Pull (Left); Posts' Matched Height (Right)

But instead, we can set the first post using the same workflow as before, then double-click the remaining posts:

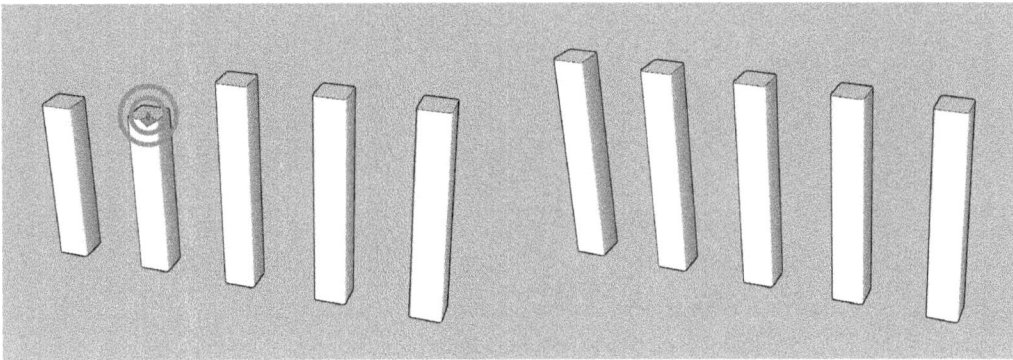

Figure 5.64: Remaining Posts are Double-Clicked (Left); All Posts Same Height (Right)

Push/Pull remembers this measurement for all directions—not just up or down. In this next example, there are two rectangular boxes on top of each other. These boxes are the exact same size, but the top box should have a 1" lip running around all sides. The scale tool could be used, but we would have to use a complicated equation to find out the scaling percentage. Instead, the top box group is activated, then the Push/Pull tool is used to pull one face out 1":

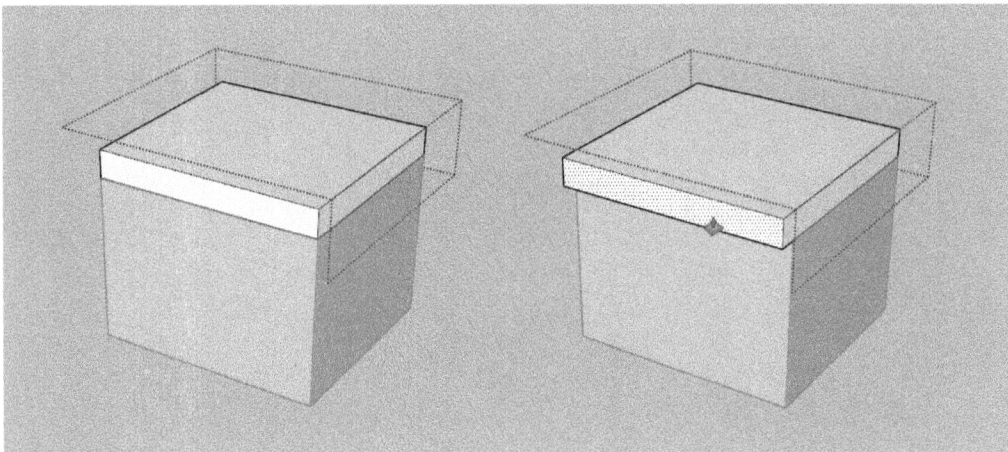

Figure 5.65: Top Box Group is Activated (Left) and One Face is Pulled Out 1" (Right)

Then, the other faces around the sides of the box can be double-clicked to extrude out 1" as well. All sides of the box are facing the same normal, meaning that all faces are oriented the same way. So, all these faces will extrude out with a double-click:

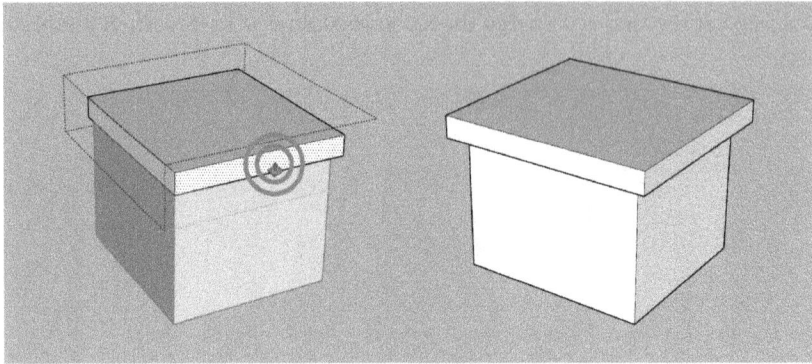

Figure 5.66: The Other Faces are Double-Clicked with the Push/
Pull Tool (Left) and Resulting Geometry (Right)

> **Note**
>
> Double-clicking workflows rely heavily on Geometry and Face orientation! We will learn more about Geometry and Face orientation later in this chapter.

The Push/Pull tool can also cut away Geometry from our models. In this first example, we can see a cylinder with a circle on the top Face. This inner circle was created with the Offset tool—we will learn more about the Offset tool later in this chapter. The inner circle can be pushed down into the cylinder to create something such as a coffee mug:

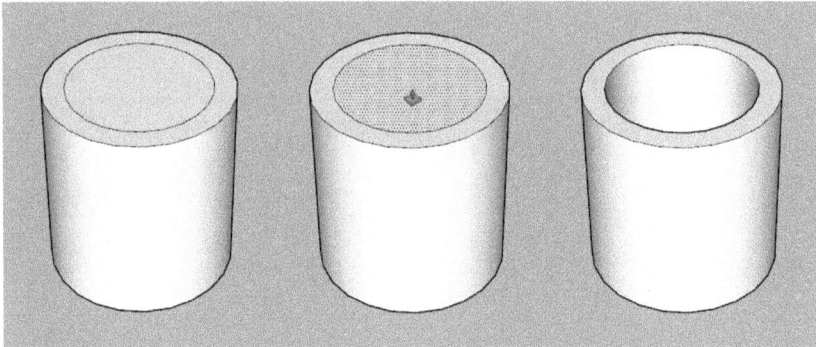

Figure 5.67: Cylinder and Offset Circle (Left); Push/Pull Active and
Picked Inner Face (Middle); Face Pushed Down (Right)

This is very different from moving the Face, which could be done with Autofold. Autofold would have folded the donut-shaped Face, but Push/Pull leaves the donut-shaped face in place and creates new, vertical faces connecting the original circle to the donut-shaped Face. We can see Move/Autofold on the left and Push/Pull on the right in the following screenshot:

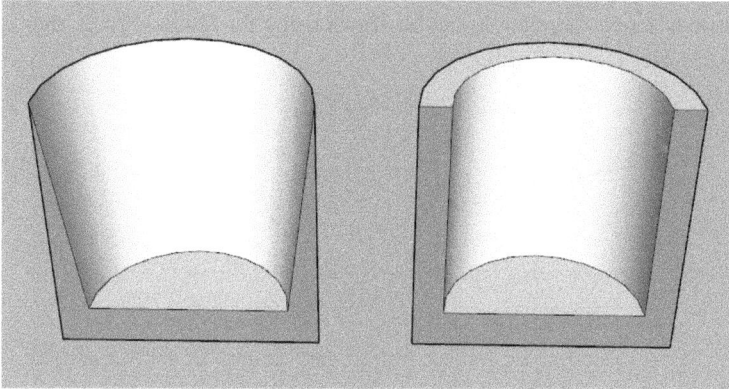

Figure 5.68: Move and Autofold (Left); Push/Pull (Right)

Geometry can only be cut with the Push/Pull tool in a perpendicular direction, so the edges of a hole will always go straight into the Geometry. However, there are great opportunities to explore with other Edit tools to see how holes can be further changed—especially with Move, Rotate, and Scale.

The Push/Pull tool can also cut complete holes through Geometry, starting on one side and going straight through the other. When cutting all the way through Geometry, the two faces must be at the exact same angle relative to one another—any other angle will not create a full hole. In this example, we see a wall and some Guides to draw a window:

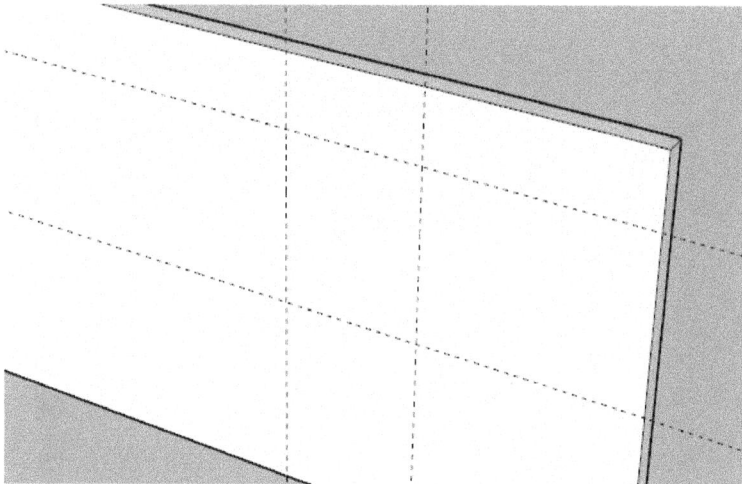

Figure 5.69: Wall with Guides Outlining a Potential Window Opening

Note

We will look at creating Guides later in this chapter.

Once the wall Group is active, a rectangle can be drawn using the Guides. Then, the Guides can be removed by going to **Edit | Delete Guides**:

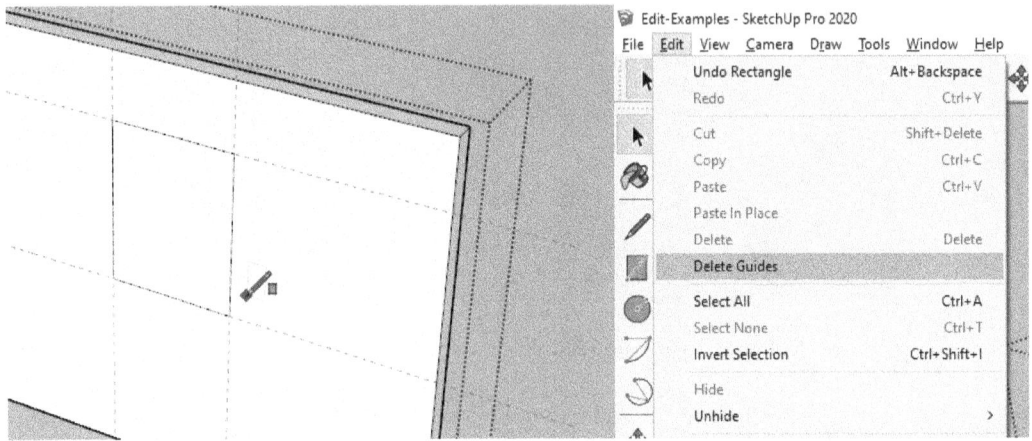

Figure 5.70: Rectangle Drawn in Wall Group (Left); Guides Deleted Using Edit | Delete Guides (Right)

The Push/Pull tool can then be activated, and the rectangle is picked. With the rectangle picked, the Geometry can be pushed into the wall:

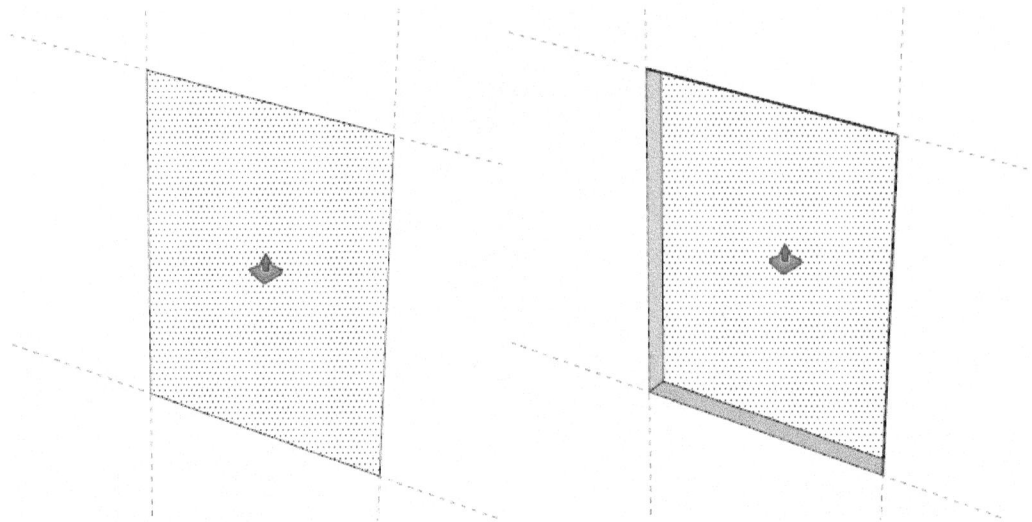

Figure 5.71: Rectangle Picked with Push/Pull Tool (Left); Rectangle Beginning to be Pushed into Wall (Right)

Any snap on the opposite Face will cause the Push/Pull workflow to create a full hole. On Face is the most common snap to see in this type of workflow, but On Edge or Endpoint can also be used:

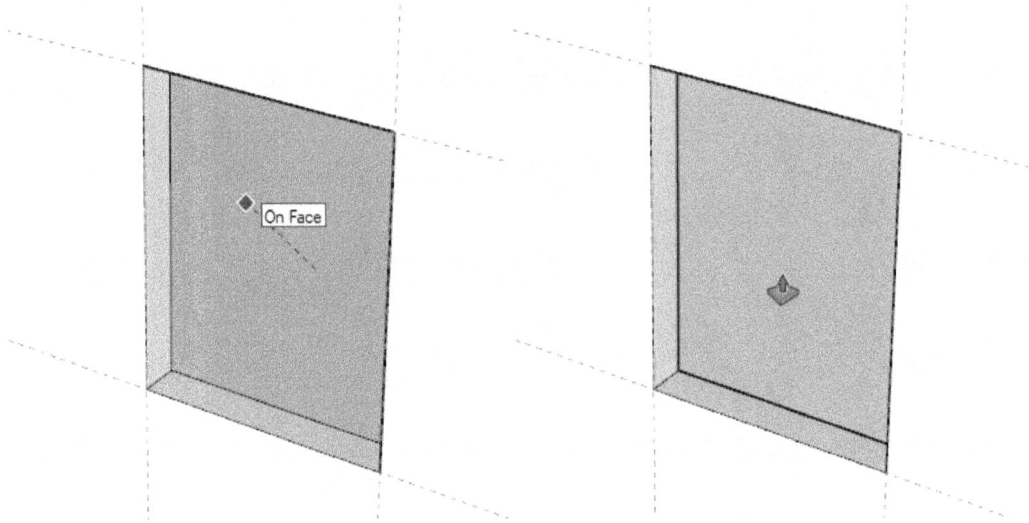

Figure 5.72: On Face Snap in the Push/Pull Tool Workflow (Left); On Face is Clicked and a Full Hole is Created (Right)

> **Note**
>
> One of the biggest stumbling points in Push/Pull workflows is that the Geometry is not at the same level of Group or Component! If the rectangle had been drawn outside the Group, a new box would have been created inside the wall and would not have interacted at all with the wall Geometry. Always make sure that you are creating Geometry for Push/Pull at the correct Object level!

At some points during a Push/Pull workflow, there might be the following warning: "**Offset Limited**." SketchUp will limit the Push/Pull tool from going over any Edge that is connected to the Geometry that the Push/Pull Tool is editing. This is a safeguard from overwriting existing Geometry or creating overlapping, complex shapes. We can see this in the following example. The shorter face is being Pulled toward the other side of the Geometry, and the offset is limited when it hits the other side of the Geometry. If it were allowed to go through the other side, then the geometry could overlap and create complex, non-watertight spaces:

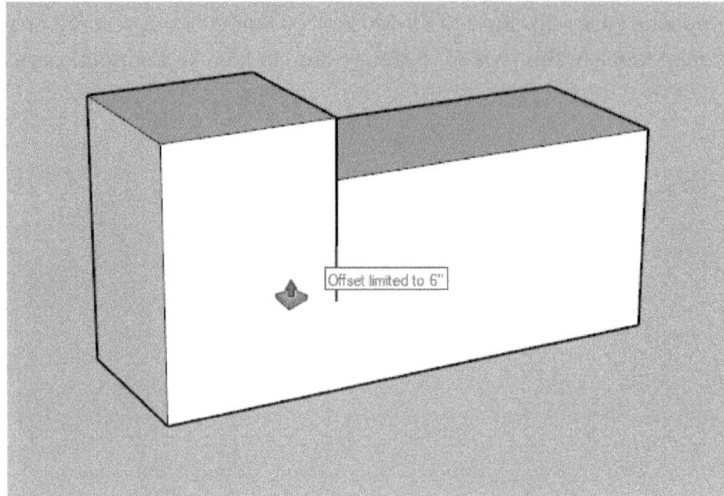

Figure 5.73: Geometry with Selected Face (Left); Offset Limited when Intersecting Same Geometry (Right)

Any distance less than the Offset limit will still work—and it might be a good reminder to Group and separate your Geometry if it is not supposed to be connected in the first place!

Geometry can also throw up errors when attempting to be edited with Push/Pull. Two common errors are "**Offset Limited to 0**" and "**Cannot Push/Pull curved or smoothed surfaces**." Let's use the same shape for both of these examples.

In the first example, we want to Pull the middle face up to match the height of one of the sides:

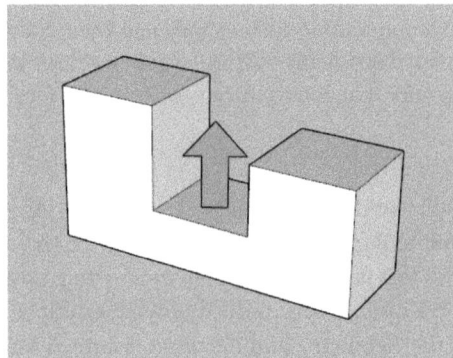

Figure 5.74: Geometry and Desired Push/Pull Workflow

However, when the Face is selected and the Push/Pull tool activated, there is an error of Offset Limited to 0 This is due to the Geometry intersecting with itself, as explained in the previous example, but it does not look like that is the case! Only when we view hidden Edges do we see the issue—the vertical face on the left was triangulated and softened to hide the diagonal Edge:

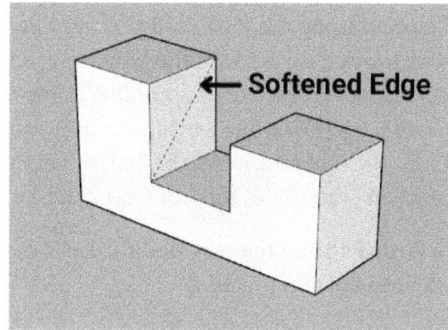

Figure 5.75: Softened Edge in the Triangulated Face

Because there was a separate Edge that was 0" away (the same as touching the Face), it could not be pulled up. It could move down, though, so SketchUp allowed the Push/Pull workflow to happen in general. In this next example, we can see when SketchUp will not allow the Push/Pull tool to even be activated, and that is when we attempt to choose a smoothed or softened Face. We can use the same Geometry, as we now know the vertical face on the left has a hidden Edge, making this a smoothed/softened Face:

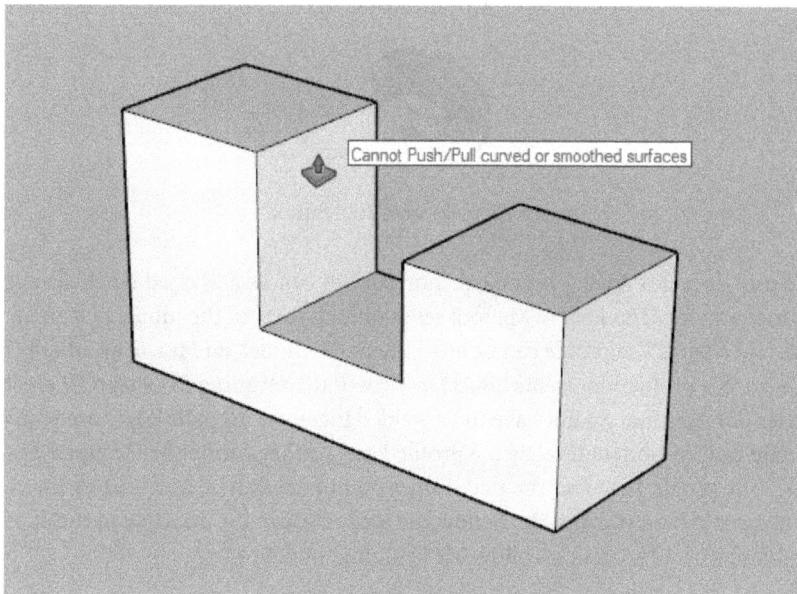

Figure 5.76: Error Message for Smoothed or Softened Face

This is the same reason why we cannot Push/Pull the outside of a cylinder or any other organic shape—there are multiple connected Faces with hidden edges, so SketchUp does not know how to move perpendicularly to the resulting Face. Even in our example, when the Faces are coplanar, the Push/Pull tool will not allow the Face to be selected.

The Offset Limit can be worked around using the Push/Pull *Ctrl* modifier key, which creates a copy of the pushed or pulled Face. As we learned earlier with the Move tool, copying a Face will leave the original Face in the same location and will create a copy according to the tool. In this case, the Push/Pull tool will do the same Push/Pull workflow while leaving the original Face in place. This can be useful to get around the Offset Limit issue, but it can also be bad for your model as it creates interior Faces and will make the Geometry non-watertight. Interior faces can be a big issue!

There are some workflows where Push/Pull and the *Ctrl* modifier key can be used appropriately, and you might find a use while you experiment and practice!

The Follow Me Tool

The Follow Me tool works very similarly to the Push/Pull tool in that it extrudes and cuts Geometry using a defined Face. However, the Follow Me tool adds a layer of complexity to the workflow because it "follows" an Edge or connected series of Edges. Where the Push/Pull tool was limited only to the perpendicular direction, Follow Me allows SketchUp to create complex 3D shapes, including sweeps, revolves, and corner detailing, such as cornices and chamfers. The Follow Me tool is the best way to quickly create curved 3D shapes using the default SketchUp Pro toolset. The Follow Me tool is represented by a curved 3D arrow inside of a swept shape:

Figure 5.77: Follow Me Tool Button

The Follow Me tool does not have a keyboard shortcut but can be activated by clicking on any of the Follow Me tool buttons. The Follow Me tool relies on two parts of the model's Geometry: a Face used as a profile and a path. The profile can be any Face in the model, and it will eventually be pulled into a 3D shape, so this profile represents a section view of the resulting 3D shape. The path can be any Edge or series of Edges that do not have to be welded together. All path Edges are required to be in the same Group or Component level as the profile Face, so they cannot be Grouped or separated from each other. The profile Face and the path Edges do not need to be touching or intersecting at any point, but this could help you and the Follow Me tool visualize the anticipated result. Geometry can be extruded/swept/revolved using Follow Me by doing the following:

1. Activating the Follow Me tool.
2. Selecting the profile Face with a click.
3. Tracing the mouse along the desired path Edge chain.
4. Clicking to finish the Follow Me workflow and create the 3D Geometry.

Additionally, if the desired shape will wrap fully around a loop of Edges, a Face bounded by those Edges can be preselected to represent the entire path Edge chain. This can be done as follows:

1. Preselecting a Face fully bounded by the desired path Edge chain.

2. Selecting the profile Face.

3. The Follow Me tool will generate a complete 3D loop or "donut."

> **Note**
>
> The *Alt* modifier key allows for a similar workflow to create fully looped 3D Geometry. However, the selection of the path Face after using the *Alt* modifier key can be tricky and may cause the Follow Me tool to create an incomplete loop. It is always suggested to preselect the path Face when attempting to make a full loop using the Follow Me tool instead of using the *Alt* modifier key.

In this first example, we will look at a simple swept shape using the Follow Me tool. We can see that there is a rectangular donut that we want to follow the drawn series of Edges:

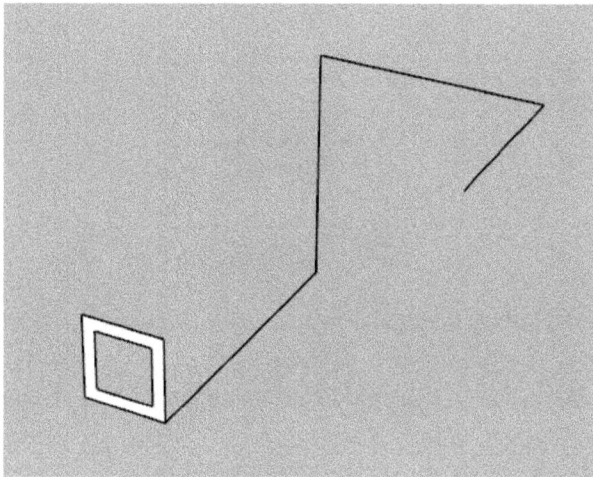

Figure 5.78: Rectangular Donut Face (Profile) and Path Edges

> **Note**
>
> This is a simplistic version of Follow Me; we will see how this interacts with more complex Geometry in the following examples.

Once the Follow Me tool is activated, we need to pick our Profile Face. After clicking the Face, we then see a red line overlay where the Follow Me tool thinks we are tracing. As long as the mouse cursor roughly follows the tracing destination along the Path Edges, the Follow Me tool will register the Path:

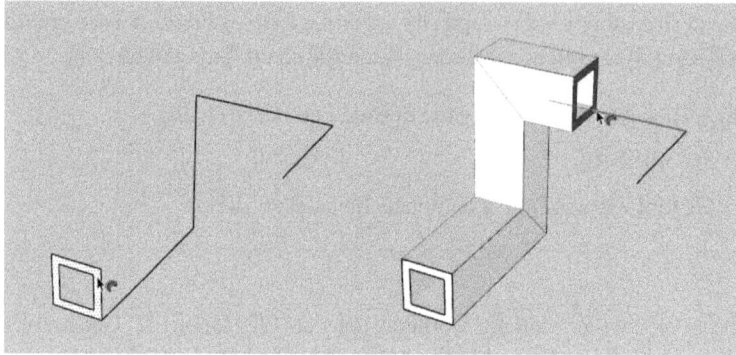

Figure 5.79: Follow Me Tool Activated and Face Selected (Left); Tracing Started (Right)

Then, we can move the mouse and click to end the Follow Me operation. In this example, the entire Path was used:

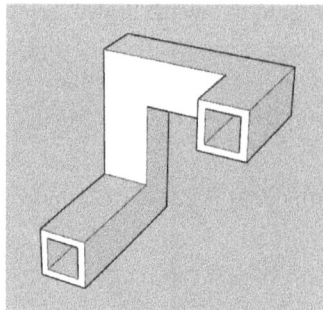

Figure 5.80: Follow Me Operation Completed Along Full Length of Path

Remember, this workflow can also be achieved by preselecting the Path, activating the Follow Me tool, then selecting the Profile:

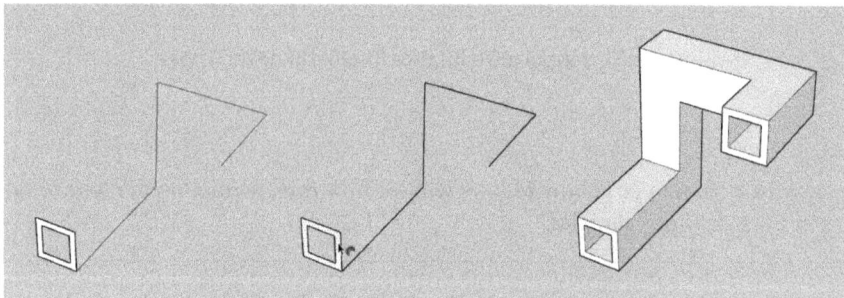

Figure 5.81: Path Preselected (Left); Follow Me Tool Activated and
Profile Selected (Middle); Completed Shape (Right)

The Follow Me tool is also a great way to create rounded 3D shapes, typically called revolved shapes. In the real world, these shapes could be made on a lathe. In SketchUp Pro, these shapes are created using the Follow Me tool and a circular path, often a circular Face. In this example, we create a bowl. Half of the bowl section has been drawn, and a circle has been created to represent the bottom of the bowl. The Path does not need to intersect with or touch the Profile, but in this case, it does:

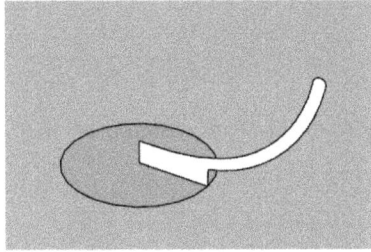

Figure 5.82: Half of a Bowl Section and Circular Bowl Bottom

In this example, we will preselect the circle that represents the bottom of the bowl—this is the Path. When a Face is preselected before the Follow Me tool is activated, then the bounding Edges that surround the Face will be used as the Path. Then, after activating the Follow Me tool, the bowl section Profile can be selected. The final bowl shape is revolved, and the bowl is completed:

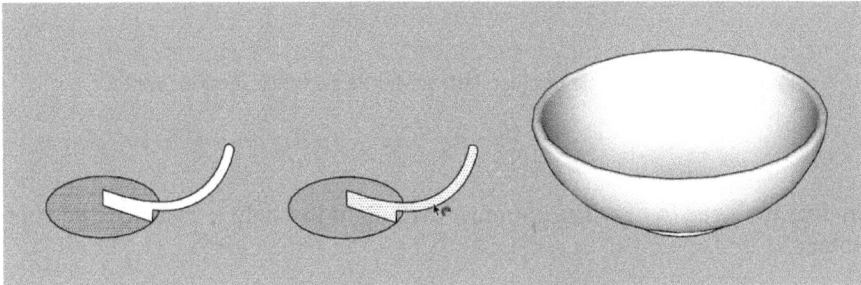

Figure 5.83: Circular Face Preselected (Left); Follow Me Tool Activated
and Profile Selected (Middle); Finished Bowl (Right)

Try to revolve and sweep shapes for yourself! As you look around, you might notice all sorts of revolved and swept shapes in your daily life. Bowling pins, water bottles, and chess pawns are all revolved shapes, as well as garden hoses, picture frames, and drawer pulls. We can see some before-and-after examples of some of these shapes in the following screenshots. We will not cover any tutorials for these shapes, but you are encouraged to try this on your own. See what kinds of shapes you can recreate using Follow Me!

First, let's look at the revolved shapes. Remember—for a fully revolved shape, you need a profile that is either half or a full section of the shape. Also, a circular Face represents the Path the Profile is revolved around, creating the final shape:

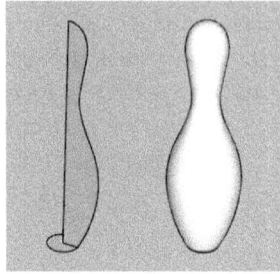

Figure 5.84: Bowling Pin Circular Path at Bottom and Half-Section Profile

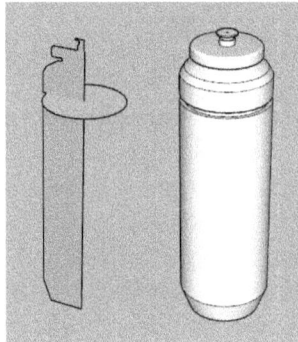

Figure 5.85: Water Bottle Circular Path at Middle and Half-Section Profile

> **Note**
>
> The Path does not need to be at the bottom—that might just help to visualize the Follow Me workflow.

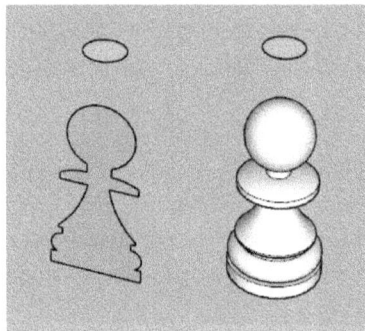

Figure 5.86: Chess Piece Circular Path Above the Geometry and Full Section Profile

> **Note**
>
> It may be easier to visualize the revolved shape in a full section. Make sure of two things:
>
> - The circular Path cannot be split into two halves—that will make the Follow Me Path a half-moon or "D" shape, not a circle!
> - Both sides of the Profile should be identical, or else it will end up with a double-revolved shape, or it might not work at all!

For swept shapes, the Path can be any connected series of Edges—it does not have to be a closed shape such as a circle. In this case, we can see the garden hose and drawer pull follow open-ended Paths, and the picture frame is a closed rectangular Path:

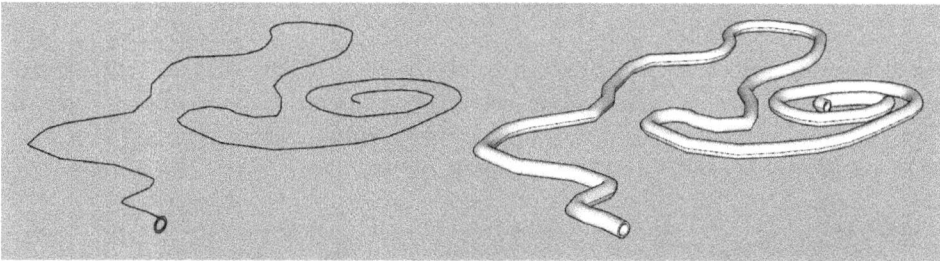

Figure 5.87: Garden Hose with Round Donut Profile and Freehand Path

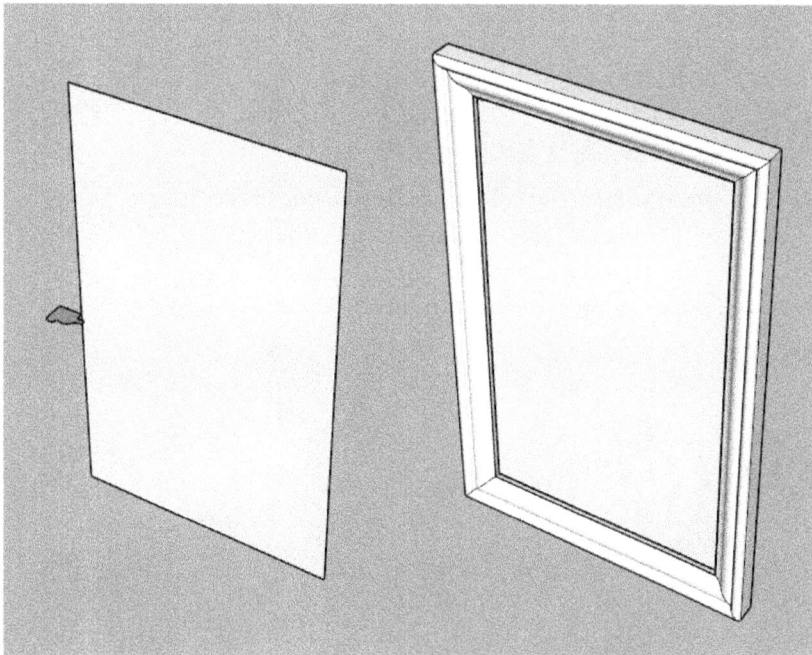

Figure 5.88: Picture Frame with Custom Section Profile and Rectangular Path

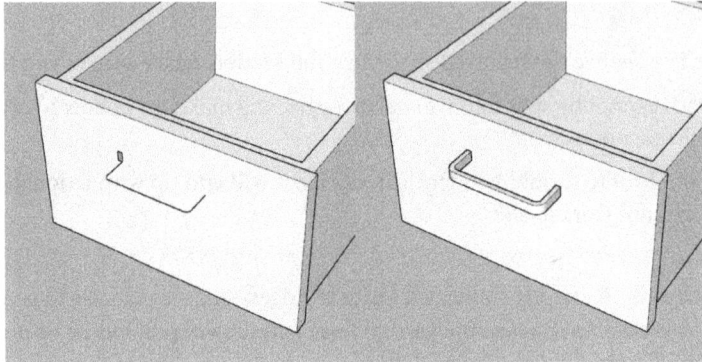

Figure 5.89: Drawer Pull with Rectangular Section Profile and Custom Path

It is tempting to use existing Geometry as the Path for a Follow Me workflow when working on projects. It is important to remember that the Follow Me tool will only recognize Geometry that is in the same Object or top level as the Profile. In this final example, we can see a countertop that will go across the tops of the cabinets. The cabinets are all separate Objects, so Follow Me does not "see" any viable Paths:

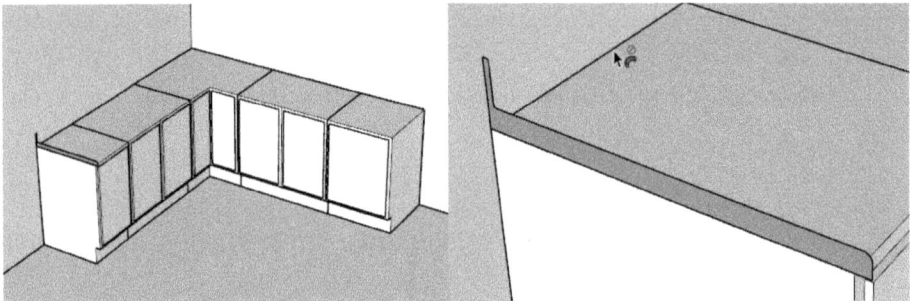

Figure 5.90: Kitchen Cabinets and Countertop Profile (Left);
Follow Me Tool Does Not Recognize Path (Right)

A path must be drawn along the top of the cabinets before the Follow Me tool can find a valid Path:

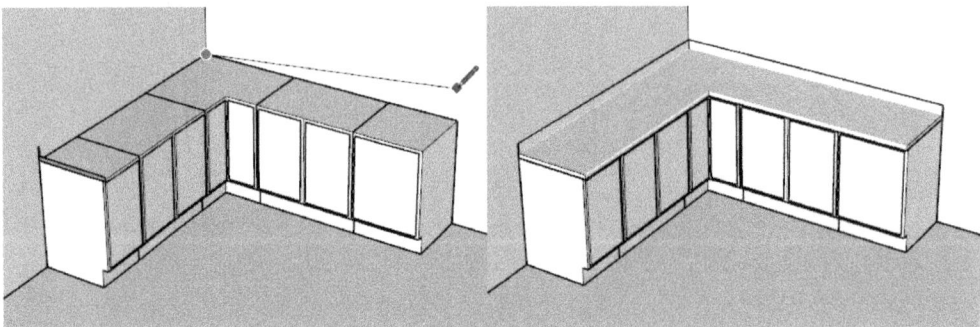

Figure 5.91: Path Drawn (Left) and Follow Me Workflow Completed (Right)

Push/Pull and Follow Me are great for creating 3D Geometry from 2D Faces. In the next section, we will look at the last Edit tool—Offset.

The Offset Tool

The Offset tool does not create 3D Geometry like the Push/Pull or Follow Me tools, but it still requires an existing Face in order to work. The Offset tool creates an offset of a Face—to make either a smaller or a larger Face that has Edges equidistant from the original Face outline. This is different from making a scaled copy, which might not make equally offset distances. The Offset tool uses *Alt* as a modifier key that allows overlap, which we will discuss later in this section. The Offset tool is represented by two arcs and an arrow showing the offset direction:

Figure 5.92: Offset Tool Button

The Offset tool can be activated by doing the following:

- Clicking on any of the Offset tool buttons
- Hitting *F* on the keyboard

The Offset tool must choose a Face to work with—it cannot offset Edges on its own. The Offset tool uses the plane of the Face to know where the new Geometry will go, although it can go on the existing Face or outside of the existing Face. Faces can be preselected in an Offset workflow, although this is not necessary. As with the Push/Pull tool, Offset will most likely not be done more than once on a single Face, and preselecting will lock in that selection until deselected by clicking with the Select tool or hitting *Esc*. Once a Face is selected, an offset distance can be clicked on the screen or can be typed into the Measurements Box. Faces can be offset by doing the following:

1. Activating the Offset tool.
2. Picking a Face to be Offset.
3. Clicking to set the offset distance, or showing the offset direction and typing a distance into the Measurements Box and hitting *Enter*.

The Offset tool is a great way to draw equidistant Geometry—that is, Geometry that is a certain distance in all directions. Let's look at this in terms of a simple box. We see that we have a box shape, but this box is solid—we would prefer it to be hollow. Deleting the top edge is the wrong workflow for many reasons—but most importantly, it would leave us with a box shape that has infinitely thin walls:

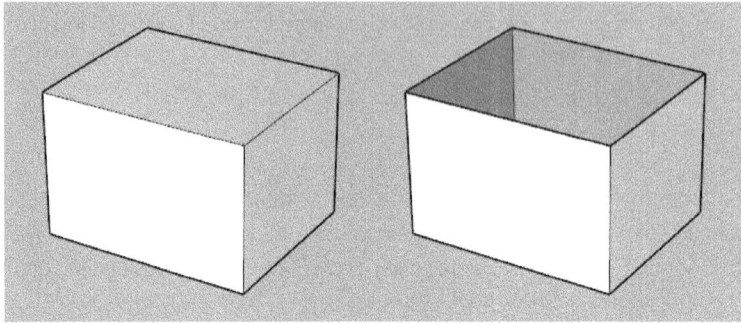

Figure 5.93: Solid Box (Left); Deleted Top Face (Right)

This box could not exist in the real world, especially if creating shop drawings or 3D printing. Instead, we need to know the thickness of the box walls—in this case, ½ inch. On the top of our box, we can then offset the Geometry ½ inch inward:

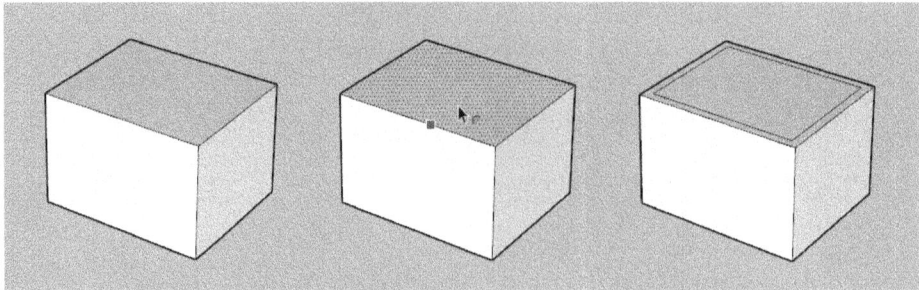

Figure 5.94: Solid Box (Left); Offset Tool Activated and Face Selected (Middle); Offset Set to ½ Inch (Right)

Then, we can use Push/Pull to push the new rectangle back into the solid box. This created ½-inch thick walls around all sides of the box, even though the box was wider than it was deep. If we had attempted to use scale, we would have ended up with thicker walls on two sides:

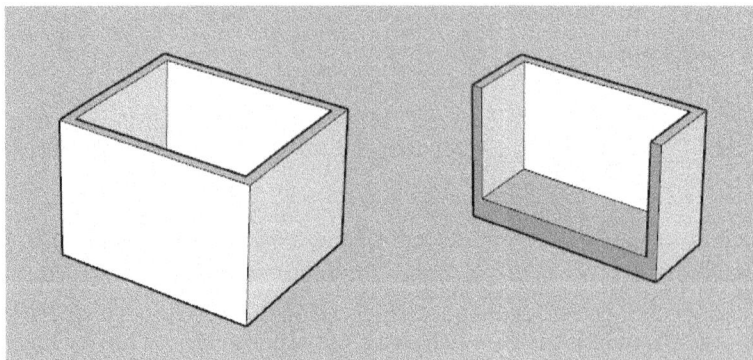

Figure 5.95: New Rectangle Pushed Down (Left); Box Section Showing ½-Inch Wall Thickness (Right)

As with Push/Pull, Offset can save a distance that can be used many times in a row on multiple Faces. In this example, we can see that there is a window that needs frames for each pane of glass. As it is drawn right now, the frames are represented by Edges, and this is the same issue as the previous example—these are infinitely thin pieces of Geometry. Instead, we need to offset into each Face the same distance:

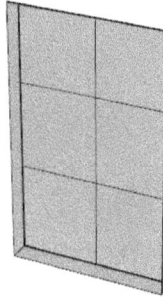

Figure 5.96: Window and Window Panes Without Frames

To begin, we can offset in one of the Faces. Then, this offset distance will be remembered by SketchUp, and by double-clicking on the other Faces, we can repeat the same offset distance:

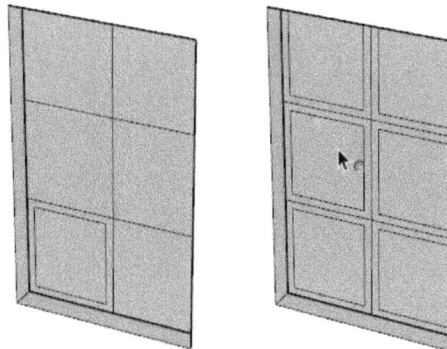

Figure 5.97: First Offset Set (Left); Remaining Faces Offset Same Distance with Double-Clicks (Right)

> **Note**
>
> Remember that the Offset tool must be able to click a Face directly—not on a Face inside a Group or Component. Even if you have to edit an Object in between Offset workflows, SketchUp will still remember the distance!

The Offset tool uses the *Alt* modifier key to allow overlapping offsets. This is much easier to visualize when overlaps are not allowed in the Offset tool's default state. In this example, we have a five-Sided Face that we are attempting to offset. The offset will be valid until we get too close to the middle when it will not show a Face as a preview. When the Face preview disappears, the offset will be overlapping, and the workflow will not create an offset Face when we click:

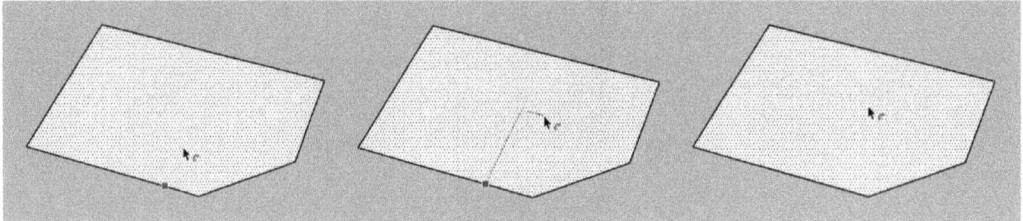

Figure 5.98: Five-Sided Face with Offset Active (Left); Preview Face
Disappears (Middle); No Offset Face Created (Right)

This is because the top and bottom edges would have crossed over each other, or overlapped, even though the left and right edges still had room to grow. This can create very confusing shapes, which is why it does not create a shape at all. The *Alt* modifier key allows this behavior, as we can see in the next example. Using the same five-Sided Face, we can begin the offset and see the preview Face disappear. Then, after hitting *Alt* on the keyboard, a Face preview appears once again, but it is a strange inverse:

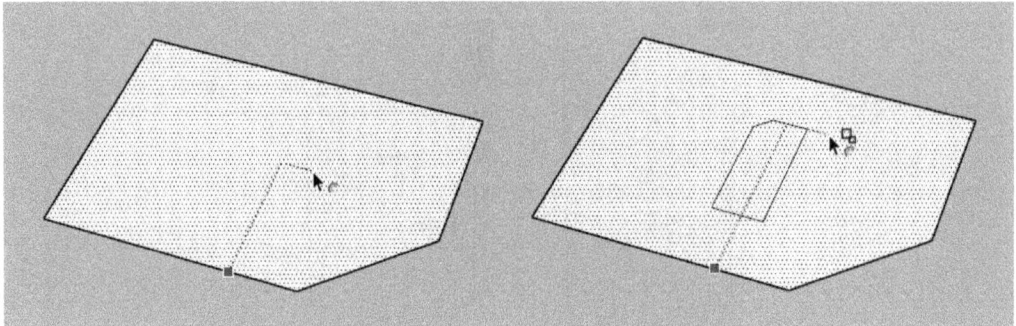

Figure 5.99: Five-Sided Face with Preview Face Missing (Left); Alt
Modifier Key Used to Show Overlapping Preview (Right)

This Face can be understood if we look at it one Edge at a time. The angled Edge is the giveaway—the Edges have all flipped around from each other—but SketchUp still connects them. In the following screenshot, we can see that each new Edge is the same distance from its original Edge, but the new Face no longer resembles the original Face:

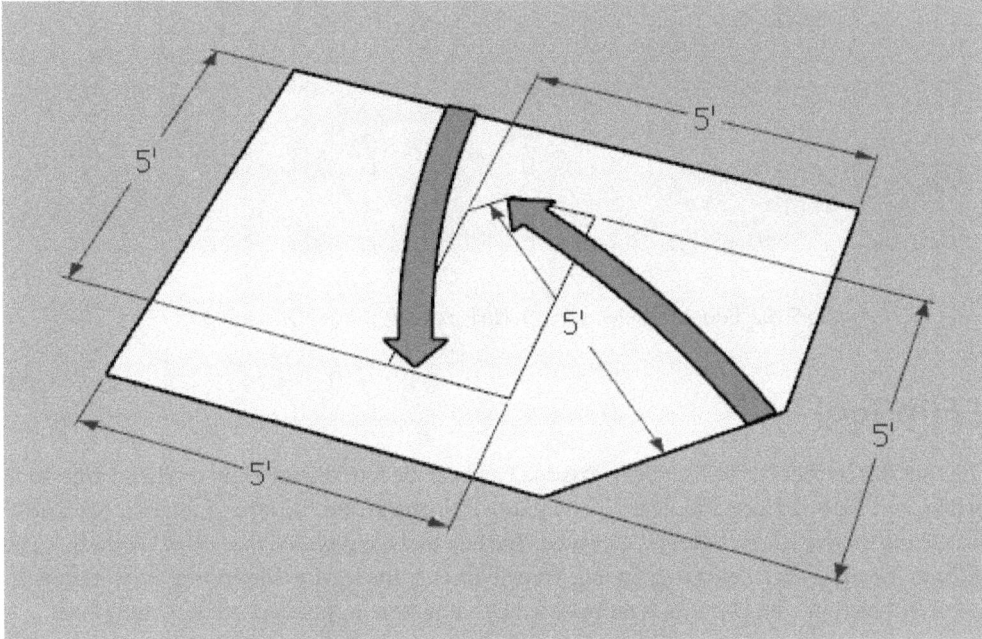

Figure 5.100: Five-Sided Face with Offset and Dimensions

SketchUp will also allow offsets of simply connected, coplanar Edges. This means that we can offset just a few Edges that are connected to a Face, or even some Edges that are not filled with a Face. In this next example, we will use the same pentagonal shape, but we only want to offset the top and left Edges. With the Select tool, the two Edges can be preselected, and then the Offset tool can be activated. The preview will only show the two Edges, and these can be offset to either side of the Geometry:

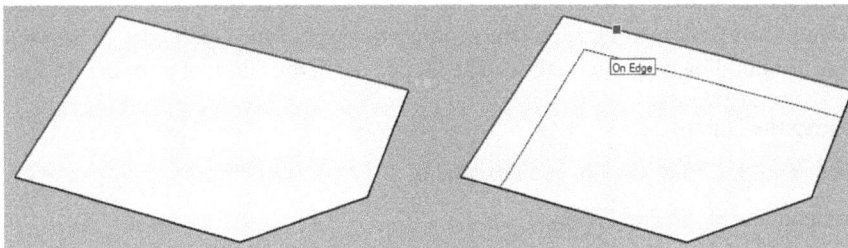

Figure 5.101: Edges Preselected with the Select Tool (Left); Edges Offset (Right)

This also means that any Edges that are coplanar and connected can also be offset. In this example, we can use the same shape, but with the Face removed. The Edges must be preselected when using the Offset tool as the Offset tool prioritizes selecting Faces. With the Edges preselected, the offset can be completed:

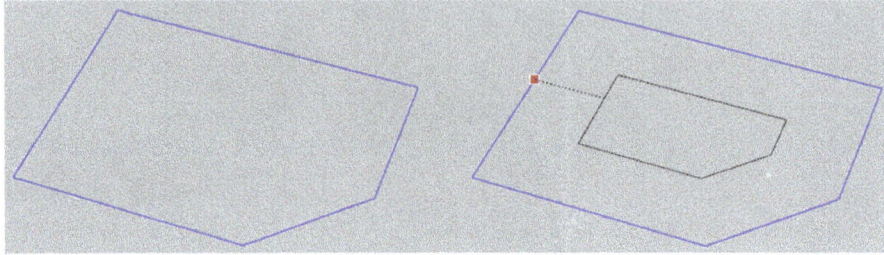

Figure 5.102: Edges Preselected with No Face (Left); Edges Offset (Right)

The Flip Tool

The Flip tool was introduced in SketchUp Pro 2023 and can be found in the Edit toolbar, Large Tool Set toolbar, and Getting Started toolbar. The Flip tool is distinct from the other Edit tools because it shows a Gizmo overlay when the tool is activated. The Flip tool has replaced the legacy "Flip Along…" workflow in the right-click contextual menu. We will discuss the legacy "Flip Along…" workflow at the end of this section. The Flip tool is represented by a house icon split by a vertical blue plane:

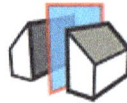

Figure 5.103: Flip Tool Button

The Flip tool will show an overlay of three Flip Planes, which are colored to the corresponding Axes—Red, Green, and Blue. When one of the Flip Planes is clicked, the model will flip along that Plane. By default, the Planes are aligned to the geometric center of the Geometry, so the Geometry will Flip inside its bounding box. Geometry can be flipped using the Flip tool by doing the following:

1. Activating the Flip tool.
2. Selecting an Edge, Face, Group, or Component.
3. Clicking on one of the Flip Planes.

Additionally, multiple Objects or Ungrouped Geometry can be flipped by preselecting and then activating the Flip tool. This can be achieved by doing the following:

1. Preselecting a selection of ungrouped Geometry or multiple Objects.
2. Activating the Flip tool.
3. Clicking on one of the Flip Planes.

In this example, a Group has been preselected, then the Flip tool has been activated to show the Flip Planes:

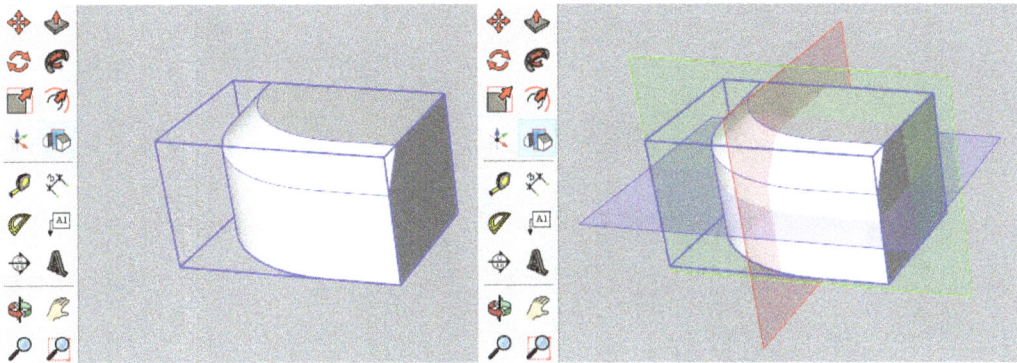

Figure 5.104: Group Preselected (Left); Flip Tool Activated (Right)

With the Flip Planes visible, the Red Flip Plane can be clicked to flip the Group along the Red Plane:

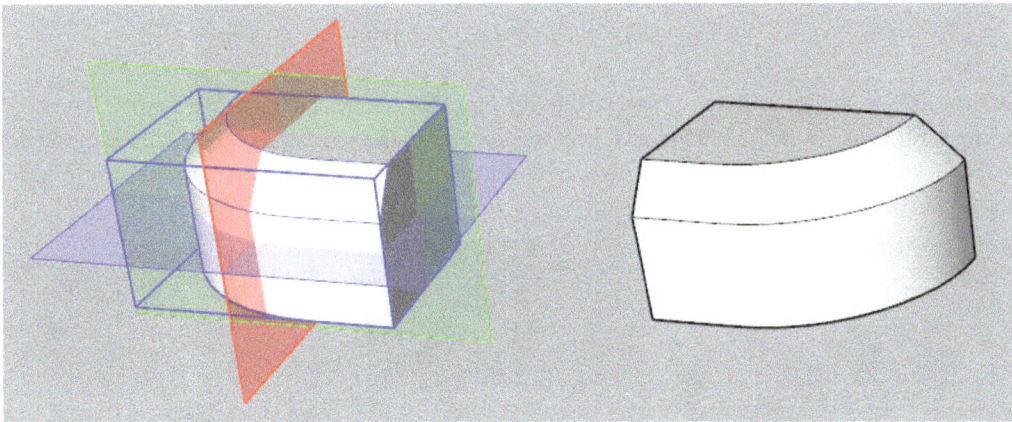

Figure 5.105: Red Plane Selected (Left); Flipped Group Deselected (Right)

> **Note**
> The Arrow Keys can be used to flip the model, with the Left Arrow flipping along the Green Plane, the Up Arrow flipping along the Blue Plane, and the Right Arrow flipping along the Red Plane.

The Flip tool will allow the Flip Planes to be clicked and dragged to move them, which changes the location of the center of the flip action, often referred to as the "Mirror Line." This allows for an Object to be flipped in context with other Geometry, which can be helpful when an Object should be moved and flipped.

In this next example, we can see a small bedroom with some furniture along the wall. If we want to flip the entire furniture arrangement in the room and align it with the other wall, we can do that by moving the Flip Plane:

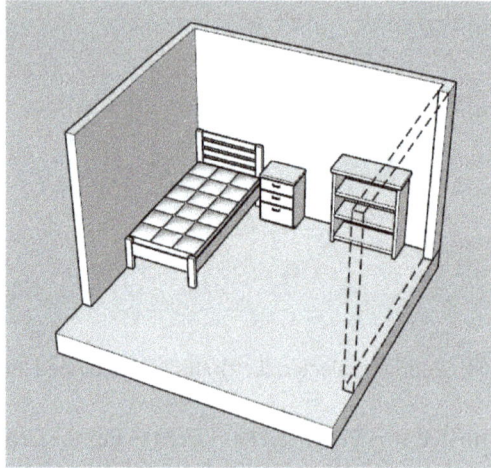

Figure 5.106: Bedroom with Furniture

With the different furniture Groups preselected, the Flip tool can be activated. We can see the Flip Plane we want to use is the Red Plane, as it is perpendicular to the back wall of the room. If the Red Plane is clicked without moving, the bed would no longer be aligned with a wall, which would not be the desired effect:

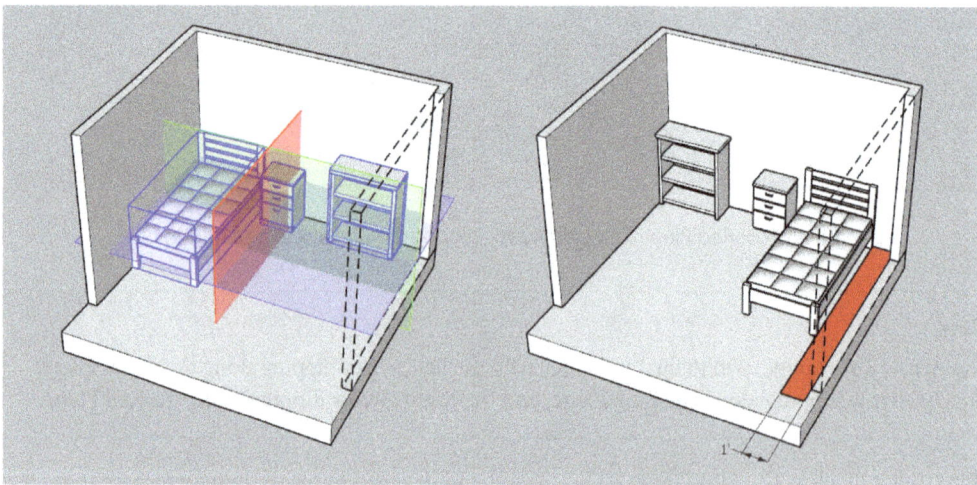

Figure 5.107: Red Flip Plane is Clicked (Left); Bed Not Aligned with Wall (Right)

Instead, we can click and drag on the Red Flip Plane to the new Mirror Line, which is in the middle of the bedroom. In this example, we can use the Midpoint Snap in the wall Group to locate the correct location. When the left mouse button is released, the Flip will happen automatically:

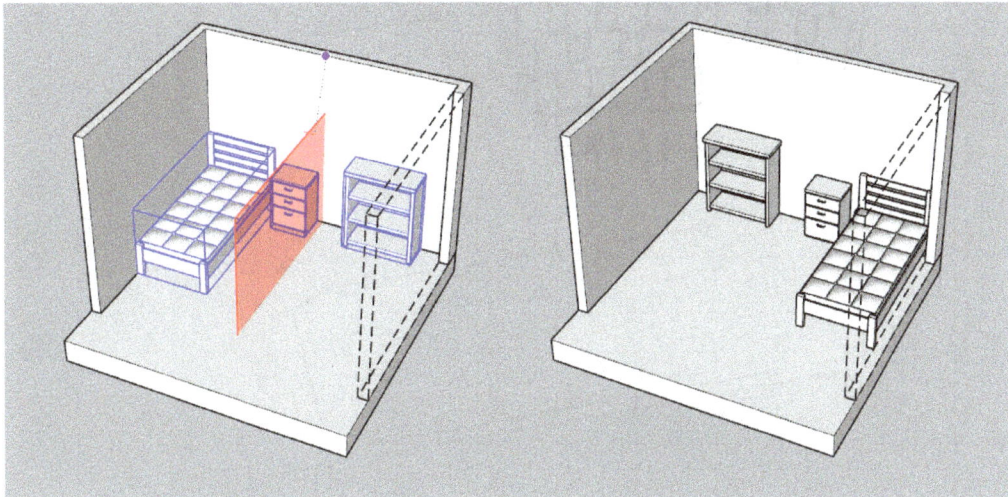

Figure 5.108: Red Plane Clicked and Dragged to Wall Midpoint Snap (Left); Mouse
Button is Released and the Geometry is Flipped Correctly (Right)

> **Note**
>
> The Flip tool is designed to snap to points such as an Endpoint, Midpoint, Intersection, or the Axes Origin. On Edge or On Face snaps will not work correctly in SketchUp Pro 2023, and the Flip workflow will not place your model Geometry in the correct location. This includes snapping to Guides, which we will discuss later in this chapter. Make sure that you snap to a specific Point when setting a different Mirror Line!

The Flip tool will show the location of the Flip Plane after the Geometry has been flipped, but the location of the Flip Plane will be reset after the workflow ends. This can help flip the Geometry back to its original location if it was done incorrectly, but it is important to remember that these Flip Plane locations cannot be saved to selections or Groups or Components.

The Flip tool can also create copies, like the Move tool and the Rotate tool. The *Ctrl* Modifier Key can be used to toggle **Flip** and **Flip as a Copy**. Flipping a copy is often more useful when moving the Flip Plane to a desired Mirror Line, and this workflow can be used to create symmetrical models. In this next example, a group of chairs has been arranged to view an event. These chairs should be flipped around the middle aisle for this space, which has been marked with a Guide in SketchUp:

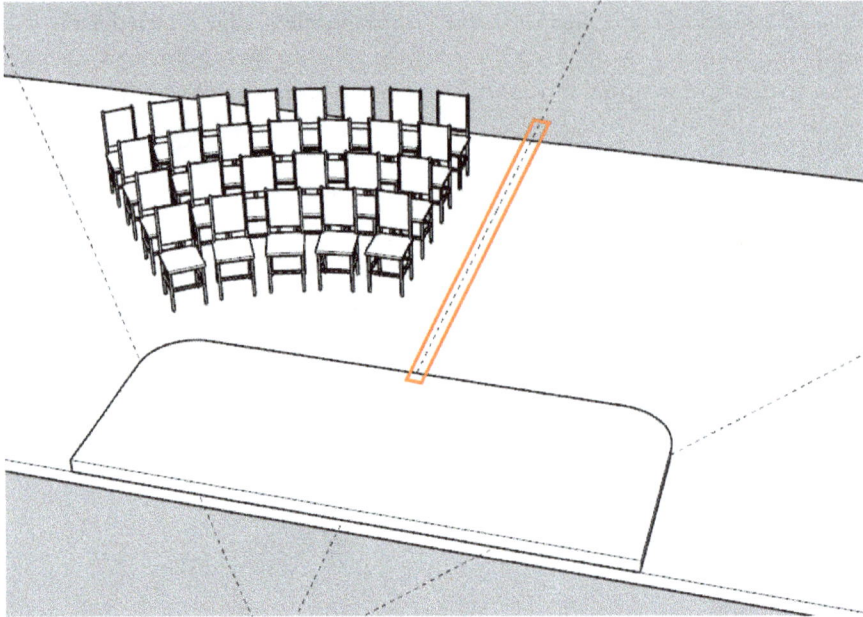

Figure 5.109: Event Space with Chairs with Middle Guide Highlighted

With the Chairs Group preselected, the Flip tool can be activated to show the Flip Planes. The *Ctrl* Modifier Key can be clicked to show a small Plus (+) Sign next to the cursor, indicating that Copy is toggled on:

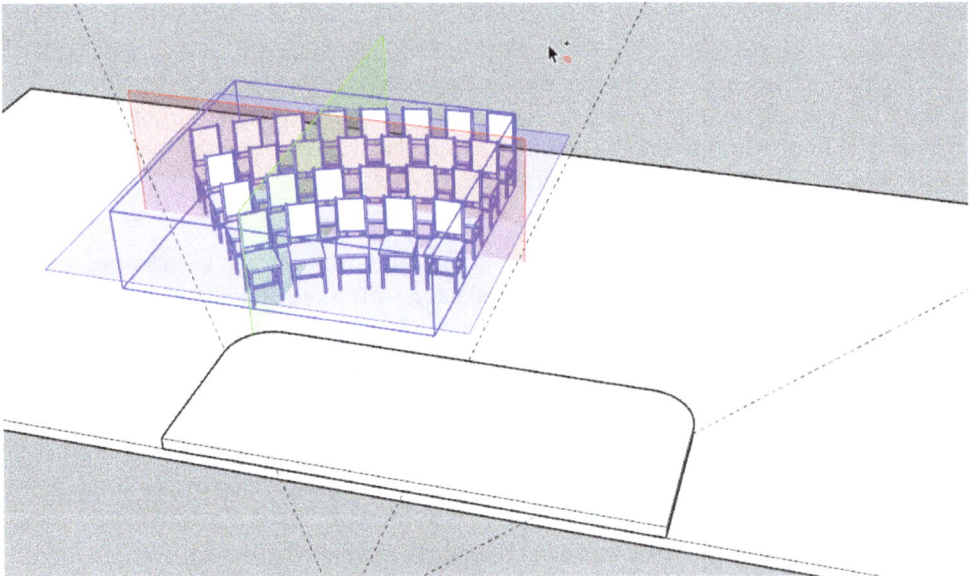

Figure 5.110: Chairs Group with Flip Planes Shown with Copy Toggled On

With Copy toggled on, the Green Plane can be clicked and dragged to snap to the Intersection of the Guide and the Floor Group, and when the mouse button is released, a copy of the chairs will be made directly across from the Mirror Line:

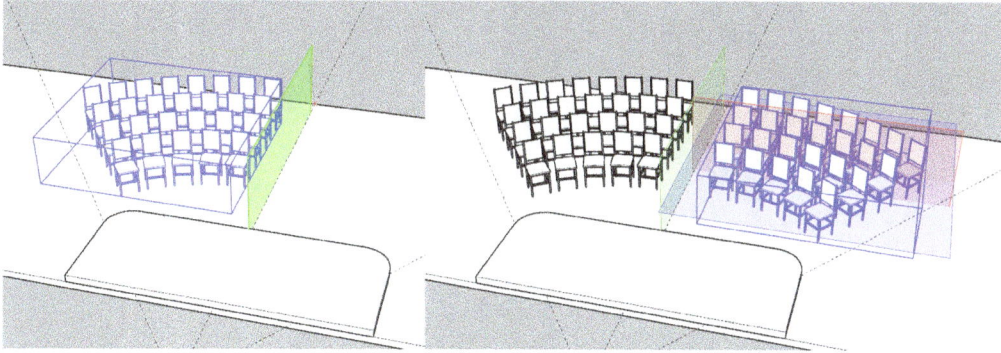

Figure 5.111: Green Plane Clicked and Dragged to Intersection (Left); Copy Created (Right)

Note

Remember—Flip requires that you pick a specific Point snap when moving a Flip Plane! Choosing the On Edge snap on the Guide would have resulted in the copy being placed in an incorrect location.

Flip Planes are automatically oriented in the same direction as the model Axes if the Geometry is not in a single Group or Component. If the Flip tool is activated and a single Group or Component is selected, then the Flip Planes will be aligned with the internal Axes of the Group or Component. In this next example, we can see the model Axes and multiple objects. When multiple Groups are selected and the Flip tool is activated, the Flip Planes are aligned to the model Axes:

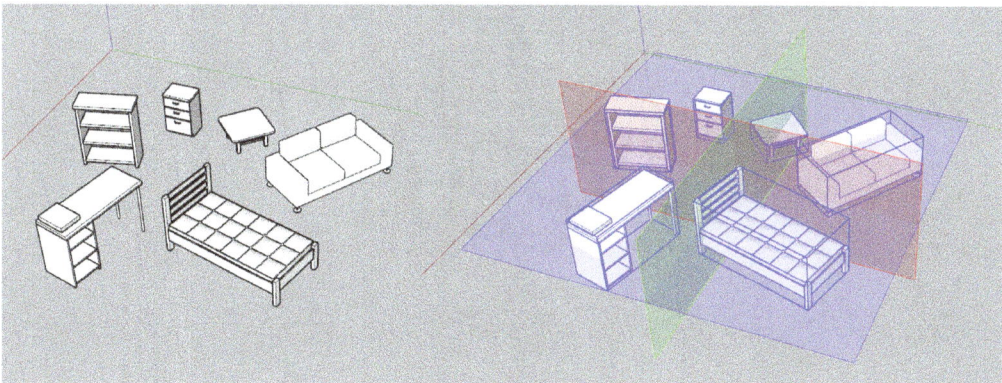

Figure 5.112: Model Axes and Objects (Left); Multiple Groups Selected and Flip Planes Shown (Right)

When only one of the Groups is selected and the Flip tool is activated, the Flip Planes align with the internal Axes for that Group. These Axes can be seen when the Group is double-clicked to activate the Group:

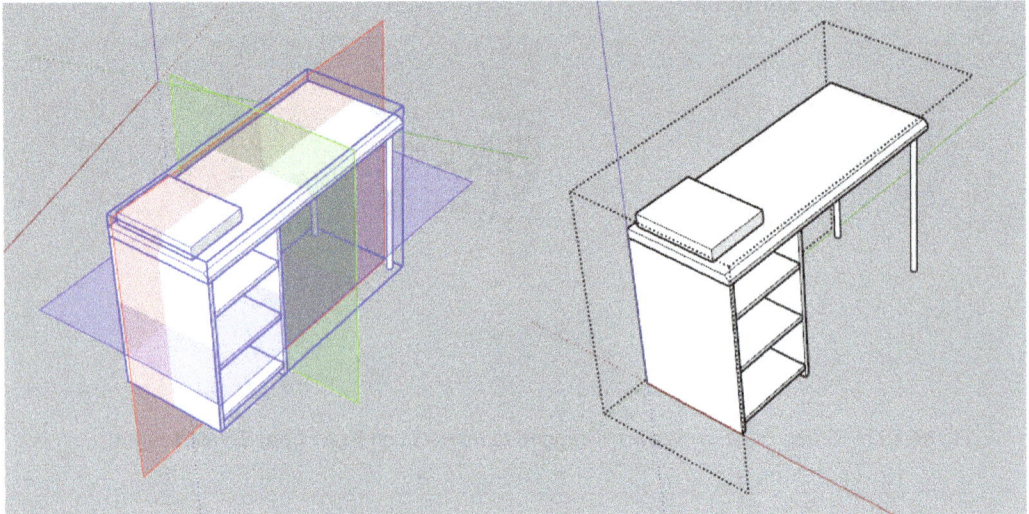

Figure 5.113: Flip Axes Aligned to Group Axes (Left); Group Activated to Show Internal Axes (Right)

When a single Group or Component is selected, the model Axes and the internal Object Axes can be toggled by clicking the *Alt* modifier key. The internal Axes will be shown by default, but when *Alt* is clicked, the Flip Planes will be re-aligned with the model Axes. In this last example, a group of chairs has been arranged to view an event. The Chairs Group should be flipped over the side aisle, which runs in the Red direction of the model Axes:

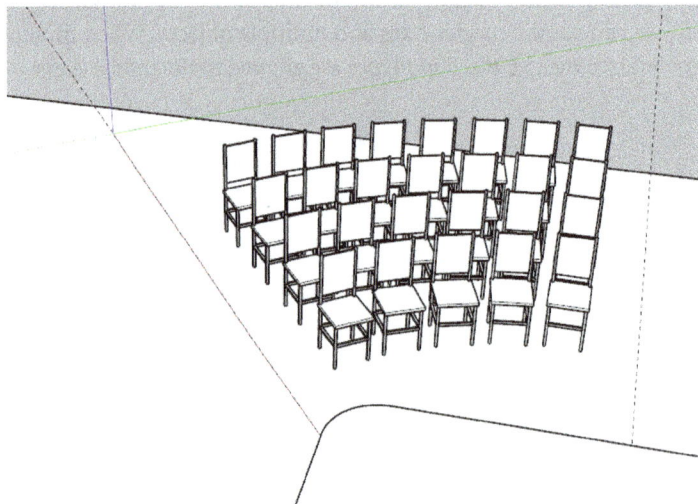

Figure 5.114: Event Space with Chairs and Side Aisle Guide Aligned with Model Axes

With the Chairs Group preselected, the Flip tool can be activated to show the Flip Planes. The *Alt* modifier key can be clicked to toggle between the Group Axes to the model Axes:

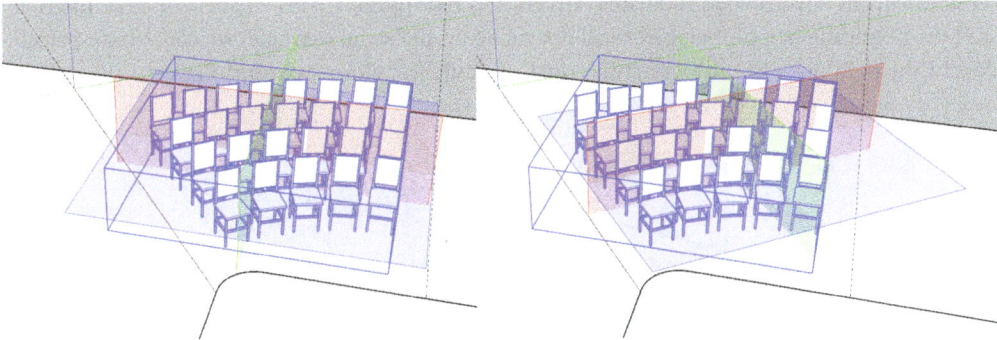

Figure 5.115: Chair Group with Flip Activated (Left); Alt Key Toggles Flip-Plane Alignment (Right)

Then, the *Ctrl* modifier key can be clicked to toggle copy, and the Flip Plane can be dragged to the intersection of the Guide and the Floor Group. When the mouse button is released, the model is flipped as a copy across the side aisle:

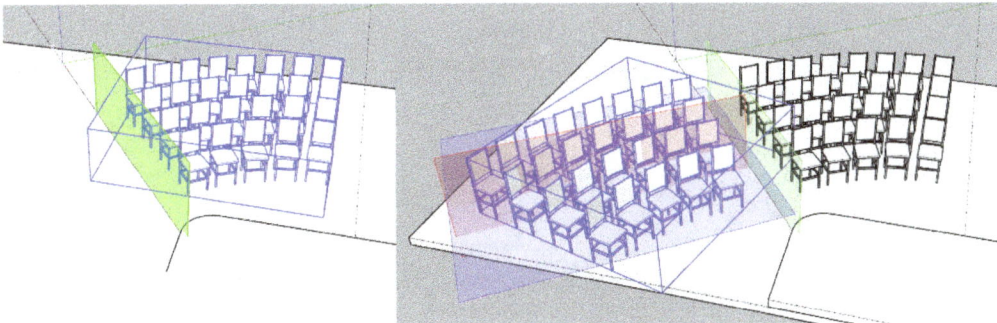

Figure 5.116: Ctrl Key Toggles Copy and Flip Plane Dragged to
Intersection (Left); Flip as a Copy Completed (Right)

The Legacy Flip Along Workflow

The Flip tool was introduced in SketchUp Pro 2023, but earlier versions of SketchUp Pro will use the "Flip Along…" workflow. This workflow can still be assigned to keyboard shortcuts in SketchUp Pro 2023, but it can no longer be added to the right-click contextual menu. The rest of this section will only work in SketchUp Pro versions before 2023.

Geometry and Objects can be flipped by selecting Geometry, then right-clicking and choosing **Flip Along | Direction**. The direction value will show different text based on the type of Geometry or Object selected, but it will always show Red, Green, and Blue options. This will behave as an internal flip, and the bounding box of the object will not move. In the same example, we could have placed the desk in the right place to start with, then right-clicked **Flip Along | Group's Green**:

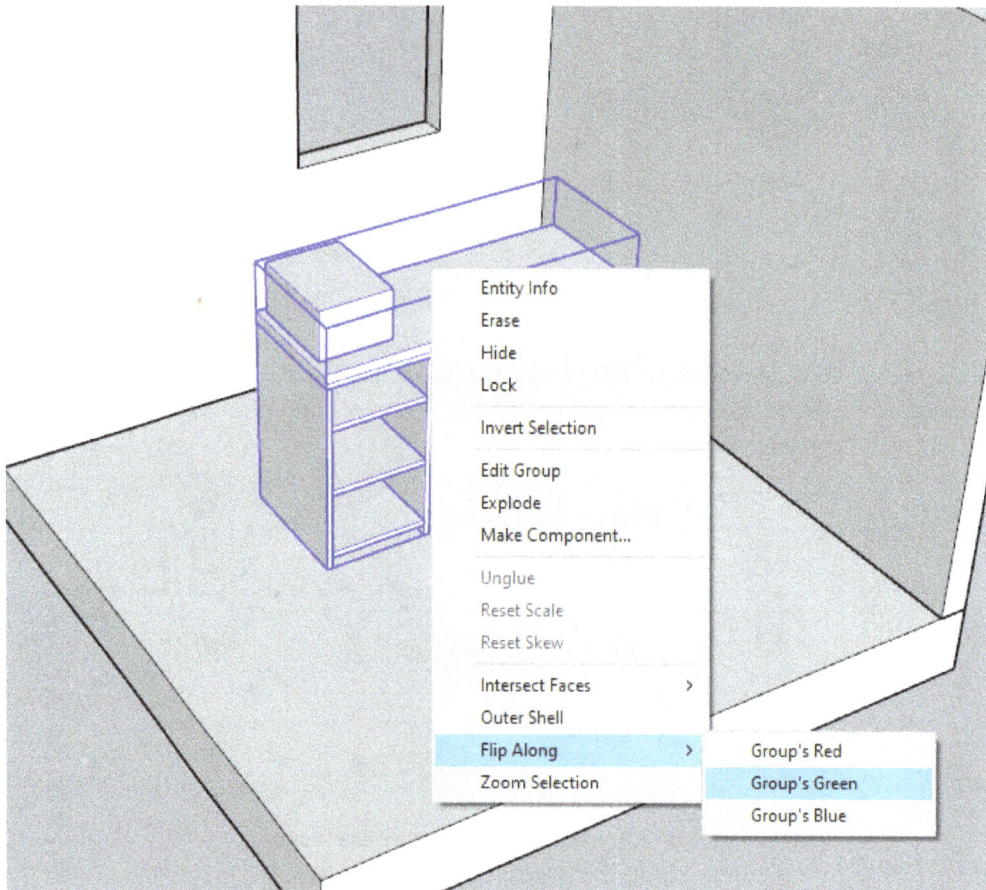

Figure 5.117: Desk Flipped using Flip Along Option

Now that we have examined all seven Edit tools, let's look at a few things to keep in mind as we practice Editing and Creating Geometry.

Editing Existing Geometry and Creating New Geometry

As you work through your projects, a common consideration is *when to create new Geometry, and when to edit existing Geometry.* This will all depend on what objects you are creating, where you are in the workflow, and what shared data or distances you might have between different objects.

The number-one thing to remember when working on your model is to separate different objects into Groups or Components—if they are not the exact same material in the real world, then they should not be stuck together in SketchUp. Of course, there are always exceptions to every rule, but this is a good one to follow 99% of the time. Even if multiple objects will eventually be glued together in the real world, they can always be Grouped in Nested Groups in your SketchUp model.

> **Note**
>
> One exception to this might be a picture frame that is made up of four connected pieces of wood. If we consider the picture frame to be one "object" for an interior design project, then it can be one connected 3D shape. But if the picture frame is being created for a woodworking project, each piece of wood should probably be modeled independently.

If we have our objects in appropriate Groups and Components, then the question of editing existing Geometry or creating new Geometry becomes easier to answer—it is obvious that we cannot "touch" the Geometry in our Groups and Components when creating new 3D models. But we should continue to use our existing Objects for our modeling whenever possible. In the following example, we have a drawer that will have an additional piece of wood attached to the front. The empty box has already been modeled as one piece and Grouped. The new piece of wood will overlap the existing box by 1 inch on all sides. We could create a new rectangle off in space, then Push/Pull it to the correct dimensions, then Move it to be next to the existing box. However, this is an inefficient workflow as we do not utilize the Geometry that we have right in front of us!

There are a few ways to utilize the existing box, but it will almost always start by tracing existing Geometry. In this case, we will trace a rectangle on the front Face of the box:

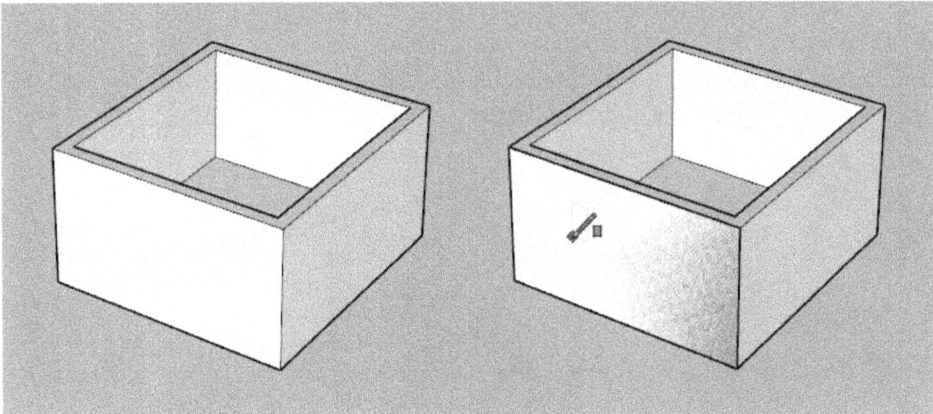

Figure 5.118: Existing Box for Drawer (Left); Rectangle Traced on Box (Right)

> **Note**
>
> This will create two Faces in different Groups that are directly on top of each other, and they visually "z-fight" or glitch, and parts of both Faces will be visible as you orbit around the model. This is expected behavior as the two Faces are on top of each other—SketchUp does not know which one should be shown first!

The new rectangle can now be used to create new Geometry. The existing Geometry was not "edited" as we did not enter the Group and edit, but we were still able to use it efficiently. Now, we can Push/Pull the rectangle out, then Push/Pull each side out 1 inch:

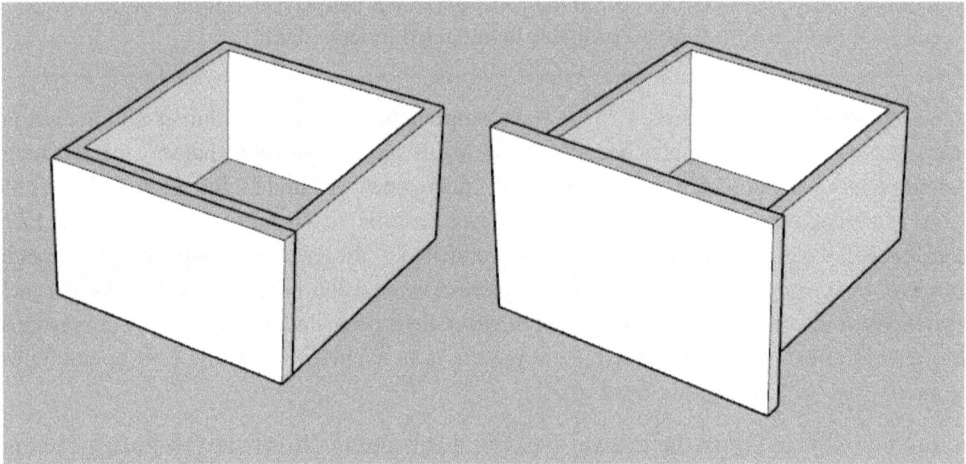

Figure 5.119: Rectangle Pulled Out (Left); Sides Pulled Out 1 Inch (Right)

Or, we could have offset the rectangle out 1 inch and erased the inner rectangle; then we could Push/Pull the newer rectangle out to the desired thickness:

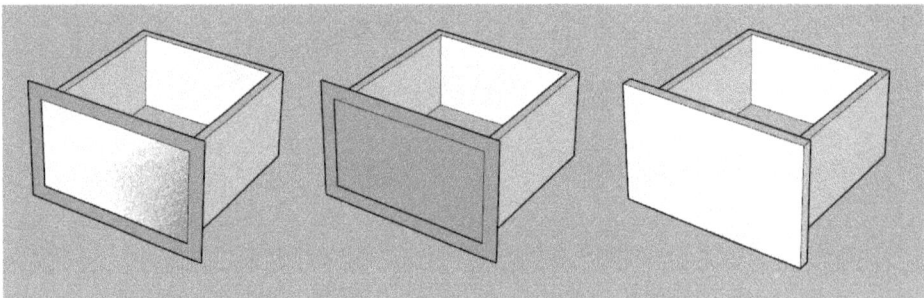

Figure 5.120: Rectangle Offset 1 Inch Out (Left); Inner Rectangle
Erased (Middle); New Rectangle Pulled Out (Right)

So, when can we really edit existing Geometry in our models? Are we always going to reference our Geometry when we work but never edit? Of course not—that is not the case. As you are working on a part or piece of a model, you will constantly be using the Edit tools, but once you complete an object, it should be Grouped or turned into a Component! It may be tempting to use Move/Copy or Push/Pull with the *Ctrl* copy modifier key to cut some corners, but it will come back to hurt your model in the end. You may end up breaking watertight Geometry or corrupting an existing bit of Geometry in the process.

Face Orientation

One immediate indication that there were too many steps in a creation/edit workflow is the SketchUp Faces. When a rectangle with the default texture is extruded using Push/Pull, it will always be white. Try this yourself—draw a new rectangle and Push it down; it will be white. Then, Undo and Pull it up—still white!

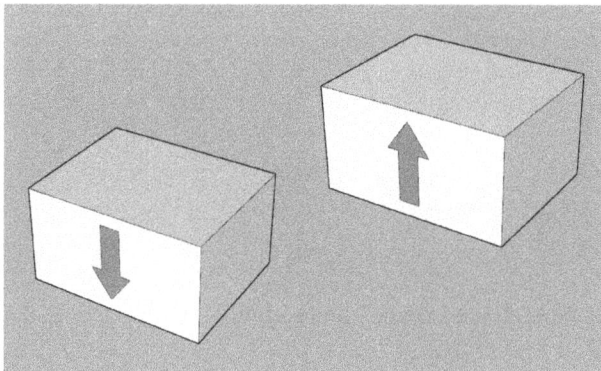

Figure 5.121: Rectangle Pushed Down and Pulled Up to Create White Boxes

All Faces in SketchUp have two sides—or Normals. These define which side of the Face is "up" or "down." When a rectangle is created, it will show either a white or a blue side. SketchUp will try to put the white side "up" or "out" whenever possible, especially when working in 3D. So, when a new shape is Pulled up using Push/Pull, SketchUp will try to show white sides "out." If we delete one of the Faces, we can see that the internal sides are all blue:

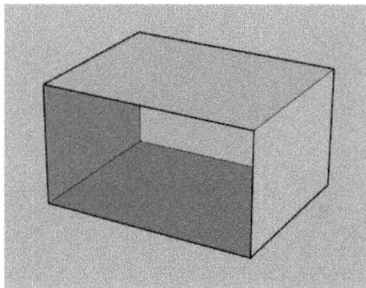

Figure 5.122: Box with Face Deleted Showing Blue Interior Faces

> **Note**
>
> The white and blue colors are part of the default texture/material in SketchUp. Coloring Faces with different textures or materials may make it harder to identify the Normal of a Face! If you have other materials applied, try using the Monochrome Face Style to view the Normals in your models! We will discuss Styles in *Chapter 7*.

If a shape is too complex, then SketchUp may have a hard time determining what is "inside" when using a 3D Edit tool. An easy example is a simple rectangle with an offset rectangle inside. These are two Faces in the same plane—very simple Geometry! However, when the inner rectangle is pushed down using Push/Pull, SketchUp does not create a closed box. Instead, SketchUp connects the outer donut Face to the inner Face as it is pushed down, causing it to act as a hole in the ground:

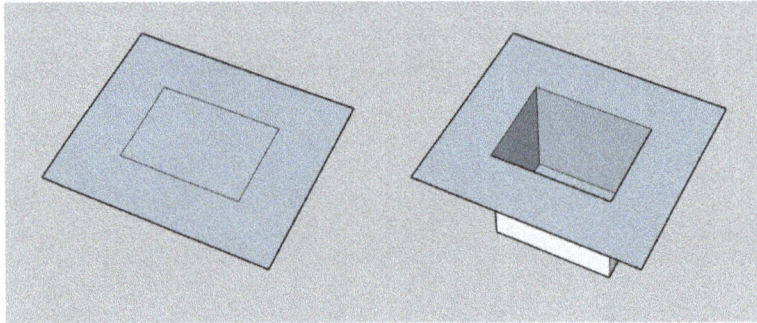

Figure 5.123: Rectangle with Offset Rectangle (Left); Push/Pull Creates a "Hole in the Ground" (Right)

SketchUp does a good job of determining that the inside of the hole should be blue, and the outside is white, but we already have non-watertight Geometry. Now, when the outer rectangle is pulled up using Push/Pull, the Geometry only gets worse! This new shape is solid, but it has a large hole in the middle and it has flipped the bottom face of the hole!

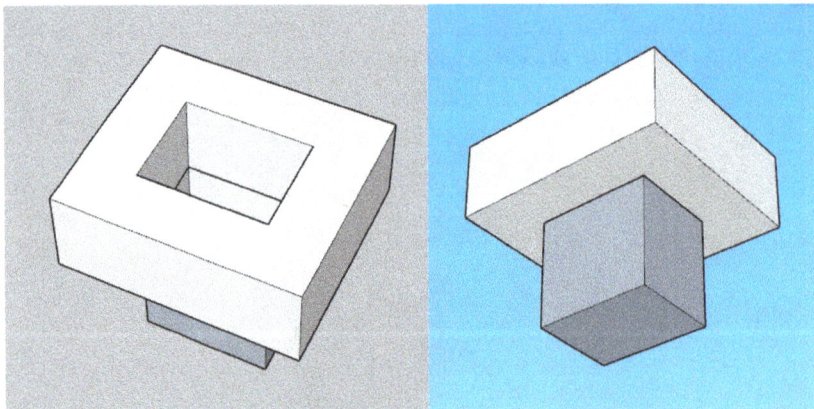

Figure 5.124: Face Extruded Up (Left); Bottom Face in Hole Flipped (Right)

The resulting shape is non-watertight and has faces that are not oriented in the right direction relating to each other. If we wanted to make this Geometry correctly, we could start with just the first, larger rectangle. This rectangle can be pulled up with Push/Pull before the Offset tool is used:

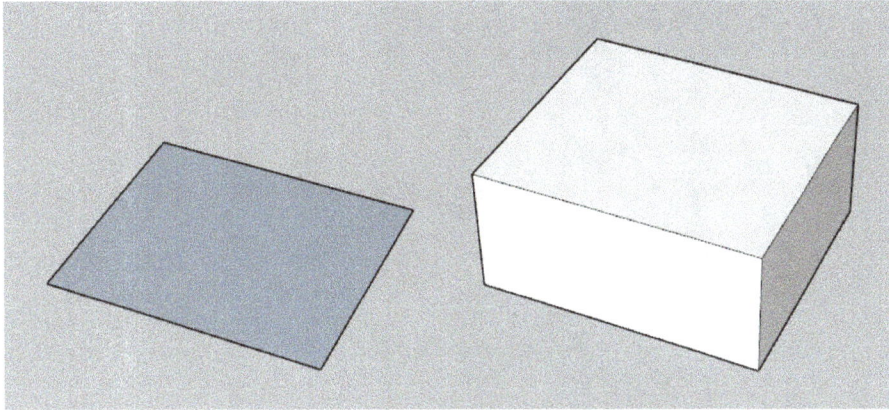

Figure 5.125: Rectangle (Left); Pulled Up (Right)

Then, the Offset tool can be used to Offset the bottom Face after we Orbit below our box. Do not forget to use the Navigation Tools to their full potential—we are in a free 3D environment after all! Then, the offset rectangle can be pulled down the appropriate distance. This creates one watertight shape with no internal Faces. Also, SketchUp is able to keep all Faces white, meaning that their Normal directions are all pointing the same way!

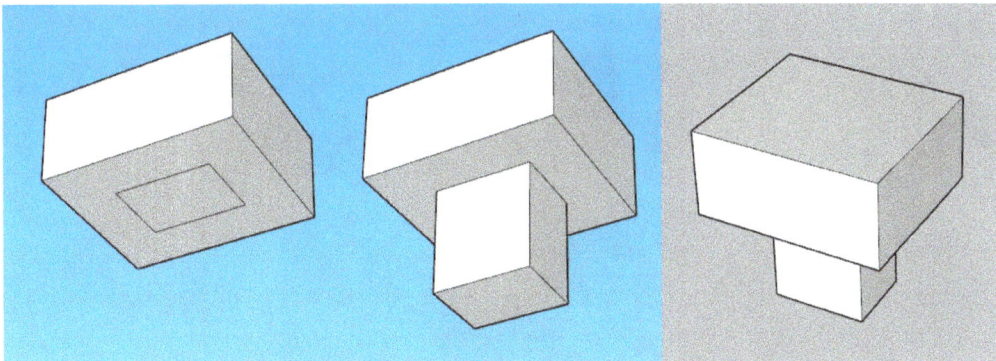

Figure 5.126: Bottom Face Offset (Left); Rectangle Pulled Down (Middle); Resulting Shape (Right)

Many workflows can be broken like this, and you will inevitably run across an instance where the Geometry does not behave correctly. We have already seen that SketchUp can Auto-Heal Faces when drawing Edges, and you may have to use this technique to fix non-watertight Geometry. In these workflows, occasionally a Face will be flipped in the wrong direction, and we might see this as a blue Face in the default texture. In this example, we can see some blue Faces were added to heal some broken Geometry:

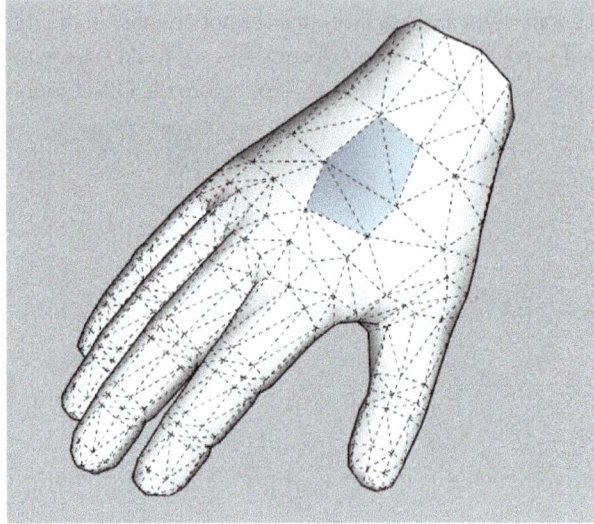

Figure 5.127: Blue Faces Used to Heal Geometry

These Faces have the right texture, so instead of trying to paint them with the Paint Bucket tool, we need to flip their Normal! This can be done by using the Reverse Faces option, which can be found with a right-click or the Edit Menu. The right-click menu is contextual based on the selection set, and it is the same menu of options that appears at the bottom of the Menu Bar Edit dropdown. With the blue Faces selected, we can go to right-click | **Reverse Faces**. The Faces will flip their Normals, and we will see the white side of the default texture:

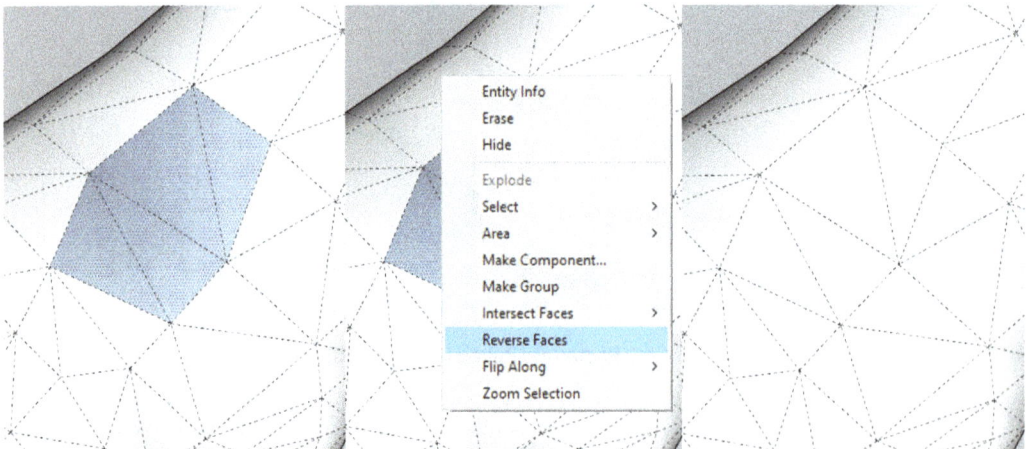

Figure 5.128: Faces Selected (Left); Right-Clicking Reverse Faces (Middle); Faces Flipped Normals (Right)

If we wanted to align all Faces with a Face we know to be in the correct direction, then we could use another option in the same menu. **Orient Faces** can only be chosen when a single Face is selected, and it will orient all other Faces to have the same Normal direction. Using the same model as before, we can see that a white Face is selected. This is the Face we want to orient to, so we want to choose one with white facing out:

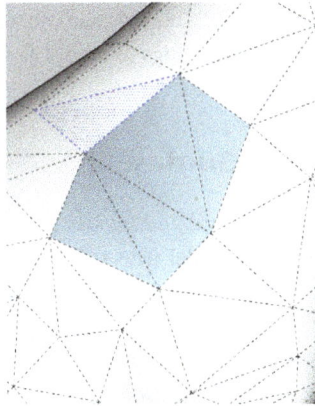

Figure 5.129: Face Selected with Correct Normal Direction

Then, we can right-click **Orient Faces**. This will go through the connected Geometry and attempt to align all Normals in the same direction. We can see that in this example, it has left the white Faces alone and has flipped the blue Faces:

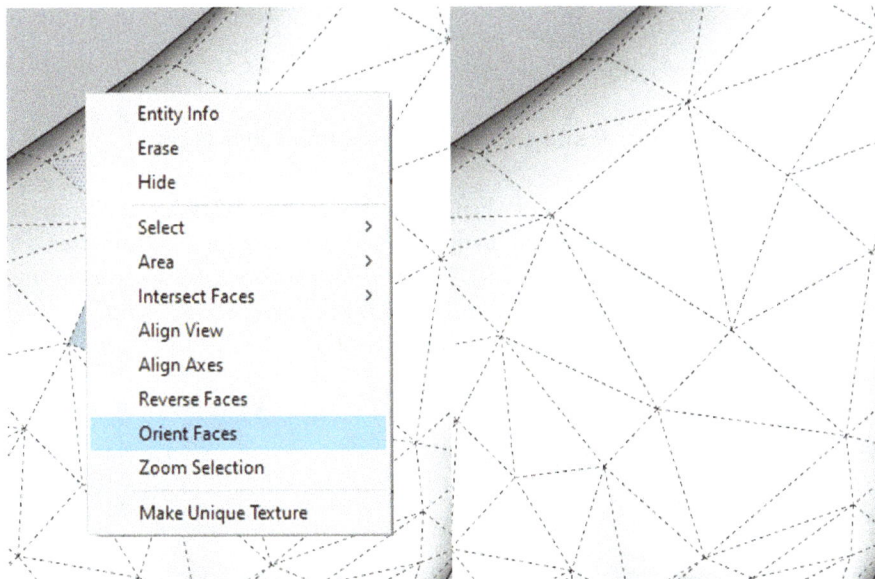

Figure 5.130: Right-Clicking Orient Faces (Left); and Blue Faces Flipped (Right)

> **Note**
>
> Why wouldn't we always use **Orient Faces**? When working with Revolves, you may have seen Geometry generate "inside out" with all blue Faces facing out. This is the perfect time to select all and use **Reverse Faces**!

It is best to keep the model watertight and all Faces oriented correctly as you move through your workflow. It is much easier to fix a series of small bugs or errors as you are working than to try to fix it all at the end!

Tape Measure and Protractor Guides

In an earlier chapter, we looked at the Axes tool and how it could be used to reorganize the Red, Green, and Blue coordinate directions in the model. This tool is grouped with two other tools in the Construction toolbar, which are used constantly in Editing workflows—the Tape Measure and Protractor tools. These tools work very similarly in that they both measure existing Geometry—Tape Measure measures distances, and Protractor measures angles. In their default workflows, they are extremely useful for quickly checking distances or angles in the model, but they do not leave dimensions or other Geometry behind. Both the Tape Measure and Protractor tools use the *Ctrl* modifier to place guides in the model, which is the most helpful addition to many modeling workflows! The Tape Measure and Protractor Buttons look like the real-world objects that they represent:

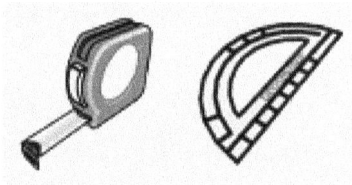

Figure 5.131: Tape Measure Button (Left); Protractor Button (Right)

The Tape Measure tool can measure distances when you click to start and then hover over the second point, and the Measurements Box will display an updated distance. The Tape Measure tool will also display the coordinates of any point in the model. The Tape Measure tool will show the distance in the Measurements Box after you click again, but only until the mouse is moved again:

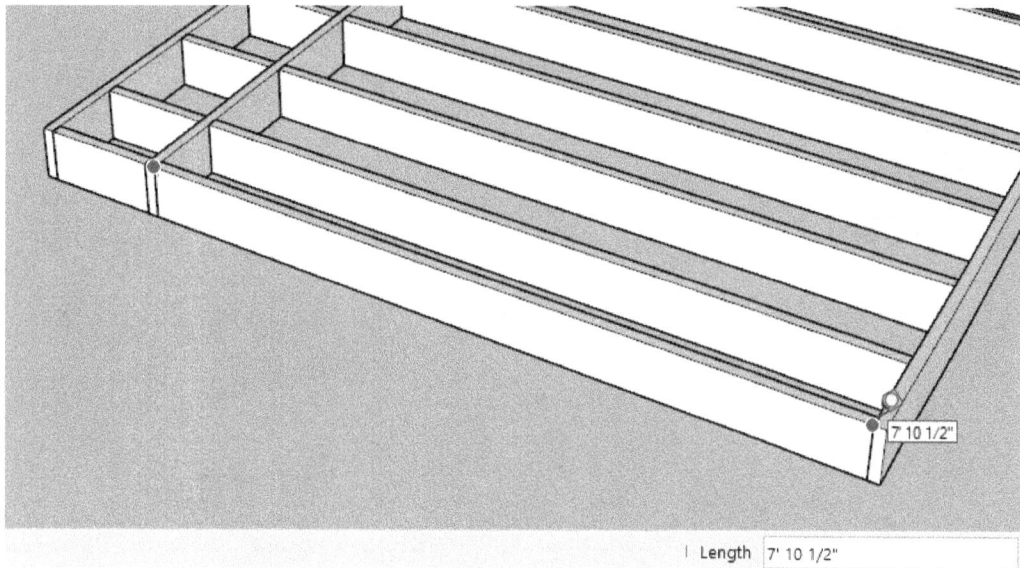

Figure 5.132: Tape Measure Tool Measuring Distance with Result in Measurements Box

> **Note**
>
> The Tape Measure tool can also be used to scale the entire SketchUp model at once. This workflow is discussed earlier in this Chapter in the *Scale Tool* section.

The Protractor tool uses a rotation gizmo like the Rotate tool, but instead of rotating Geometry, it measures the angle between the two vectors created by the center point and the first specified point, then a second point. The Protractor tool will also display a running updated angle in the Measurements Box, but it will also leave the angle in the Measurements Box after the mouse is moved:

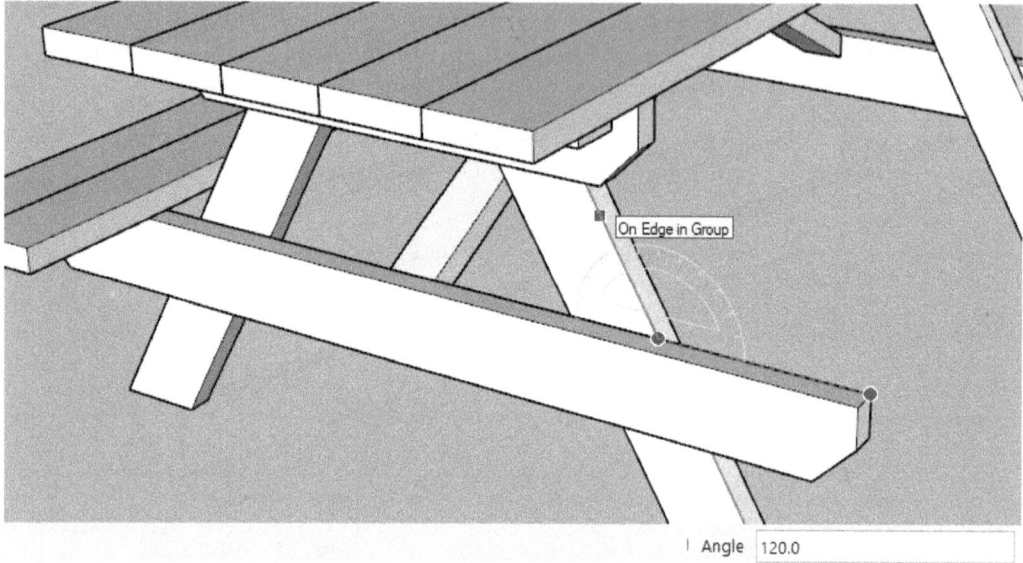

Figure 5.133: Protractor Tool Measuring Angle with Result in Measurements Box

The Tape Measure tool and Protractor tool provide an even more helpful function when modeling—they both create Guides. Guides are dashed lines, not model Geometry, that specify a vector through the model. As Guides represent a vector, they are infinite, going past the Geometry in both directions.

> **Note**
> Tape Measure Guides can be created to be a specific, finite distance. This will be examined at the end of this section.

Guides can be selected, moved, and deleted from a model. Guides can also be included Groups and Components and can be hidden and shown, like regular model Geometry. They can also be inferenced or snapped to when creating and editing Geometry, which is why they are used in so many workflows! Guides allow for data to be set at specific distances or angles away from existing Geometry, which new Geometry can be drawn to. This allows existing Faces and Edges to not be messed up by drawing lines or other shapes that can intersect with and break existing Geometry. In this example, we can see a rectangle that should have a circle drawn on its Face. The Circle should be 10 inches from the left side and 12 inches from the bottom. We could draw some lines on the Face 10 inches from left to right then 12 inches up, resulting in extra Geometry on our Face. Also, the bottom Edge of the rectangle will be broken in two!

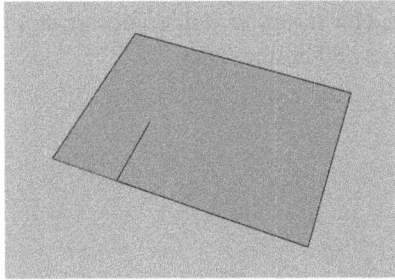

Figure 5.134: Rectangle with Lines Drawn and Broken Bottom Edge

Instead, we can place Guides using the Tape Measure tool. We can create these Guides by doing the following:

1. Activating the Tape Measure tool.
2. Activating the Create Guides *Ctrl* modifier Key. The Tape Measure Cursor will show a dashed line and a plus (+) sign representing that Guides will be created.
3. Clicking on an existing Edge that will be parallel to the created Guide.
4. Showing a distance and typing in the offset value and hitting *Enter*.

We can see this workflow twice in our example. We know the offset distances for our Guides—10 inches and 12 inches—and we know the parallel Edges. With the Tape Measure tool active and the Create Guides option turned on, we can click on the left Edge of the Rectangle. Showing the Red Direction, we can type 10 and hit *Enter*. This creates a parallel guide on top of the rectangle:

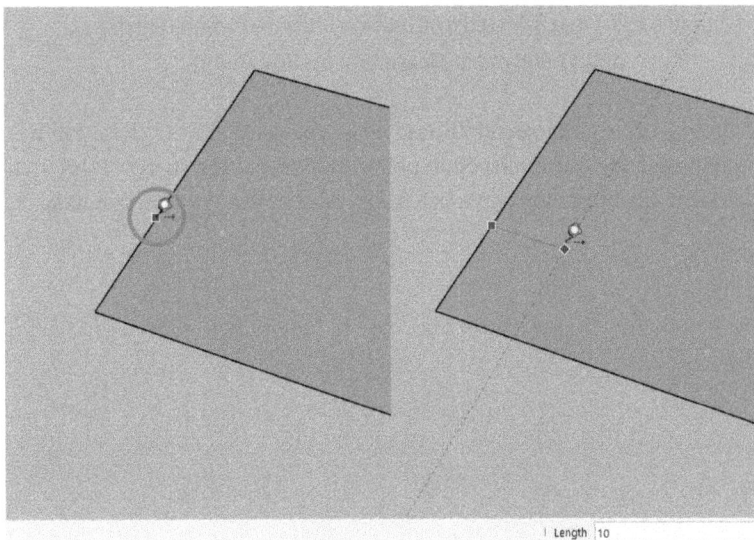

Figure 5.135: Edge Selected (Left); Guide Shown in the Red Direction
and 10 Entered in Measurements Box (Right)

The same workflow can be repeated for the other Guide, clicking on the Bottom Edge and showing the Green Direction, then typing 12 and hitting *Enter*:

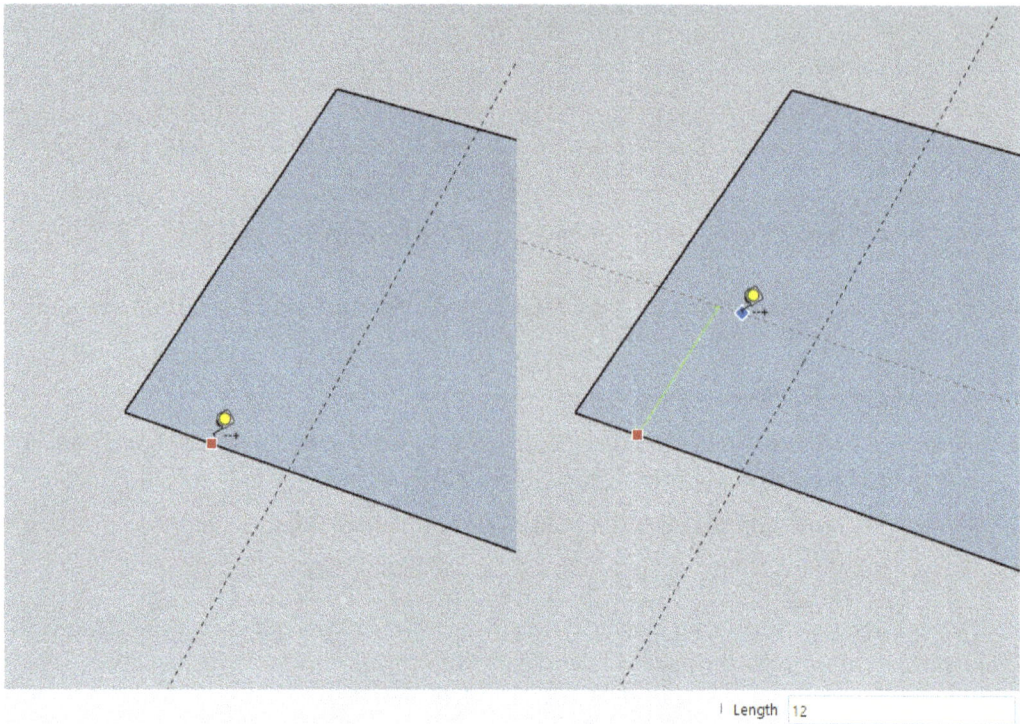

Figure 5.136: Edge Selected (Left); Guide Shown in Green Direction
and 12 Entered in Measurements Box (Right)

These two guides now create an intersection point where the center of the circle can be drawn. The Circle Tool can be activated, and the intersection point can be set as the center of the circle. Then, the Guides can be erased with the Eraser tool, or selected and deleted with the Selection tool and *Delete* key:

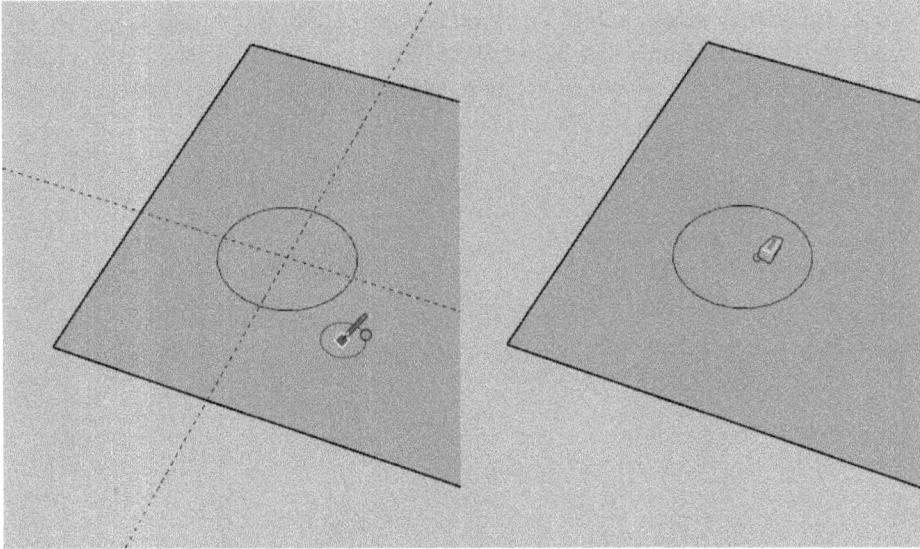

Figure 5.137: Circle Drawn on Intersection (Left); Guides Erased with Eraser Tool (Right)

Guides can also be deleted from the model using **Edit | Delete Guides**. However, this will delete all Guides from the model, including Guides in Groups or Components and hidden Guides. So, be careful when using this option.

The Protractor tool creates guides in the same manner. We will use a similar workflow to what we used in an earlier chapter when we looked at folding Geometry using the Move tool. We see that we have a rectangle that is split in two, and we want to move the bisecting Edge up in the blue direction to make a tent shape. The distance is unknown, but we know that the sides of the tent should be 30 degrees. We can use Protractor guides to help us Move this Geometry into the correct location. First, the view should be orbited so that we can see the horizon in the distance. This will put the Protractor tool's gizmo into the Green orientation, and will create Guides going up the model, instead of flat on the rectangle:

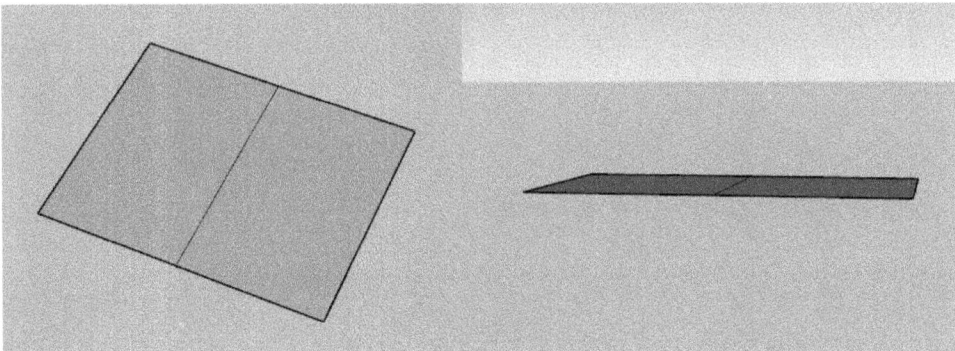

Figure 5.138: Rectangle with Bisecting Edge (Left); View Oriented to See Horizon (Right)

Now, we can activate the Protractor tool and turn on the *Ctrl* modifier key option to create Guides. The Protractor tool will also show a dashed line and a plus (+) sign to represent that this option is turned on. To create our Guides, we should first pick one corner of the rectangle with the Protractor Gizmo. This will set one point of the Guide, and the center point of the angle:

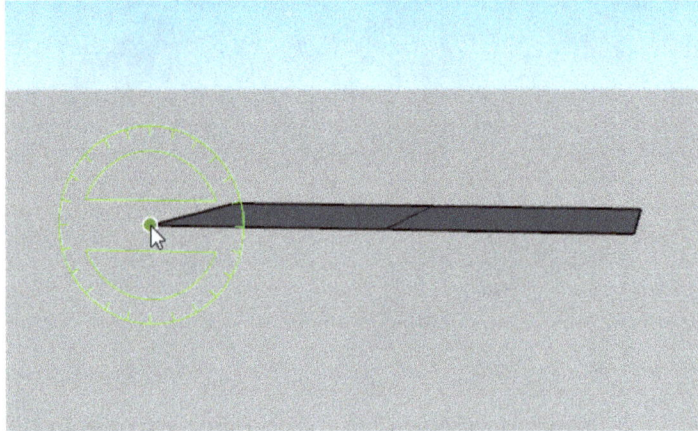

Figure 5.139: Protractor Gizmo Showing Green Direction and Picking Endpoint of Rectangle

Then, the first point of the angle should be set by clicking on the other end of the rectangle. This sets the base angle to be flat across the rectangle's Face:

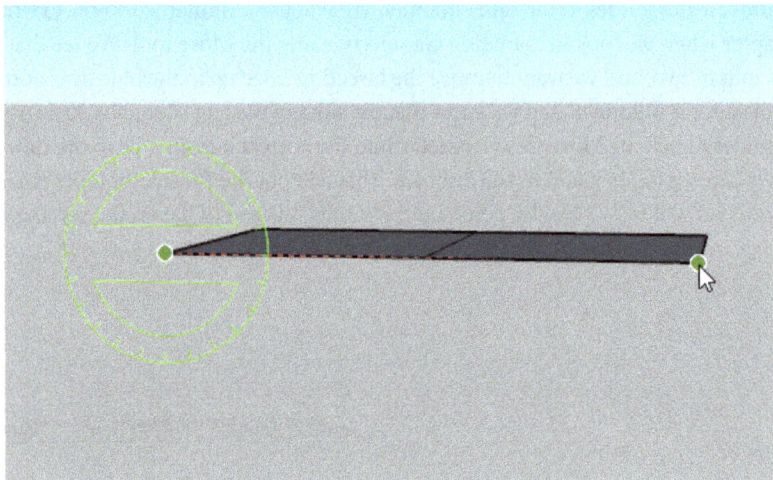

Figure 5.140: Second Point Set on Rectangle

Finally, we can specify the degree angle by showing the direction (up), typing our desired angle amount into the Measurements Box—30—and hitting *Enter*. This will create a guide that goes up and is 30 degrees from the Edge of the Rectangle:

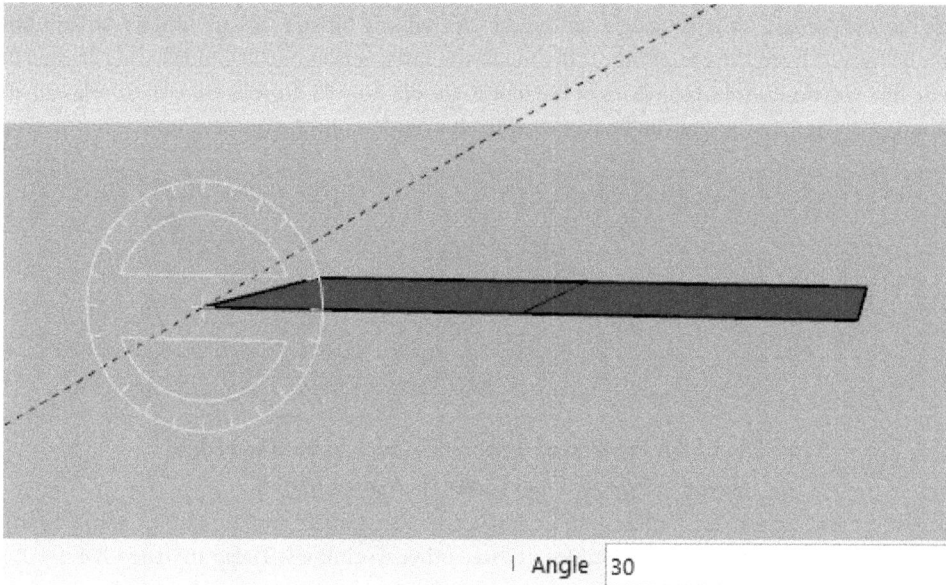

| Angle | 30 |

Figure 5.141: Angle Specified and Guide Created

We can repeat this workflow on the other side of the rectangle, once again specifying 30 as the degree value, and we will then have two Guides creating a triangular shape above one end of the rectangle. This does not create a Face as the Guides are not Edges, so no new Geometry is created:

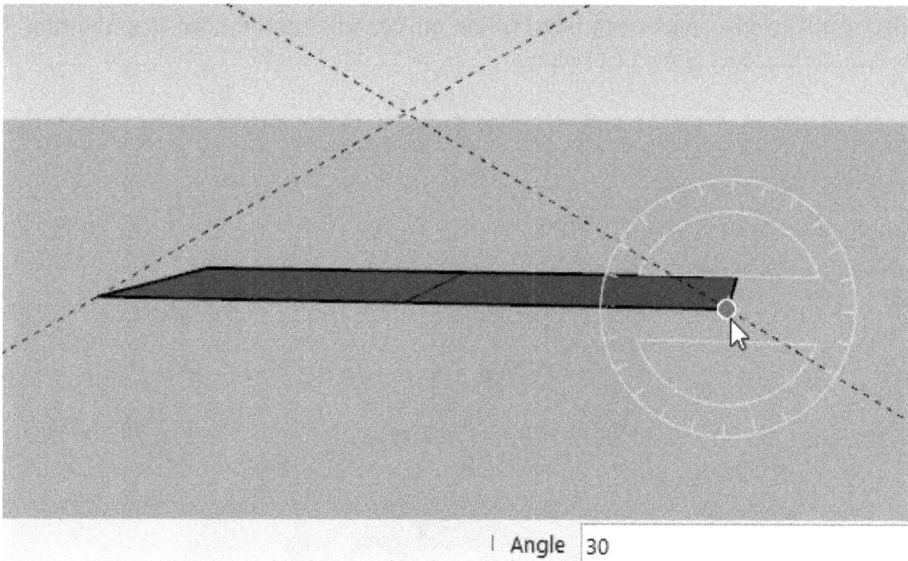

| Angle | 30 |

Figure 5.142: Workflow Repeated on the Other Side to Create a Second Guide

Now, the bisecting Edge in the rectangle can be preselected, the Move tool can be activated, and the Edge can be moved from the endpoint of the line to the intersection of the Guides. This creates the tent shape that was desired, and the sides of the tent shape are now 30 degrees from their original, flat angle. Then, if desired, the Edges can be removed from or hidden in the model:

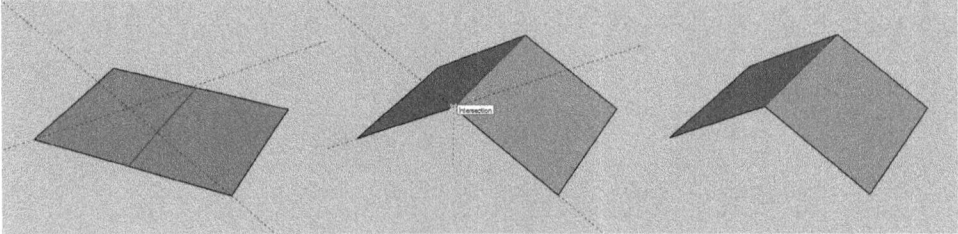

Figure 5.143: Edge Preselected (Left); Move Tool Activated and Edge
Moved to Intersection (Middle); Final Shape (Right)

The Tape Measure tool can also create Guides in three other workflows. These are listed here:

- Duplicating an Edge, Guide, or Axis
- Through Two Points
- Short Guide with a Specified Distance and Direction

Duplicating an Edge, Guide, or Axis is fairly simple. With the Tape Measure tool active and Create Guides toggled on, double-click on any Edge, Guide, or Axis with the On Edge Snap showing. This will create a Guide that goes along that vector:

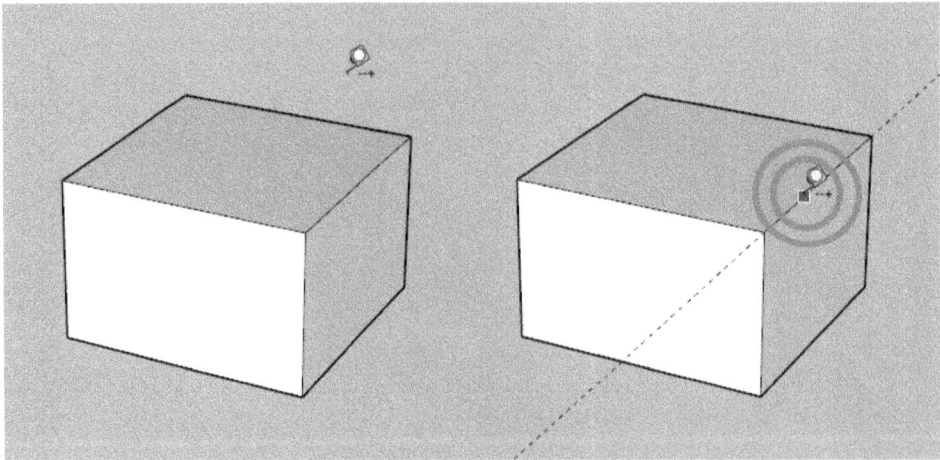

Figure 5.144: Tape Measure Tool and Create Guides Toggled On (Left);
Edge Double-Clicked and Guide Created (Right)

It is easier to see when a guide should be created on an existing Edge. It could transfer the head high of a door opening to an adjacent window or it could "extend" a line to help locate an intersection point. But when would an Axis or another Guide need to be duplicated? In the case of an Axis, remember that the Axis tool can move Axes around a model! If you are about to move Axes but would like to retain one or more of those vectors, then duplicating the Axis with a Guide makes sense. As for a Guide on another Guide, remember that Guides can be Grouped or inside of Components. If a Guide is in a Group that is about to be moved, but you would like to keep the vector data in place, the Grouped Guide can be duplicated with a double-click, and then the Group can be moved.

Creating Guides through two points is also fairly simple. With the Tape Measure tool active and Create Guides toggled on, click on any Inference or Snap in the model, and then on any other Inference or Snap. This will create a Guide running through those two points:

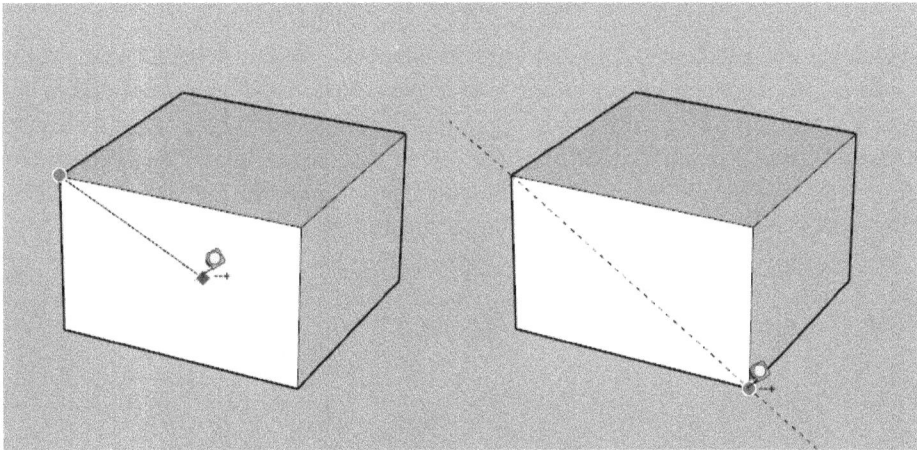

Figure 5.145: Tape Measure Tool and Create Guides Toggled On with
Point Selected (Left); Second Point Creates Guide (Right)

This workflow will not be activated if the Snap that is first selected is the Midpoint or On Edge Snap. This will activate the standard Guide creation workflow, which is to create a parallel Guide. Using Endpoint, On Face, or Intersection will activate this workflow.

Short Guides with Specified Distance and Direction Guides are started with the same workflow as through two points Guides. A Snap or Inference is first selected, then a distance and direction are specified. This can be done by clicking in space in a model that is not connected to a Snap or Inference, such as randomly clicking in one of the Axes' directions. However, it is more common to show the direction and type a distance into the Measurements Box. This creates a short Guide that can be snapped to along its length, but also two Guide Points. This is a different type of Snap that behaves like an Endpoint Snap, but because it is not on an Edge it is referred to as a Guide Point. This workflow might be used when a parallel guide cannot be created for the specified location. In this example, we see that there is a diamond shape, and we want to find the point 8 inches to the right of the diamond endpoint:

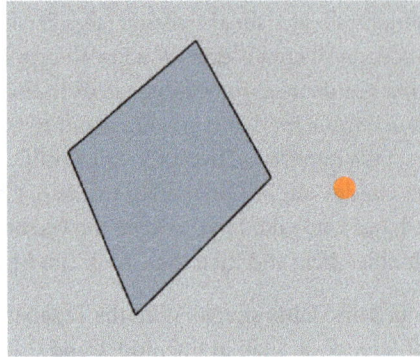

Figure 5.146: Diamond and Desired Reference Point Location

We could draw a line along the Red Direction using the Line Tool, but this would then interfere with our Diamond and the Geometry that we want to draw next. Instead, we can use a short Guide. To begin, we activate the Tape Measure tool and toggle on Create Guides. Then, we pick the endpoint on the diamond to start creating a Guide. The Tape Measure can then show the Red Direction with the Guide preview, and we can type 8 into the Measurements Box and hit *Enter*. A short Guide has been created with a Guide Point exactly 8 inches from the diamond:

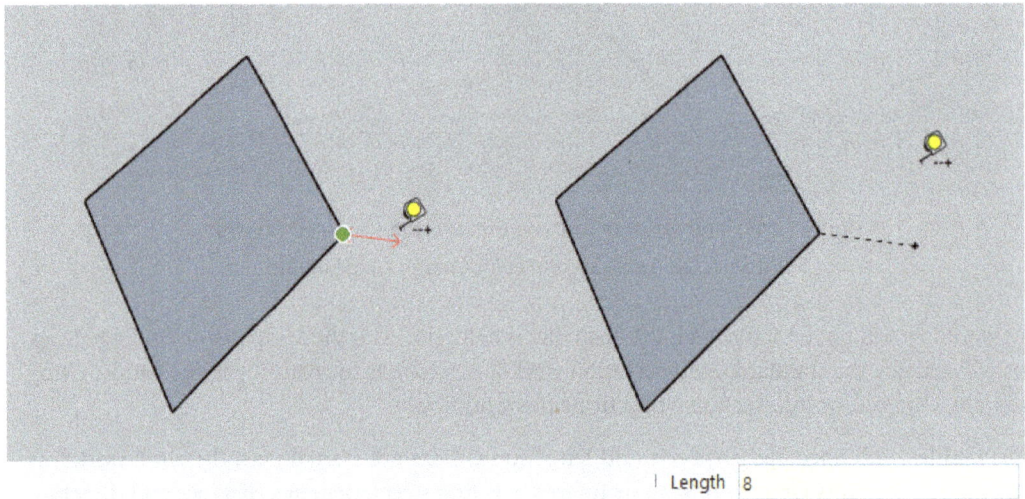

| Length 8

Figure 5.147: First Point Set and Direction Shown (Left); 8 Inches Entered
into Measurements Box and Completed Guide (Right)

Guides are an essential part of many workflows as they can be used freely knowing that the model Geometry will not be corrupted, and the Guides can all easily be deleted at once. Experiment with Guides in your own models and get used to using them—they are great tools that can be used in a variety of workflows.

Summary

In this chapter, we discussed the Edit tools and their fundamental workflows and modifier key uses in SketchUp Pro. The Edit tools allow for unlimited possibilities in SketchUp Pro, even without using extensions or plugins. Using the Edit tools effectively in their basic forms and with their different options allows for a greater understanding of the model you are attempting to create and requires you to stay on top of ensuring your model is watertight and oriented the correct way! Finally, we looked at creating Guides in our model with the Tape Measure and Protractor tools.

In order to take our models to the next level, we will next discuss view options in SketchUp Pro, examining how to best navigate our model while hiding, showing, and looking at our objects.

Part 2 – Views, Animations, and Materials

After introducing the basics of SketchUp Pro in *Part 1*, *Part 2* focuses on how to manipulate the unique visual styles in SketchUp Pro. We will explore camera and view options, including how to show and hide Geometry and Objects, and how to change the view styles to edit Groups and Components. We will also explore Parallel Projection, Perspective, and Two-Point Perspective options, and more detailed options to change the Drawing Area camera view. We will end this part with a deep dive into SketchUp Materials, including SketchUp Material Libraries, how to import and create Texture Materials, how to edit Materials, and how to apply Materials to your SketchUp models.

This part has the following chapters:

- *Chapter 6, Camera Options*
- *Chapter 7, View Options*
- *Chapter 8, Materials*

6
Camera Options

In this chapter, we will examine 3D camera options in SketchUp Pro. We will discuss the Perspective, Two-Point Projection, and Parallel Projection views as we look at example Geometry. We will also analyze the available Camera Tools to introduce different ways to place the SketchUp camera and navigate the model.

The following topics are covered in this chapter:

- Camera View Options:
  - Perspective
  - Two-Point Perspective
  - Parallel Projection (Orthographic)
- Camera Tools:
  - Position Camera
  - Walk
  - Look Around

Camera Options – My Geometry is Not Straight!

You may have noticed while working in SketchUp (or even while looking at the screenshots in this book) that Geometry often appears to lean or to get larger or smaller depending on the camera angle. We know that moving the camera is not changing the size of the Geometry, so what is happening?

The default view in SketchUp is the Perspective camera and it will distort how Geometry looks, depending on where it is on the screen. This is an attempt to mimic the way we see the real world, but it does not always come across perfectly in SketchUp or other 3D modeling software.

The Perspective view is often the best way to work while modeling in SketchUp, as it provides a sense of depth and scale in the model. The Two-Point Perspective option is great for static Scenes, such as when creating a rendering. The Parallel Projection view provides a sense of scale throughout the model while removing any sense of depth:

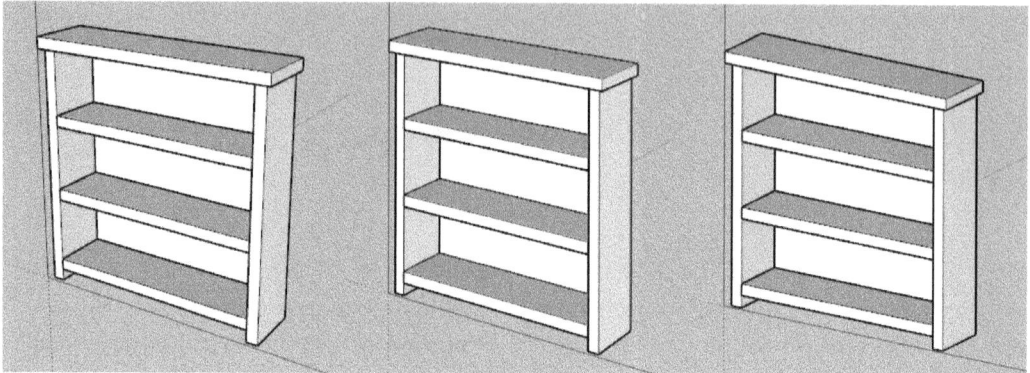

Figure 6.1 – Perspective View (Left); Two-Point Perspective View (Middle); Parallel Projection View (Right)

The three camera views can be selected by going to the **Camera** menu dropdown and clicking a view option:

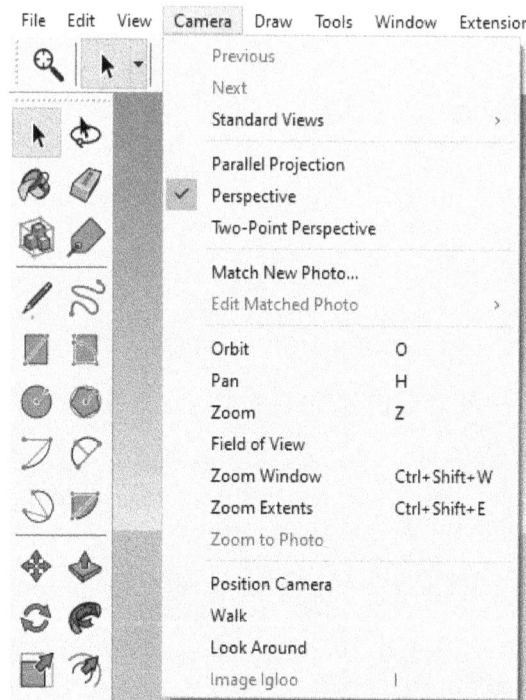

Figure 6.2 – Camera Menu Dropdown with Perspective Selected

In this section, we will look at the three views: **Perspective**, **Two-Point Perspective**, and **Parallel Projection**. We will begin to understand how the views are displayed and also provide some best practice tips for when to use each view.

Perspective View

As mentioned in this section's introduction, the Perspective view is the default camera setting in SketchUp Pro. The Perspective view is immediately recognizable because no two lines appear parallel!

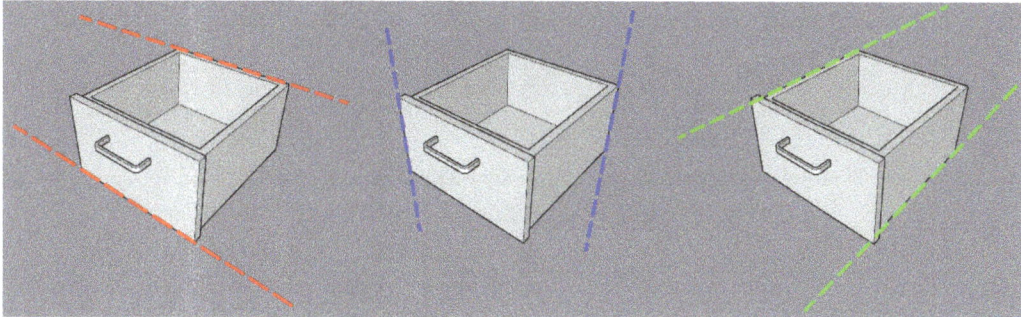

Figure 6.3 – The Perspective View Showing No Parallel Lines along Standard Axes

The horizontal lines on the Geometry appear to "point" or "vanish" to the horizon, and vertical lines appear to be slanted:

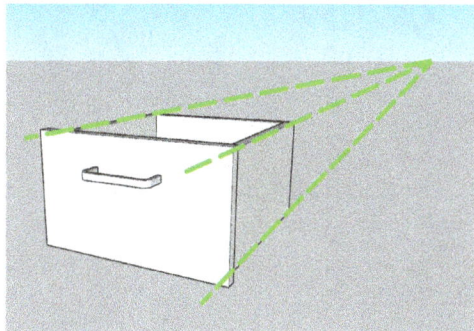

Figure 6.4 – Horizontal Lines Appear to Vanish to the Horizon

The vertical lines will appear to be slanted "in" or slanted "out," depending on the camera's location relative to the Geometry. This all depends on the location of the camera relative to the Geometry and the Field of View. We talked about the Field of View in *Chapter 1, Beginning with SketchUp Pro*, and we will talk about it in more technical terms later in this chapter. Vertical lines in a model will look mostly vertical when the camera is far away from the Geometry, but the vertical lines will slant more when the camera is close to the Geometry:

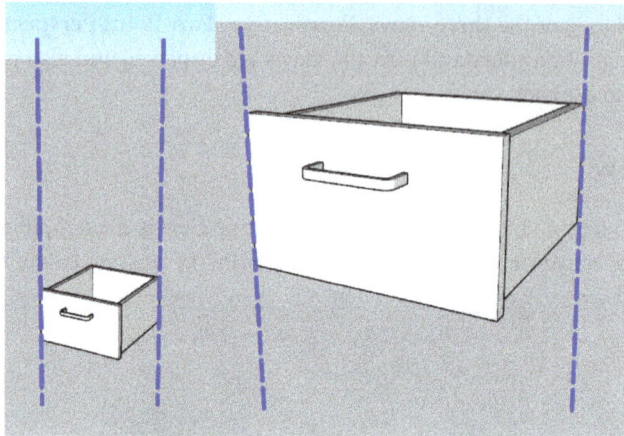

Figure 6.5 – Vertical Lines Mostly Straight when Camera is Far (Left);
Vertical Lines Slant when Camera is Close (Left)

Additionally, Geometry on the sides of the screen will appear to slant much more than Geometry that is in the middle of the screen, even if the camera remains the same distance from the Geometry. This is achieved through a Pan, which we also discussed in *Chapter 1, Beginning with SketchUp Pro*:

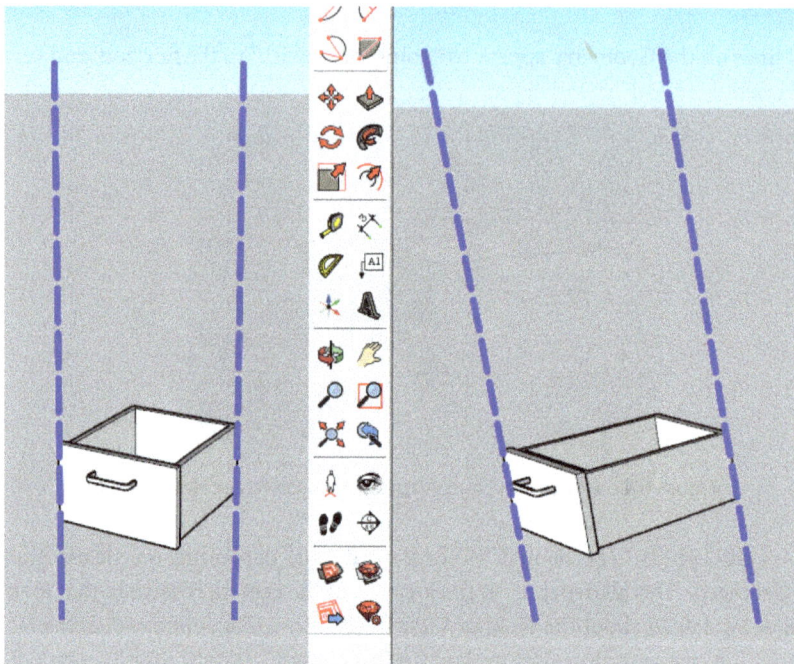

Figure 6.6 – Vertical Lines Slanted at Sides of Screen (Left);
Vertical Lines Straight at Middle of Screen (Right)

The Perspective view is the default setting in SketchUp as it allows for easy size comparisons concerning distance. Geometry that is the same real-world size will appear larger or smaller, depending on the camera's relative position, and we can use that to know what objects are closer or farther away in our model. It helps us build a sense of 3D space – and that can better explain our model! In this example, we can see a row of columns that appear to get smaller from left to right. However, we know that these columns are the same size, and the Perspective view is showing us that the row of columns is going away from us:

Figure 6.7 – Columns Appear to Get Smaller from Left to Right (Left); Perspective View Shows Depth (Right)

While this view does provide a sense of depth, it can be very difficult to understand a sense of relative scale. The two boxes in the following example might be the same size – we cannot easily tell without some clues or dimensions:

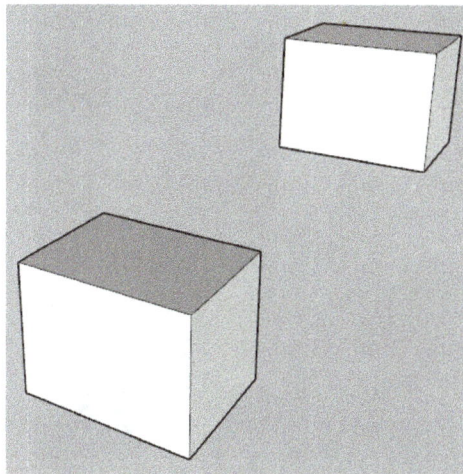

Figure 6.8 – Two Boxes of Unknown Scale

However, when we put some scale Objects – such as standard SketchUp people – or some dimensions on the model, we can see that these boxes are not the same size. The Perspective view does not do a good job of clearly showing the scale – at least not from this angle:

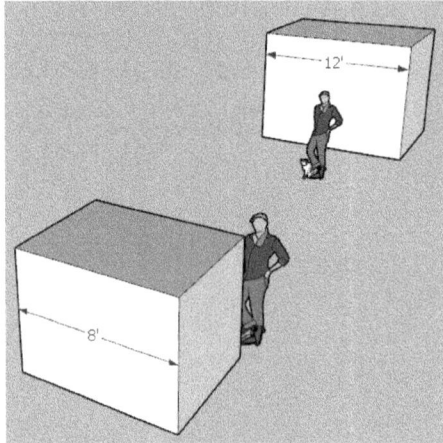

Figure 6.9 – Two Boxes with SketchUp People for Scale and Dimensions

The vertical slanting or distortion is part of the way SketchUp wants you to view SketchUp models – it mimics the camera distortion that we see in most photographs. If you look at this photograph of a building, you can see the vertical lines are leaning – this is common camera distortion:

Figure 6.10 – Photograph Showing Camera Distortion on Vertical Lines

Many photographers often edit their photos to re-align the vertical lines to better represent real-world objects, as seen in the following figure:

Figure 6.11 – Photograph with Vertical Lines Adjusted

This is a good representation of our next view in SketchUp – the Two-Point Perspective view.

The Two-Point Perspective View

The Two-Point Perspective view is very similar to the Perspective view, but it does not have a vertical vanishing point. That means that there will never be any distorted vertical lines – they will always be straight up and down and will all be parallel.

We can see a Perspective view of some boxes on the left of this example. You can easily see the vertical distortion, especially on the sides of the screen. On the right-hand side of this example, we can see a Two-Point Perspective view, which corrects the distortion in the Perspective view and shows all vertical lines as parallel:

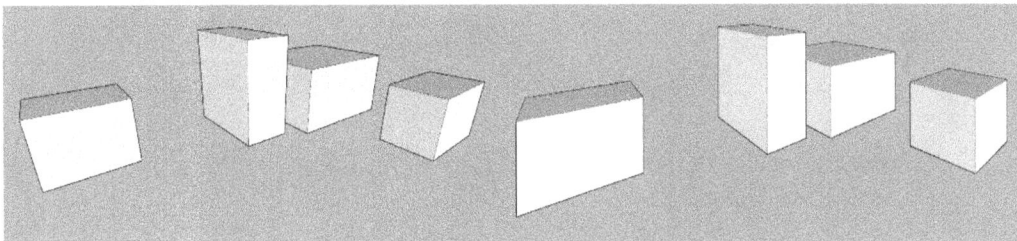

Figure 6.12 – Perspective View (Left); Two-Point Perspective View (Right)

The Two-Point Perspective view is great to use when setting up a rendering or Scene. A common workflow is to use the Perspective view while modeling and finding a good camera angle (using Orbit and Pan) and then changing the view to Two-Point Perspective. After saving this view, a Scene can be created. We will discuss Scenes in a later chapter:

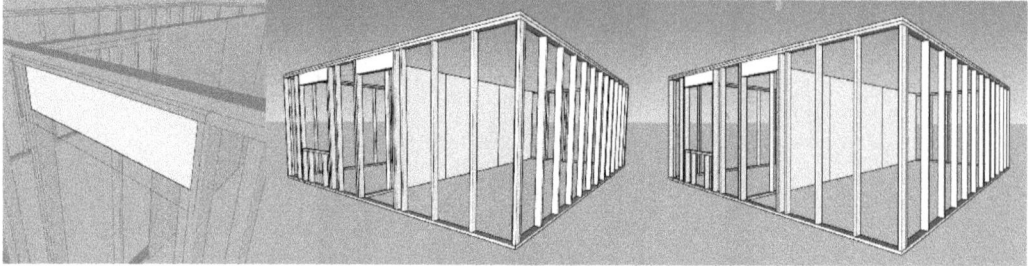

Figure 6.13 – Modeling in Perspective (Left); Finding a Good Camera
Angle (Middle); Two-Point Perspective (Right)

Note

Orbiting while in Two-Point Perspective will automatically set the view to the standard Perspective view. This is one reason why Two-Point Perspective is often set for static views – such as Scenes for renderings.

In the standard Perspective view, it can be difficult to understand the vanishing points. Vanishing points are where the horizontal and vertical lines "point," or where they would visually meet if they were infinitely extended. In SketchUp Pro, we have guides and Axes to help us visualize where vanishing points are in the drawing!

In this example, we can see that the Field of View has been changed so that we can see where the Red Axis and Green Axis meet the horizon line. These two points are our horizontal vanishing points!

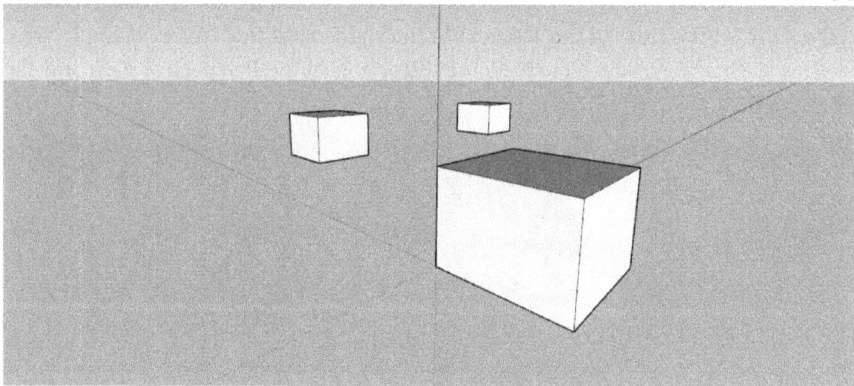

Figure 6.14 – The Field of View Changed to Show Vanishing Points

So long as the Geometry has been drawn on the Red and Green Axes, the horizontal lines will appear to point toward the vanishing points no matter where the Geometry is relative to the Origin, where the Axes meet. The following example shows that the horizontal lines on the boxes appear to point toward the vanishing point – and the guides on the right confirm the direction of the horizontal lines:

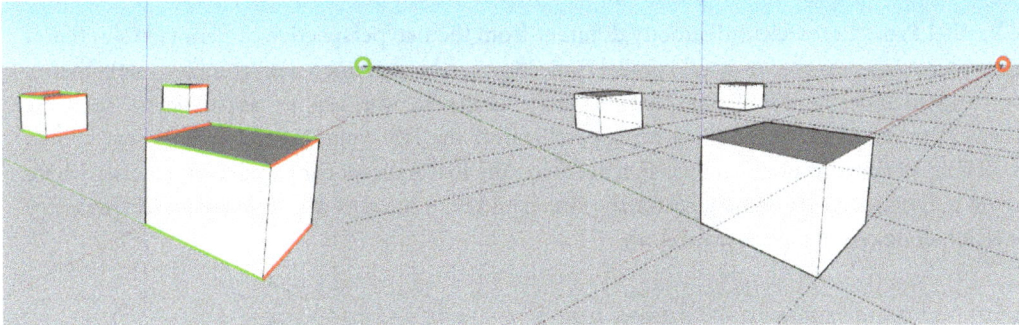

Figure 6.15 – Horizontal Lines on Boxes (left) and Guides Point to Vanishing Points (Right)

The Standard Perspective view also has a third, vertical vanishing point, depending on where the camera is located. This point is either straight up or straight down and it is where the Blue Axis disappears. This is what accounts for the distortion in the vertical lines. If we view the preceding model in the Perspective view, we will see the guides converging at the vertical vanishing point, but if we look at it in the Two-Point Perspective view, we will see that we have no vertical vanishing point. All vertical lines are parallel in the Two-Point Perspective view, so they can never meet!

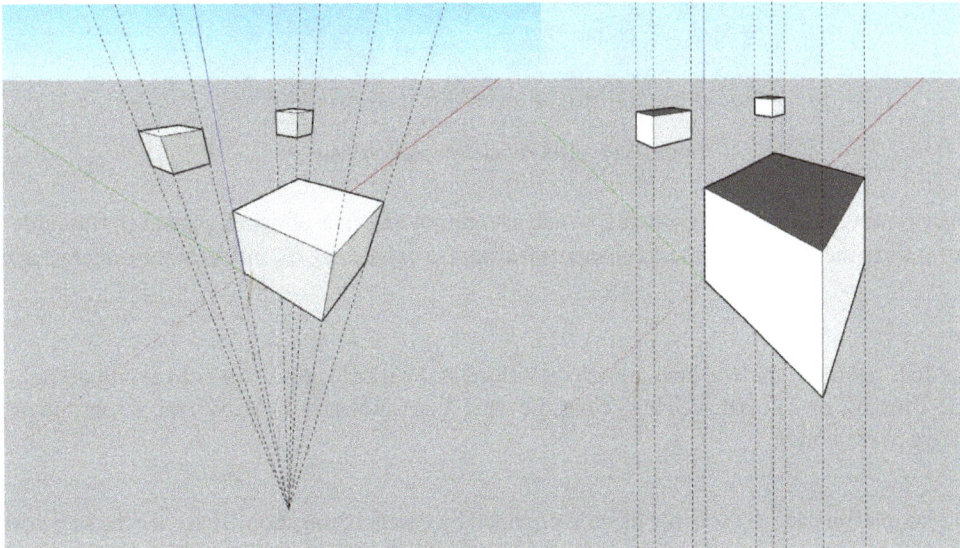

Figure 6.16 – Vertical Vanishing Point in Perspective (Left); No Vertical
Vanishing Point in Two-Point Perspective (Right)

The Two-Point Perspective view was named this due to the two perspective vanishing points that can be seen – both of which are horizontal.

The Parallel Projection View

The Parallel Projection view is distinctly different from the two perspective camera views in that it does not have any perspective at all! Parallel Projection can be explained with complex mathematics but is easier to understand as all lines that are parallel in the real world appear as parallel in SketchUp. This takes the idea of the vertical lines being parallel from the Two-Point Perspective view and applies it to all lines in any direction. All lines that are along the Red Axis appear parallel – no matter where they are in the model. The same goes for the Green and Blue Axes, as well as any parallel Geometry that is not along one of the standard Axes:

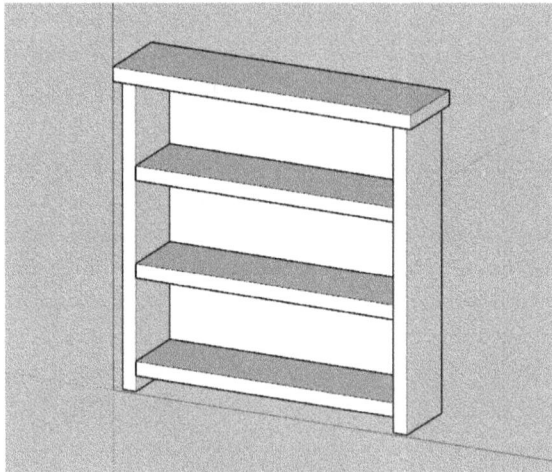

Figure 6.17 – The Parallel Projection View

This view removes perspective elements, which also remove a sense of depth. However, this adds a sense of scale to the model – all model elements will appear as true size regardless of the camera angle.

> **Note**
> This does not mean that the lines are always viewed as the true length – that is an axonometric view. The sense of scale is that all lines along the same Axis will always show the same proportions relative to each other.

In this example, we can see the sense of scale provided by Parallel Projection. These are the same two boxes as before, which were difficult to distinguish in the Perspective view. In the Parallel Projection view, we can see that the box on the right is larger, although we can no longer tell how far away the two boxes are from each other:

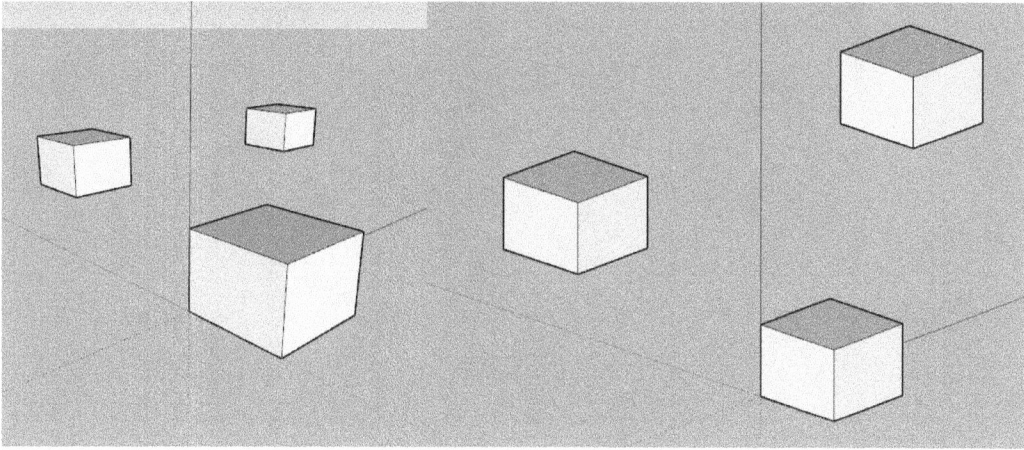

Figure 6.18 – The Perspective View of Boxes (Left); The Parallel Projection View of Boxes (Right)

This scale allows for a model to pass a quick eye test – if two lines look the same length, then they are. Remember, the lines must be on the same Axis to be compared this way! In this example, we can see some windows on a wall. In the Perspective view, it is difficult to see if the windows are all the same width – the depth of the view makes the windows appear smaller as they get farther away from the camera:

Figure 6.19 – Wall with Windows in the Perspective View

In the following figure, we can see the wall and windows in the Parallel Projection view. In this view, we can immediately see that the fourth window is larger than the others. This was not immediately apparent in the Perspective view, but in Parallel Projection, the other windows are the same size:

Figure 6.20 – Wall with Windows in the Parallel Projection View

The relative size of the windows will stay the same, no matter which way the camera is orbited around the model. In the following figure, we can see that the angle of the camera has changed in the Parallel Projection view. The wall appears to be shorter and longer, but this is just the rotation of the camera. The important thing to note is that the proportion of the windows to each other and the wall has not changed:

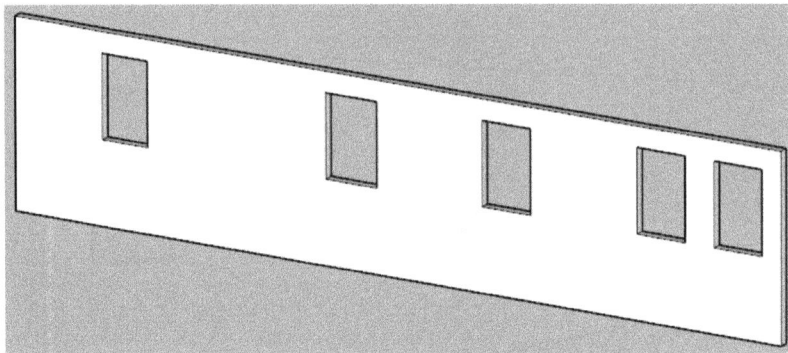

Figure 6.21 – The Camera Orbited to See the Wall from Different Angles

Parallel Projection views are often used while modeling small parts or while modeling for mechanical or industrial design. With small parts, it is not as necessary to have a sense of depth, and a sense of scale is much more important. Parallel Projection can also be helpful while modeling drawings of any size, and some designers prefer to model in Parallel Projection at all times. Parallel Projection can be especially helpful when trying to view a planning model of an architectural file, for instance. This would be an example of an orthographic view.

Orthographic Views

Orthographic views are a type of Parallel Projection where the camera is directly aligned with one of the standard Axes. Orthographic views can be created in SketchUp by setting the camera to the Parallel Projection view and then selecting one of the standard views from the Views Toolbar or **Views > Standard Views** from the menu dropdown:

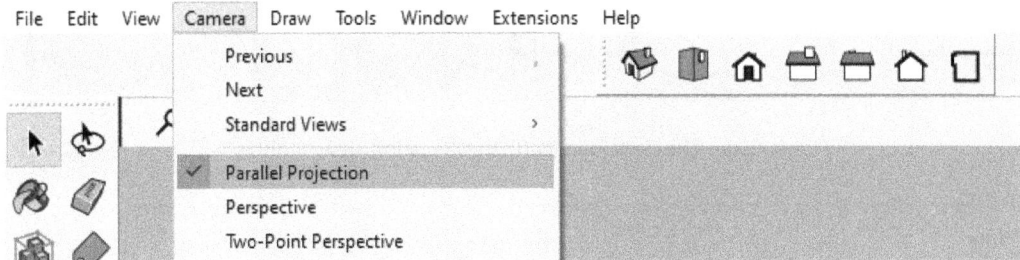

Figure 6.22 – Parallel Projection Set in the Camera Menu Dropdown
(Left); Standard Views on the Views Toolbar (Right)

This is a great way to create a quick plan, elevation, or section view of a SketchUp model. These are standard views in most design industries, especially in architectural and mechanical design. In the following example, we can see that a SketchUp model has been completed and is ready to have drawings created for it:

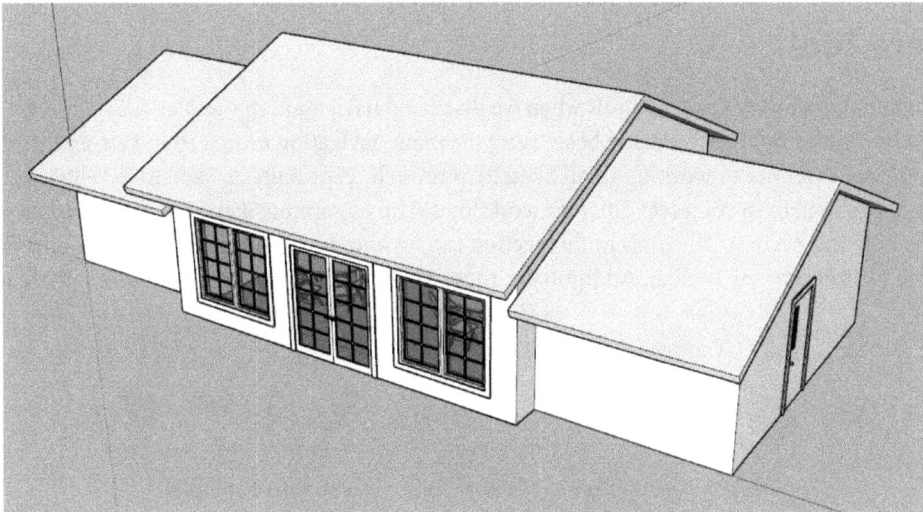

Figure 6.23 – Finished SketchUp Model

For a quick check, or to export some sketching views from within SketchUp Pro, some Orthographic views can be created. In the following figure, we can see the Floor Plan and Elevation views:

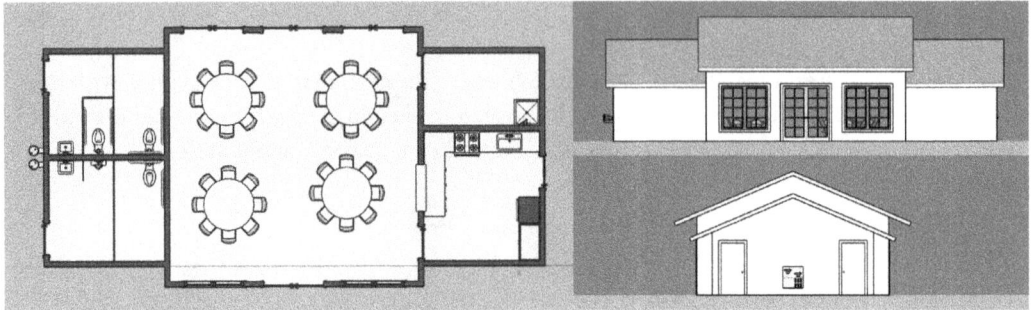

Figure 6.24 – The Floor Plan (Left); the Elevation View (Right)

> **Note**
>
> Combining Orthographic views with Section Planes in a Scene is a great way to create a Floor Plan view that can be quickly opened! We will discuss Section Planes and Scenes in *Chapter 7*.

Switching between Perspective, Two-Point Perspective, and Parallel Projection or Orthographic views is a great way to understand how a SketchUp model is coming together! In the next section, we will look at some Tools that will help us explore our SketchUp models; Position Camera, Walk, and Look Around.

Camera Tools

We discussed the primary Camera Tools when we discussed navigation in *Chapter 1*, *Beginning with SketchUp Pro*. You should have already been using the main navigation tools; Orbit, Pan, and Zoom. These tools are essential to working in all SketchUp models, especially as SketchUp relies on the position of the camera to complete different workflows. The remaining SketchUp camera tools will be reviewed in this section. The tools in this section can be found on the Large Tool Set Toolbar but not on the Getting Started Toolbar. Additionally, these tools can be found on the **Camera** Toolbar:

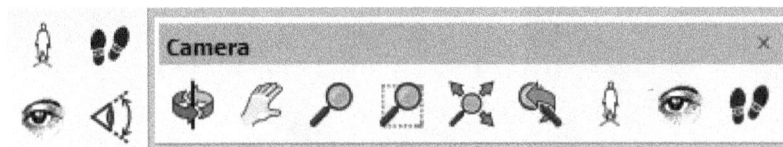

Figure 6.25 – Getting Started Toolbar (Left); Camera Toolbar (Right)

Position Camera

The Position Camera Tool is an excellent way to set the camera in the exact location that you want while modeling or while setting up a Scene. The Position Camera Tool requires two things to set the camera location: a Position and a Height Offset. The Position is where in the Drawing Area a "person" would be standing to view the surrounding model elements, and the Height Offset represents the Eye Height of the "person" in the model. A person isn't created in the model – it is only represented by the Position Camera Tool settings.

The Position Camera Tool button looks like an outlined person standing on a red X:

Figure 6.26 – The Position Camera Tool Button

The Position Camera Tool can be activated by doing the following:

1. Clicking the **Position Camera** button.
2. If desired, type in a new **Height Offset** into the Measurements Box.
3. Clicking anywhere in the Drawing Area to set the Position.

> **Note**
> The default height for the Height Offset in the Measurements Box is 5' 6". This is a nominal height for the eye height of an average person.

When the Position Camera workflow has been completed, the Look Around Tool is immediately activated. From there, the Look Around Tool and the Walk Tool can be used to explore the model. The Position Camera is a fairly simple tool to use in concept, but when used in tangent with the Look Around and Walk Tools, it can create a powerful exploratory workflow.

Before we look at the Look Around Tool, let's look at a quick example of when the Position Camera tool can be used to help us model and explore a SketchUp model. In this example, we have a small building, and the next step in the modeling workflow is to update the kitchen cabinets. In this case, we could begin to hide the Geometry and use the Section Tool to cut away the Geometry. We will discuss the Section Plane Tool in Chapter 7:

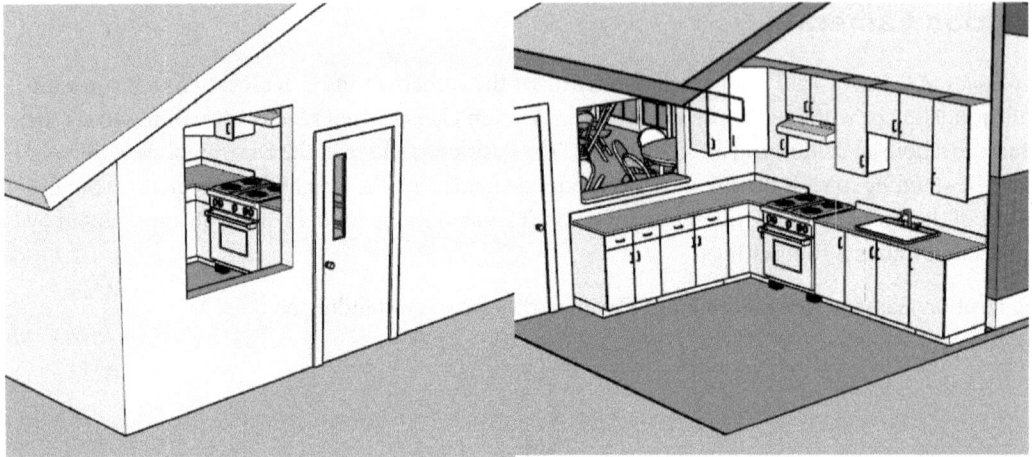

Figure 6.27 – A Small Building with Kitchen (Left); Hidden Geometry and Section Cuts (Right)

Instead of creating the view shown in the preceding figure, with multiple Section Cuts and hidden Geometry, we can use the Position Camera to jump into the model quickly and precisely for this edit. In this case, the camera has been orbited around so that the floor of the kitchen is visible through one of the window openings. There is no window yet in this model, so the floor Geometry can be clicked. With the Position Camera Tool active, the floor can be clicked on through the window opening:

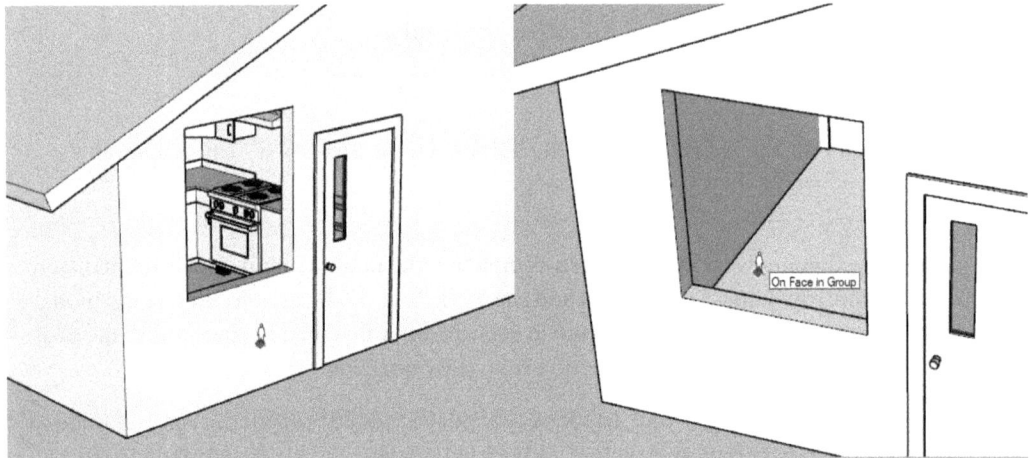

Figure 6.28 – Position Camera Tool Active (Left); Kitchen Floor Clicked through Window Opening (Right)

When the floor is clicked, the camera will automatically move inside the building and it will appear as though we are standing in the kitchen with the camera placed at the default 5' 6" eye height:

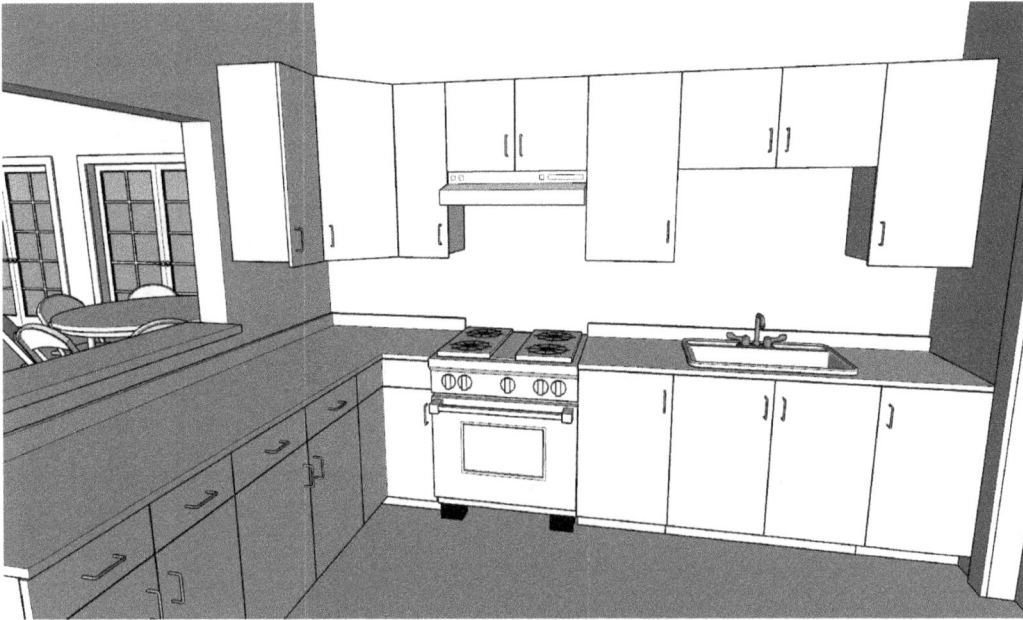

Figure 6.29 – Camera at 5′6″ in the Kitchen

This is a perfect location to work on the kitchen sink or the countertops, but what if the toe kicks of the kitchen cabinets needed to be edited? In that case, the Eye Height should be much lower and closer to the kitchen floor – for example, at 6 inches. In the following example, we have the same starting view, but **Height Offset** has been set to 6″ before the Position Camera Tool is clicked:

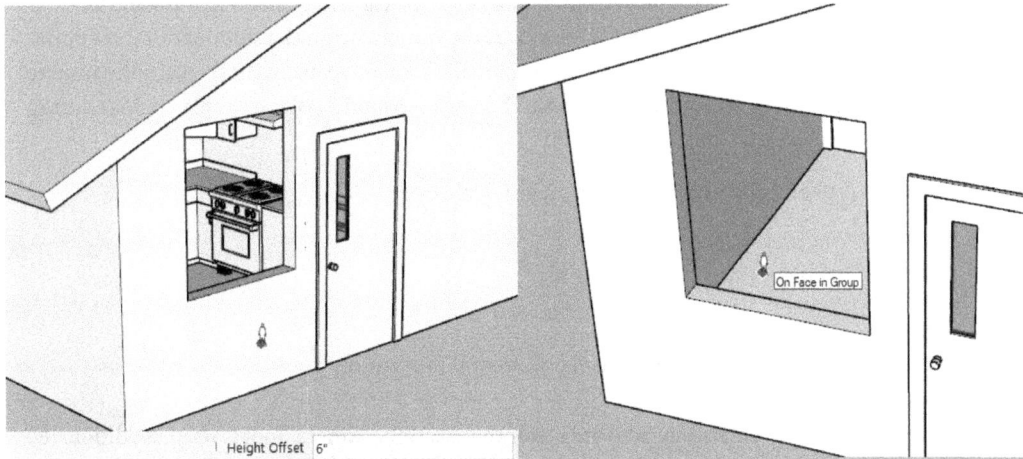

Figure 6.30 – The Position Camera Tool is Active and the Height Offset is 6″ (Left);
the Kitchen Floor Clicked through the Window Opening (Right)

Now, when the floor is clicked, the camera will still be moved into the building but will be placed 6" above the floor. This is a much more appropriate height to work on the toe kicks of the kitchen cabinets:

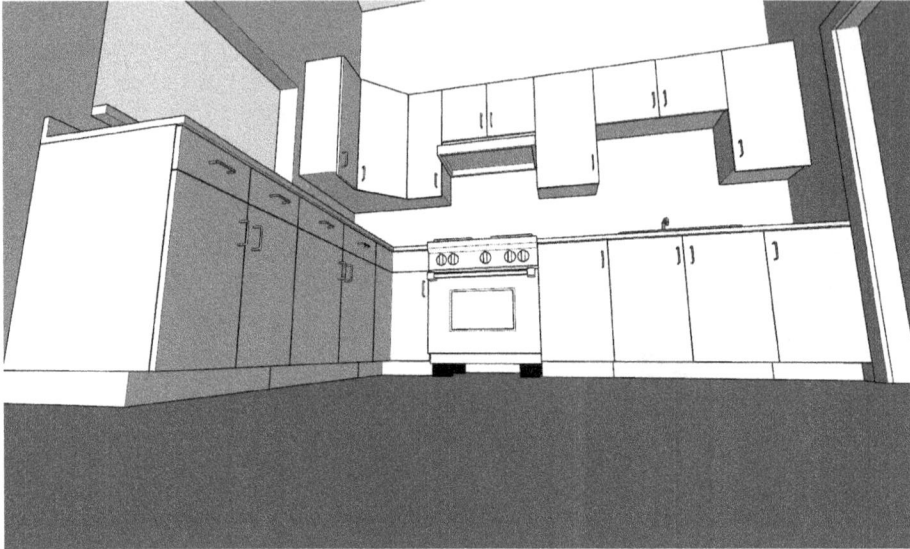

Figure 6.31 – Camera at 6" in the Kitchen

Look Around

The Look Around Tool is a perfect companion to the Position Camera Tool. The Look Around Tool works opposite to the Orbit Tool. The Look Around Tool keeps the camera position in the same location but moves the target point, while the Orbit Tool moves the camera around a stationary target point. This is exactly what we want after using the Position Camera Tool as we just went through the trouble of placing the camera in the exact location we need! The Look Around Tool is a simple click-and-drag tool; this can be done using the left mouse button.

The Look Around Tool button is represented by a grayscale eye:

Figure 6.32 – The Look Around Tool Button

The Look Around Tool is automatically activated after the Position Camera Tool is used. Additionally, the Look Around Tool can be activated by clicking the Look Around Tool Button. When active, the target point can be moved by clicking and dragging on the screen with the left mouse button. In the following figure, we can see that the Look Around Tool has been used to look further to the right of the room to see the rest of the kitchen cabinets:

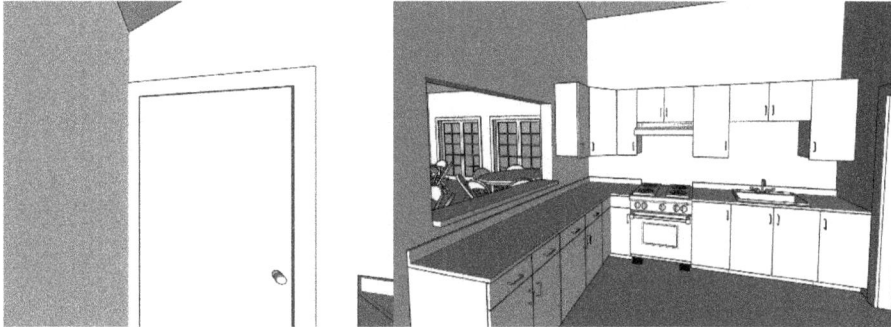

Figure 6.33 – Original Camera View (Left); the Look Around Tool
Active and Dragged to Show the Correct View (Right)

The Look Around Tool should be used at any point when the camera is in the correct location but when the camera might be pointing in the wrong direction. This is often after the Position Camera Tool has been used, and also after the Walk Tool has been used in model exploration.

Walk

The Walk Tool fits in with the Position Camera and Look Around Tools when it comes to model exploration. The Walk Tool allows the camera to be moved around the same camera height that is currently set in the model. This can be set using the Position Camera Tool or can be the current camera height after panning and orbiting around the model.

> **Note**
>
> The Walk Tool is a great way to make Scenes for creating an Animation in SketchUp Pro. We will discuss Scenes and Animation in *Chapter 7, View Options*.

The Walk Tool has three modifier keys that change the way the Walk Tool can be used. The *Ctrl* key can be held to activate a "run," which moves the camera more quickly along the standard Walking limitations. The *Shift* key can be held to change the orientation of the camera movement. Instead of being locked in the horizontal plane, the *Shift* key locks the camera movement into the perpendicular plane – that is, up and down and panning left and right. The *Alt* key can be held to remove collision detection, which is something we will discuss later in this section.

The Walk Tool Button looks like two footprints:

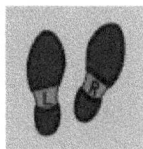

Figure 6.34 – The Walk Tool Button

The Walk Tool can be activated by doing the following:

1. Clicking any of the **Walk Tool** Buttons.

2. Clicking the left mouse button and holding to walk:

    -   Dragging up or down walks the camera forwards or backward

    -   Dragging left or right turns the camera to the left or right

    -   Holding *Shift* and dragging up or down moves the camera up or down

    -   Holding *Shift* and dragging left or right pans the camera left or right

The Walk Tool is a great way to explore models that are at a human scale – that is, humans will exist within a designated space. Architectural and landscape designs and some art designs are the perfect applications for the Walk Tool. The Orbit Tool is a much better solution for mechanical, industrial, or woodworking design when the finished product is a single object or something that would fit inside a room. The Walk Tool allows us to explore space in SketchUp Pro by simulating what it might look like to walk through a space.

In this first example, we can see that we have an architectural building. This is the same example that we used for the Position Camera Tool. In this case, we will still use the Position Camera Tool to place the camera at 5' 6" inside the kitchen:

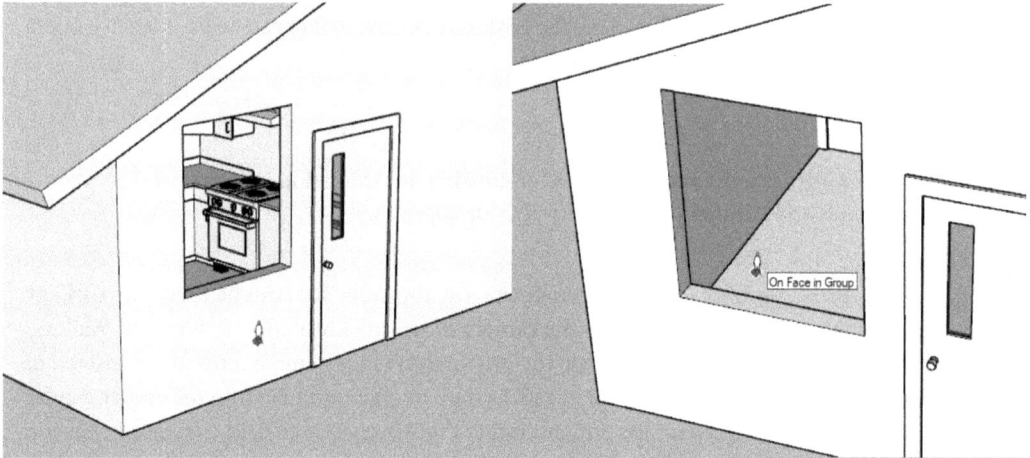

Figure 6.35 – Architectural Building (Left); the Position Camera
Tool Used to Place a Camera in the Kitchen (Right)

We have also seen how to use the Look Around Tool to explore the space from a stationary position. In the following figure, we can see how the Look Around Tool can be used to see more of the kitchen:

Figure 6.36 – The Look Around Tool Used to See Multiple Views of the Kitchen

The Walk Tool adds a dynamic element to exploring space since we can explore by walking around! With the Walk Tool active, we can drag on the screen to walk through the kitchen door opening and into the next room. This is a great opportunity to introduce one great feature of the Walk Tool: collision detection. The Walk Tool will not allow the camera to walk through Geometry in the model, but it will instead "bump" into the Geometry as if it were solid. If we attempted to walk through the wall of the kitchen instead of through the door opening, we would simply stop – we won't be able to get through!

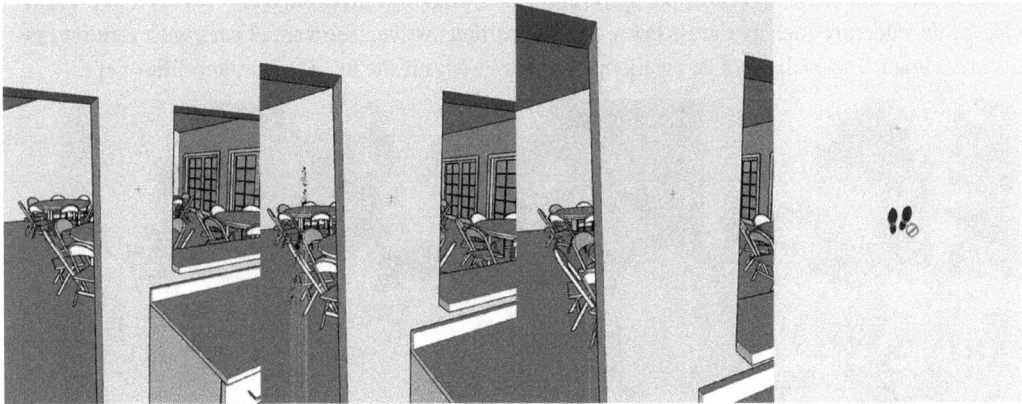

Figure 6.37 – Using the Walk Tool to Attempt to Go Through a Wall

Note

This does not happen when using Pan or Orbit, which will allow you to go right through and into the Geometry in your SketchUp model. This collision detection is even more helpful in making it feel like the camera is inside the spaces that you are exploring!

The *Alt* modifier key can be used to get through a wall or any other Geometry while using the Walk Tool. The *Alt* modifier key turns off collision detection, which can help in short instances, but collision detection is usually a great feature to keep on!

Figure 6.38 – Using the Alt Modifier Key to Move Through a Wall

One reason to keep collision detection on is that SketchUp Pro has pre-programmed tolerances for keeping the Eye Height a standard height from what SketchUp perceives as the surface being walked on. That means that the Walk Tool can walk up ramps, stairs, and even mesh surfaces created to represent landscape architecture such as a grass lawn. In this example, we can see a small park with some stairs and a grass lawn. The Walk Tool can walk up the stairs without the need for any modifier keys:

Figure 6.39 – The Walk Tool Walks Up Stairs

The stairs represent small increases in the surface height, but the Walk Tool can also be used to move over sloped surfaces. In the following example, the Walk Tool has automatically adjusted the camera height to be the same distance above the Geometry, even though the Geometry is sloping up!

Figure 6.40 – The Walk Tool Walking Over a Hill

The Walk Tool can also be used in larger spaces that need to be explored from a "flyover" perspective. Different than the Orbit Tool, the Walk Tool can be used to fly through a space, instead of around it. In this example, we have a small city block, and we want to see what the block looks like as we move through the street in a more aerial view:

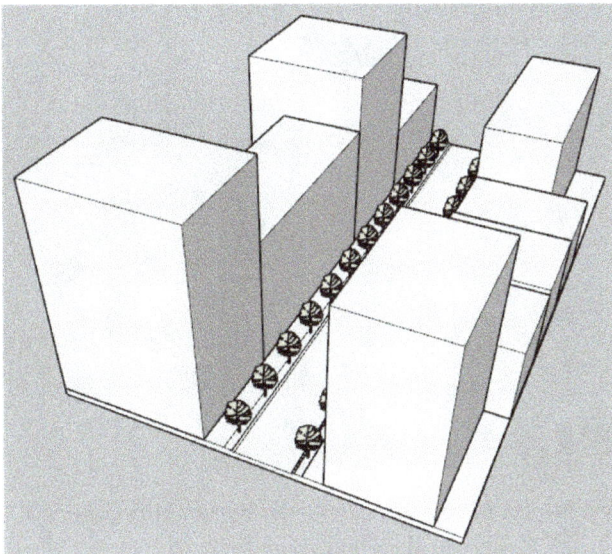

Figure 6.41 – Model of a City Block

If we were to use the Orbit Tool, we would see that we quickly collide and go through or behind the buildings on either side of the block:

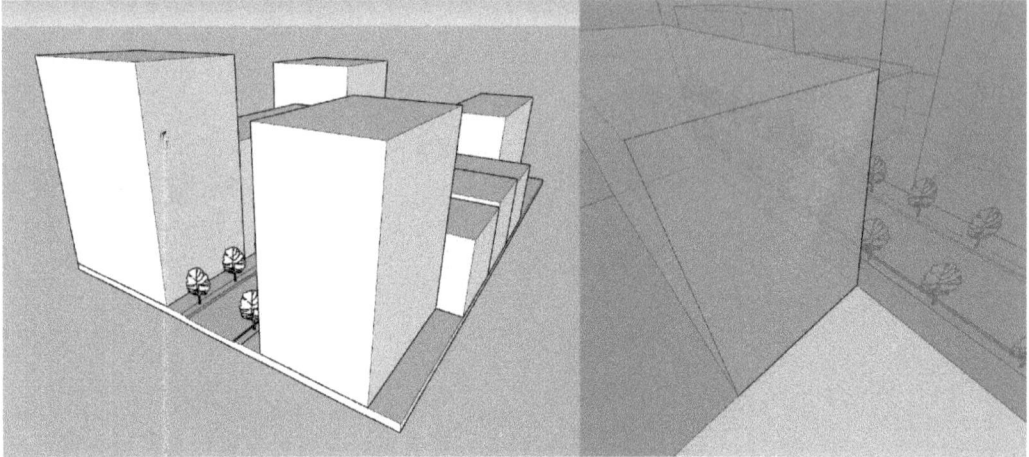

Figure 6.42 – The Orbit Tool Moves the Camera Behind (Left) or Inside of Buildings (Right)

The Walk Tool can instead be used to fly directly down the street. In the following example, we start at the ground level. To change the camera level, we could use the Position Camera with the Height Offset set to a much higher value than the standard Eye Height. However, because we don't know the exact height that we need, we can use the *Shift* modifier key to first move the camera vertically by clicking and dragging up the screen with the left mouse button:

Figure 6.43 – The camera at the Ground Level (Left); Using the SHIFT
Modifier Key to Move the Camera Vertically (Right)

Now that the camera is at an appropriate height for this example, we can release all modifier keys and start to move the camera down the street:

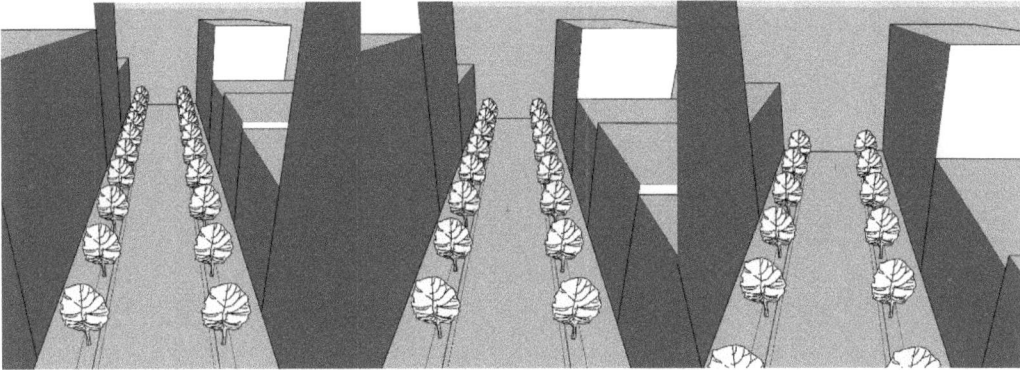

Figure 6.44 – Camera Moving Down the Street with the Standard Walk Tool

> **Note**
> Remember – if you want to move quickly, use the *Alt* modifier key to activate the Run option! Running is faster than Walking, but it is also harder to control. So, be careful!

Field of View

The Field of View tool was added to the Camera section of the Large Tool Set Toolbar in SketchUp Pro 2023, but it cannot be found on the Camera Toolbar. We discussed the Field of View Tool in *Chapter 1, Beginning with SketchUp Pro*.

Using a combination of the Position Camera, Look Around, and Walk Tools will allow you to explore your model in camera views that represent real-world viewing angles. This can be very helpful when working in enclosed interior spaces or exterior spaces.

Summary

This chapter is the first step toward understanding how to view your SketchUp model. You will most likely spend most of your time in the Perspective view rather than the Two-Point Perspective or Parallel Projection view, but there are some specific instances where different view options are necessary. Becoming familiar with the Camera Tools is essential – especially Orbit, Pan, and Zoom. While Position Camera, Look Around, and Walk may not be used as often as the main Camera Tools, they can also be extremely helpful when setting up views for Scenes.

We talked about setting up views for modeling and exporting in this chapter a lot, and we will continue with those ideas in the next chapter. In the next chapter, we will discuss the View options in SketchUp Pro.

7
View Options

In this chapter, we will examine the various options to change the view in SketchUp Pro. We will begin with Styles, including the Edge, Face, and Component Styles. Then, we will discuss Shadows and Fog, including how this can be set by a Geo-location. Styles can be edited to have any combination of these settings, and we will look at editing workflows for Styles.

We will also discuss sections that allow for dynamic views of a SketchUp model. Finally, we will review how to save these view options into snapshots, known as Scenes, which can be animated!

The following topics are covered in this chapter:

- Styles:
  - Edge Styles
  - Face Styles
  - Component Styles
  - Creating and Editing Styles
- Shadows and Fog
- Sections
- Scenes:
  - Animation

Styles

Styles in SketchUp Pro refer to a collection of display settings and options. These display settings primarily focus on Edges and Faces, and through built-in templates and custom settings, SketchUp can achieve hand-sketched, painted, technically drafted, or even semi-realistic appearances for a SketchUp model. These Styles are not post-production rendering – they are the visual representation of the live SketchUp model!

Many of the Style options can be found in the **View** dropdown on the Menu Bar:

Figure 7.1 – The View dropdown on the Menu Bar

Certain Style options also are found on individual toolbars (such as the **Styles** toolbar) and in the Style panel, which can be docked in the **Default Tray** panel:

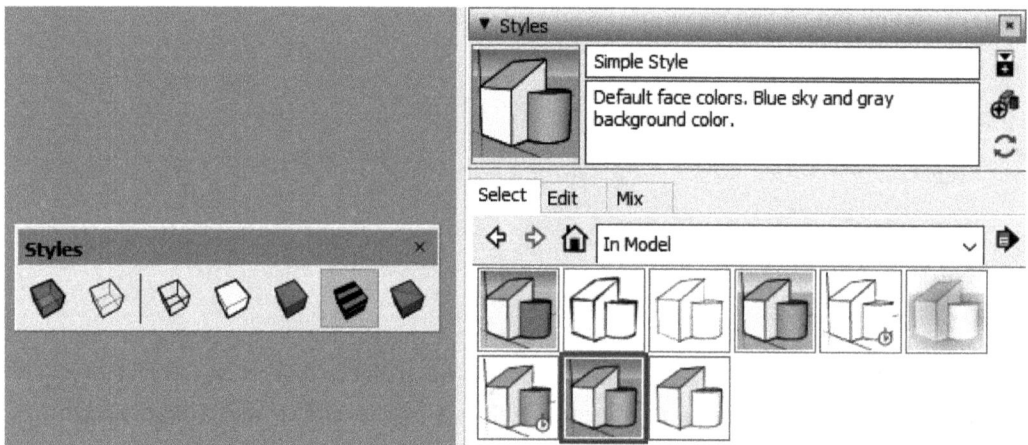

Figure 7.2 – The Styles Toolbar (Left) and the Styles Panel in the Default Tray (Right)

We will not discuss the Style Builder in this book, but we will discuss how to edit Styles in the *Creating and editing Styles* section of this chapter. This chapter will focus on the basics of Styles and will expand on how to further edit and save our SketchUp views with Sections, Scenes, and Animation.

Edge Styles

As mentioned in this section's introduction, the Edge and Face Styles are what make SketchUp unique and recognizable. We have looked at how SketchUp can create precise, rigid models that can be used for manufacturing or construction documents in the previous chapters. However, these drawings can often seem very formal and confusing for some people to understand, and a sketch of the project might be the perfect view for them! When we sketch with a pen or pencil, the lines are never perfectly straight; they may overlap or have darker endpoints. Let's look at the following figure as an example:

Figure 7.3 – Default SketchUp Style (Left) Pencil Sketch SketchUp Style (Right)

In the figure on the left, we can recognize the Geometry as being computer-generated. The lines point exactly to the vanishing points, the corners are all crisp and connected, and the lines are all a consistent color. In the figure on the right, the drawing looks more like a traditional pencil and paper sketch. The lines wobble, the corners overlap in some cases and do not connect in others, and the lines have varied thicknesses and colors. When we zoom into the figures, we can see this in even more detail:

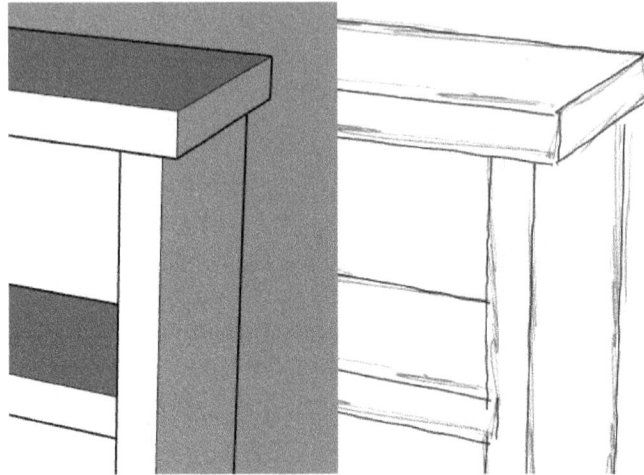

Figure 7.4 – Default SketchUp Style Detail (Left) Pencil Sketch SketchUp Style Detail (Right)

Before we can begin to edit or create custom Edge Styles like the Style we just looked at, we need to understand some of the basic options of Edge Styles. All of these **Edge Style** options can be toggled in the **View** | **Edge Style** dropdown in the Menu Bar:

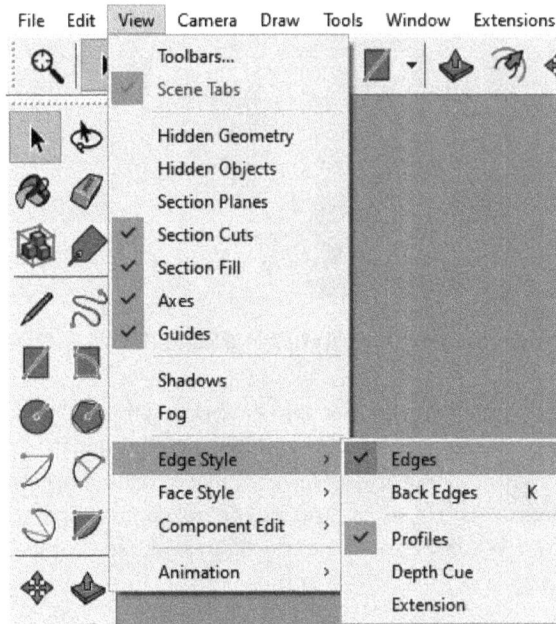

Figure 7.5 – Menu Bar Dropdown for Edge Styles

Let's look at the Edge Style options in detail.

Edges

All Edge Style options can be toggled, even the Edges themselves! If all Edge Style options are toggled off, including the **Edges** option, the model will look like it has no Edges. Of course, we can assume that there are Edges in the model if we can see Faces:

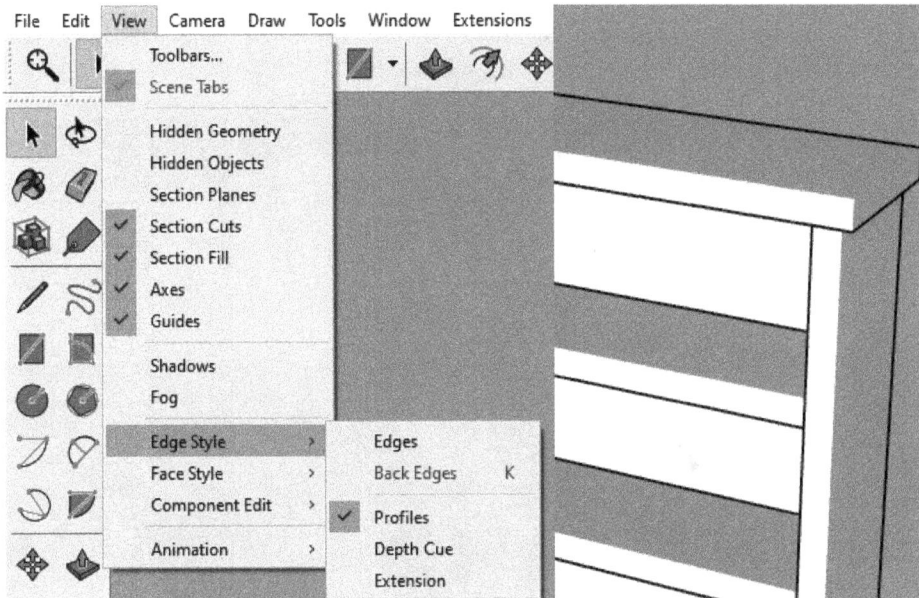

Figure 7.6 – Edges Toggled off in the View | Edge Style | Edges

The Edges can still be selected and edited even while toggled off. Changing the Style only updates the visual appearance of the Geometry; it does not hide or lock the Geometry:

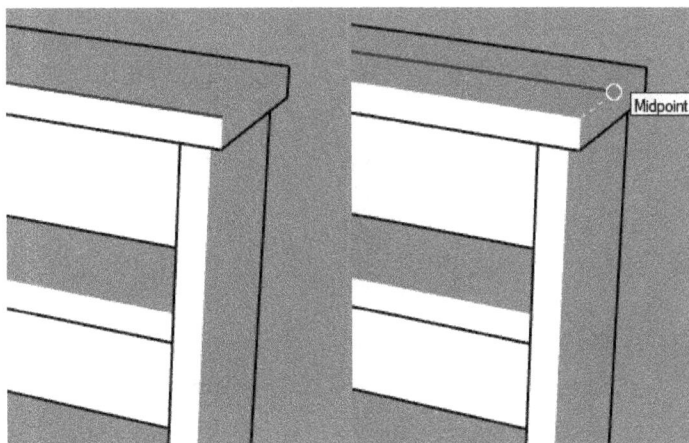

Figure 7.7 – Invisible Edge Selected (Left) Invisible Edge Copied (Right)

Turning off Edges in the **Edge Style** menu may be helpful when creating renderings or graphics, but it is not advised while modeling. Remember that it is often helpful to hide specific Geometry while modeling.

Back Edges

Back Edges are the Edges that are hidden from the view of the camera by Faces. Back Edges can be either partially or fully covered by a Face or multiple Faces:

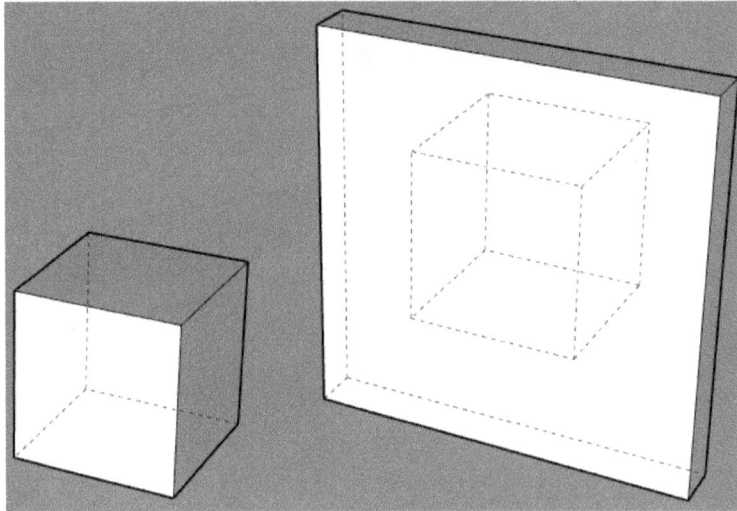

Figure 7.8 – Back Edges of a cube (Left); Back Edges partially hidden by other Geometry (Right)

Back Edges act differently than the X-ray Face Style because they appear as dashed for the Edges that are not currently visible. In the X-ray Face Style, all Edges appear to be solid, and it can be more helpful to see the Back Edges as dashed:

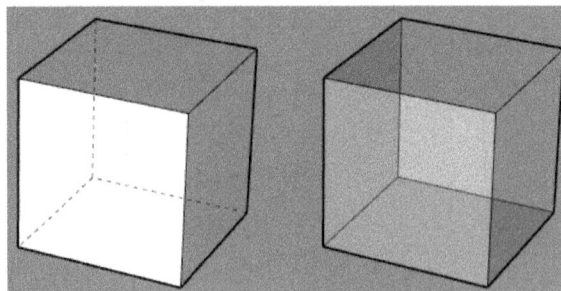

Figure 7.9 – Back Edges (Left); X-ray Face Style (Right)

Back Edges have an advantage over the X-ray Face Style because they retain full-color Face materials. We will talk more about the X-ray Face Style and some of its benefits in the *X-ray* section of this chapter.

> **Note**
> Back Edges and the X-ray Face Style can't be on at the same time – activating one of these options automatically deactivates the other. It is up to you to find the right time to use each option!

Profiles

Profiles are one of SketchUp's most distinctive Edge Style options. Profiles create a bolder line around connected Geometry, including Grouped and Ungrouped Geometry:

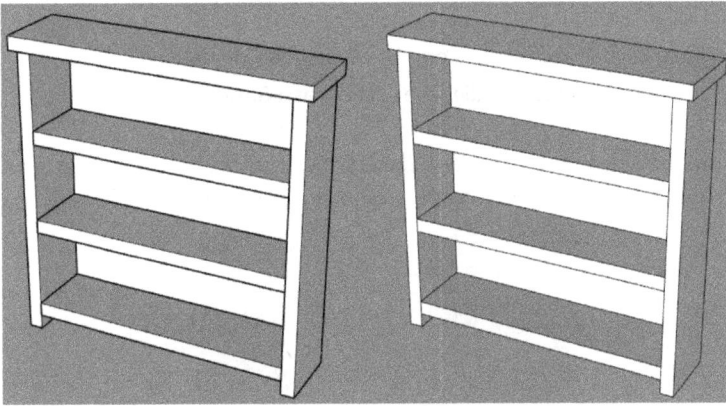

Figure 7.10 – Profiles On (Left) and Profiles Off (Right)

SketchUp automatically assigns the bold highlight to the Edges that create the Geometry's outline, and these Edges will be updated when the camera angle changes:

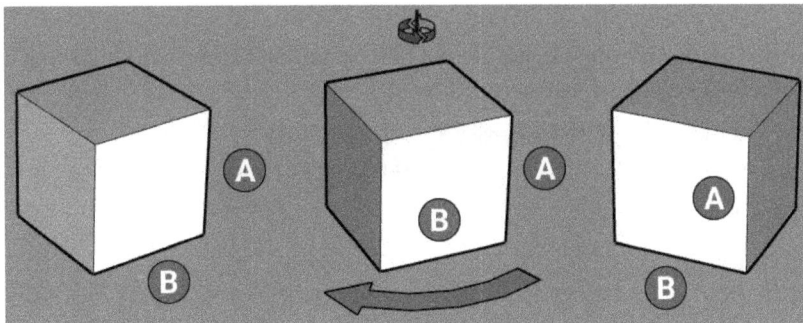

Figure 7.11 – Edges A and B are in the Profile (Left); Orbit has been
used (Middle); Edge A is no longer a Profile (Right)

Profiles help us quickly identify what Geometry is connected, which quickly helps us identify Grouped Geometry and establish depth perception in a model. In the following figure, we can quickly tell that the cylinder on the right is Grouped separately from the box because it has a profile along its bottom Edge:

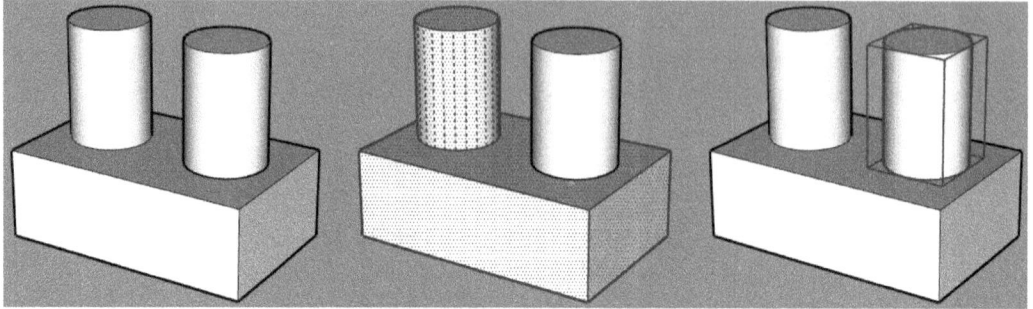

Figure 7.12 – Two Cylinders on a Box (Left); the Cylinder and Box are
Connected (Middle); the Cylinder as a Group (Right)

Profiles are not additional Edges in the model – SketchUp assigns the bold highlight to existing Edges. This is why Profiles occasionally extend into the Geometry, as seen in the following figure:

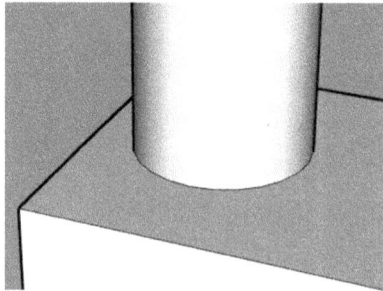

Figure 7.13 – The Profile Extends into the Connected Geometry

Profiles can be used to identify when Edges are not fully bounding Faces. In the following figure, we can see a rectangle with an Edge drawn across it:

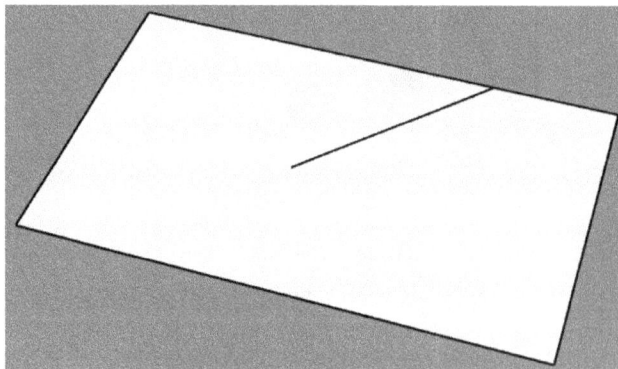

Figure 7.14 – Rectangle with a "Hanging" Edge

The Edge does not go fully across the rectangle, so it is represented as a bold Profile. This is because the Edge is "hanging" and is not defining the boundary of a Face. If the Edge is connected to the other side of the rectangle, the Edges have split the rectangular Face in half, creating two new Faces. Both Edges will be thin as they are not Profiles – they are interior Edges to the Geometry:

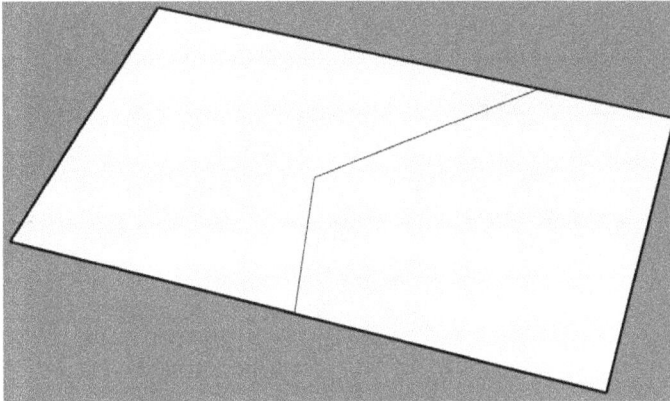

Figure 7.15 – The Edge is Connected to the Rectangle with the Line Tool and the Edges are Thin

This is a great way to find "hanging" Geometry in your SketchUp models as you begin to create more complex models.

Depth Cue

Depth Cue bolds Edges, depending on how close the Geometry is to the camera. Edges that are closer to the camera are bolder than Edges that are further away:

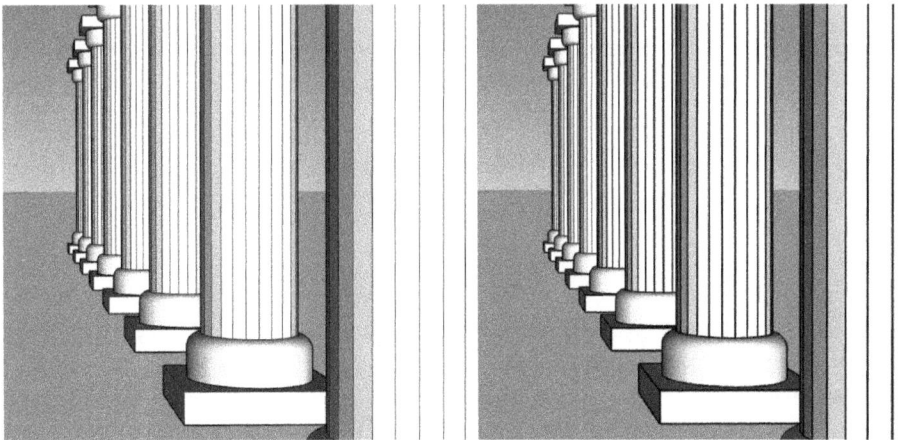

Figure 7.16 – Depth Cue Off (Left); Depth Cue On (Right)

Depth Cue can be used with the **Profile** option, but the Depth Cue may be bolder than the Profiles.

Extension

Extension mimics a hand drafting technique where the drafter would purposely extend and bold the ends of line segments to create more impactful corner conditions. This helps to highlight sharp corners in the drawing and ensures that the hand drafter can create strong, connected lines. SketchUp includes the **Extension** option for creating a similar effect. The lines are not extended past the corners; they are only visually represented that way. This may be confusing to new SketchUp Pro users!

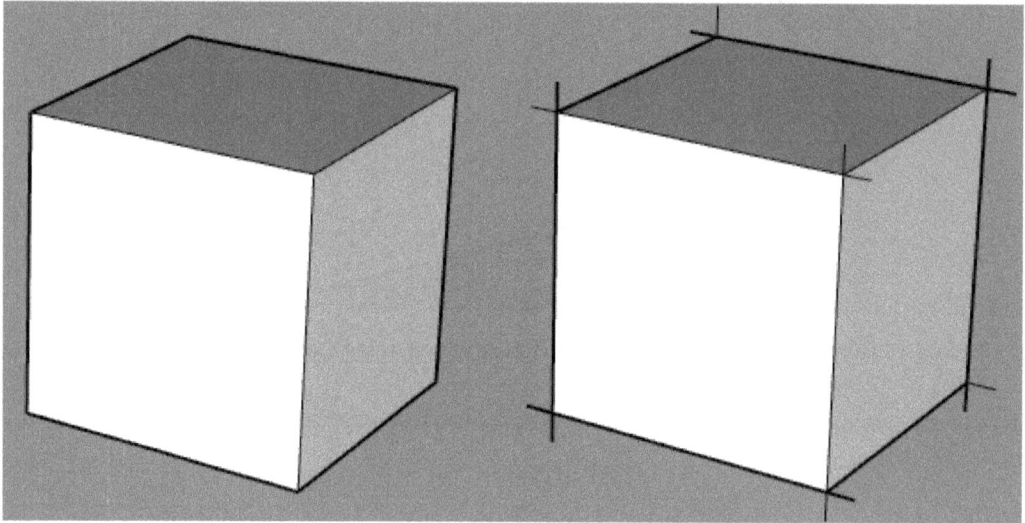

Figure 7.17 – Extension Off (Left); Extension On (Right)

> **Note**
>
> The Extension is relative to the screen size, not to the length of the line or location in the model. All Extension lines will appear to be the same length as represented by the total number of pixels.

As with all Edge options, detailed settings will be explored in a later chapter.

Face Styles

The Face Styles in SketchUp complement the Edge Styles in giving SketchUp its unique look and feel. These Face Styles are often toggled during modeling to help us understand complex Geometry, as well as create unique designs for exporting renderings and images. Faces cannot be toggled off like Edges, but the **Wireframe** option is essentially the same thing. The **Face Style** options can be toggled in the **View | Face Style** dropdown in the Menu Bar:

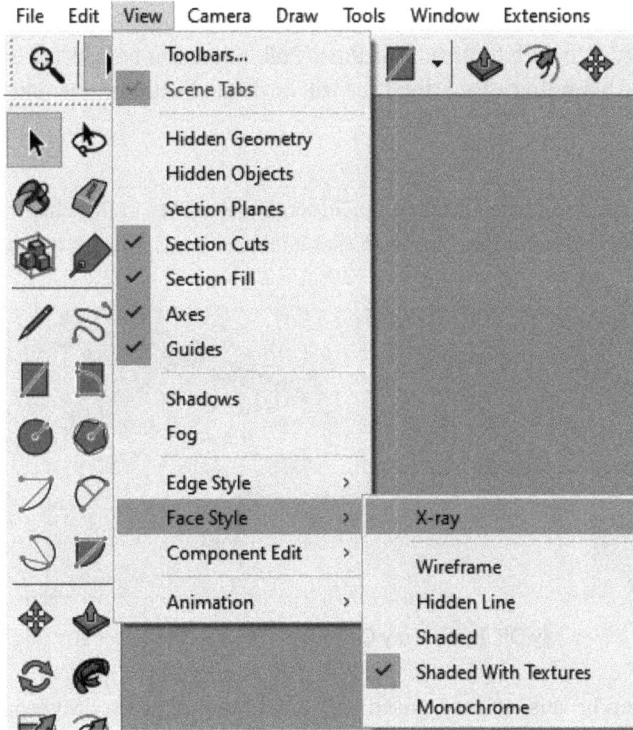

Figure 7.18 – Menu Bar Dropdown for Face Styles

Let's look at the Face Style options in detail.

Wireframe

Wireframe is not the first option in the dropdown, but we will discuss it first as it is the most unique. Wireframe essentially hides the Faces in the model, which allows for all Edges to be shown completely. When Wireframe is toggled on, no Faces can be selected. Additionally, toggling on **Hidden Geometry** does not show the Faces as being hidden; they are completely turned off:

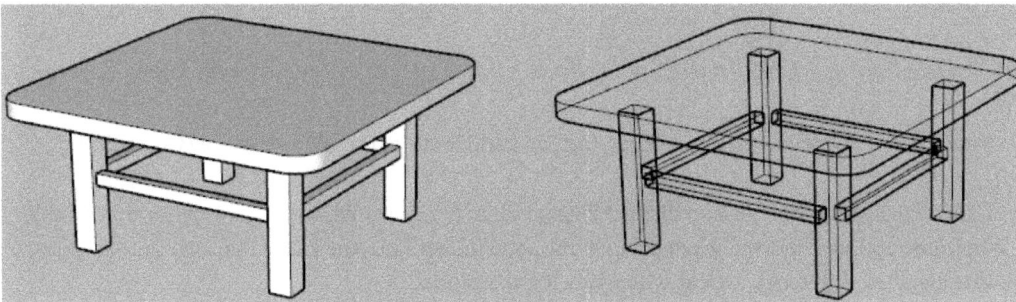

Figure 7.19 – Shaded Face Style (Left); Wireframe Face Style (Right)

Wireframe is rarely used in modeling except in very specific workflows. Faces cannot be manipulated while using Wireframe, even with tools such as Push/Pull. Edges can be selected easier, but if Edges are deleted that may be bounding a Face, the Face will be deleted without warning.

X-ray

We mentioned the X-ray option when discussing Hidden Edges earlier in this chapter. X-ray performs a similar function to the Hidden Edges option in that it allows the user to see through the Geometry in the model:

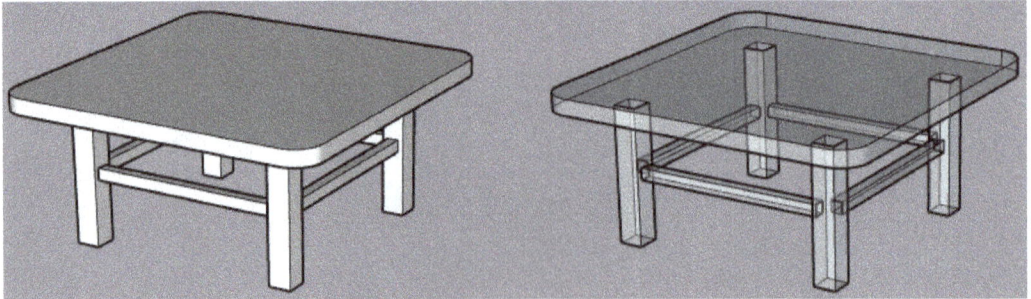

Figure 7.20 – X-ray Off (Left); X-ray On to See Through the Geometry (Right)

X-ray is a unique option because it can be used with any of the other Face Style options, except for Wireframe as wireframe completely hides the Faces.

X-ray dims the appearance of the Faces in the model, including any materials or textures on Faces. This is a visual reminder that X-ray is on:

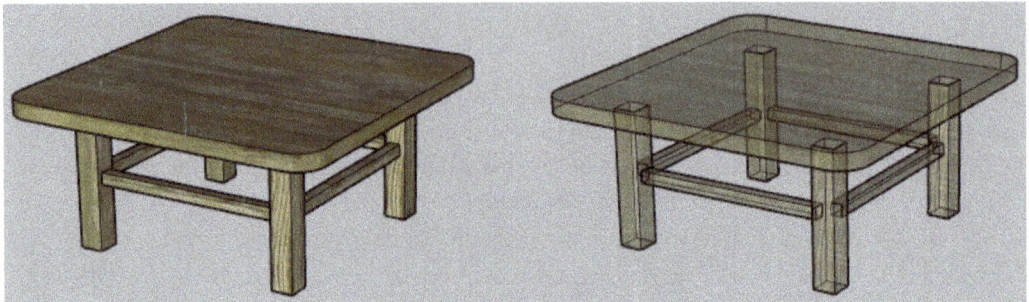

Figure 7.21 – Shaded with Texture (Left); Materials and Textures Dimmed with X-ray On (Right)

The Hidden Line option has the least distinct change when X-ray is used, and it may appear to resemble the Wireframe option. However, when using X-ray and Hidden Line, the Edges that are covered appear to be dimmer. The difference is clear when viewing the model:

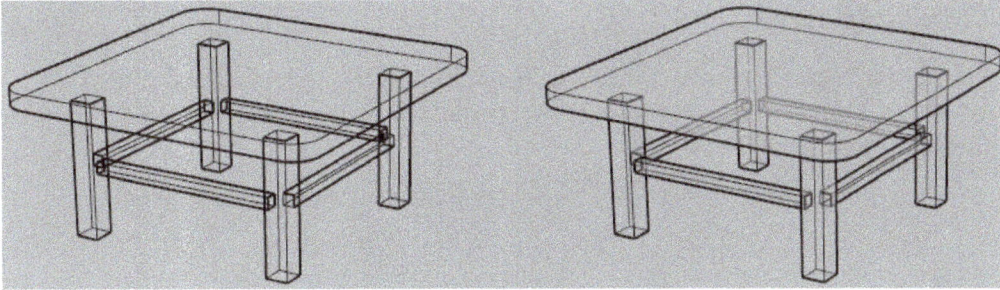

Figure 7.22 – Wireframe (Left); Hidden Line and X-ray Show Dimmer Edges (Right)

Viewing the model from below allows the Hidden Line color to be contrasted with the sky color:

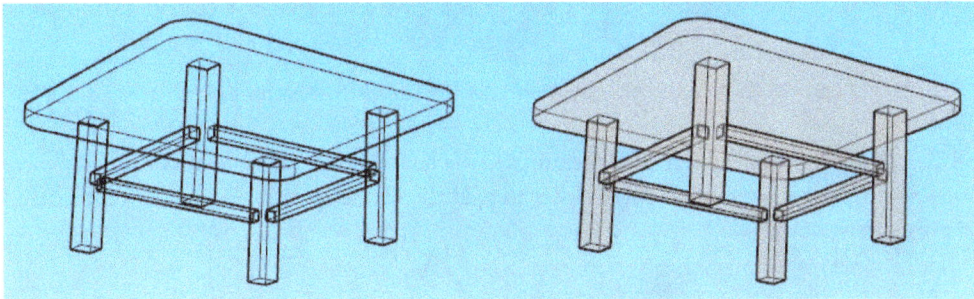

Figure 7.23 – Wireframe (Left); Hidden Line and X-ray (Right)

Hidden Line

Hidden Line uses the current Style background color to replace the Face color for all Geometry. Transparent materials will have their colors updated but will retain their transparency:

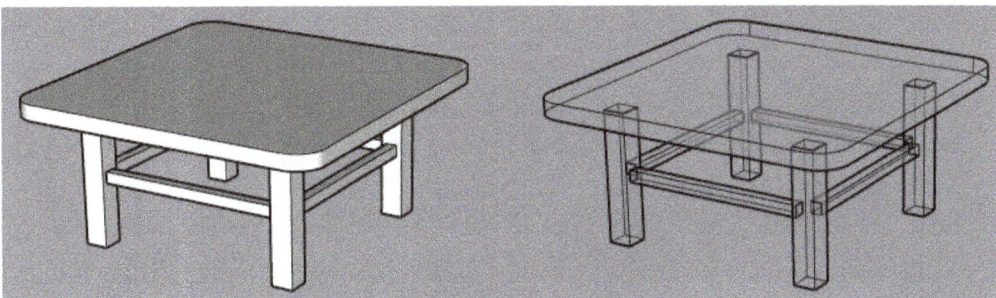

Figure 7.24 – Shaded (Left); Hidden Line (Right)

As mentioned before, Hidden Line will render the background color even when the Geometry is viewed against the sky. The background color can be changed in the Style panel in the Modeling tab, which we will discuss in the *Creating and Editing Styles* section of this chapter.

Shaded with Texture

The Shaded with Texture Face Style is the most common and recognizable Face Style in SketchUp. The default SketchUp colors for new Faces (white/gray and blue/gray) are recognizable in the Shaded with Texture option:

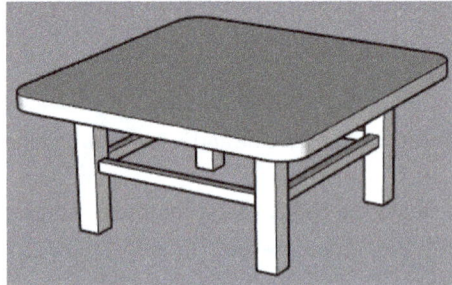

Figure 7.25 – Shaded Face Style with Default SketchUp Material Colors

We will discuss how to apply materials and textures in the next chapter. Textures can be used to move the visual appearance of a model one step closer to realistic, whereas materials are often considered to be representative colors:

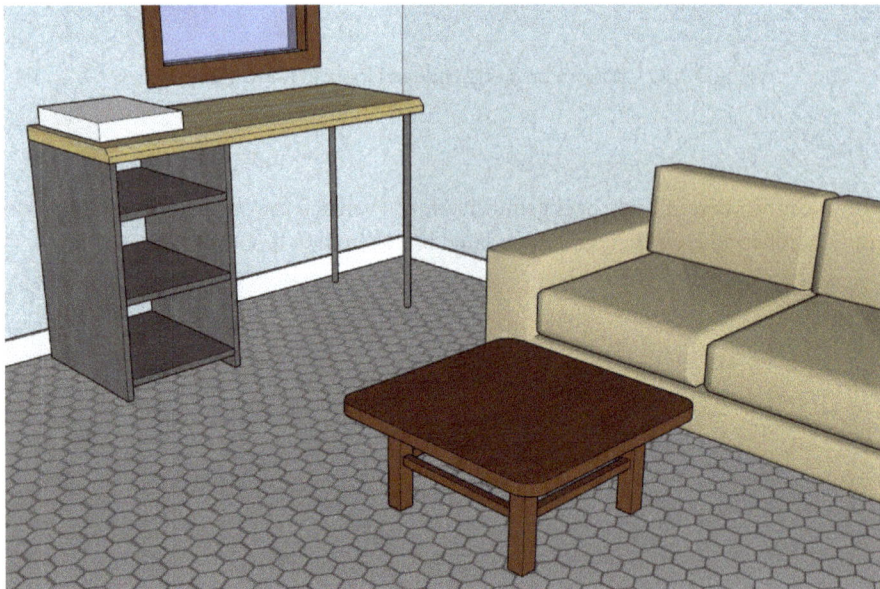

Figure 7.26 – Shaded with Texture Face Style used to Identify Geometry

When using the Shaded with Texture option, the SketchUp Geometry can be represented in several ways, which can help us immediately identify the Geometry.

Shaded

The Shaded Face Style simplifies the model's appearance when textures are applied to Faces. The Shaded Face Style renders all faces as solid colors, even while using textures. SketchUp identifies the overall color appearance of the texture and uses that color as a substitute:

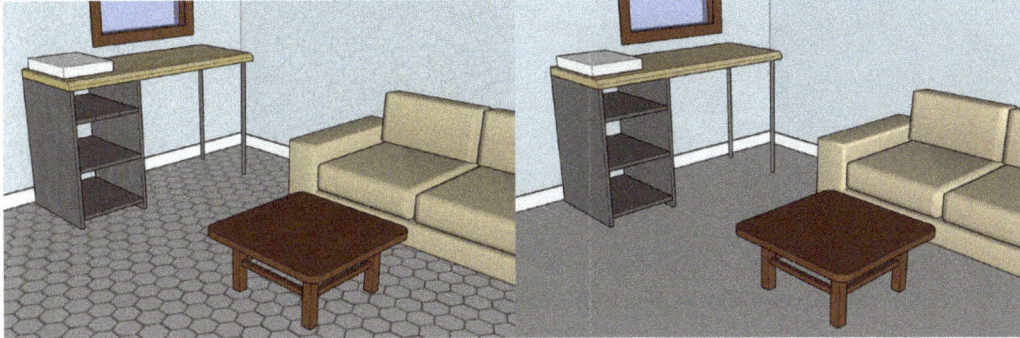

Figure 7.27 –Shaded with Texture Face Style (Left); Shaded Face Style Represents Textures as Colors (Right)

The Shaded Face Style can be used occasionally to simplify a model for image export, but it is always recommended to work in the Shaded with Texture option when applying materials and textures. It can be easy to apply a texture in the Shaded Face Style and have it represented as a color; when the Shaded with Texture Face Style is toggled on, the texture may have been used incorrectly. In the following figure, we can see an exaggerated example of this dilemma:

Figure 7.28 – Shaded (Left); Shaded with Texture (Right)

> **Note**
>
> Sorry to have to show you that! All of the Textures used are included with SketchUp Pro: Mosaic Hexagonal Tile, Water Pool Light, White Subway Tile, Brick Antique 01, Brick Tumbled, Fencing Lattice Natural, Polished Concrete New, Roofing Scalloped, and Vegetation Bark Maple.

Monochrome

The Monochrome Face Style renders all materials and textures as the default SketchUp material. This can help us view the Geometry as a uniform material, as it can help us identify flipped Geometry:

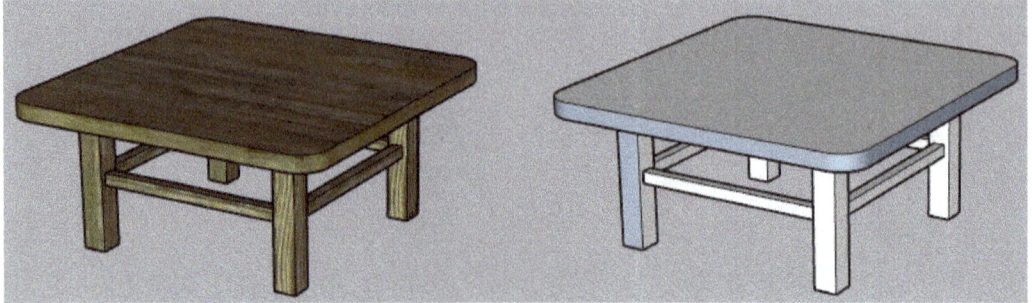

Figure 7.29 – Shaded with Texture (Left); Monochrome Showing Flipped Faces (Right)

The Monochrome Face Style can also help us export images – either to sketch over by hand or in a photo editing software or to simply see the model as pure Geometry:

Figure 7.30 – Room Rendered as Shaded (Left); Room Rendered as Monochrome (Right)

Component Styles

The Component Style options can be used while editing a model. These Component Styles can be found in the **View** dropdown in the Main Menu:

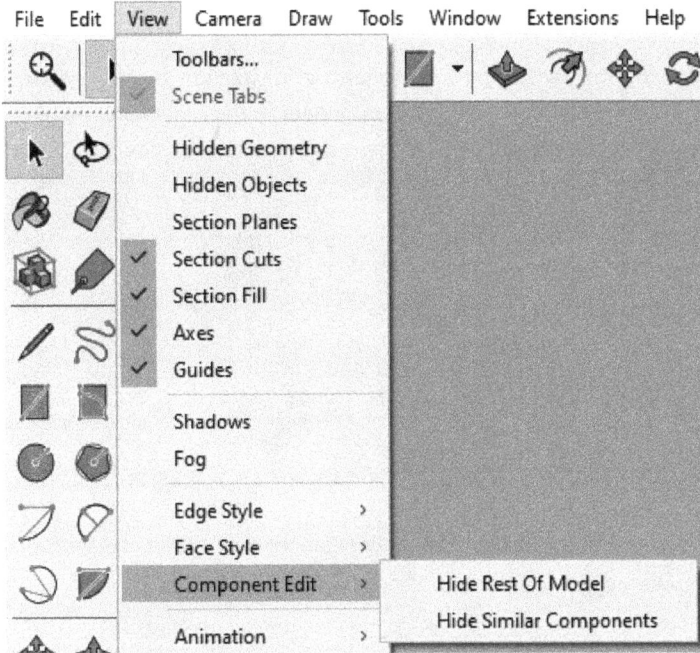

Figure 7.31 – Component Dropdown in the Main Menu

Let's look at the options under **Component Edit** in detail.

Hide Rest of Model

The **Hide Rest of Model** option will completely hide all other Geometry when the Component is edited. Other instances of the same Component will remain visible as grayed-out objects. The **Hide Rest of Model** view option will work when editing Groups or Components, although copied Groups will be completely hidden as Groups are not instances:

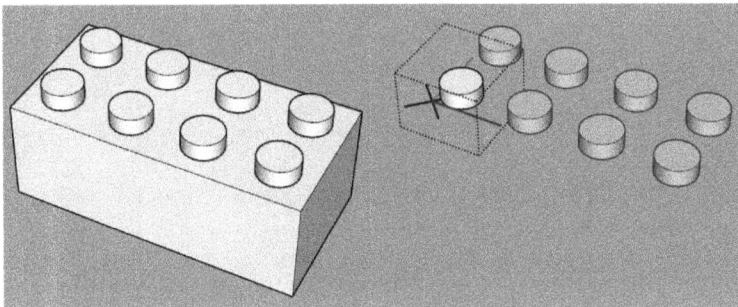

Figure 7.32 – Object with Cylinder Components (Left); Cylinder
Component Edited and other Geometry Hidden (Right)

Hide Similar Components

Hide Similar Components only works on Components as it isolates Component instances. When a Component is edited, all other instances of the Component in the model are hidden:

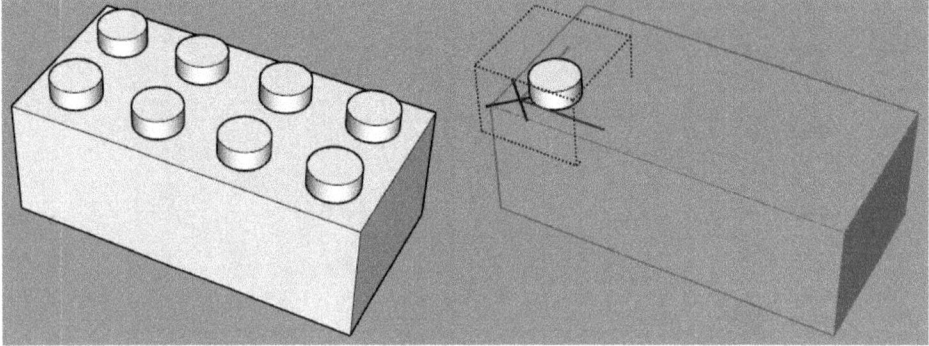

Figure 7.33 – Objects with Cylinder Components (Left); Cylinder
Component Edited and Component Instances Hidden (Right)

This workflow can be especially helpful when viewing many Component instances in sequence, such as a railing. In the following example, we can see that many Components make up the vertical balusters in the railing. Only one Component needs to be edited to update all Component instances, but it is hard to see the baluster from the side to complete the workflow:

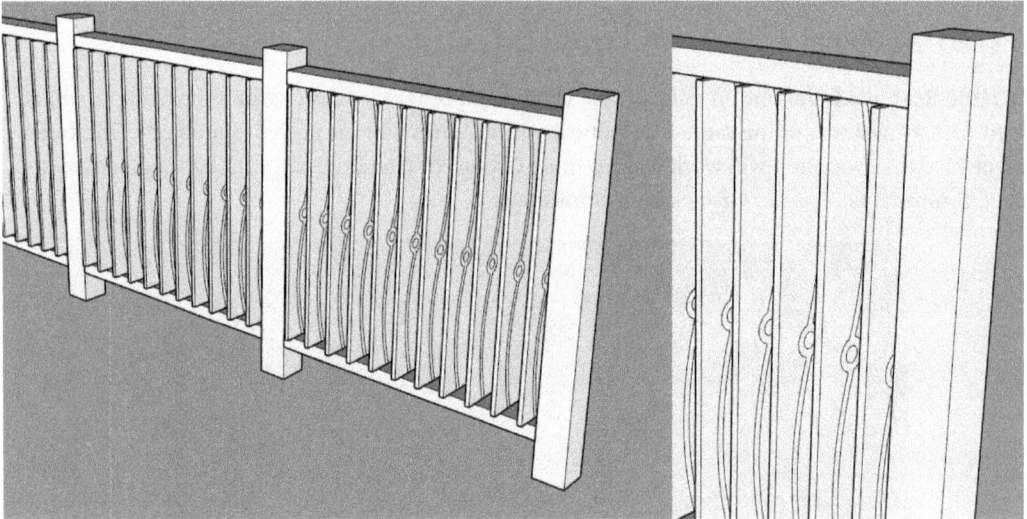

Figure 7.34 – Railing with Vertical Balusters (Left); Difficult Camera Angle to Edit (Right)

With **Hide Similar Components** toggled on, one Component is isolated and the edit can be performed with an easier camera angle:

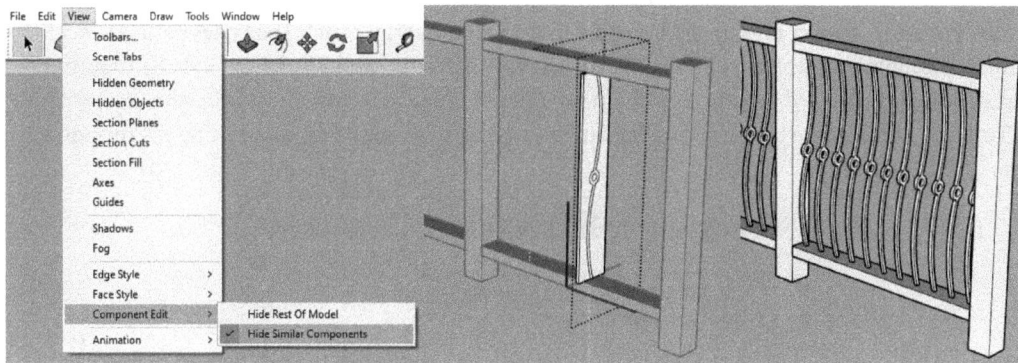

Figure 7.35 – Hide Similar Components Toggled On (Left); Edit
Completed (Middle); Finished Component (Right)

Hide Rest of Model and **Hide Similar Components** can be used together to completely hide all Geometry in a model, including all Component instances:

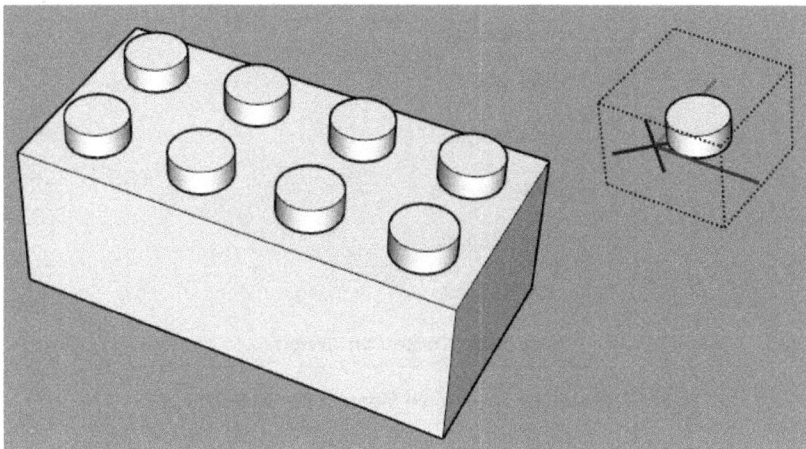

Figure 7.36 – Objects with Cylinder Components (Left); Hide Rest of
Model and Hide Similar Components Used Together (Right)

The Style options are a great way to change the way you can interact with the Geometry while modeling and exporting your SketchUp Pro models.

Creating and Editing Styles

Now that we understand the settings that can be changed in the **View** dropdown in the Menu Bar, we can dive a little deeper into the Styles panel in the **Default Tray** panel. The Styles panel allows for unlimited combinations of Edge, Face, Background, Watermark, and Modeling Style Settings to be created, edited, and saved in your SketchUp models. To create a new Style, it is best to start with a similar Style. Styles come preloaded in SketchUp Pro, and they can be accessed using the dropdown menu in the **Styles** panel:

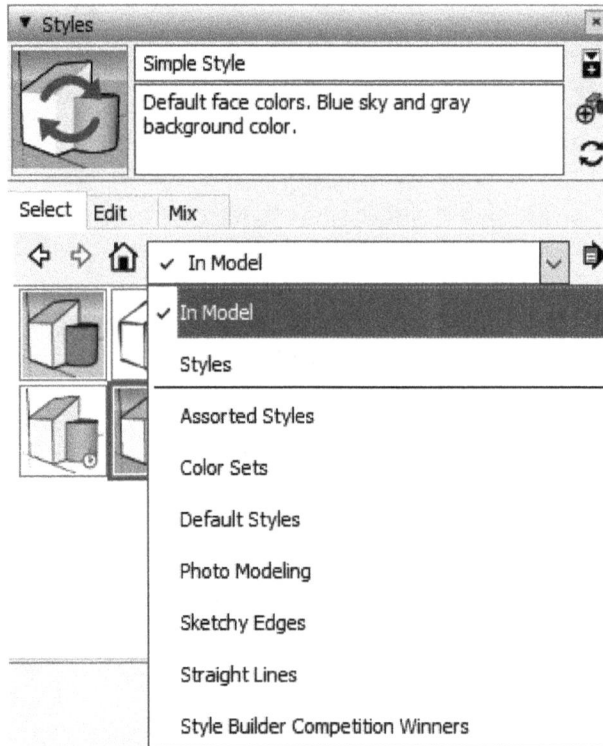

Figure 7.37 – Styles Dropdown Menu in the Styles Panel

Once a Style has been selected, it will be added to the **In Model** tab of the Styles panel, which can be selected in the dropdown or by clicking on the Home button next to the dropdown:

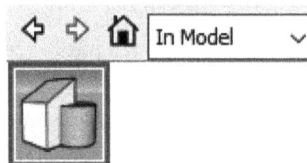

Figure 7.38 – The In Model Tab and Home Button

With the Style currently selected, the **New Style** button can be clicked. This will copy the current style and add it to the **In Model** tab. Now, this new Style can be Edited without the original Style being overwritten. Styles can be Edited by changing any of the Edge, Face, or Component settings in the **View** dropdown Menu. To update the current Style with these changes, you can click the **Update Style** button, which is located underneath the **New Style** button in the Styles panel:

Figure 7.39 – The New Style Button and Update Style Button

More precise editing can also be done to Styles. The **Styles** panel has three tabs below the current Style image: **Select**, **Edit**, and **Mix**. The **Select** tab is the default tab and will allow you to choose one of the default or preloaded Styles. The **Edit** tab is where most of the editing can be done to the current Style, including more options than are available in the **View** dropdown on the Menu Bar:

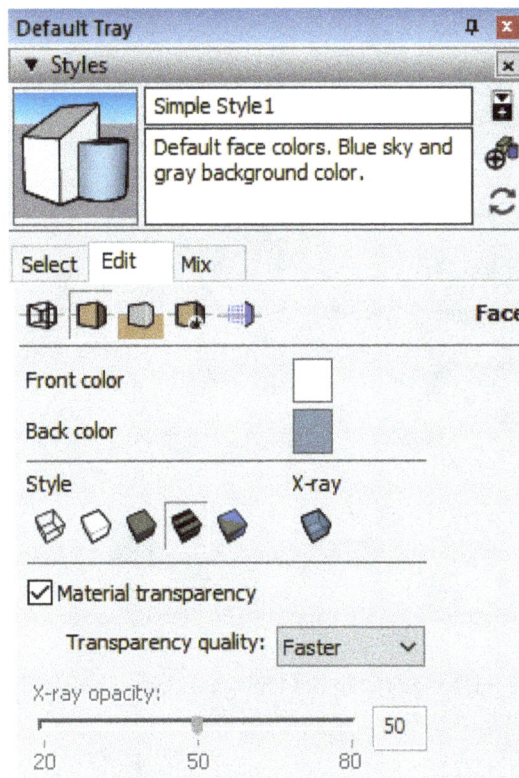

Figure 7.40 – Styles Panel Edit Tab for the Current Style

When the **Edit** tab is selected, there will be five icons at the top of the tab. These icons represent the five Style Settings groups: Edge, Face, Background, Watermark, and Modeling:

Figure 7.41 – The five Style Settings Groups – Edge, Face, Background,
Watermark, and Modeling (Left to Right)

Each of these tabs has different options for editing parts of the Style. Edge and Face are the most familiar tabs, containing many of the options found in the **View** Menu Bar dropdown. However, more advanced settings are also available. We will not discuss every setting at length in this chapter, but you can explore the options in depth in your own models!

The Edge Settings include the options from the **View** dropdown and include the option to tweak the values for **Profiles**, **Depth Cue**, **Extension**, and **Endpoints**:

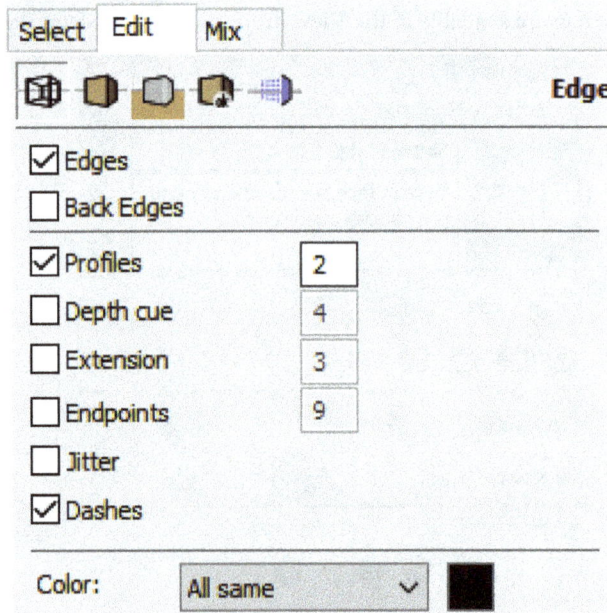

Figure 7.42 – The Edge Style Edit Tab Options

The Face Settings also include the options from the **View** dropdown represented in button form. These are the same buttons that can be found on the Styles toolbar:

Figure 7.43 – The Face Style Edit Tab Options

Importantly, the Default Material Front and Back Colors can be updated in the Face Settings. These colors can be updated by clicking on one of the colored boxes next to the option name. When the colored box is clicked, SketchUp will open a **Choose Color** Dialog Box. A color can be chosen from **Color Wheel**, **HSL** (Hue, Lightness, Saturation), **HSB** (Hue, Saturation, Brightness), or **RGB** (Red, Green, Blue):

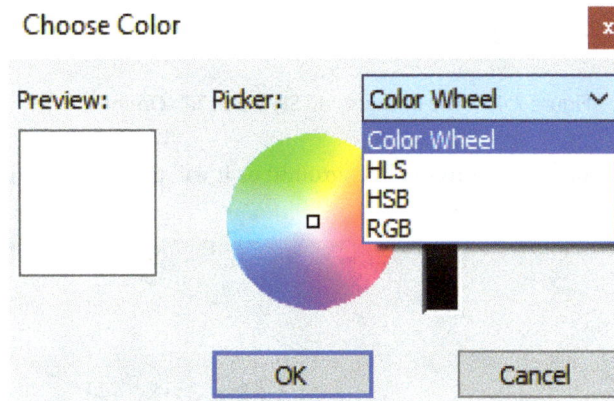

Figure 7.44 – The Choose Color Dialog Box with the Dropdown Open

Once a new color has been chosen, it will update the color in the Style. The Undo and Redo tools do not remember changing colors in the Styles Edit tab, so it is suggested that you change these colors in a copy of a Style, never in the original Style itself!

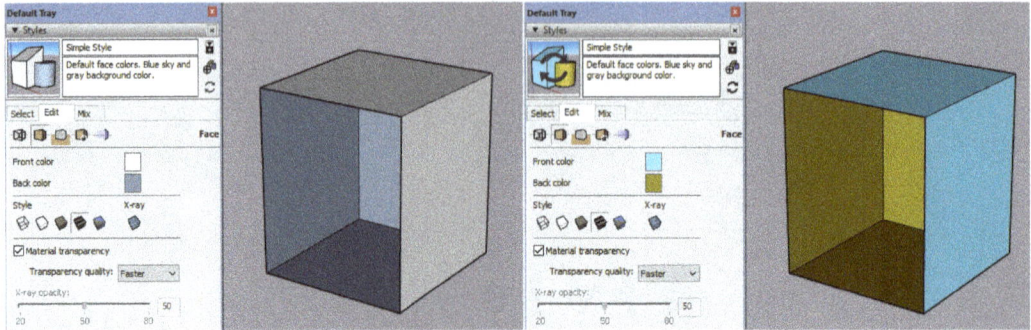

Figure 7.45 – Default Colors (Left); Updated Colors (Right)

The third section is the Background Settings. This section includes the color options for **Background**, **Sky**, and **Ground**:

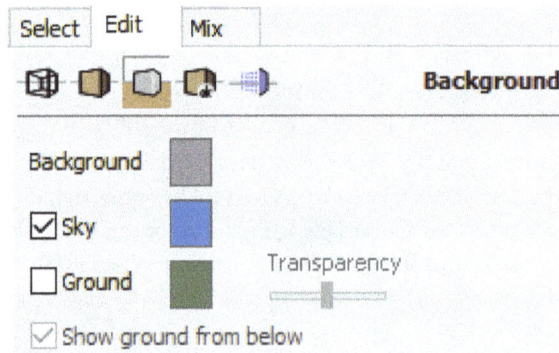

Figure 7.46 – The Background Style Edit Tab Options

The Ground is distinctly different from the Background as it will partially obscure Geometry that extends into the negative Blue direction:

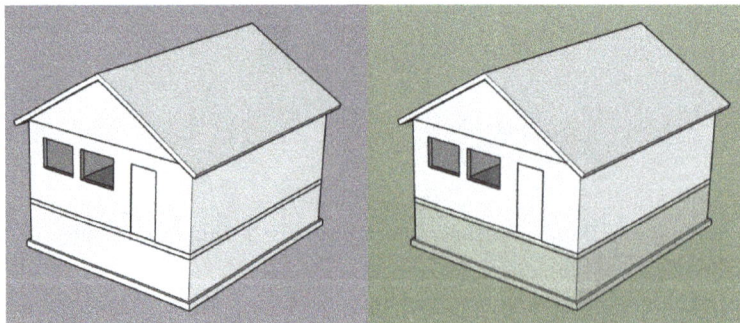

Figure 7.47 – Ground Off (Left); Ground On (Right)

The fourth section is the Watermark Settings. This option is a great way to add a background or overlay to the SketchUp model – especially when the Style will be used to export images or animations:

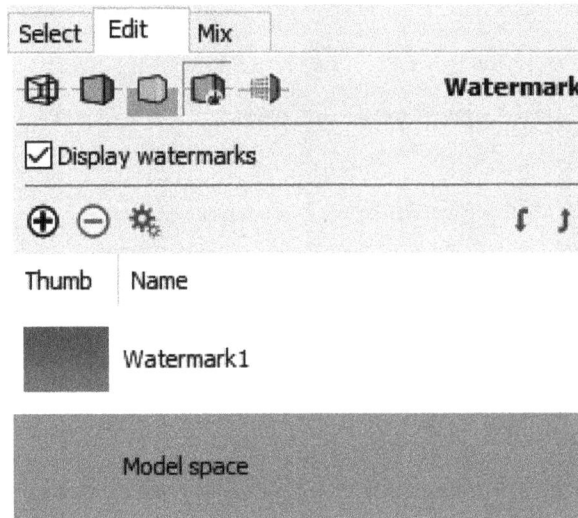

Figure 7.48 – The Watermark Style Edit Tab Options

Watermarks can be added by clicking the **Add Watermarks** button. Once clicked, the **Create Watermark** dialog box will walk through the process of creating a Watermark. Watermarks can be edited on each section of the **Create Watermark** dialog box, first to pick an image, then to choose a **Background** or **Overlay**:

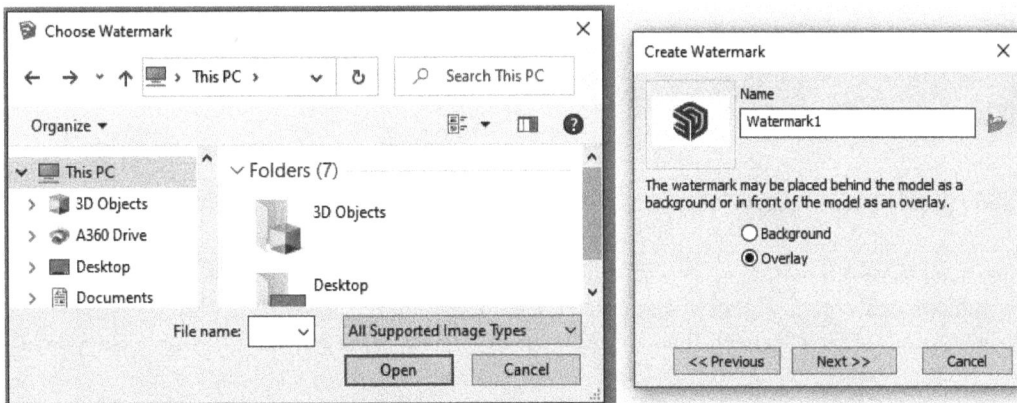

Figure 7.49 – Create Watermark Dialog Box Steps 1 and 2 (Left to Right)

The final two steps are to create a mask, if appropriate, then set the position of the image:

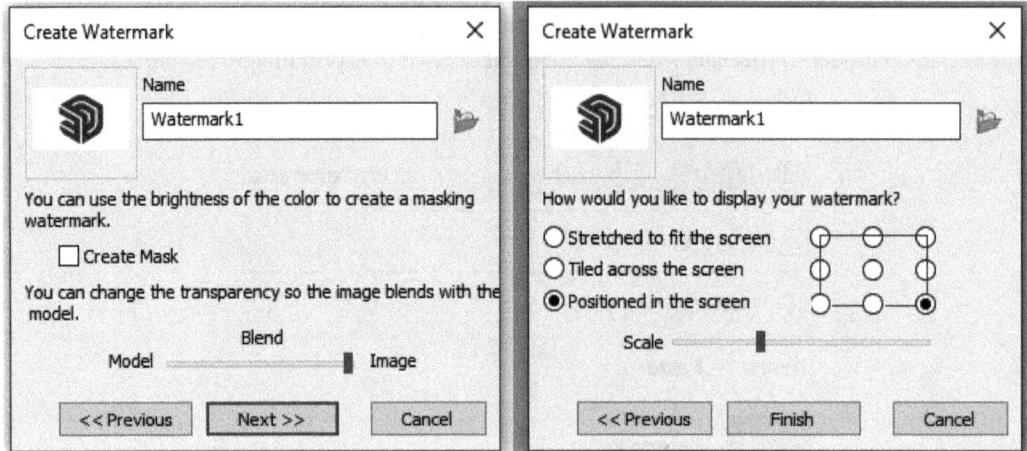

Figure 7.50 – Create Watermark Dialog Box Steps 3 and 4 (Left to Right)

In the following figure, a paper texture has been applied as an overlay to make the model appear as if it has been drawn on paper:

Figure 7.51 – Paper Texture Image (Left); Paper Texture Overlay Applied as a Watermark (Right)

In this example, we can see that a company logo has been included in the Style as a Background element in the bottom-right corner. This can be exported as part of the animation to reduce the time needed for post-processing, and to effectively brand an Animation video as being created by a certain company:

Figure 7.52 – Company Logo Image (Left); Company Logo Background Applied as a Watermark (Right)

In this final example, we can see that a black and white image has been used to create a watermark mask of the project. Masks can be made in any shape from any image, but in this example, it has been used to create a round vignette around the model:

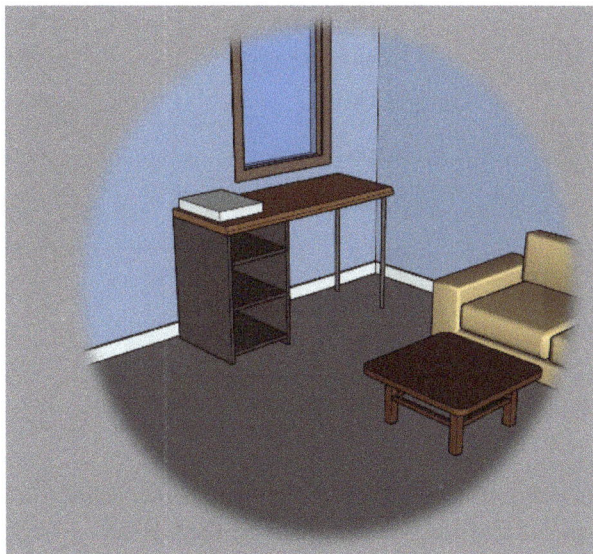

Figure 7.53 – Round mask image (left); round mask overlay applied as a mask Watermark (right)

The fifth and final section is the **Modeling Settings** section. This section includes all of the Style options for modeling elements, including the Selection colors, the Section colors, and the Guides colors.

> **Note**
>
> The Axes and Direction and Other Reference Colors can be updated in **Preferences | Accessibility**.

Additionally, this section includes the remainder of the settings found on the **View** dropdown in the Main Menu, as well as the Section options:

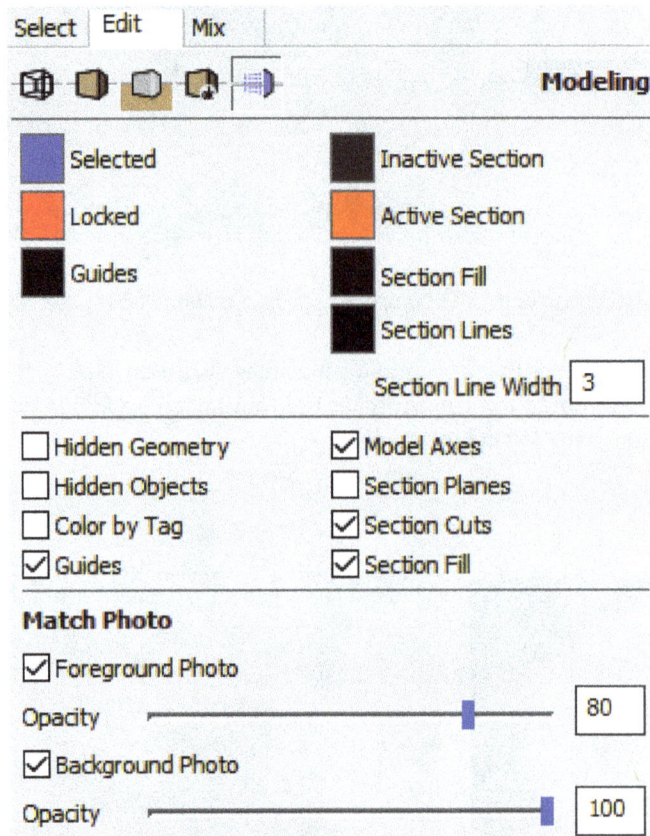

Figure 7.54 – The Modeling Style Edit Tab Options

These five Settings sections can be meticulously edited to create the perfect Style for any SketchUp model, but that process can be tedious. The **Mix** tab can be used to speed up this process, by allowing preloaded and already created Styles to be sampled to create new Styles. The process should still begin with the closest available Style being selected on the **Select** tab, and a copy of that Style being created with **Add Style**:

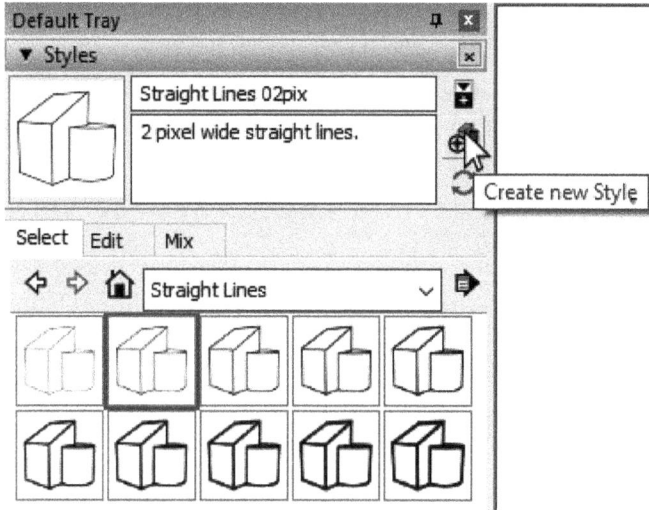

Figure 7.55 – Style Selected and Add Style on the Select Tab

Then, the **Mix** tab can be selected. This will split the Style panel into top and bottom sections, with the top section showing the five Style Edit Option tabs and the bottom showing a copy of the **Select** tab:

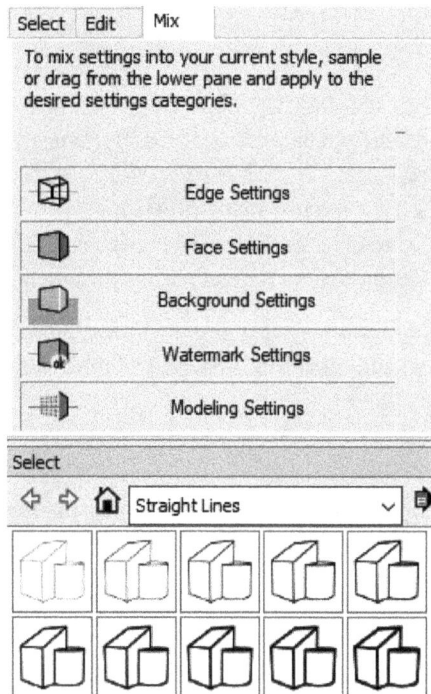

Figure 7.56 – The Mix Tab Selected Showing the Top and Bottom Sections

The top section is used to update the current style by sampling the Styles from the bottom section. This can be done by dragging and dropping the styles into the corresponding boxes above. Additionally, a Style can be clicked in the bottom section with the Eye Dropper icon; then, a Paint Bucket icon will appear when hovering over one of the Style Edit options in the top panel. By clicking on an option in the top panel with the Paint Bucket, the Style settings will be applied:

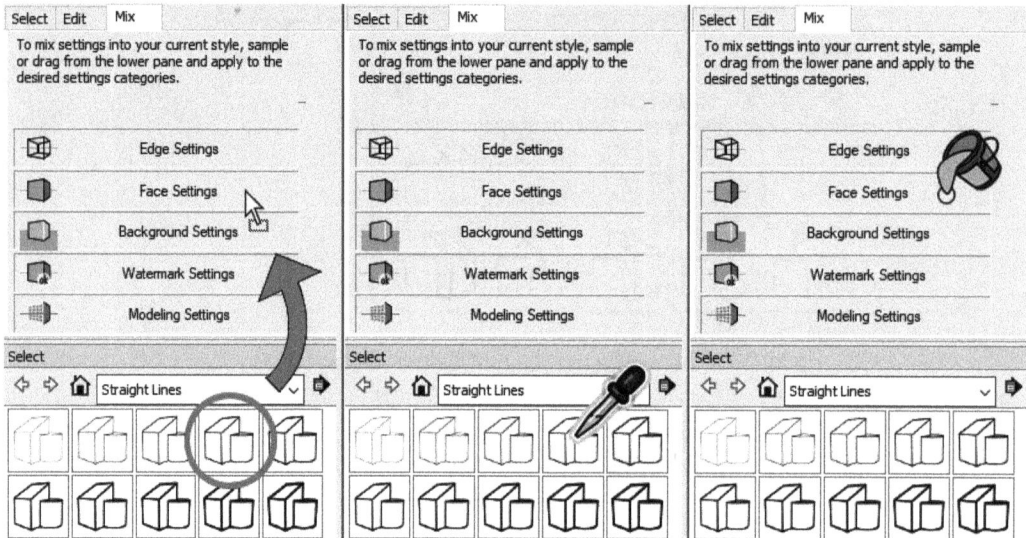

Figure 7.57 – Drag and Drop (Left); Eye Dropper Icon in the Bottom
Panel (Middle); Paint Bucket Icon in the Top Panel (Right)

The **Mix** tab can be used with the **Edit** tab to quickly experiment with combinations of Styles, and fine-tune those experiments to quickly and effectively create the perfect Style for your SketchUp model. Try different combinations for your model but remember to experiment with new styles!

In the next few sections, we will look at other settings that change the way your SketchUp model can be viewed – including saving views in Scenes. But first, we will look at Shadows and Fog.

Shadows and Fog

Shadows and **Fog** are two **View** options that can be toggled on in the Menu Bar dropdown:

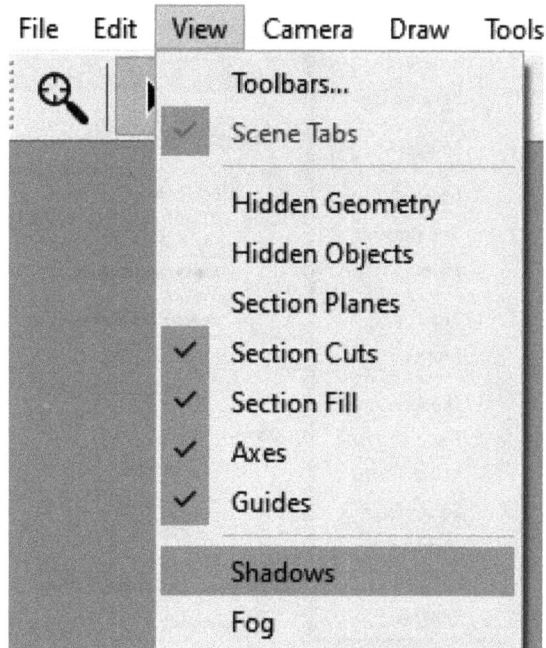

Figure 7.58 – Shadows and Fog in the Menu Bar Dropdown

While both of these options can be toggled in the **View** Menu Bar dropdown, additional **Shadows** options can be found in the **Shadows** toolbar. This toolbar is not shown by default and must be turned on by accessing **Views | Toolbars** and checking **Shadows**:

Figure 7.59 – The Shadows Toolbar

Additionally, **Shadows** and **Fog** both have panels that can be found in the **Default Tray** panel. These panels can be turned on by checking the respective boxes in **Window | Default Tray** in the Menu Bar:

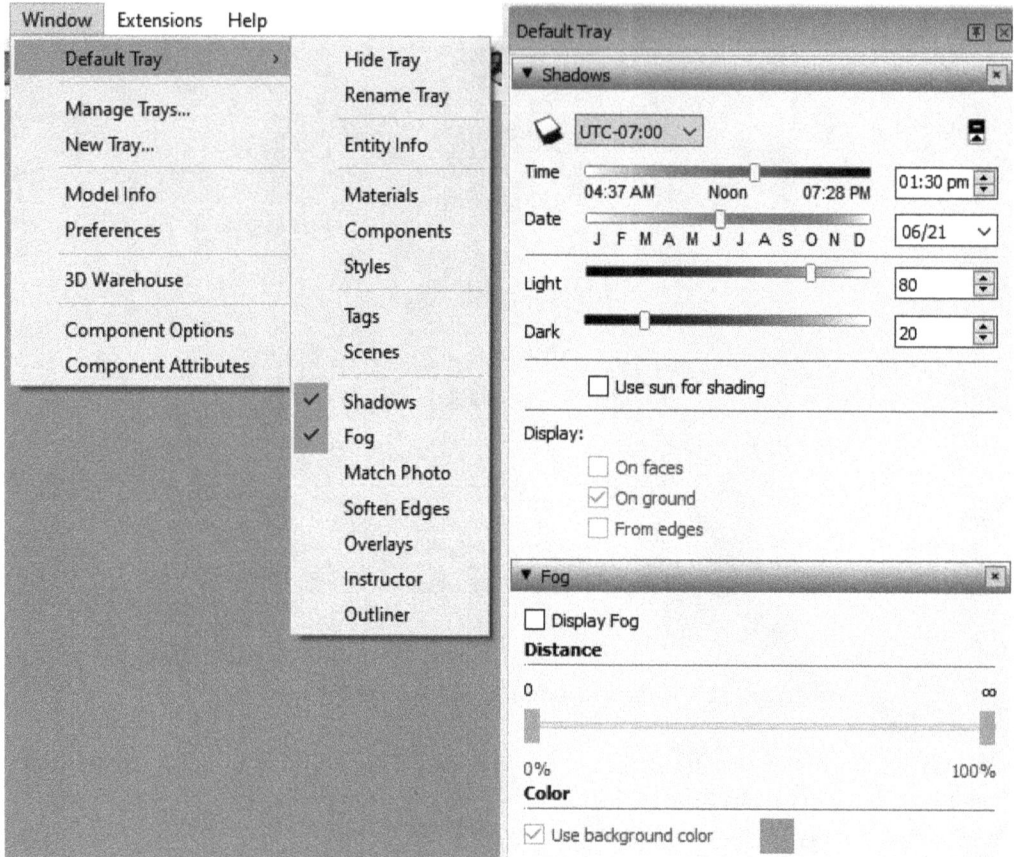

Figure 7.60 – Shadows and Fog Checkboxes in the Menu Bar (Left);
Shadows and Fog Panels in the Default Tray (Right)

We'll examine **Shadows** and **Fog** in greater detail in the following subsections.

Shadows

Shadows in SketchUp Pro are cast from an imaginary light source that represents the Sun. There are no physical light sources in the standard toolset for SketchUp Pro – light sources for rendering must be added by using extensions, plugins, or additional rendering software. We will focus on the built-in Shadows in this chapter.

The **Shadows** toolbar and **Default Tray** panel have the same settings, with the **Default Tray** panel having additional and more precise options. We will use the **Default Tray** panel for the examples in this section. To see all Shadow options in the panel, click on the black and white button to Show Details:

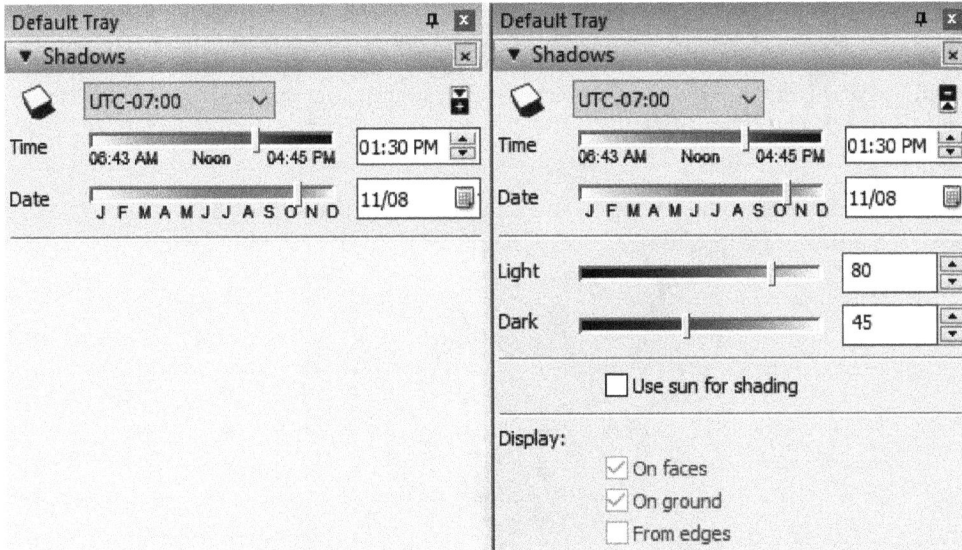

Figure 7.61 – Shadows Default Tray Panel with Details Hidden (Left);
Details Shown After Clicking the Button (Right)

There is only one button on the Shadows panel, and that is to toggle Shadows on and off. This button can be difficult to identify as it is fairly nondescript:

Figure 7.62 – Shadows Button with Shadows Off (Left); Shadows Button with Shadows On (Right)

Shadows are defined by setting the time of day (Time), time of year (Date), and location. The time of day is probably the most expected setting, as this is something we observe daily. The Sun rises in the east and sets in the west, so shadows point in the opposite direction – toward the west at the beginning of the day, the north in the middle of the day, and the east at the end of the day:

Figure 7.63 – Shadows Pointing West at 9 A.M. (Left); Shadows Pointing
North at 12 P.M. (Middle); Shadows Pointing East at 3 P.M. (Right)

> **Note**
>
> This is the case for locations in the northern hemisphere, which is the default setting for SketchUp. We will discuss Geo-location for setting locations in the southern hemisphere later in this section.

The cardinal directions of west, east, north, and south align with the SketchUp Pro Axes by default. Positive Green is true north, positive Red is east, negative Green is south, and negative Red is west:

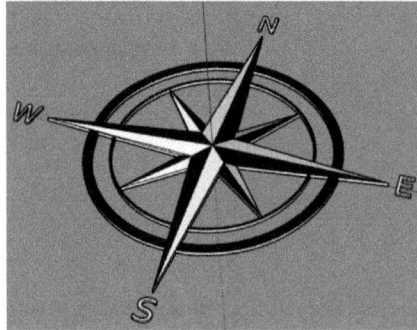

Figure 7.64 – Positive Green is True North, Positive Red is East,
Negative Green is South, and Negative Red is West

> **Note**
>
> By default, true north points along the positive green axis. This can be changed by adding an extension to SketchUp. It is recommended to rotate the model and update the Axes to accurately locate the model relative to true north.

Time

The **Time** slider can be used to change the time of day in SketchUp Pro. The **Time** slider will start at sunrise and will end at sunset, depending on the Location and Date. Time can be changed with the slider or typed into the box next to the slider and can be finely adjusted by clicking the arrows in the text box:

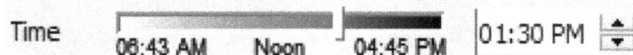

Figure 7.65 – Time Slider and Text Box in the Shadows Dialog Box

> **Note**
>
> Time can be set outside of daytime hours by typing in the text box. This will cause the SketchUp model's Geometry to turn a dark gray, representing nighttime. The Sky color does not update based on the time of day in SketchUp Pro without adding an extension.

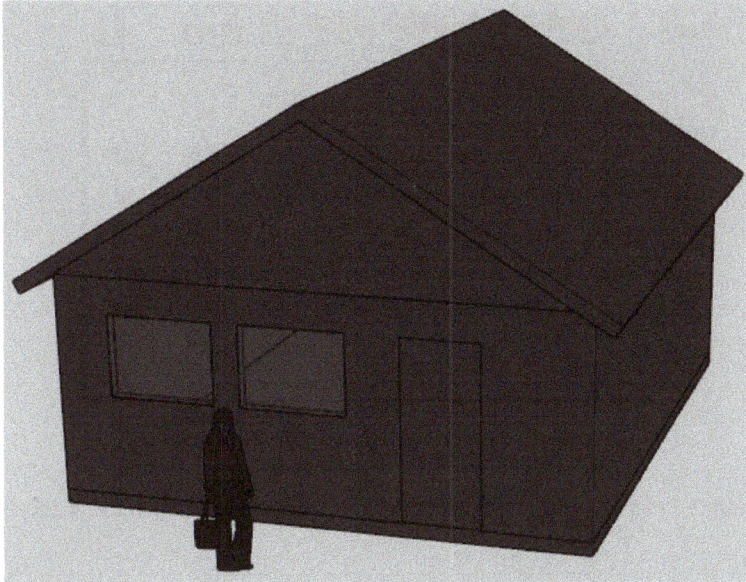

Figure 7.66 – Time Set to 11 P.M.

Adjusting the Time can help prepare shadow studies for your model throughout the time of day, it can also be used to try and get the best-looking shadow for one still image:

Figure 7.67 – Time Set at Different Times of Day to Create a Shadow Study

Date

In the real world, the tilt of the Earth's axis changes the length of shadows, depending on the time of year. SketchUp understands this phenomenon and provides a time of year slider that changes the path of the Sun, and therefore the height and angle of the shadows. The letters listed below the **Date** slider represent the months in English: January, February, March, and so on. Similar to the **Time** setting, the date can be typed in the text box or selected using the calendar button:

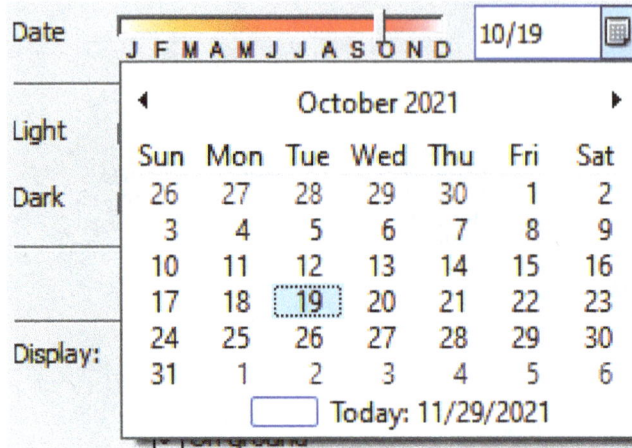

Figure 7.68 – Date Slider and Text Box with the Calendar Dropdown Open in the Shadows Panel

As a rule of thumb, shadows are shorter in the summer and longer in the winter. This can be important to understand for shadow studies or when trying to create the best still image:

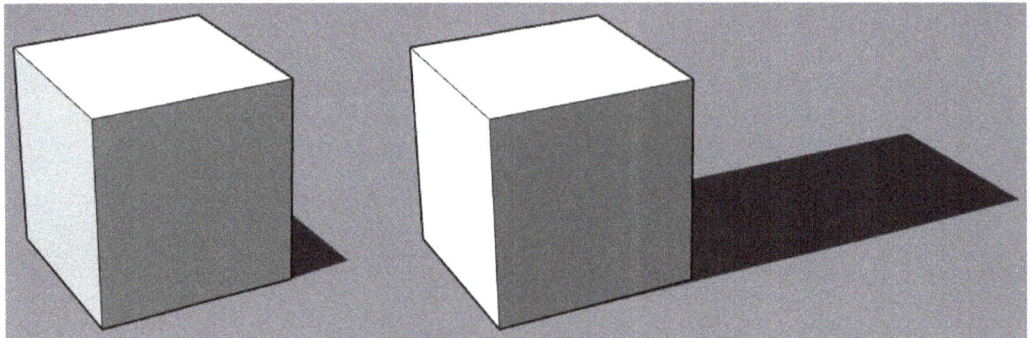

Figure 7.69 – Shadows at 12 P.M. Noon in the Summer (Left) and in the Winter (Right)

Adjusting the Date will also change the sunrise and sunset times, as these vary throughout the year. This will update the options on the Time slider as appropriate.

Location

The final setting to get the most accurate shadows for a SketchUp Pro model is the Location. SketchUp sets a default location in Boulder, CO, USA when using SketchUp Pro in English. This Location can be updated to any location on Earth. There are two ways to update the location: either by adding GPS coordinates or selecting a location on a map. Setting the Location can be done by opening the **Window | Model Info** dropdown in the Menu Bar, then selecting the **Geo-location** menu:

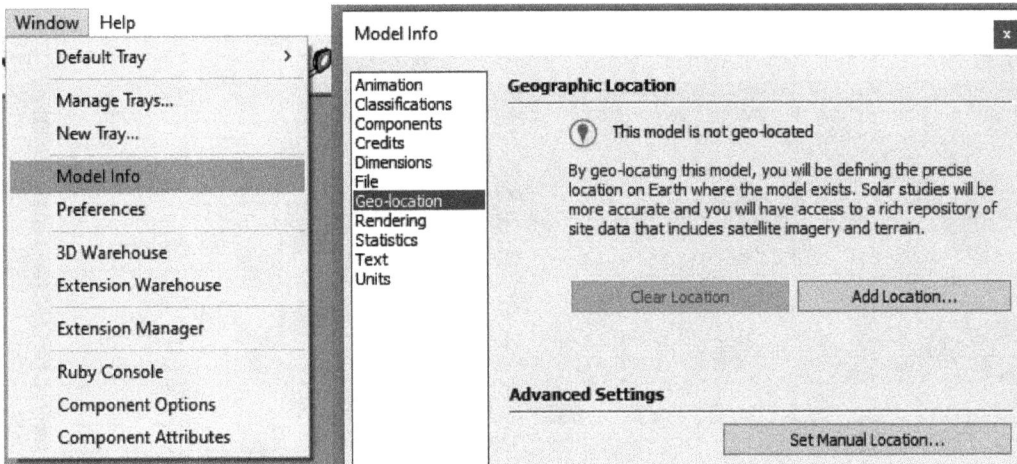

Figure 7.70 – Window | Model Info (Left); the Model Info Dialog Box with the Geo-location Menu (Right)

The Geographic Location section of the Geo-location menu will allow a physical location to be set by picking a region of topography on a map. This topography can be brought into the model during this process as well. The menu will show if the model has been geo-located:

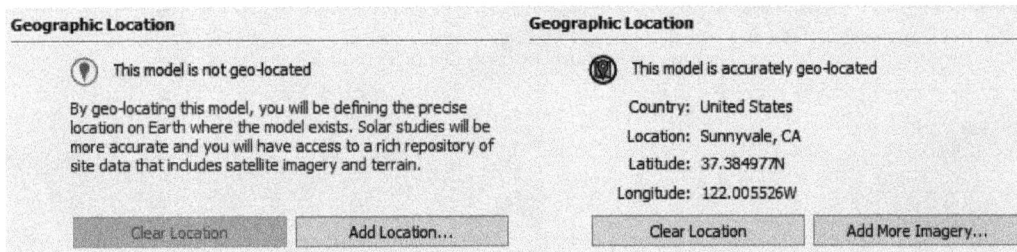

Figure 7.71 – The Model is not Geo-located (Left); The Model is Geo-located (Right)

When **Add Location** is clicked, SketchUp Pro will open the **Add Location** dialog box. **Add Location** will allow you to choose a specific location on the map and select a region of the map to be imported into the model. The center of the rectangular region will be placed at the Origin in the SketchUp model.

The **Add Location** dialog box also contains some specific tools to help you find a specific location:

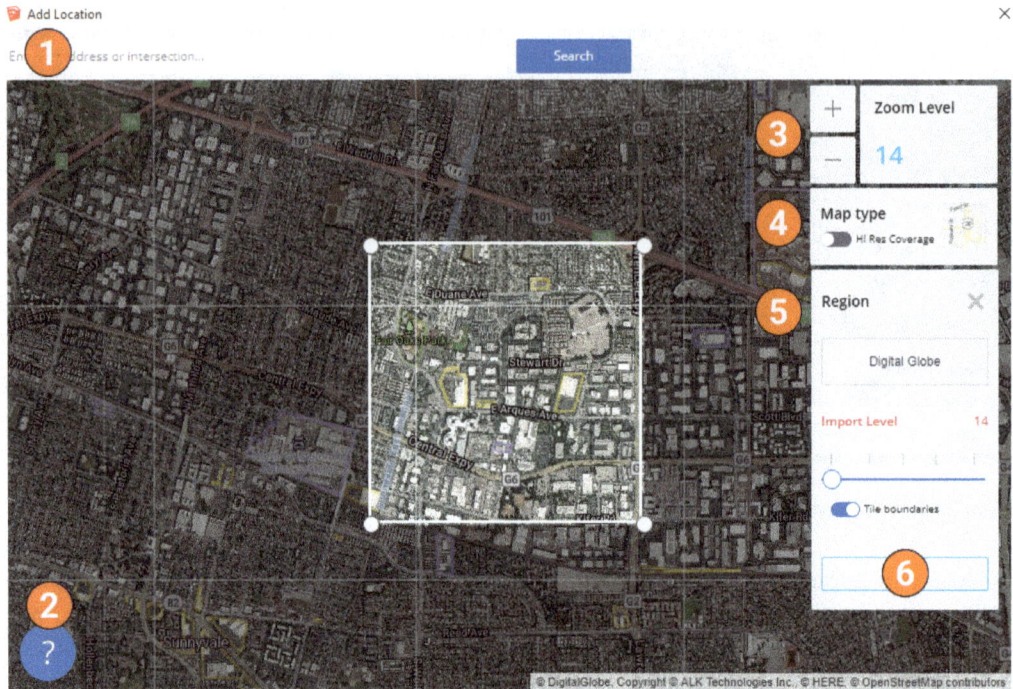

Figure 7.72 – Add Location Dialog Box

These are as follows:

1. Search Bar.

2. Add Location Help/Tour.

3. Zoom Buttons & Zoom Level.

4. Map Type (Hi-Res Coverage).

5. Region (including multiple imagery providers).

6. Import button.

> **Note**
>
> Multiple imagery providers contribute to the map and region Terrain, but Google Maps is no longer contributing to the **Add Location** option in SketchUp Pro.

Once the Location has been set and the Region has been selected, the **Import** button can be used to import the Terrain into the SketchUp model. Once the Terrain has been added to the SketchUp model, it will be placed on two Tags with the names **Location Snapshot** and **Location Terrain**:

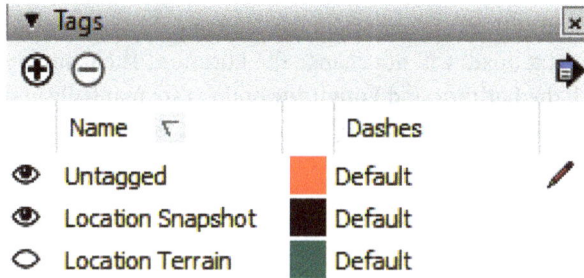

Figure 7.73 – Location Snapshot and Location Terrain Tags in the Tags Panel

These Tags can be toggled off in the **Tags** panel if you do not wish the Terrain to be included in the final model. We will discuss Tags in *Chapter 9, Entity Info, Outliner, and Tags Dynamically Organize Your Models*.

If Terrain is not important for the SketchUp model, an easier solution is to **Set Manual Location** under **Advanced Settings**:

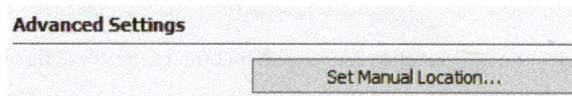

Figure 7.74 – Set Manual Location Button in the Model Info Geo-Location Menu

When **Set Manual Location** is clicked, the **Set Manual Geo-location** dialog box is opened. This Dialog Box prompts you to add **Latitude** and **Longitude** coordinates. The default location of Boulder, CO, USA, and its corresponding coordinates will appear if a Location has not been set already:

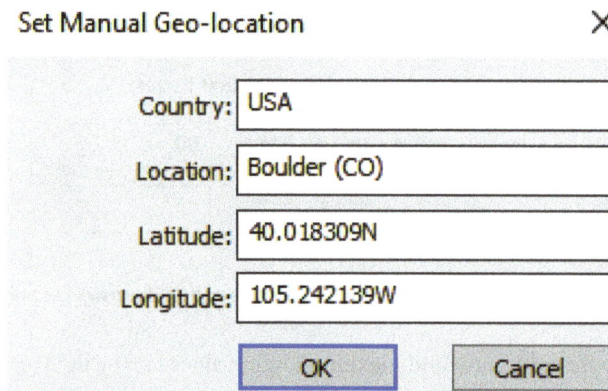

Figure 7.75 – The Set Manual Geo-location Dialog Box with Default
Location, Latitude, and Longitude Coordinates

The **Latitude** and **Longitude** options are important when setting the Location manually, as updating the **Country** or **Location** text boxes will not change the Location. The **Country** and **Location** text boxes should be updated if the **Latitude** and **Longitude** options are manually updated to not confuse anyone viewing the Geo-Location settings. These text boxes can be set to any value, including made-up locations!

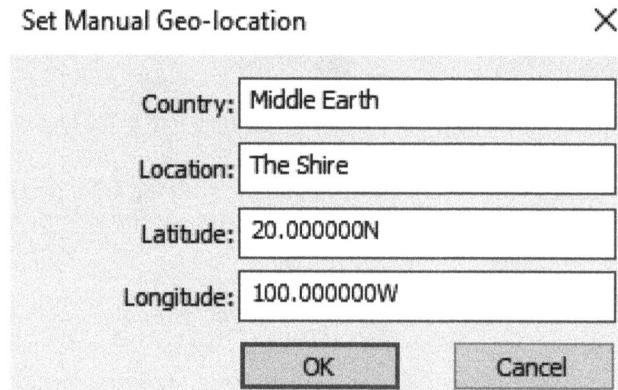

Figure 7.76 – Model Info Geo-location Menu with Incorrect Country and Location Text

By setting a Location, Date, and Time, a SketchUp model can provide precise Shadow information for any model, which is extremely helpful in architectural projects. For most models, simply updating the Time and Date sliders will be enough for you to find an appropriate Shadow that shows off the model!

Other Options

The remaining options in the Shadows panel can change the appearance of the Shadows in the model, but it is recommended to use the default options unless a specific appearance is desired.

The **Light** and **Dark** values can be changed using sliders or text boxes:

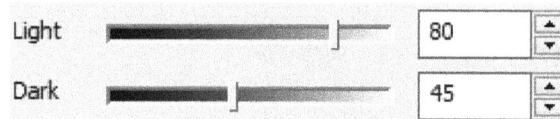

Figure 7.77 – Light and Dark Sliders and Text Boxes in the Shadows Default Tray

These values will update the minimum and maximum light values for the light and Shadows:

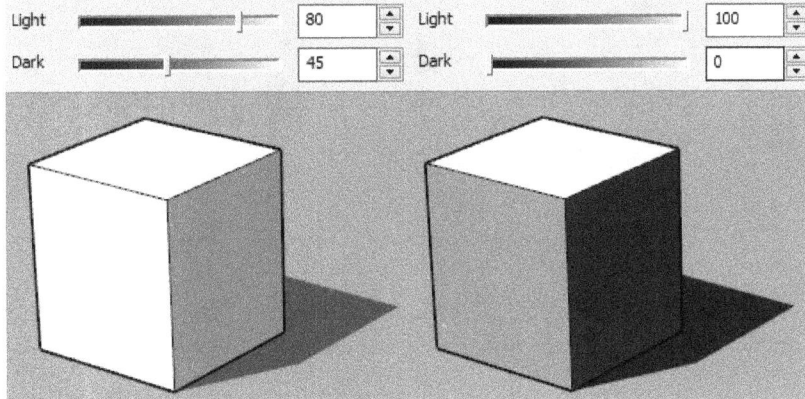

Figure 7.78 – Light and Dark Default Settings (Left); Light set to 100 and Dark set to 0 (Right)

The **Use sun for shading** option can be toggled on so that the Geometry creates shadows on its Faces without casting a Shadow on other Geometry, even with the Shadows toggled off. This can show the Sun's direction without you hiding any objects in shadow:

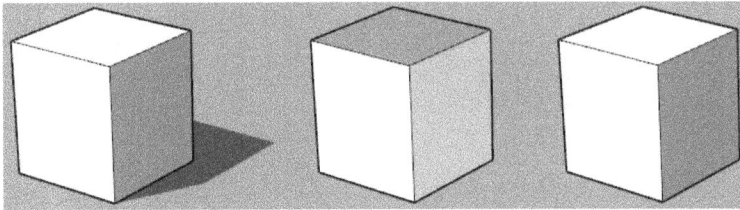

Figure 7.79 – Shadows On (Left); Shadows Off (Middle); Shadows
Off with Use Sun for Shading Toggled On (Left)

Finally, the Display options allow for different Geometry to be understood to cast and receive Shadows. On Faces casts shadows on the Faces of other Geometry:

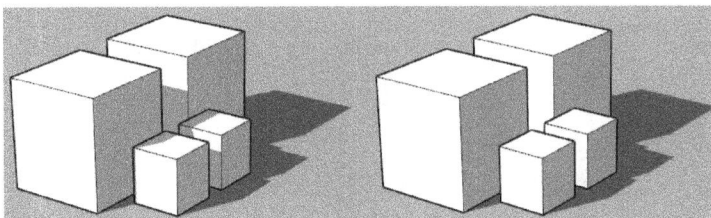

Figure 7.80 – On Faces On (Left); On Faces Off (Right)

On Ground casts the Shadows on the ground plane of SketchUp (at 0 in the Blue Axis) even though there is no ground Geometry in the model:

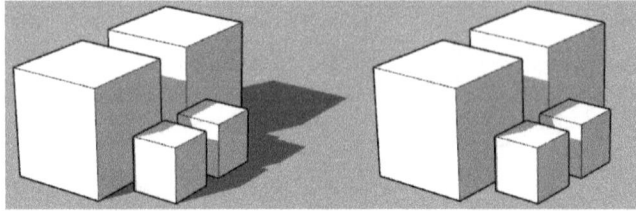

Figure 7.81 – On Ground On (Left); On Ground Off (Right)

This option is often turned on in basic SketchUp models but can be turned on when using advanced models with custom Terrain. On Faces must be turned on so that the Shadows are cast onto the custom Terrain:

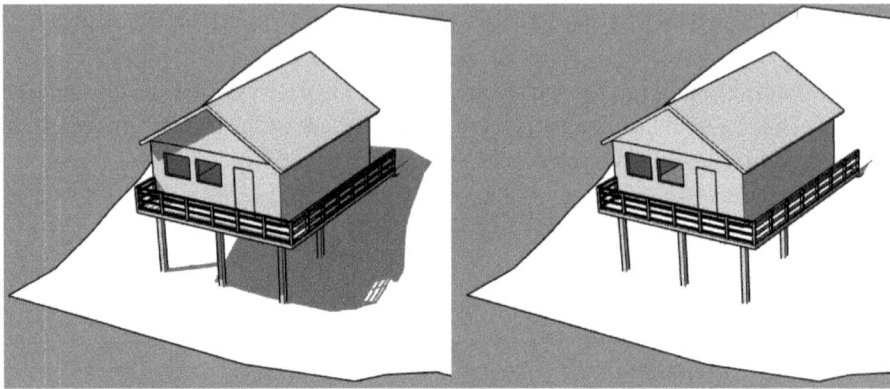

Figure 7.82 – Additional Shadow Cast On Ground On (Left); On Ground Off (Right)

On Edges allows for Shadows to be cast from Edges instead of only Faces. Technically, Edges are infinitely thin Geometry in SketchUp, so they wouldn't cast a Shadow. However, in some SketchUp models, Edges can be used to represent thin Geometry, so Shadows should be cast:

Figure 7.83 – From Edges Off (Left) From Edges On (Right)

Shadows do not have to be complex in your SketchUp model. There are many options to set up and tweak, but in most cases, the biggest impact to the SketchUp model can be to simply toggle Shadows on and leave it at that!

In the next section, we will look at another option that can have a big impact: Fog.

Fog

Compared to **Shadows**, **Fog** is a relatively simple option in SketchUp Pro. **Fog** does not have a toolbar in SketchUp Pro, but it does have a panel in the **Default Tray** panel:

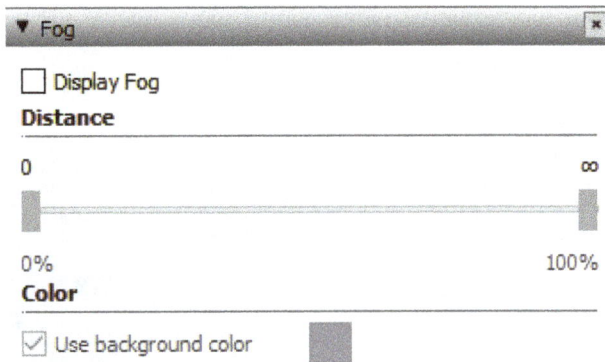

Figure 7.84 – The Fog Panel in the Default Tray Panel

The **Fog** panel has a **Display Fog** checkbox in place of a Fog button. Displaying Fog in SketchUp creates a gradient of color that obscures Geometry that is further away from the camera until it fades into the background color. In the following figure, we can see that when Fog is turned on, the trees further away from the camera eventually fade into the background color:

Figure 7.85 – Fog Off (Left); Fog On (Right)

Fog is a tool that can be used to provide atmosphere and depth to a SketchUp model. The Fog cannot be used to place patches of Fog in specific locations to prepare a rendering, as it is camera-dependent and not placed in specific locations. Zooming, Panning, and Orbiting around a model will change how the Fog appears. In the following example, the camera is zoomed out and the Geometry is obscured by the Fog:

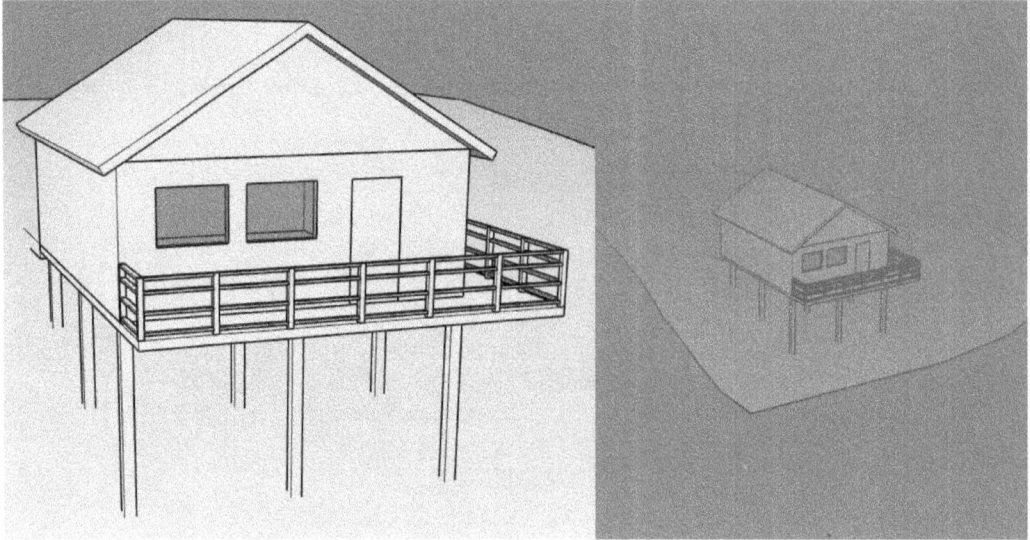

Figure 7.86 – Camera Closer to Geometry (Left); Camera Zoomed Out (Right)

Moving the camera can change how much Geometry is obscured and updating the **Distance** slider can also change how the Geometry is obscured. The slider changes exponentially from 0 to infinity and sets the distance where Geometry begins to be obscured by Fog, and where it is completely obscured by Fog. The slider contains two grips – the 0% grip and the 100% grip:

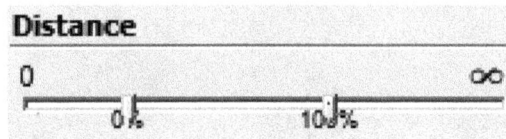

Figure 7.87 – Fog Distance Slider

There are no units on the slider, so the best way to adjust the model is to set the camera in the desired location before sliding the sliders. In the following example, we can see what happens when the 0% slider grip is moved closer to 0. The Fog begins to generate right at the camera location and continues into the distance:

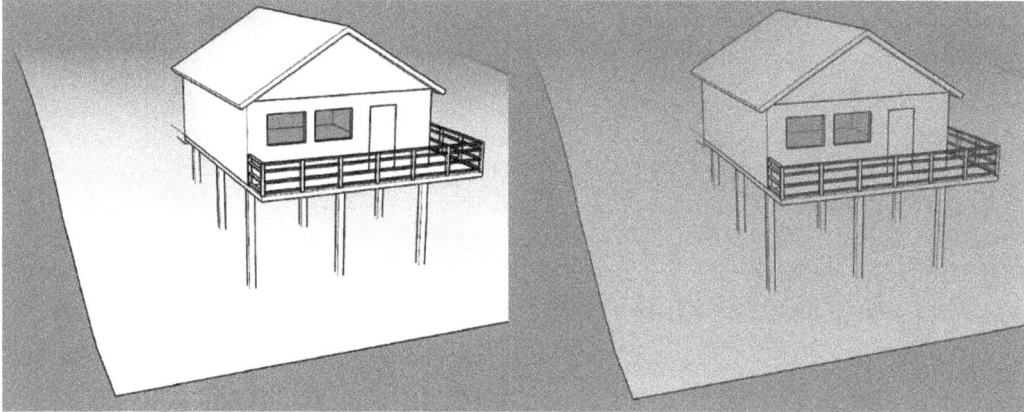

Figure 7.88 – Fog with Default Settings (Left); 0% Fog set to 0 (Right)

In the following example, we are changing the 100% slider grip distance closer to infinity. This makes the transition to fully obscured less severe:

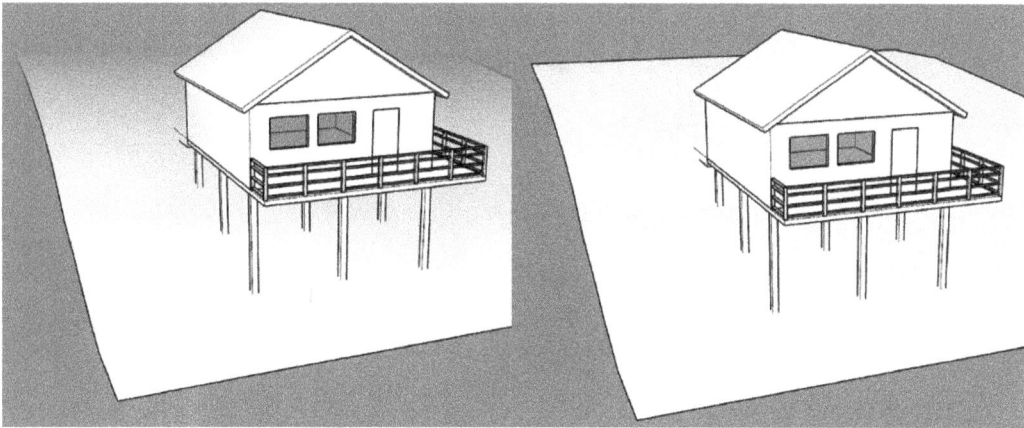

Figure 7.89 – Fog with Default Settings (Left); 100% Fog set Closer to Infinity (Right)

> **Note**
> Changing the 100% slider grip to infinity will effectively turn Fog off. It is impossible to set a gradient to smoothly transition until infinity!

The final option in the Fog panel is to change the color of the Fog. This option changes the background color of the Style but only when Fog is turned on. When **Use background color** is checked, the Fog will appear to be the background color in the Style. When **Use background color** is unchecked, SketchUp will allow you to choose a new color by clicking on the colored box:

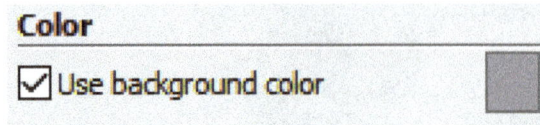

Figure 7.90 – Color Options with Checkbox and Colored Box

A new color can be selected using the **Choose Color** dialog box. This is the same dialog box that is used to pick colors for Styles, which we looked at in the *Creating and editing Styles* section of this chapter. Once a new color has been chosen, it will update the background color and Fog appropriately:

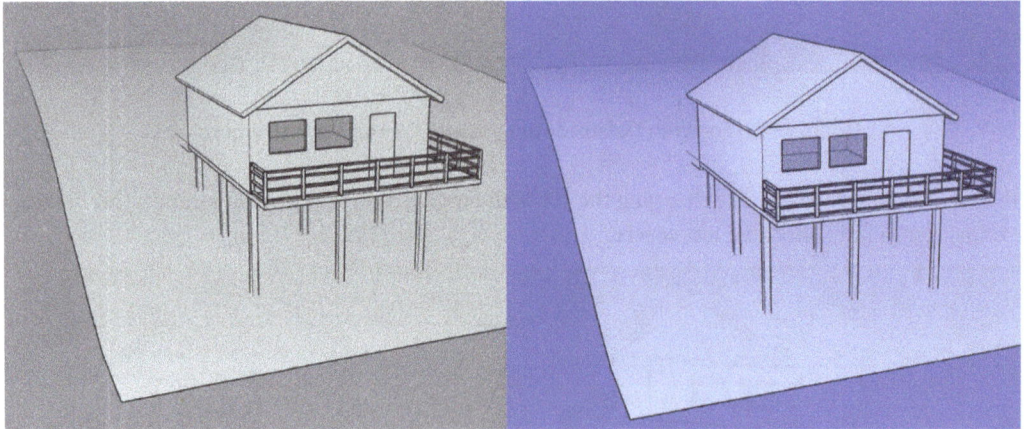

Figure 7.91 – Fog with Background Color (Left); Fog with a Specific Color Chosen (Right)

Fog is especially useful when using custom Terrain as it can limit the distance that the camera will render. When Fog is off, it can be easy to see the edge of the Terrain, but when Fog is turned on, it crops the view in a smooth transition, which can be less distracting to the viewer:

Figure 7.92 – View of Terrain with Fog Off (Left); View of Terrain with Fog On (Right)

Using Fog and Shadows in a model can drastically change its appearance. This can take a model with simple color materials and add a level of detail that helps us understand the view in a way that we might understand when viewing a photograph – with the Shadows and Fog showing areas of depth under, behind, and far away:

Figure 7.93 – SketchUp model with Shadows and Fog Off (Left); Shadows and Fog On (Right)

Now that we understand how to add recognizable depth using Shadows and Fog, we will look at ways to view our SketchUp models in a way that can only be done digitally – by using Sections!

Sections

SketchUp Pro allows you to cut the model along a 2D plane to see inside the model. This can be done by placing a Section Plane! Sections are common tools used in professional drawings to show how the inside of a part or building works, and SketchUp allows this Section functionality in real time in working models. Sections do not require you to Move or Hide any Geometry to see inside the model – the Section Plane tool takes care of that automatically. Active Section Planes hide all Geometry on one side of the Section Plane in the current model, Group, or Component, while showing the Geometry on the other side.

The Section Plane tool can be found on the Large Tool Set with the Camera tools, but all of the Section View options can be found in the **View** dropdown in the Menu and the Section toolbar. The Section Plane tool is represented by a section callout, which is a circle with the view name inscribed and a section arrow pointing to the right:

Figure 7.94 – Section Plane Tool

The Section tools can be found on the Large Tool Set, the **View** dropdown in the Main Menu, and the **Section** toolbar:

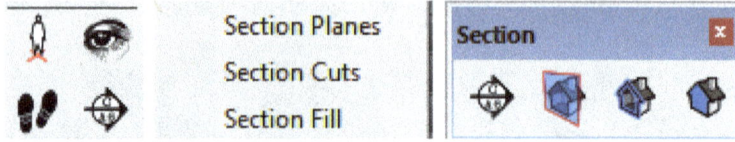

Figure 7.95 – The Section Plane Tool on the Large Tool Set (Left); the Section View Options in the View Menu (Middle); Section Toolbar (Right)

Section Planes can be added by clicking the **Section Plane** button in the Large Tool Set or the **Section** toolbar. When the **Section Plane** button is clicked, SketchUp will show the Section preview, which is a rectangle with Section callouts at the corners showing the Section direction:

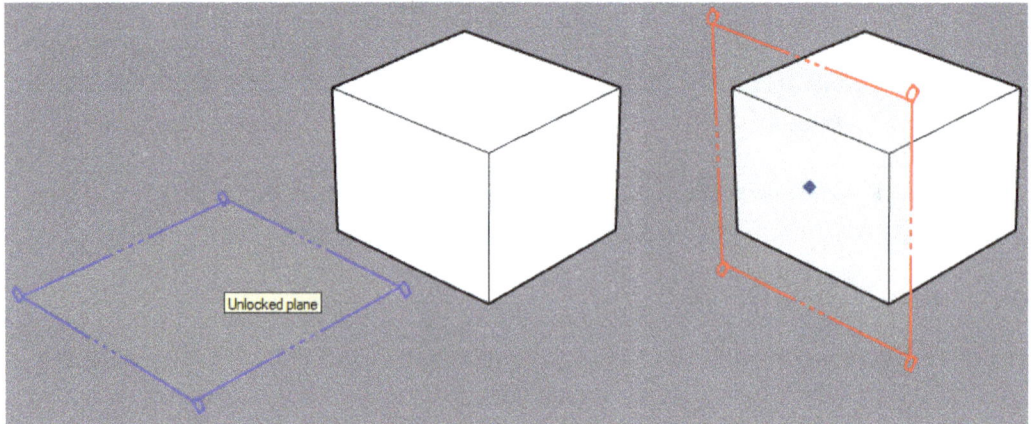

Figure 7.96 – Section Tool Clicked (Left); Section Preview (Right)

The Section preview will be colored Red, Green, or Blue if it is perpendicular to that Axis direction. Additionally, the Section preview will be colored Magenta when it is inferencing a Face that is not on a standard Axis:

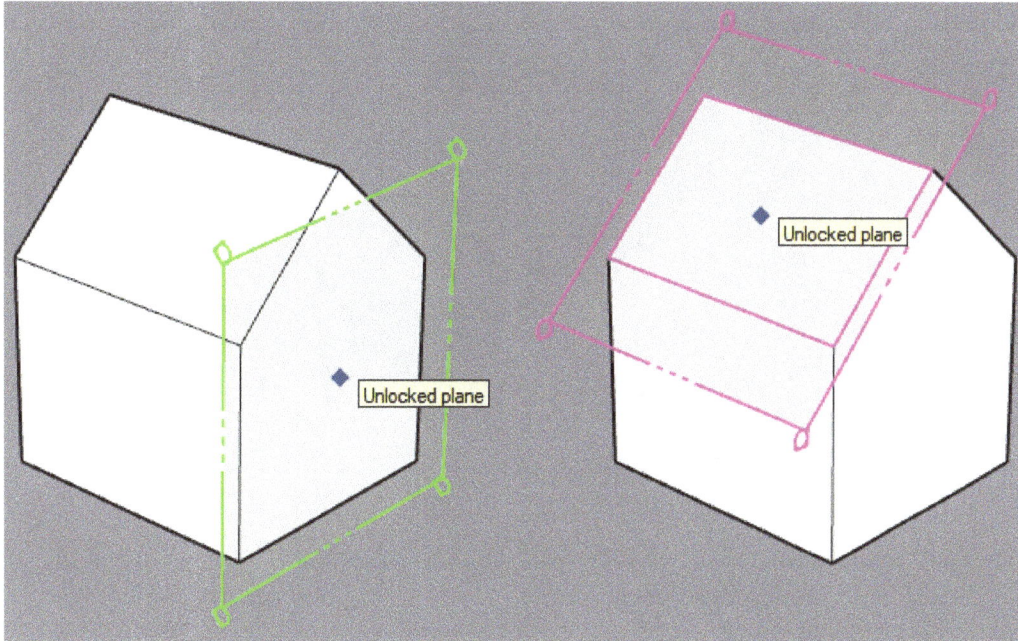

Figure 7.97 – Section preview showing the Green direction (left); Section preview inferencing Face (right)

Once the Section preview is in the correct location, the Geometry can be left-clicked. This will prompt the **Name Section Plane** dialog box, which allows the Section to be named and labeled with a symbol number. This option can be disabled with the checkbox at the bottom of the dialog box:

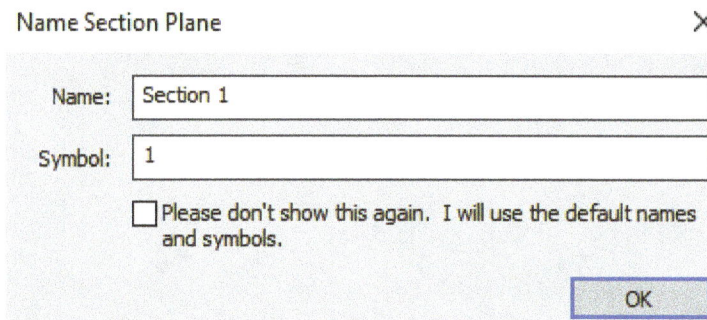

Figure 7.98 – Name Section Plane Dialog Box

Once the name has been updated and the **Name Section Plane** dialog box has been closed by clicking the **OK** button, the Section Plane will be created. Sections in SketchUp are automatically created to cut all Geometry in the active model, Group, or Component. So, if no Group or Component is currently active, the Section Plane will cut all Geometry in the model:

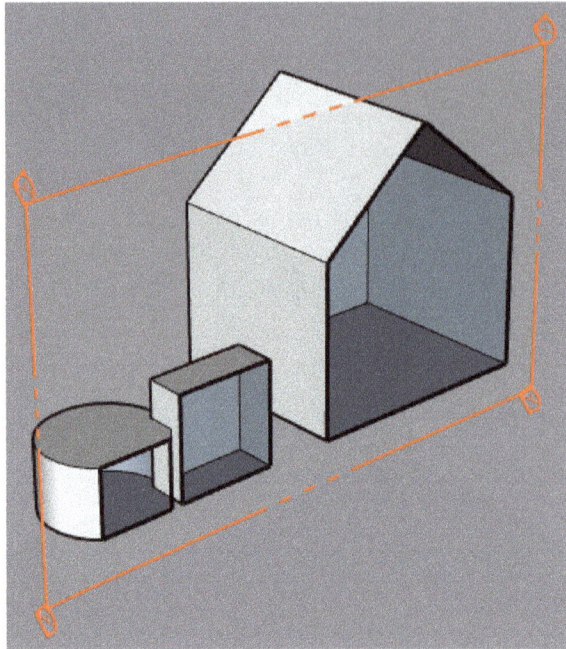

Figure 7.99 – Section Plane Created and Cutting all the Geometry in the Model

Section Planes that are created in an active Group or Component will only cut the Geometry within that Object, so the Section Plane will appear relatively smaller and will only cut the Object:

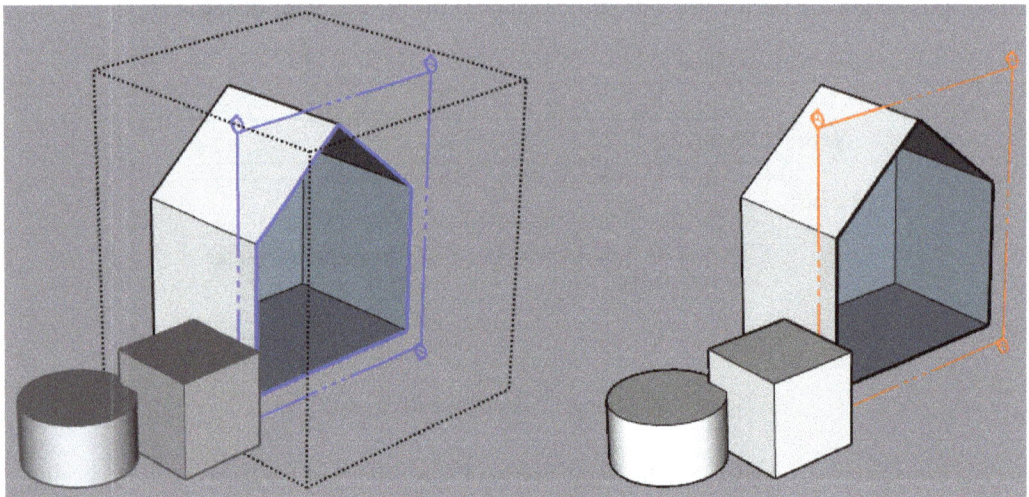

Figure 7.100 – Section Plane Created in an Active Group (Left);
Section Plane Only Cuts Group Geometry (Right)

Multiple Section Planes can be created in a model, but only one Section Plane can be active at a time. When a new Section Plane is created, it is automatically set to active and will deactivate the active Section Plane. Section Planes can also be manually deactivated by right-clicking the Section Plane and choosing the **Active Cut** option:

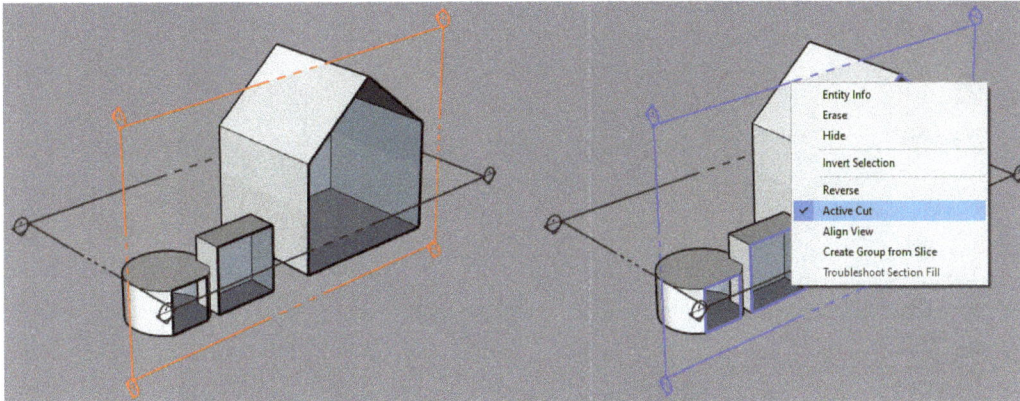

Figure 7.101 – Multiple Section Cuts with One Active Section (Left);
Manually Deactivating Section Cut in Right-Click Menu (Right)

> **Note**
>
> There can be multiple active Section Planes if they are created in different Groups or Components. But there can only be one active cut at each level – for instance, in the model with no Objects active.

Section Planes can be modified with the Move and Rotate tools. Section Planes are created directly on top of the Face that is clicked during the creation process. Typically, Section Planes can show more Geometry when they are moved toward or away from the Face that was originally selected. In this workflow, the Move tool has been activated, and the Section Plane has been selected:

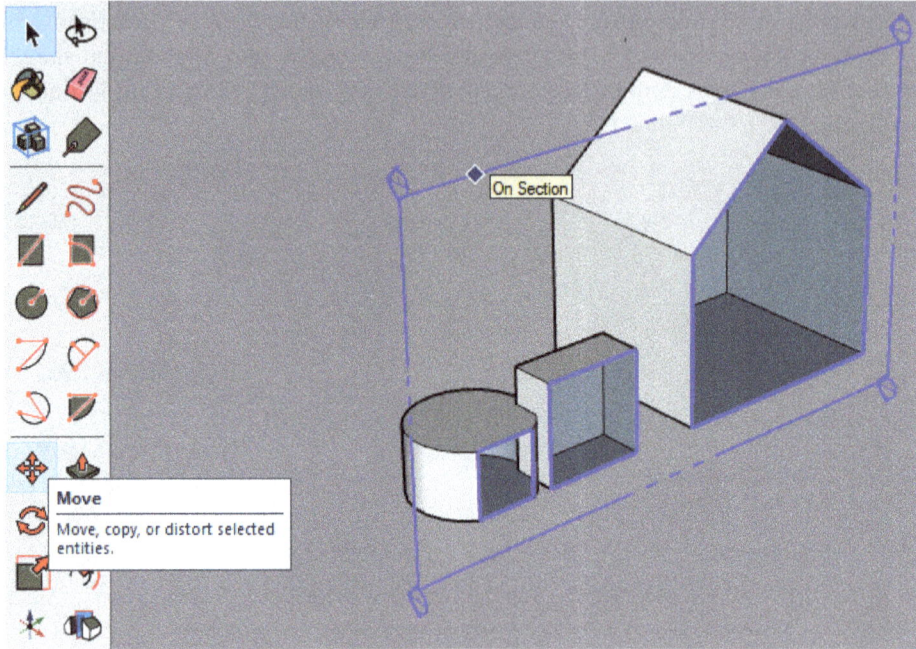

Figure 7.102 – Move Tool Activated (Left); Section Plane Selected (Right)

Then, the Section Plane can be moved along the perpendicular Axis. The SketchUp model will automatically update with a preview of the Section Plane. This can help you place the Section Plane in a good position:

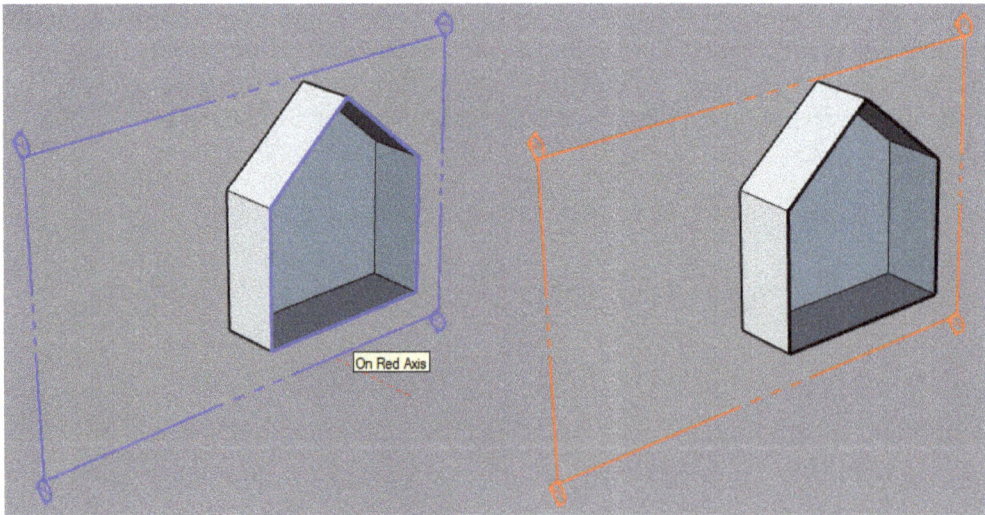

Figure 7.103 – Section Plane Preview (Left); Section Plane Moved and Deselected (Right)

Additionally, the Rotate Tool can be used to create unique Section views within a Model. If a Face is not in the model that is at the desired angle for the Section Plane, the Rotate tool can be used to update the Section angle. In the following example, a Section Plane has been created by cutting through the corner of the room. A Section Plane must first be created on one of the walls:

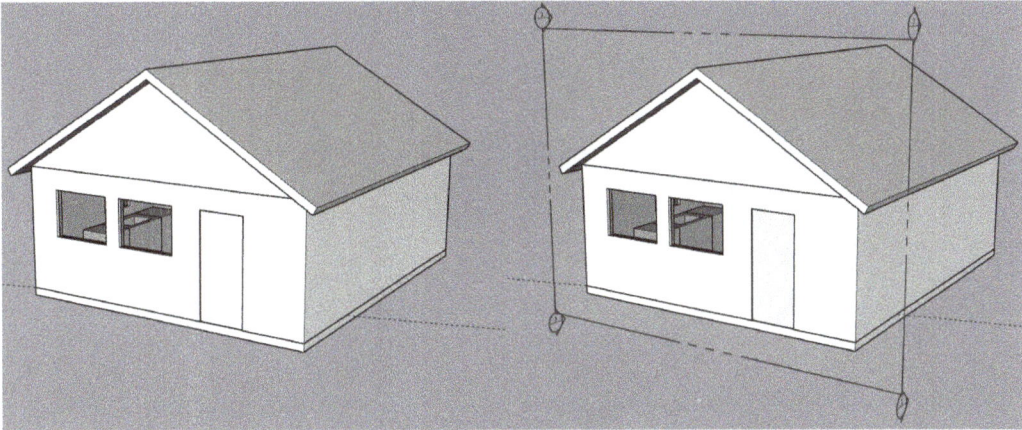

Figure 7.104 – Room with the Desired Section Angle (Left); Section Created on Wall (Right)

Then, the Rotate tool can be activated with the Section Plane pre-selected. The Rotate Basepoint can be set by left-clicking, and the Rotate angle must be entered into the Measurements Box. The Section Preview will update to show the new angle for the Section Plane:

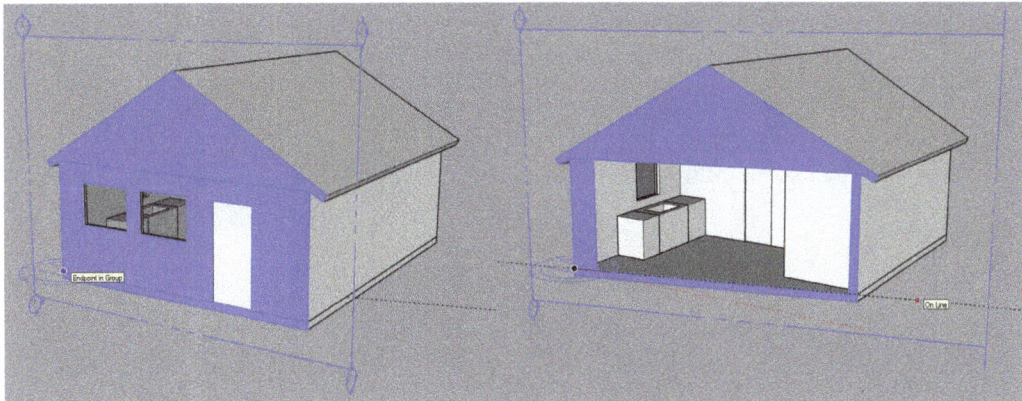

Figure 7.105 – The Rotate Tool Activated and the Basepoint Set (Left); the Section Preview Updated (Right)

When the Rotate workflow is complete, the Section Plane will be set at the updated angle even though there is no Face at that angle in the model:

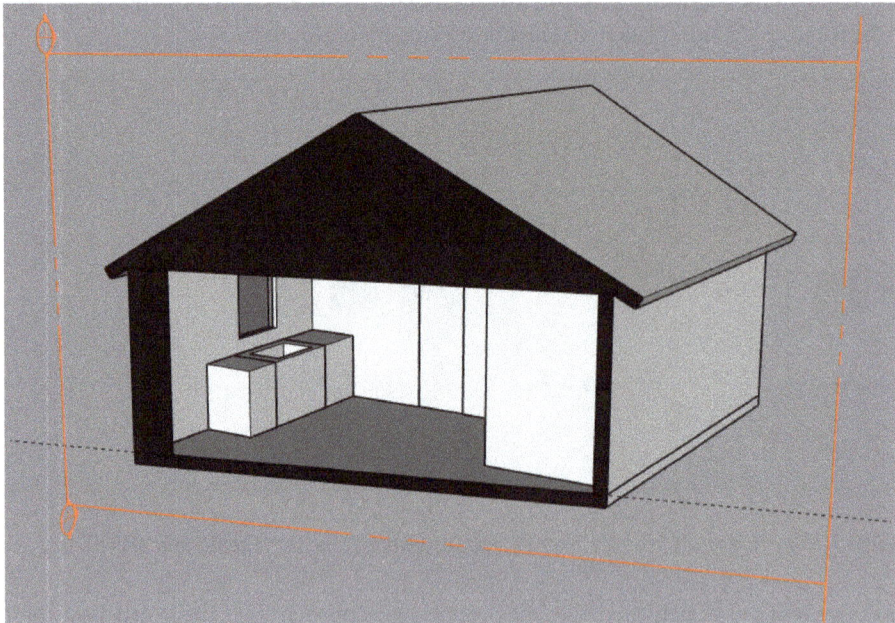

Figure 7.106 – Section Plane Rotated and Deselected

There is one final edit workflow, which is to reverse the Section Cut. This workflow is the same as rotating the Section Plane 180 degrees. With the Section Plane selected, right-click and choose **Reverse**:

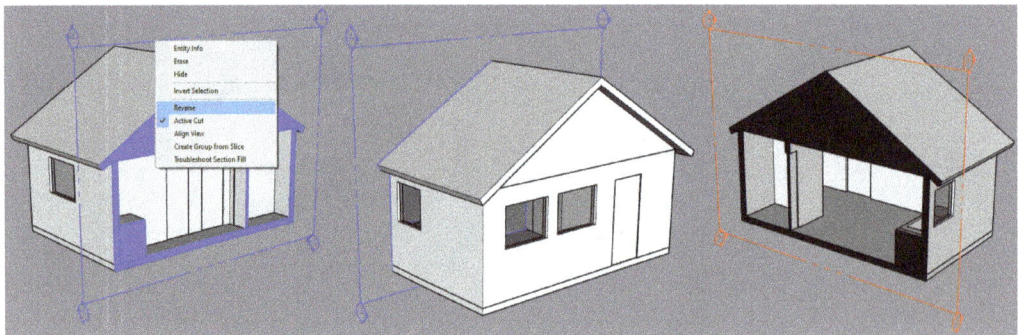

Figure 7.107 – Section Plane Selected (Left); Right-Click > Reverse
(middle); Section Plane reversed and deselected (right)

Section View Options

Three Section View options can be found in the **Section** toolbar and the **View** dropdown in the Menu. These are **Section Planes**, **Section Cuts**, and **Section Fill**:

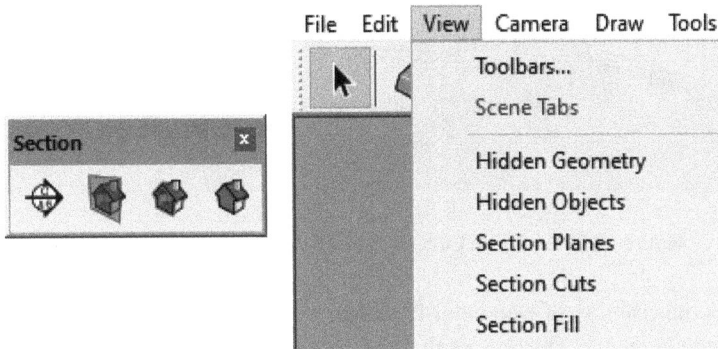

Figure 7.108 – Section Planes, Section Cuts, and Section Fill in the
Section Toolbar (Left) and the Menu > Dropdown (Right)

Section Planes toggles the visibility of the Section Planes themselves. This does not turn off the active Section Cuts in the model. This option is best used once the Section Planes are in the correct location, and the model is being prepared for images or animations to be exported:

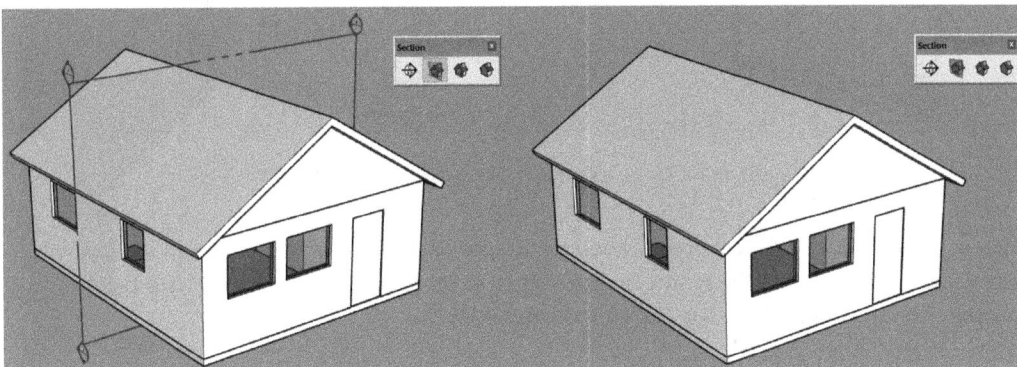

Figure 7.109 – Section Planes On (Left); Section Planes Off (Right)

Section Cuts toggles the visibility of all Active Section Planes in the model. This is a quick way to turn the Active Section Plane when modeling to work with Geometry on both sides of the Section Plane:

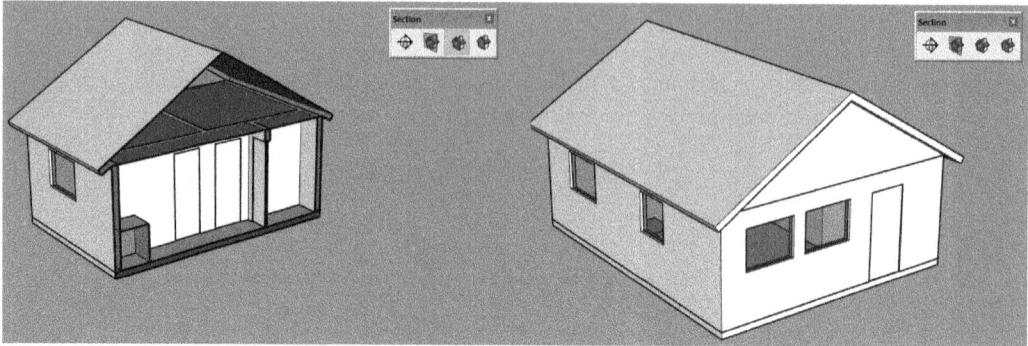

Figure 7.110 – Section Cuts On (Left); Section Cuts Off (Right)

Section Fill toggles the visibility of Geometry to appear as solid if the Section Plane cuts through a profile of connected Geometry. This means that the cut Geometry may be solid/watertight or that there are missing Faces above or below the Section Plane. So long as the Faces are connected in a loop, the Section Fill option will fill the profile:

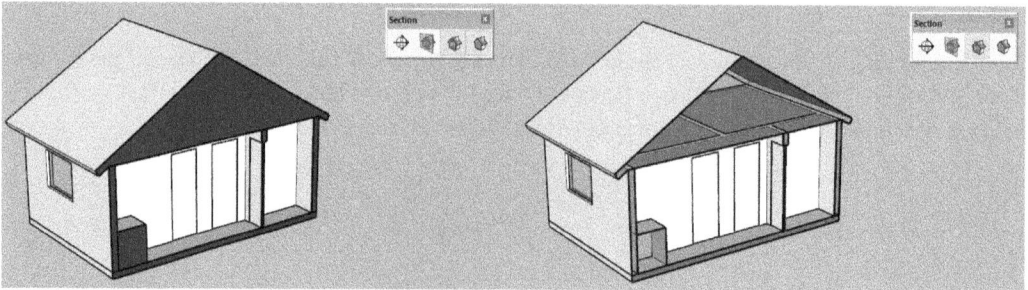

Figure 7.111 – Section Fill Off (Left); Section Fill On (Right)

To better understand what the cut profile describes, let's look at the following example. In this example, we are using a basic box, but the same principles apply to more complex Geometry. This box has six Faces, four on the sides (A-D), a top (E), and a bottom (F):

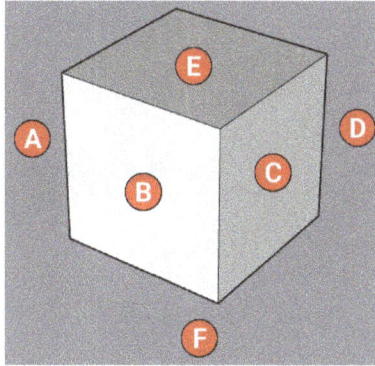

Figure 7.112 – Box with Six Sides (A-F)

With a Section Plane cutting horizontally through the box, sides **A, B, C,** and **D** are cut with the Section Plane. The top Face, **E,** is completely hidden, and the inside of the bottom Face, **F,** is shown at the bottom of the box. Toggling on **Section Fill** shows that SketchUp understands that Faces A-D create a connected loop, and a closed profile at the Section Plane cut:

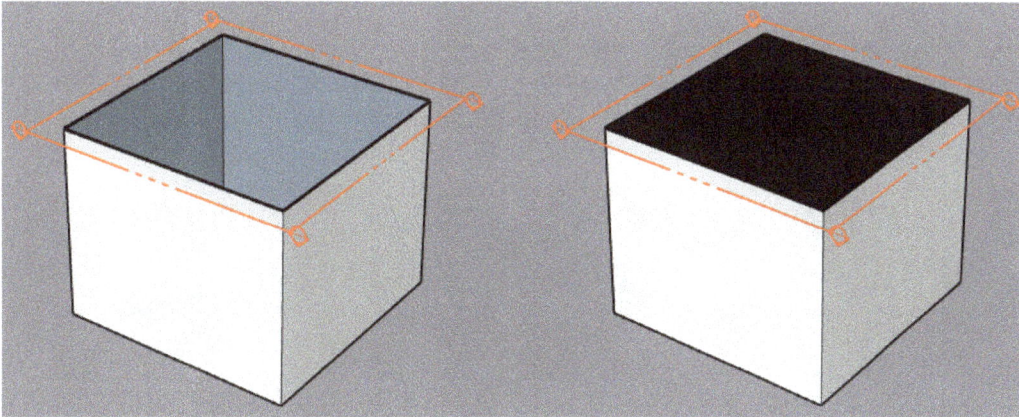

Figure 7.113 – Box with a Horizontal Section Plane Cut (Left); Section Fill Toggled On (Right)

If the top Face (**E**) or the bottom Face (**F**) is deleted, then the box will no be longer solid or watertight. However, with the Section Plane active and Section Fill toggled on, we can see that sides A-D still create a closed profile at the Section Plane cut, so the Section Fill shows as solid:

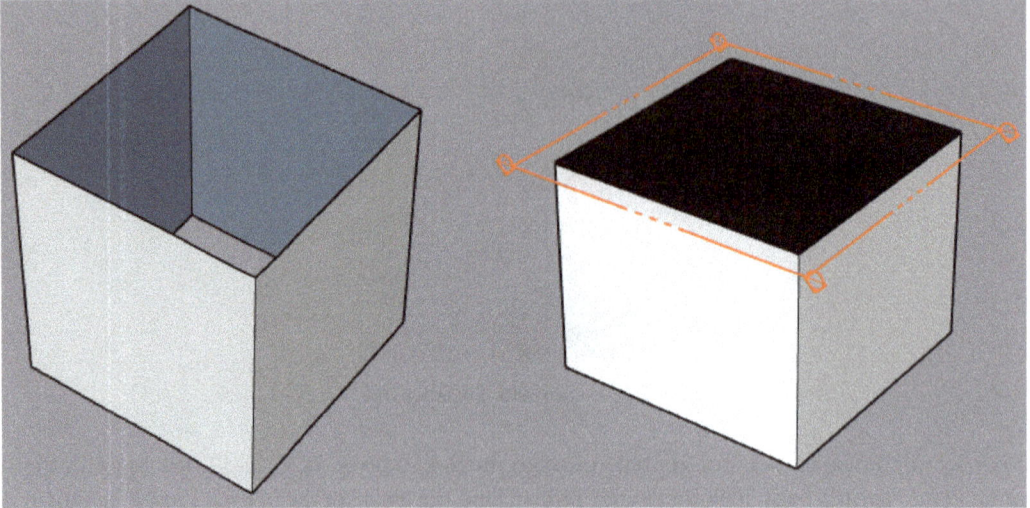

Figure 7.114 – Box with Sides E and F Deleted (Left); Section
Plane Active and Section Fill Toggled On (Right)

To have the Section Fill fail, the closed profile must be interrupted, so one of the sides of the box must be deleted. In the following example, we are starting with all six Faces and deleting only side **B**. Then, with the Section Plane active and Section Fill toggled on, we can see that the Section Fill does not fill the cut portion of the Box. This is because the cut profile is open and not closed in a loop:

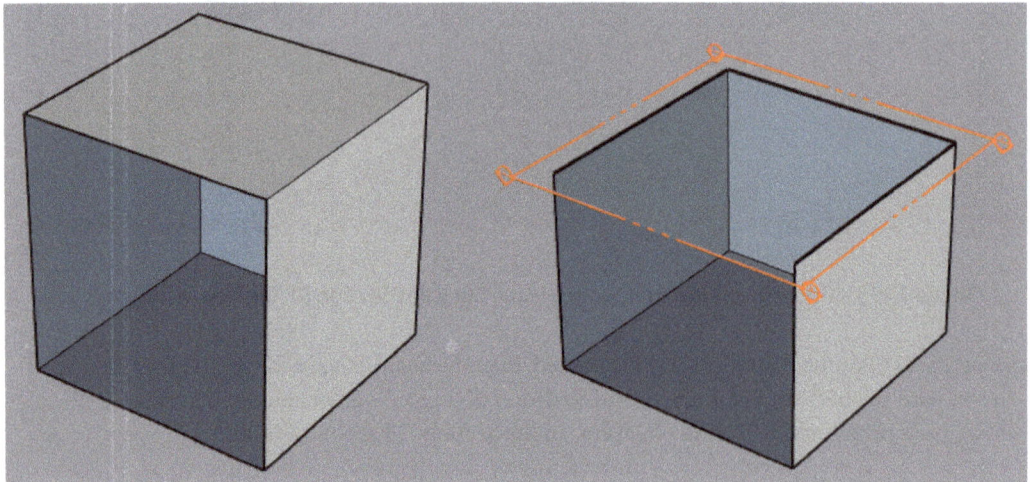

Figure 7.115 – Box with Side B Deleted (Left); Section Plane Active and Section Fill Toggled On (Right)

Section Fill can also fail if the Geometry appears to be in a connected loop, but the Geometry is separated by one or more Groups or Components. In this last example, we can still see the basic box, but sides **B** and **C** have been Grouped while Faces **A**, **D**, **E**, and **F** remain as ungrouped Geometry. When two of the sides are Grouped, it breaks the connected Faces and there is no longer a closed profile at the Section Plane cut:

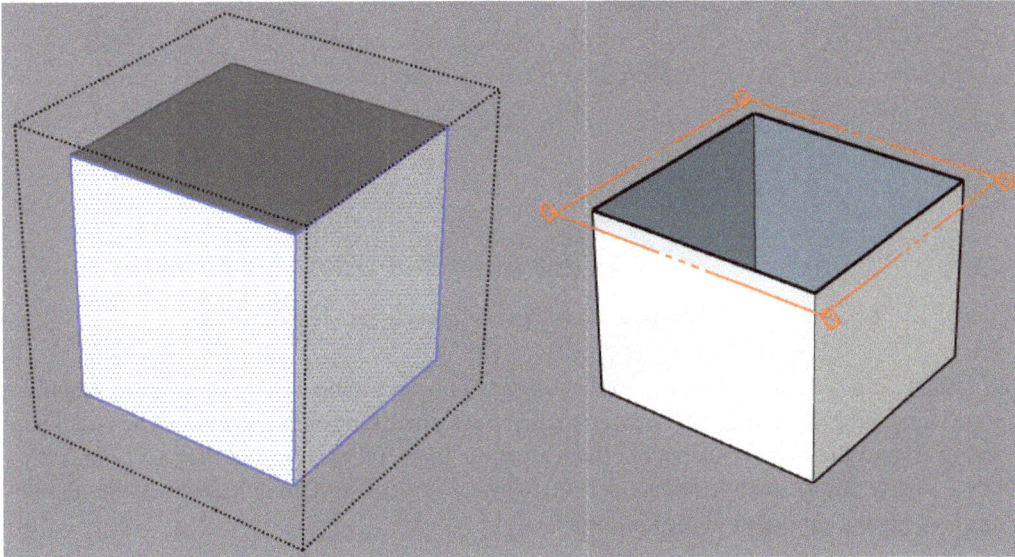

Figure 7.116 – Box with Sides B and C Grouped (Left); Section
Plane Active and Section Fill Toggled On (Right)

> **Note**
> The Section Fill color is set to dark by default. Additionally, **Inactive Section**, **Active Section**, and **Section Lines** are all set with default colors as well. All of these colors can be updated in **Style | Edit** in the **Styles** panel in the **Default Tray** panel.

In this example, we are looking at a very common use for the Section Fill option – filling a Plan Section Plane cut in an architectural model. A Plan view is a horizontal cut through a building that allows a 2D view to be created that can show the relationship between rooms, furniture layout, or fire egress:

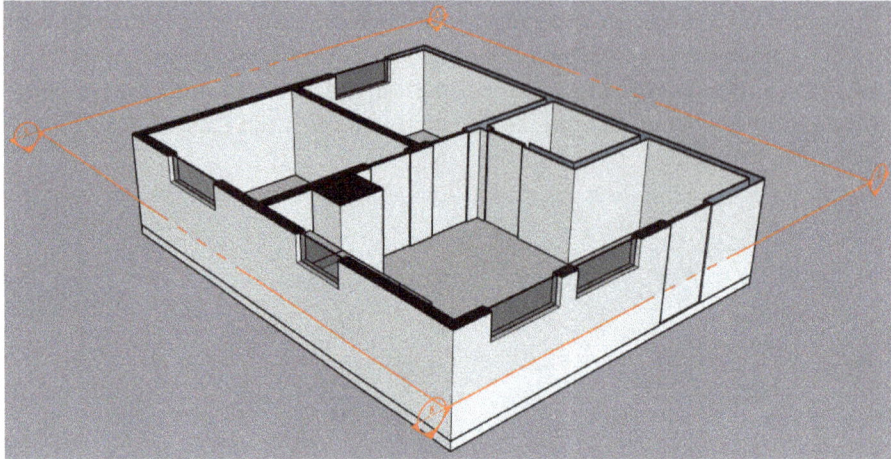

Figure 7.117 – Architectural Model Cut Horizontally to Create Plan View

In this case, we can see that Section Fill has been toggled on, and some walls, the furniture, and the doors are being filled, but some wall Geometry hasn't been filled. One of the Faces must be Grouped or missing around the wall perimeter, but there are more than 50 Faces around the wall Geometry, and this is a fairly simple architectural model. Models may be even more complicated with hundreds of Faces creating the cut profile at the Section Plane cut.

In this example, we are turning on the **Troubleshoot Section Fill** option by selecting the Section Plane, right-clicking, and choosing **Troubleshoot Section Fill**:

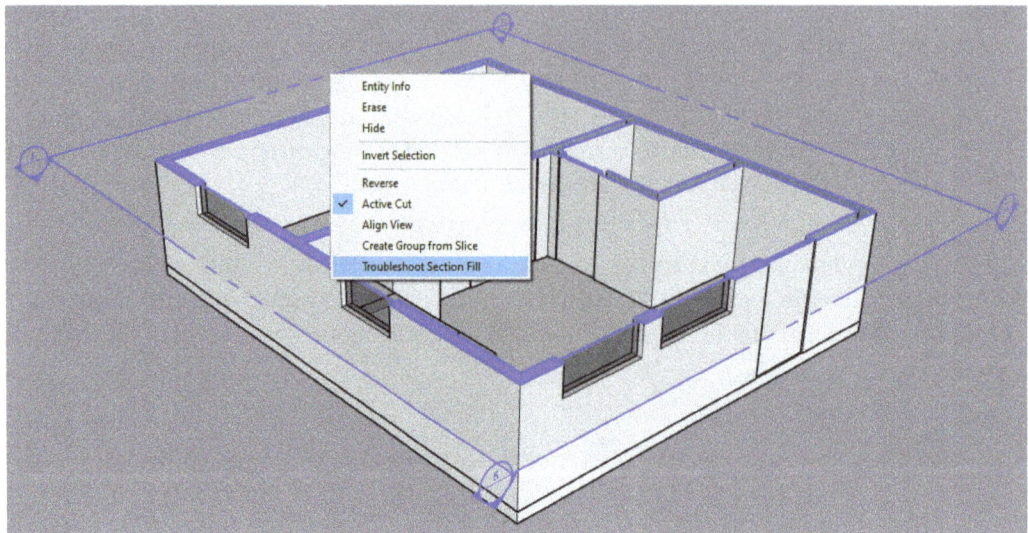

Figure 7.118 – Section Selected and Right-Clicked and Troubleshoot Section Fill Selected

With **Troubleshoot Section Fill** toggled on, red circles will appear where SketchUp believes that the profile has been interrupted. In this example, we can see that one of the doors has accidentally Grouped some wall Geometry in a Group, and it has removed one of the Faces from the connected loop of Faces:

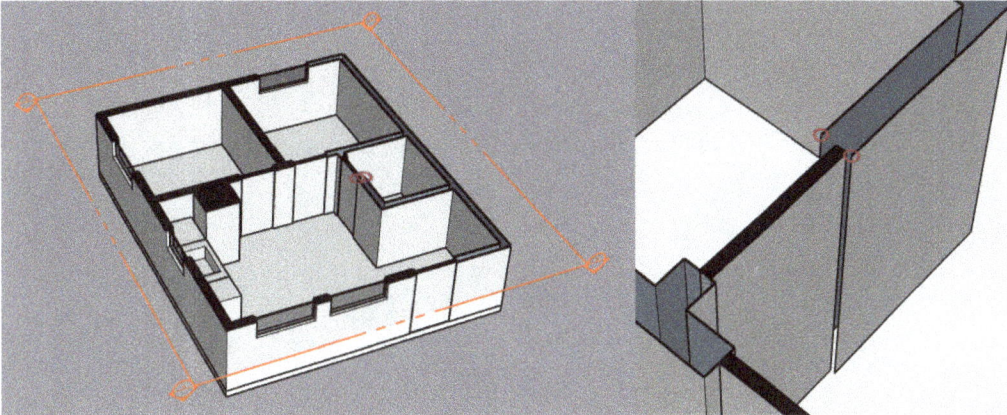

Figure 7.119 – Troubleshoot Section Fill Red Circles (Left); Wall Geometry Isolated with Missing Face (Right)

This Face can be redrawn with the Rectangle tool; by doing this, the loop of connected Faces will be restored. Now, with the Section Plane active and Section Fill toggled on, we can see the Section Fill in the walls. Troubleshoot Section Fill is still toggled on in this example, but it can be toggled off if there are red circles in the Geometry that are not meant to be part of a closed profile:

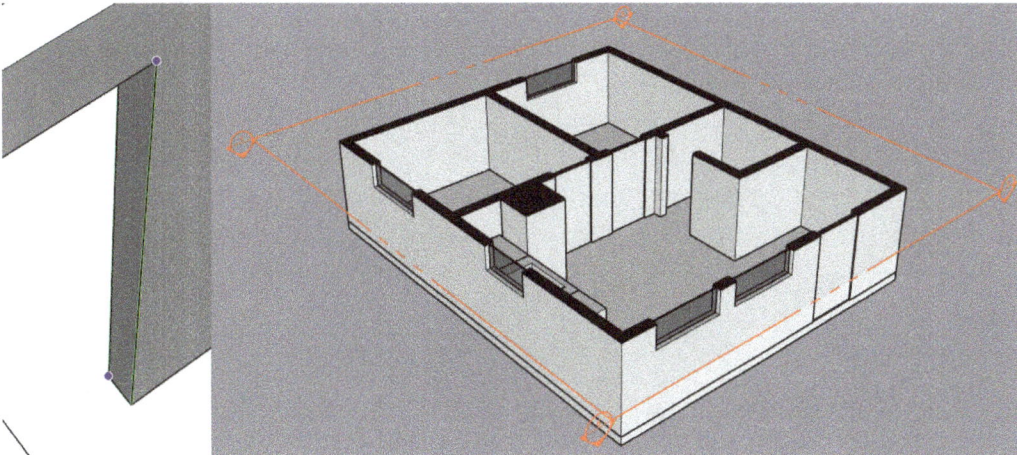

Figure 7.120 – Wall Geometry Redrawn with the Rectangle Tool (Left);
Section Plane Active with Section Fill Toggled On (Right)

The final Section Plane feature we will look at is **Create Group from Slice**, which can also be found in the right-click menu of a Section Plane. We will use the same architectural model for this example:

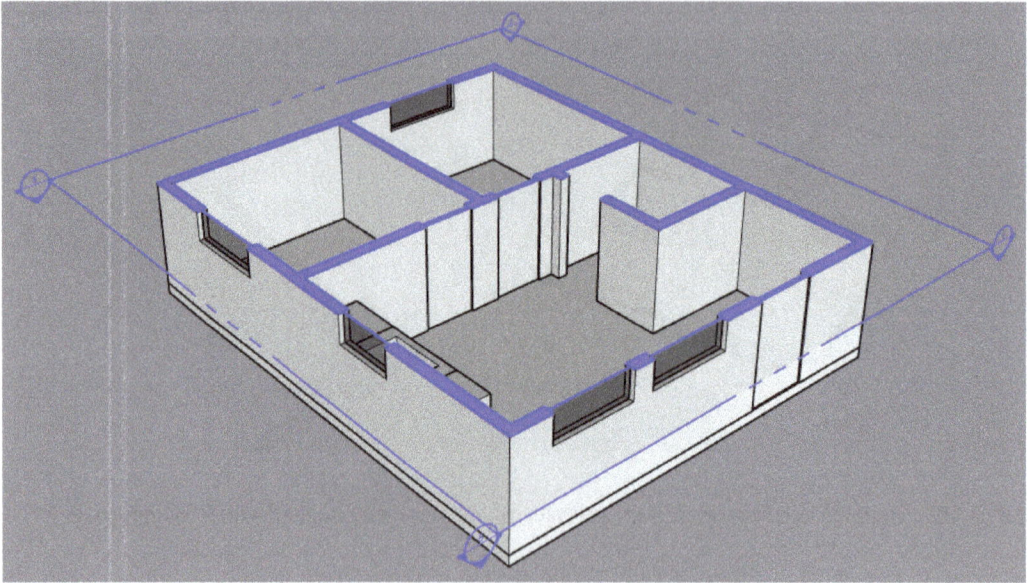

Figure 7.121 – Architectural Model with Section Plane Active

With the Section Plane selected, we can right-click and choose **Create Group from Slice**. Any Geometry that is sliced by the Section Plane will be used to create an Edge along the Section Plane, and the new Edges will be added to a Group. This will create a 2D representation of the 3D model along the Section Plane:

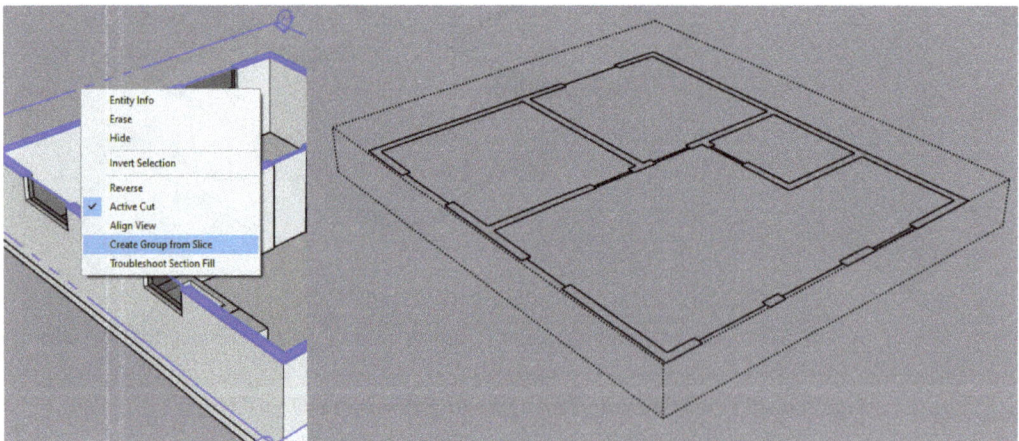

Figure 7.122 – Right-Clicking Create Group from Slice (Left); Grouped New Edges Isolated (Right)

This Group of Edges can be used as a background for 2D drawings of Plans or Sections in technical or construction documents, or to create new Geometry for the SketchUp model. This workflow can also be used to add new Edges to the SketchUp model if the Group is exploded.

The remaining **View** options are some of the most important – Scenes and Animation. Scenes and Animations can be used to save our View and Camera options so that we can save precise views and quickly toggle between saved settings.

Scenes

Scenes can be understood as View snapshots, which save the View and Camera options. Scenes will be shown in the **Scenes** tabs that appear at the top of the Drawing Area when a Scene is created. Additionally, more Scene options can be found in the **Scenes** Panel in the **Default Tray** panel:

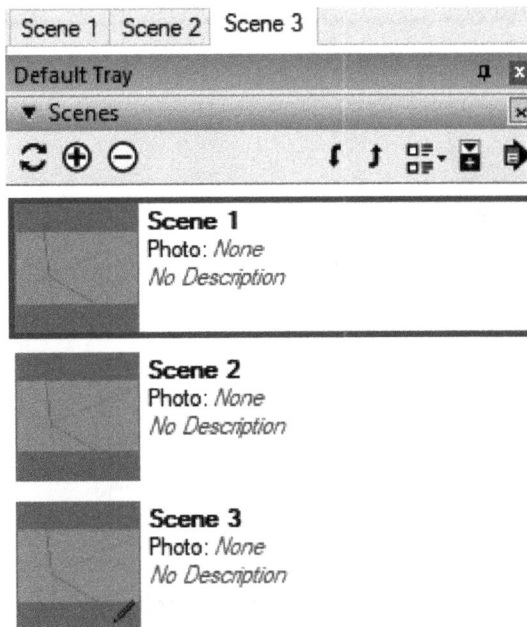

Figure 7.123 – The Scenes Tab and Scenes Panel in the Default Tray Panel

> **Note**
>
> The Scene tabs can be toggled off in the **View** dropdown in the Menu Bar if you are interested in preserving the Drawing Area space:

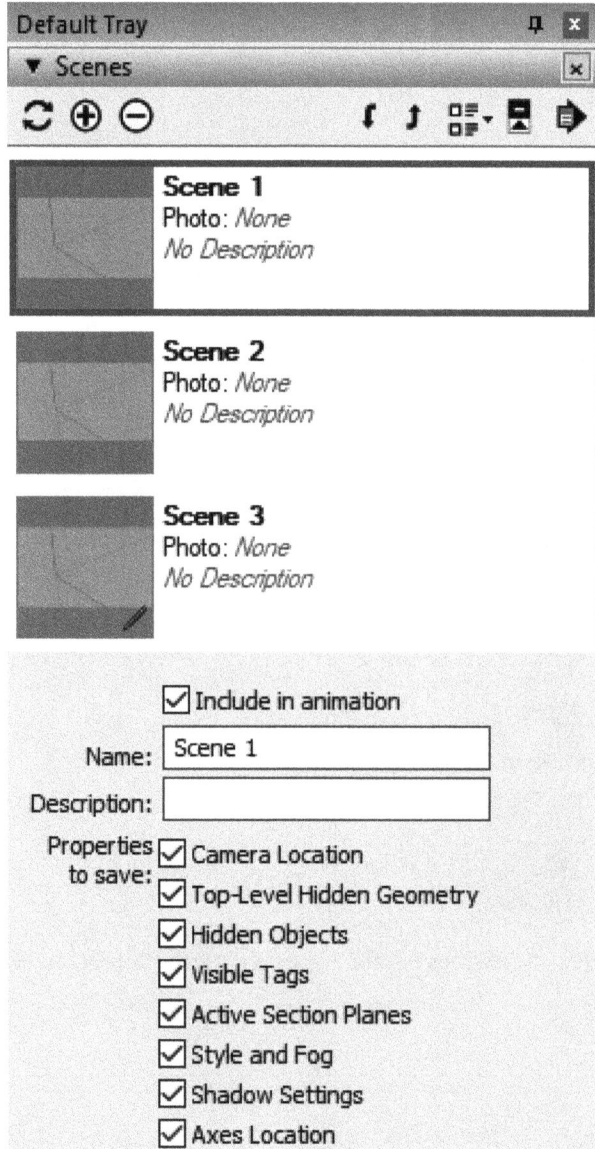

Figure 7.124 – The Scenes Panel in the Default Tray Panel with Show Details Toggled On

The Scene Tools can be found in the **View | Animation** area in the Menu bar dropdown:

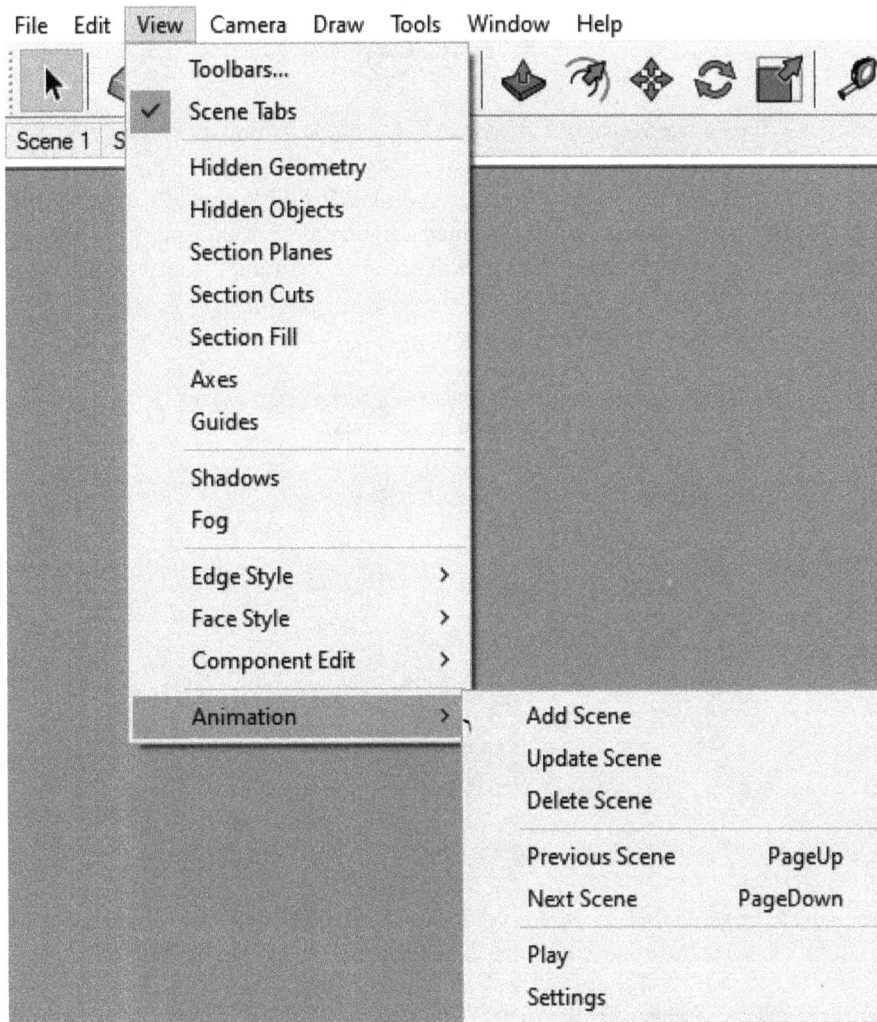

Figure 7.125 – Animation Flyout in the View Menu

It may be confusing to see Scene tools located in the Animation flyout, but we will see later in this chapter how Scenes and Animation are closely connected in SketchUp Pro.

Update Scene is represented by two circular arrows and **Add Scene** and **Delete Scene** are represented by plus and minus signs on the Scenes panel in the **Default Tray** panel, respectively:

Figure 7.126 – The Update Scene, Add Scene, and Delete Scene Buttons on the Scenes Panel

Add Scene and **Delete Scene** do exactly what they sound like! **Add Scene** adds a Scene using the currently set view and camera options, including camera location and current Style. If the View options do not match the current Style options, SketchUp will prompt you to add the settings in a new Style:

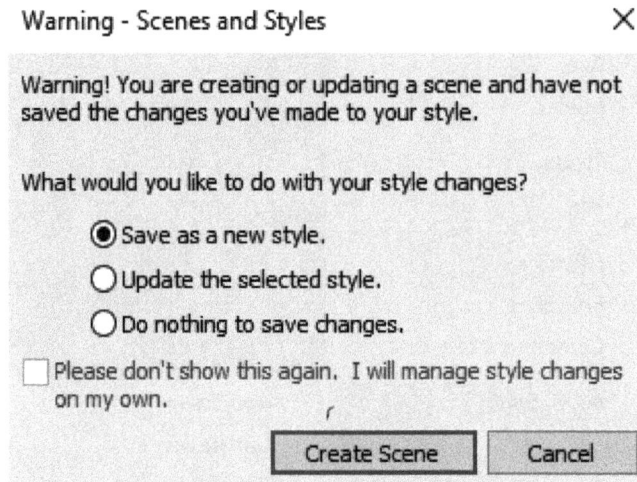

Figure 7.127 – Warning – Scenes and Styles Dialog Box

Styles do not need to be saved for each Scene, but it can be useful if the settings are going to be used to create multiple Scenes in the project, or if the View options would be useful for further modeling.

When a Scene is added, it will appear in the Scenes tabs and the Scenes panel. The Scene tab only shows the Scene name, while the **Scenes** panel in the **Default Tray** panel will include a thumbnail and name, and an optional Matched Photo and description. The Scene name can be updated by right-clicking on the Scene and choosing **Rename Scene**, or by changing the **Name** text box with **Show Details** toggled on. When a Scene is selected, the description can also be edited in the **Details** section of the Scenes panel:

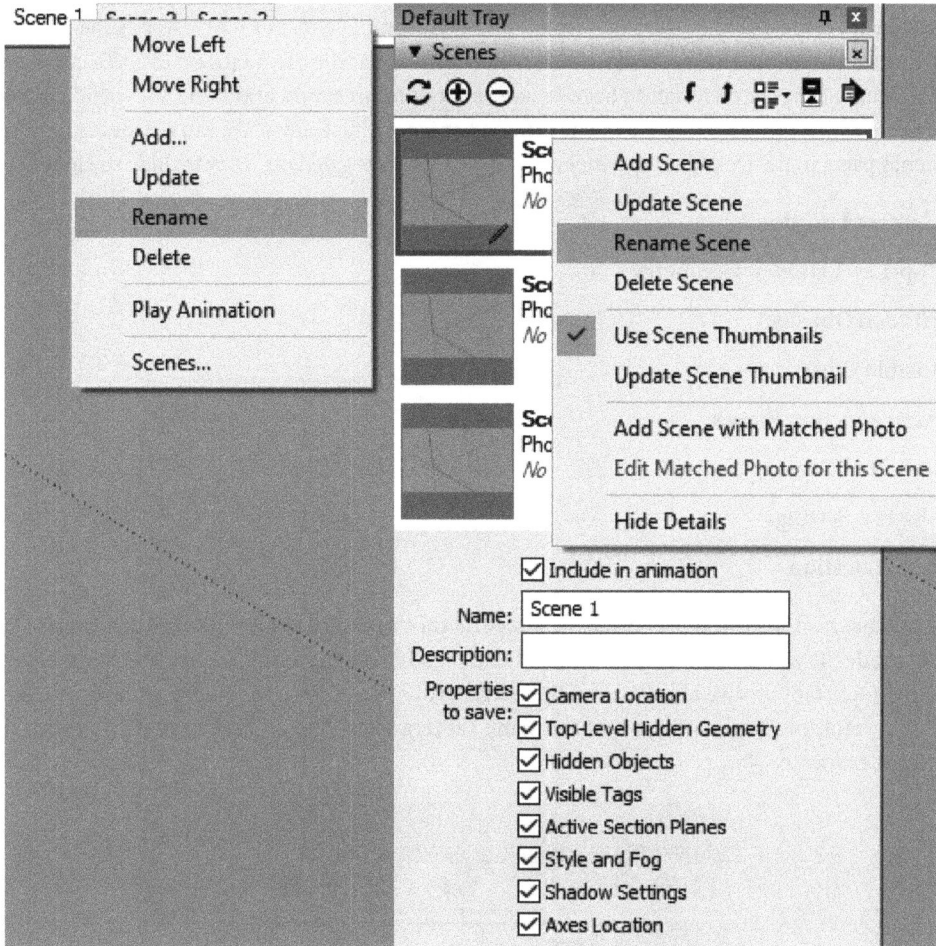

Figure 7.128 – Right-Clicking to Update Name (Left); Details with Name and Description Text Boxes (Right)

Scene Thumbnails can also be updated by right-clicking, which can be useful if additional Geometry has been added to the model.

> **Note**
> Adding, Deleting, and Updating Scenes are not the same as editing the SketchUp model. The Undo and Redo tools do not remember Adding, Deleting, or Updating Scenes – so be careful!

Updating Scenes essentially has the same function as Adding a Scene, but it overwrites the previous Scene settings of the selected Scene. So, be careful when updating Scenes if you meant to add a new Scene instead! Adding Scenes and Updating Scenes saves a lot of information about the View and Camera options. These properties can be selected to be included or ignored in Scenes. We can see these properties in the Scenes panel in the Default Tray panel with **Show Details** toggled on. They include the following:

- **Camera Location**

- **Top-Level Hidden Geometry**

- **Hidden Objects**

- **Visible Tags**

- **Active Section Planes**

- **Style and Fog**

- **Shadow Settings**

- **Axes Location**

Checking or unchecking these properties will depend on the workflow and desired outcome of the SketchUp model. If you are working on a model and creating multiple still Scenes for image export, it would be important to save almost all of the Scene properties. In the following example, we have a model with multiple distinct Scenes of the same camera view. These Scenes are named with the default Scene names 1 – 4:

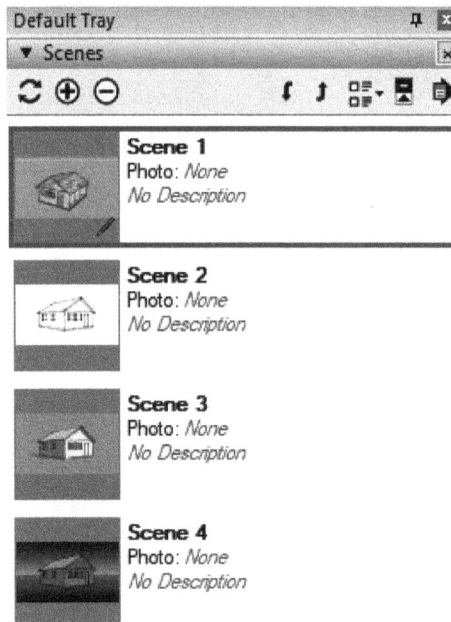

Figure 7.129 – SketchUp Model with four Scenes

Each of these Scenes is meant to represent a specific Style, even though they all have the same camera angle. In this case, it is important to set a Scene of the modeling environment. In this case, Scene 1 has X-ray on, and there are no other stylized options such as Shadows or Fog:

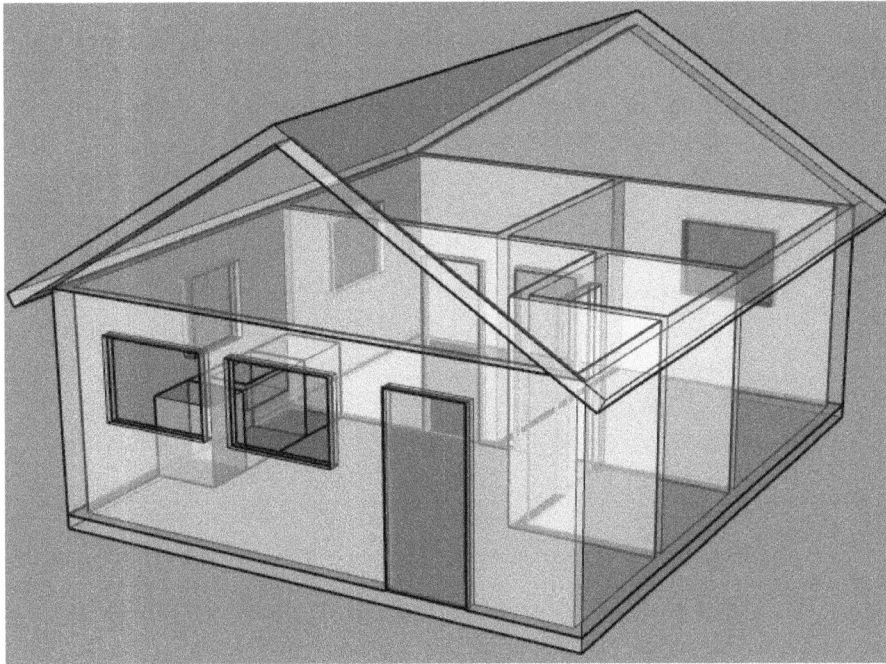

Figure 7.130 – Scene 1 is used for Editing the model

Scene 2, Scene 3, and Scene 4 all have the same camera angle but have drastically different Styles. These Styles differ from Edge Styles, Face Styles, Shadows, and Fog. In this case, it is useful to have multiple Scenes to quickly switch between these Styles so that the image can be exported for different uses. Scene 2 might be used as a background for hand-coloring studies, Scene 3 might be used as a web graphic for marketing, and Scene 4 might be used to prepare a post-production night render:

Figure 7.131 – Scenes 2, 3, and 4 save different Style options

It may be frustrating that Scene 1 returns to the same camera location each time it is selected to return to modeling. In this case, **Camera Location** could be unchecked for Scene 1 and the **Update Scene** button could be clicked. Now, the Scene will no longer return to the original camera view while modeling, which can save time and frustration! Along the same lines, **Hidden Objects, Visible Tags**, and **Active Section Planes** are often useful to have unchecked for editing Styles. **Axes Location** might be a great option to have set differently between multiple Scenes for editing, especially in projects with Geometry that are not aligned to the world Axes. It will depend on the project you are working on, and it will take some trial and error to get used to!

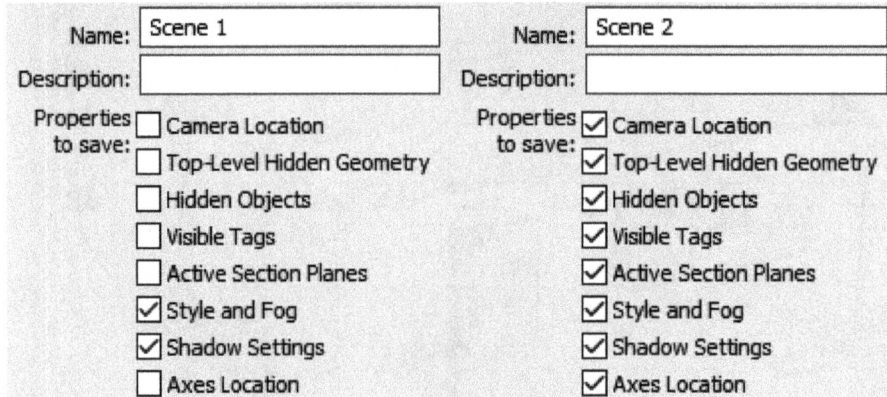

Name: Scene 1	Name: Scene 2
Description:	Description:
Properties to save: ☐ Camera Location	Properties to save: ☑ Camera Location
☐ Top-Level Hidden Geometry	☑ Top-Level Hidden Geometry
☐ Hidden Objects	☑ Hidden Objects
☐ Visible Tags	☑ Visible Tags
☐ Active Section Planes	☑ Active Section Planes
☑ Style and Fog	☑ Style and Fog
☑ Shadow Settings	☑ Shadow Settings
☐ Axes Location	☑ Axes Location

Figure 7.132 – Scene 1 Properties for Editing (Left); Scene 2 Properties for Image Export (Right)

Scene settings are especially critical to understand in Animation workflows. In the next section, we will discuss how Scenes and Animation work together in SketchUp Pro.

Animation

SketchUp supports physical cameras in the Drawing Area environment, but for this book, we will focus on the Animation settings built in for Scenes. If you are interested in working with more complex cameras in SketchUp Pro, I suggest checking out SketchUp Extensions online.

Animations in SketchUp Pro are created by interpolating between the Scenes in the SketchUp model. Interpolation is a term used in animation, which is sometimes also called inbetweening or filling in frames between keyframes in the Animation. Many animations created in other software use interpolation with complex formulas for changing the Geometry, but SketchUp uses a much simpler solution. SketchUp simply moves the camera between the two Scenes with a smooth motion to create points for the animation. This smooth motion is called a Scene Transition.

> **Note**
> Animation is very difficult to show in still images. Try this for yourself to see how SketchUp smoothly transitions between Scenes!

Once Scenes have been created in the model, the Animation can be viewed. Scenes will only be included in the Animation if the **Include in Animation** checkbox is checked for the Scene in the **Scenes** panel in the **Default Tray** panel. The Animation can be played by doing the following:

1. Right-clicking on the Scene tabs and choosing **Play Animation**.

2. Choosing **View** | **Animation** | **Play** from the Menu Bar dropdown.

When the Animation is played, it will begin on the Scene that was most recently selected. The SketchUp **Animation** dialog box will appear while the Animation is playing. The **Animation** dialog box will contain **Pause** and **Stop** buttons:

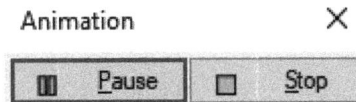

Figure 7.133 – Animation Dialog Box

Hitting **Pause** will keep the dialog box open, with the option of restarting the Animation by hitting **Play**. Hitting the **Stop** button will stop the Animation and will close the dialog box. No tools may be used while an Animation is playing or paused, and the Animation dialog box must be closed before you can continue to work in the SketchUp model.

Scenes can be moved to change the order of an Animation. Scenes can be moved like so:

1. Right-clicking on a Scene tab and Choosing **Move Right** or **Move Left**.

2. Clicking on the **Move Scene Down** or **Move Scene Up** buttons at the top of the Scenes panel in the **Default Tray** panel:

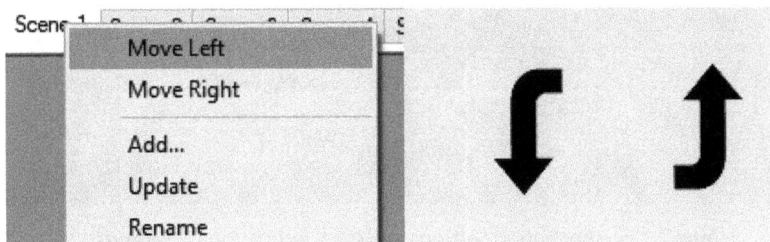

Figure 7.134 – Right-Clicking the Scene Tab (Left); the Move Scene Down
and Move Scene Up Buttons on the Scenes Panel (Right)

> **Note**
> Scene names are not considered in the animation! You must order them correctly on the Scenes Tab or Scenes panel in the Default Tray for the Animation to play in the right order!

SketchUp Pro has two adjustable settings for Animations. These Animation Settings can be found in the **View** | **Animations** | **Settings** dropdown in the Menu Bar dropdown:

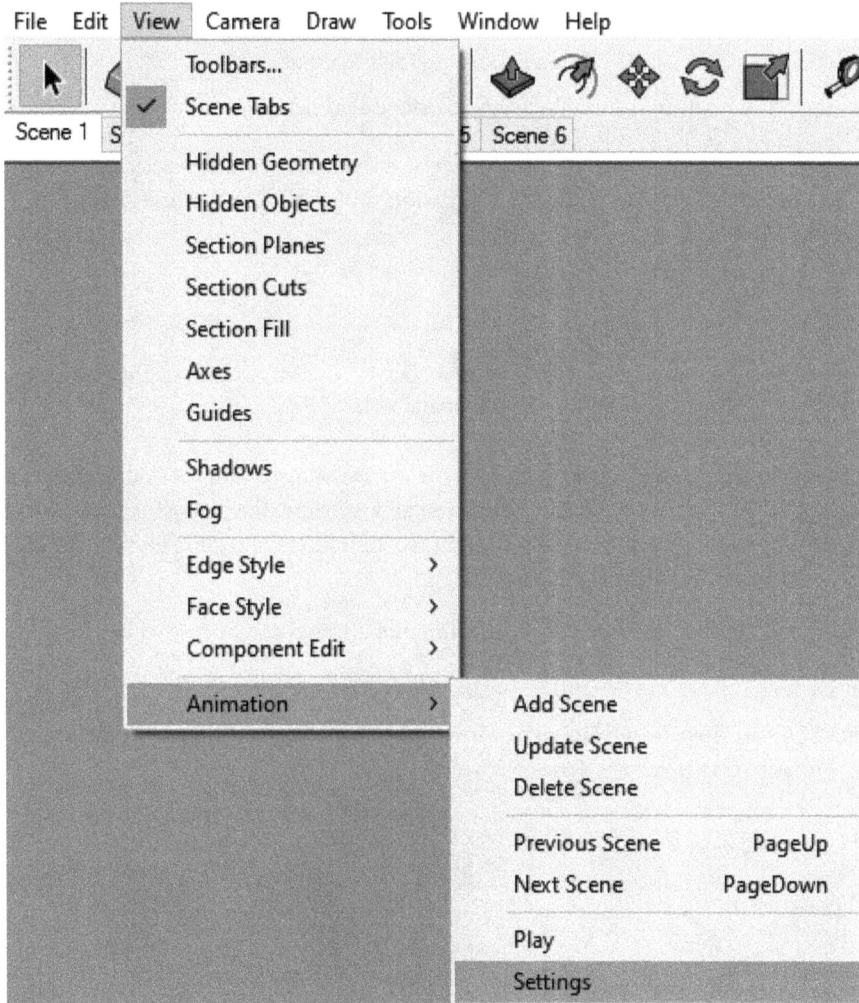

Figure 7.135 – Animation Settings in View > Animations > Settings

The Animation Settings are located in one of the tabs in the **Model Info** dialog box, which we discussed when looking at Geo-location. We can see the two adjustable settings here – **Scene Transitions** and **Scene Delay**:

Model Info - House 2

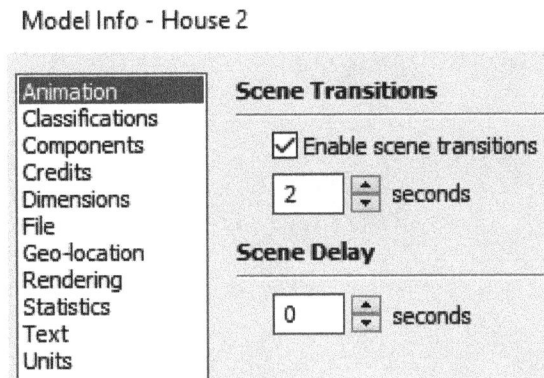

Figure 7.136 – The Animation Tab in the Model Info Dialog Box

Scene Delay sets the time that the camera should pause on each Scene. This may be a desired effect if you would like to allow the viewer to spend more time on the Scene itself, rather than having constant motion in the Animation. In this case, **Scene Delay** can be set for any number of seconds up to 100. If no pause is desired for the Scenes, **Scene Delay** should be set to 0.

> **Note**
>
> When **Scene Delay** is set to 0, the SketchUp Pro Animation sequence will appear to pause briefly at each Scene. This slight pause only occurs in SketchUp and will not be present if the Animation is exported.

Scene Transitions sets the time that the camera has to move from one Scene to the next. This setting does not precisely set the speed of the camera, but SketchUp will calculate the relative speed of the camera based on the camera locations in the corresponding Scenes. If two Scenes are created where the camera is relatively close together, the camera will appear to move slowly. If the Scenes are created with the camera far apart in the Scenes, the camera will appear to move quickly.

Scene Transitions may take some trial and error to set up in your model correctly. If a smooth animation is desired for a fly-around or walkthrough, it will be beneficial to create the Scenes at regular intervals. In the following example, we can see a representation of where eight Scenes were created for a fly-around animation:

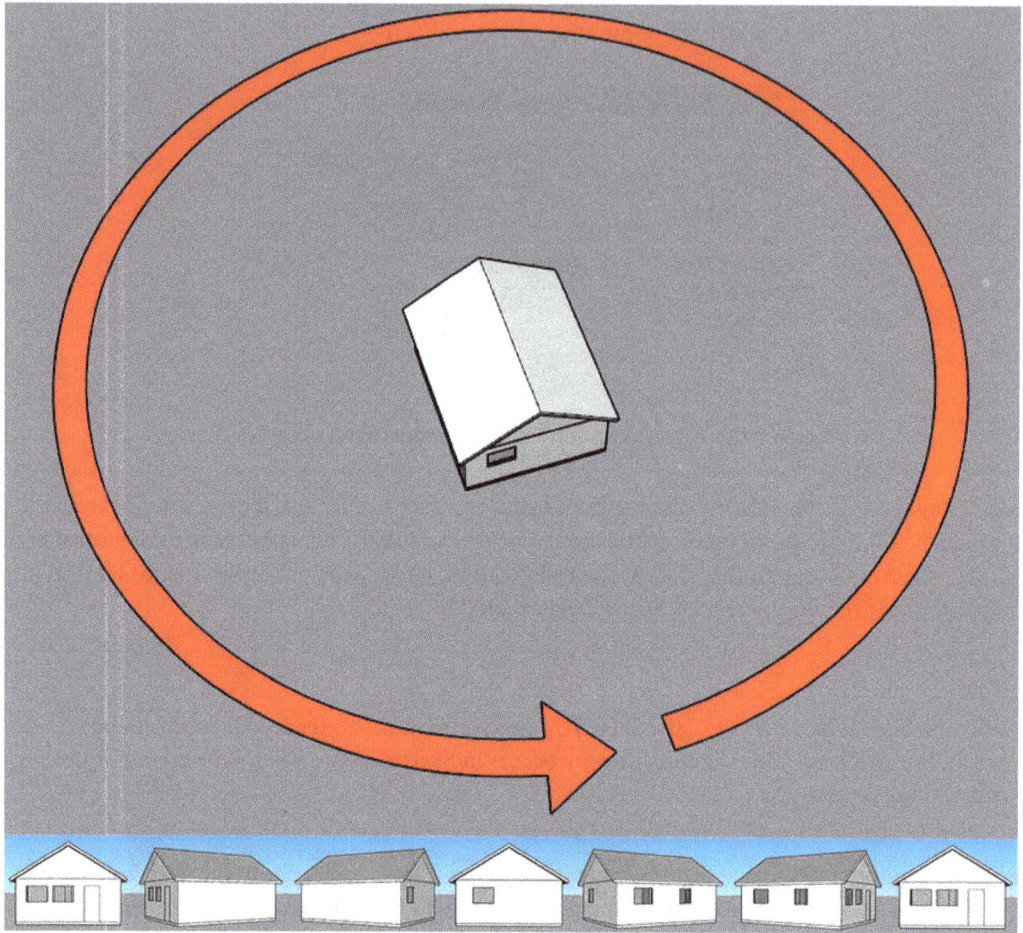

Figure 7.137 – Flyaround Animation from Above (top) Animation Storyboard (bottom)

If the Scenes were created at irregular intervals, the animation would appear to speed up and slow down between each Scene. In the following screenshot, we can see poor spacing for the Scenes for this fly-around:

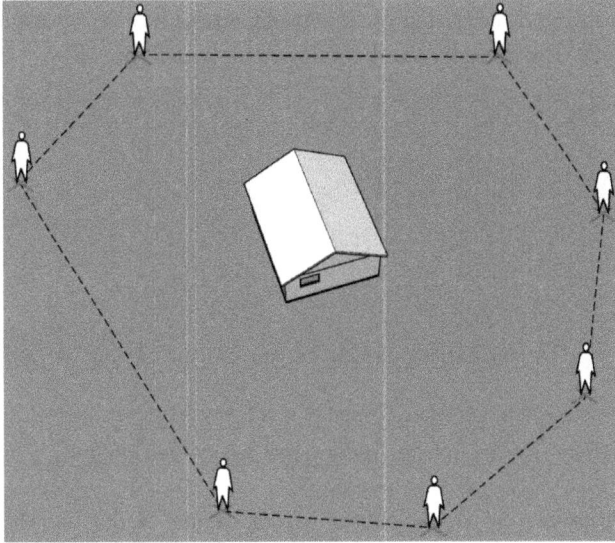

Figure 7.138 – Fly-around Animation with Poor Scene Locations

In the following figure, we can see excellent Scene locations for this fly-around:

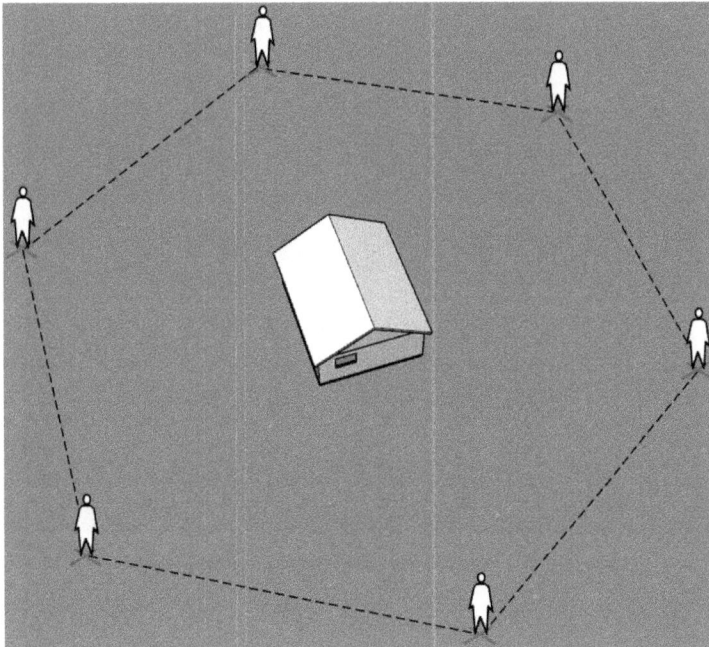

Figure 7.139 – Fly-around Animation with Good Scene Locations

In this case, the cameras were placed using the **Position Camera** tool on a polygon to get the exact location and camera angle. The polygon was then hidden for each Scene:

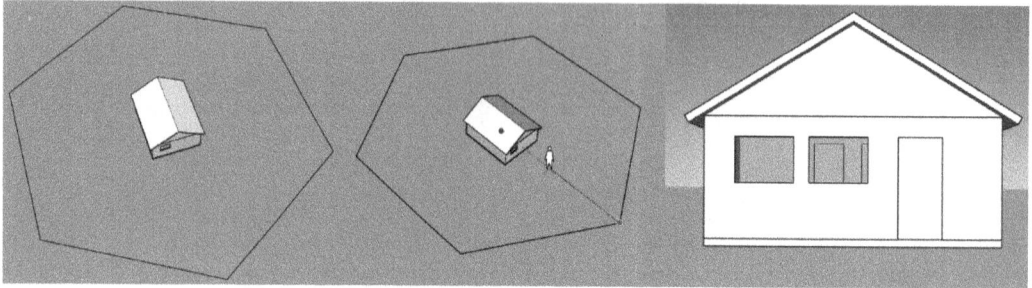

Figure 7.140 – Polygon Drawn in the Model (Left); Position Camera Tool
Used (Middle); Polygon Hidden and Scene Added (Right)

The same principle applies when creating a walkthrough of a model. In the following example, we can see poor Scene locations and good Scene locations. Even though the example in the second image has more Scenes, they are spaced closer together to create a more consistent camera speed. The Scene Transition time can always be adjusted to update the time spent between each Scene:

Figure 7.141 – Poor Scene Locations for Walkthrough

Figure 7.142 – Good Scene Locations for Walkthrough

> **Note**
> **Scene Transitions** can be toggled off using the checkbox in the **Model Info** dialog box. This may be helpful when modeling so that there is no delay when switching between Scenes!

Camera movement in Scene Transitions is the only smooth transition between Scenes in an Animation. Unfortunately, Animations in SketchUp Pro do not smoothly transition between the other Properties in Scenes, including the Styles, Hidden Objects, or Visible Tags. During an Animation, or when a new Scene is selected, all other Scene Properties will instantaneously change at the beginning of the Scene Transition. Then, the smooth camera transition will follow.

Animations can be exported from SketchUp into image or video files. Image files will produce a series of images that can act as slides for the Animation. SketchUp natively supports exporting to the JPEG, PNG, BMP, and TIFF image formats. SketchUp natively exports to one video format, .mp4. Animations can be exported by going **File** | **Export** | **Animation...** in the Menu Bar dropdown:

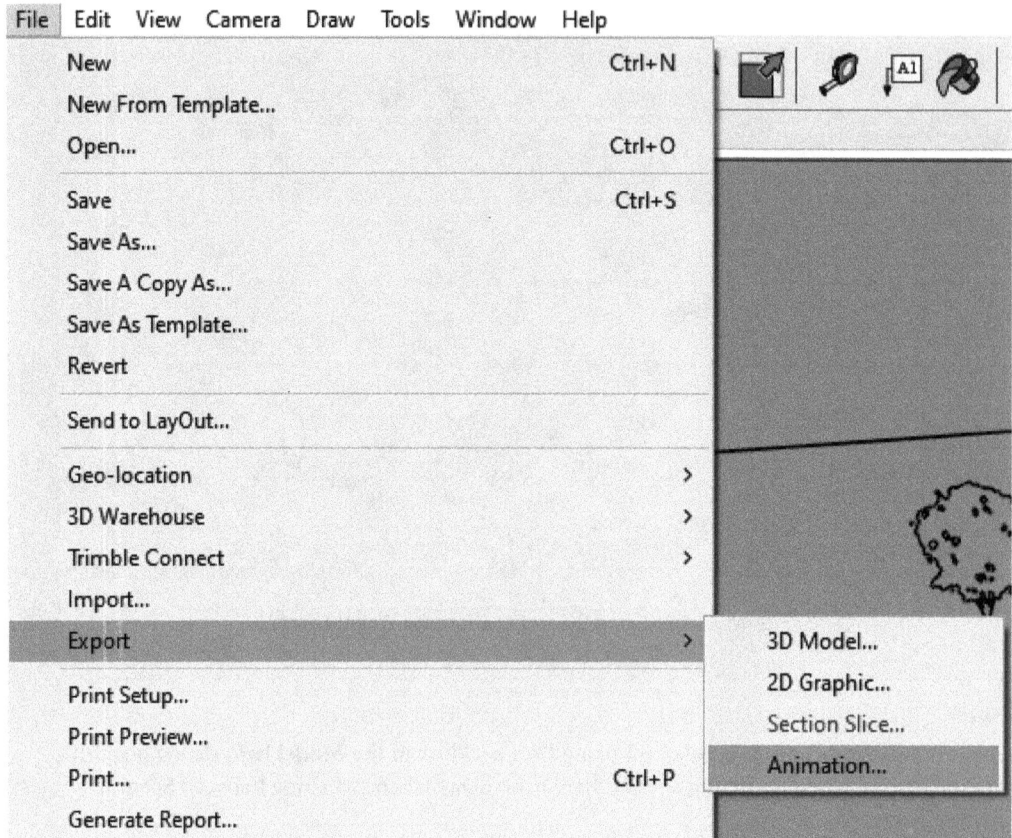

Figure 7.143 – Animation Export in File | Export | Animation

> **Note**
> By choosing **2D Graphic…** instead of **Animation…,** you can export single images of the model. This is great for individual Scenes!

SketchUp will automatically calculate the length of the animation based on the Scene Transition duration and the number of Scenes. SketchUp exports 24 frames per second of video time. In the following example, we can see that this model has four Scenes, **Scene Transitions** has been set to 2 **seconds**, and **Scene Delay** has been set to 0 **seconds**:

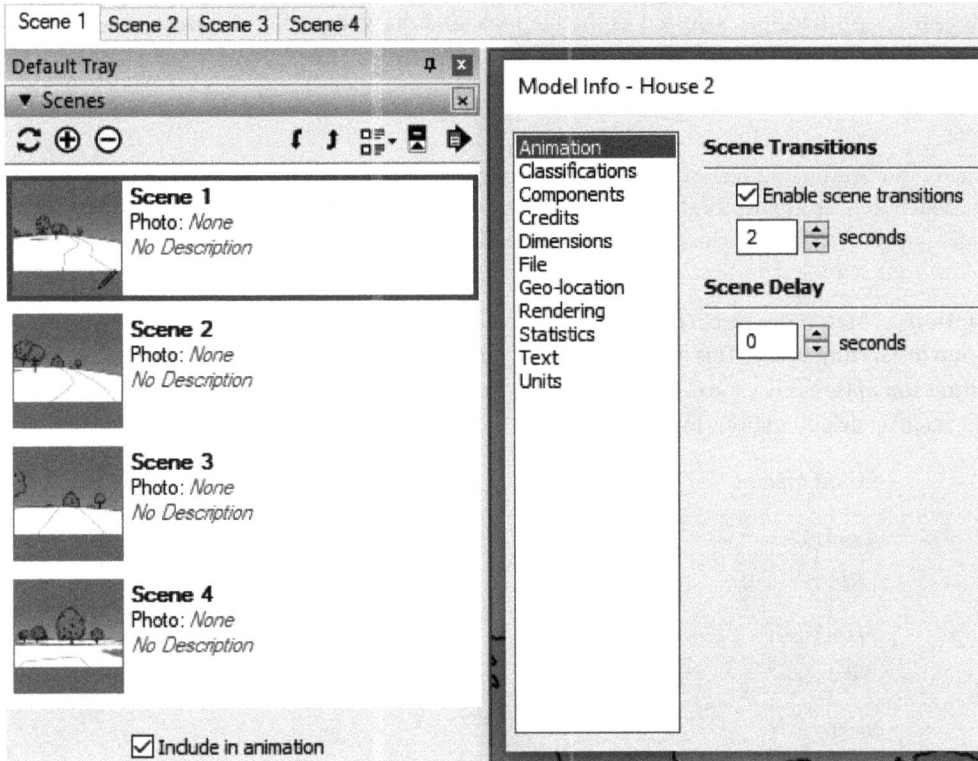

Figure 7.144 – Model with four Scenes, Scene Transition Set to
2 Seconds, and Scene Delay Set to 0 Seconds

When this model is exported to an Animation, the **Export Animation** dialog box opens:

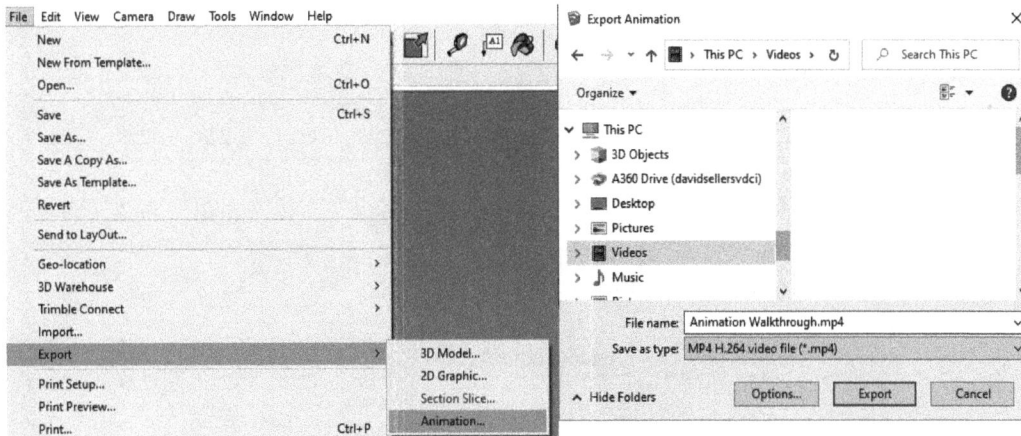

Figure 7.145 – File | Export | Animation… is Clicked (Left); Export Animation Dialog Box (Right)

In this case, `.mp4` has been selected as the file type, and the **Videos** folder has been selected as the destination.

> **Note**
>
> When exporting images, make sure that you select an empty folder for the destination! SketchUp Pro does not create a folder for the Animation images, and they will all be placed in the destination folder. If your Animation is longer than 1 minute, that could be thousands of images!

The **Options…** button can be clicked to set more precise options, such as **Resolution**, **Frame Rate**, and **Loop to Starting Scene**. This dialog box can be toggled to always show using the **Always prompt for animation options** checkbox. Additionally, the **Restore Defaults** button can be clicked to return to the SketchUp default options for Animations:

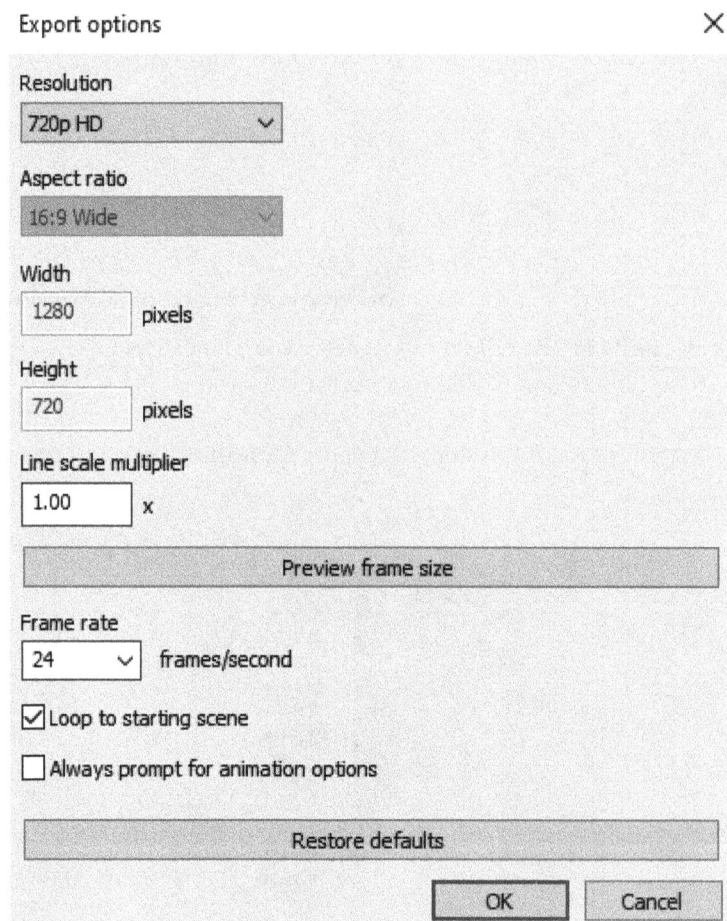

Figure 7.146 – The Export Options Dialog Box

In this example, the default options have been used. Because there are four Scenes, and **Loop to Starting Scene** has been checked, there will be four transitions or 8 total seconds of Animation time. 8 seconds times 24 frames a second equals 192 total images. SketchUp adds one more image for the starting Scene, so we end up with 193 total images. After clicking the **Export** button, we will see the **Exporting animation…** dialog box, which shows this number and a progress bar for the export:

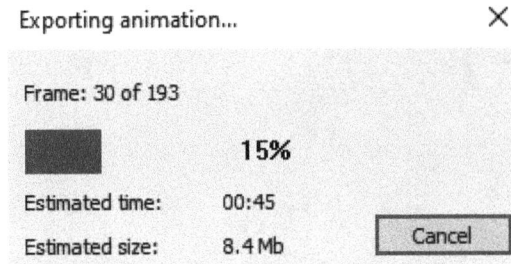

Exporting animation...   ✕

Frame: 30 of 193

                              15%

Estimated time: 00:45
Estimated size: 8.4 Mb   Cancel

Figure 7.147 – The Exporting animation… Dialog Box

Animation and Scenes are the best way to showcase your SketchUp models with a distinct SketchUp look and feel! Make sure to update Scenes with the right Properties checked and include only the Scenes you want in your Animations.

Summary

In this chapter, we discussed the View Tools and their importance when it comes to assisting with modeling and showcasing your SketchUp models! With custom settings for Edge, Face, and Component Styles, you can create unique and dynamic combinations to bring your SketchUp models to life! Using the Style Tools effectively can tell different stories using the same model – only by updating the way SketchUp shows us the Geometry.

We also discussed Shadows and Fog, which can provide depth and a sense of realism to your SketchUp models. Try using Geo-location to find a place that you have been and see how well the Shadows match up with your model!

Sections are powerful tools for modeling but can also be used in exporting Scenes and Animations. Scenes help save Style information from one Scene to the next but can also help save camera views that can lead to dynamic images and animations.

We now understand how to see our models with the Camera tools, and how to view the models with Styles and Scenes. In the next chapter, we will discuss Materials and Textures.

8

Materials

In this chapter, we will look at Materials in SketchUp Pro. Materials in SketchUp Pro are applied to Faces and can be represented by colors or image textures. We will discuss how to use the Paint Bucket tool and the Materials Panel to select and apply Materials to Objects. We will also look at the Modifier Keys to better understand how to apply Materials to update existing Materials. We will briefly look at creating Materials from colors and image imports, and we will finish the chapter by looking at advanced texture Material editing techniques.

The following topics are covered in this chapter:

- Materials:
  - Colors
  - Textures
- The Paint Bucket Tool:
  - Applying Materials
  - Sampling Materials
  - Modifier Keys
- The **Materials** Panel:
  - Creating and Editing Materials
  - Managing Materials in Your Model
- Advanced Texture Material Editing:
  - Texture Position
  - Projected Textures

Materials

Materials in SketchUp Pro are applied to Faces. Materials can represent a solid color or image texture. Materials can be painted onto Faces using the Paint Bucket tool found on the Large Tool Set. Before we learn about all of the Paint Bucket options and ways to add Materials to your SketchUp Models, let's look at the difference between color and texture Materials.

Colors

Color Materials are represented by a single color for the entire Face. Color Materials can also include an Opacity amount to make the color translucent. A color Material may appear to look different, depending on the orientation of the Geometry, because of how SketchUp shades a Model to provide depth, even when Shadows are turned off. In this next example, a box has all Faces colored the same shade of orange, but the Faces appear to be different shades:

Figure 8.1 – Orange Box Appears to Have Different Color Materials

This box has been painted with the same Material on all sides. It is important to keep this in mind when working with color Materials, as different camera orientations can drastically change the anticipated color for a scene or image export.

Textures

Texture Materials include an image as part of the Material. This image can be a seamless texture and applied across a large surface area, or a standalone image for a single purpose. In the following example, we can see that a seamless brick texture has been used to tile the Material across an entire wall, and a standalone image has been used as the Material on a sign:

Figure 8.2 – A Brick Texture Material on a Wall and a Sign Texture Material on a Sign

When working with texture Materials in SketchUp there are very few rules that SketchUp will enforce. We can see in the following figure that the Materials have been switched, and SketchUp would not acknowledge this as a problem:

Figure 8.3 – A Sign Texture Material on a Wall and a Brick Texture Material on a Sign

SketchUp does not assign certain texture Materials to a tile and others to be standalone. All texture Materials will tile or be cropped based on the image size and Geometry size, so be careful when using texture Materials!

Texture Materials can also be colored and have opacity, but black-and-white patterns can also be used as textures. Also, if the image contains transparency, such as in a PNG image file, the texture Material will render that portion of the Face as 100% transparent. We will discuss how to create and edit texture Materials later in the *Create and Edit Materials with the Materials Panel* section.

A common misconception is that a texture Material should always be used instead of a color Material. It can often be more beneficial to use a color Material for your SketchUp Model, especially if the object is not a featured part of the design. Let's take a look at some examples of why color Materials may be more appropriate to use than texture Materials.

The Object is Not the Feature of the Scene

It can be very helpful to use a mix of color Materials and texture Materials in your SketchUp Models. In this example, there is a feature object for the scene, and it is fully painted with texture Materials. There are also objects in the background, and these objects are not meant to draw any attention to themselves. If they are fully textured, they might draw the eye of the viewer and could detract from the feature object.

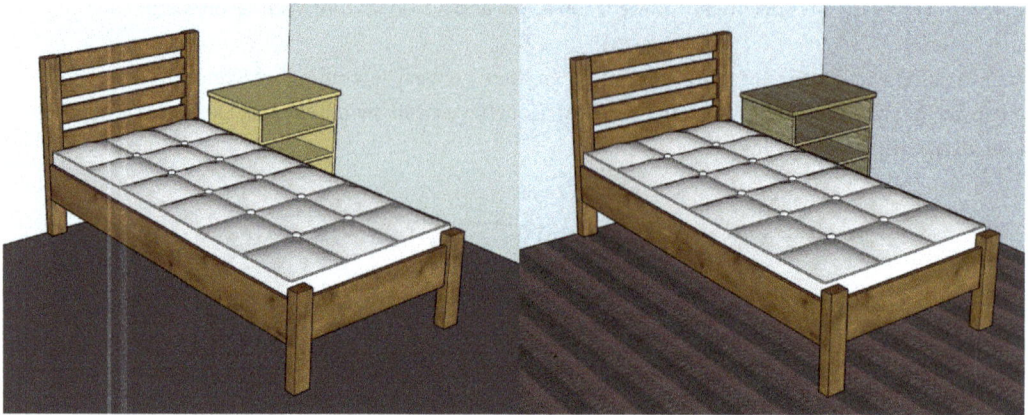

Figure 8.4 – A Feature Object Fully Textured (Left) and All Objects Fully Textured (Right)

This can also be helpful when working with smaller objects that can easily be represented by a color because the amount of texture provided would not improve the overall impression:

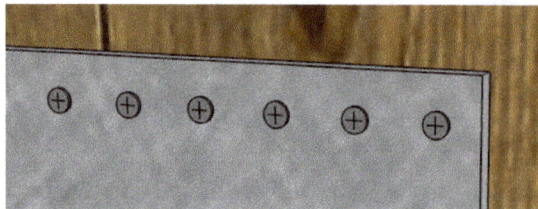

Figure 8.5 – Small Screws Colored Grey

And finally, if objects are far in the distance, they do not need to be textured, as the texture would be too small to see in the desired view:

Figure 8.6 – Distant Objects Painted with Color Materials

> **Note**
> Remember that the **Shaded and Shaded with Textures** options will show only colors and colors and textures respectively. This can be very helpful if, at some point, you want to see all colors but do not want to swap out the textures in your Model.

Textures are Difficult to Scale, Position, and Project

Texture Materials in SketchUp are created using an image file, and that image must be set up to represent real-world dimensions. The scale of an image is critical to get right when using texture Materials, as an incorrect scale can be noticed immediately by the viewer, and even a slightly incorrect scale can be noticed subconsciously. In this example, a brick material has been painted at an incorrect scale, and it makes the scene very difficult to understand:

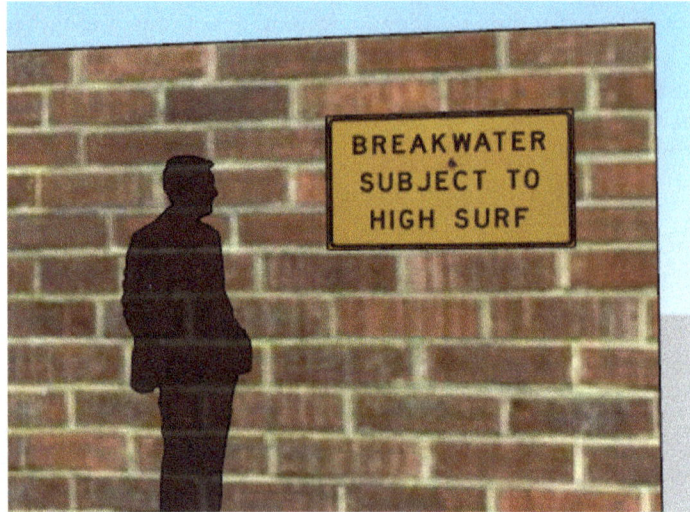

Figure 8.7 – A Brick Texture Material at an Incorrect Scale

Texture Materials must also be positioned to best represent real-world Geometry. In some cases, multiple texture Materials might need to be used to appropriately represent how an object would physically appear. In the next example, we can see a wood beam has been textured appropriately using multiple texture Materials and two beams that have Materials incorrectly positioned. Incorrect positioning can be far worse than a simple color Material representing the Geometry.

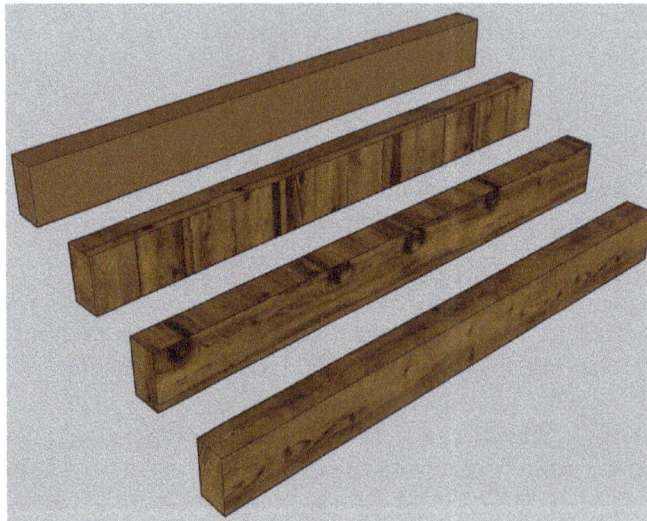

Figure 8.8 – Beams with a Color Material (Left), Incorrect Positions
(Middle), and a Correct Texture Position (Right)

Finally, texture Materials might need to be Projected. This will always be the case with non-planar Faces, as SketchUp Pro does not have a built-in tool to recognize the best way to apply the texture Material. If a large texture Material needs to be placed on complex, non-planar Geometry, the Material will appear to scatter across the Geometry Faces.

Figure 8.9 – A Scattered Texture Material on Complex, Non-Planar Geometry

This can be fixed with a Projected Texture Material, although in some cases, multiple Projected Textures may be required. This can be a time-consuming process, and it might not be necessary if the object can be represented by a solid color:

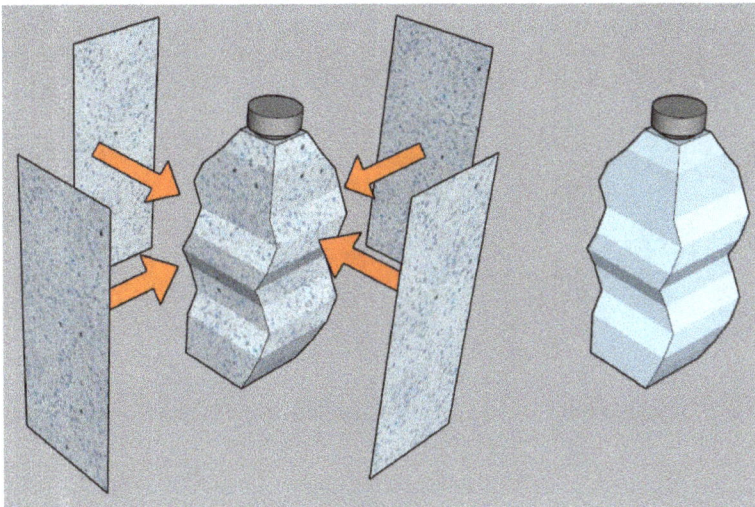

Figure 8.10 – An Object with a Projected Textures (Left) and an Object with Color Materials (Right)

> **Note**
>
> There are plugins and extensions that will assist with placing texture Materials on complex, non-planar Geometry. We will not discuss any specific extensions in this chapter, but it can be helpful to look online for the best tool for your SketchUp Model!

We will discuss workflows to appropriately set the scale, position, and projection of texture Materials later in this chapter. These are important workflows to understand, but they are not necessary for every object in a Model. That can be very time-consuming!

Colors Provide for Client Interpretation

In some cases, using a specific texture Material may provide an incorrect impression to a client or project shareholder. Texture Materials are meant to represent real-world Geometry more so than color Materials, and clients may understand the texture representation as a fully designed element. Color hue, wood grain, pattern spacing, and even grout lines can be misunderstood by clients, especially if texture Materials are presented early in the design process. It may be best to use color Materials to represent Geometry until specific expectations have been provided by a client.

Textures Tile at Scale

Textures will always tile if the Geometry is larger than the image. This is often the case with Materials in SketchUp Pro. Images used in texture Materials can often be referred to as seamless textures, meaning that there is no distinct seam between the edges of the image when it is tiled. In this next example, we can see a seamless texture and a non-seamless texture:

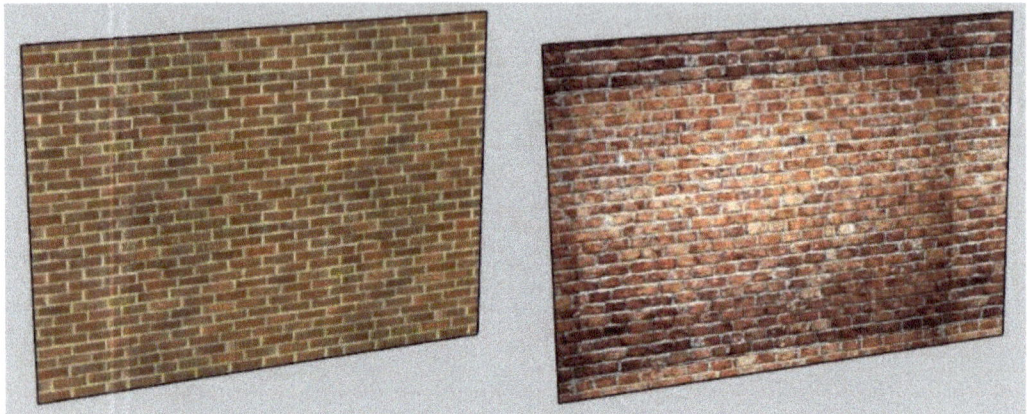

Figure 8.11 – A Seamless Brick Texture (Left) and a Non-Seamless Brick Texture (Right)

> **Note**
>
> SketchUp curates a limited number of Components in the 3D Warehouse of Seamless Texture Images. These can be found in the **Building Products | Finishes** section of the 3D Warehouse. Seamless Texture Images can also be found online using an image search engine, such as Google Images.

Even if a texture image is seamless, it may still be recognizable as a tiled texture, especially if many tiles are required to cover a large Face. In this case, it might be better to use a representative color Material so that the large Geometry does not appear to be so computer-generated:

Figure 8.12 – A Tiled Texture (Left) and a Color Material (Right)

Color Materials Reduce File Size

Using color Materials instead of texture Materials in a Model can reduce the overall file size of it. This may be required when working on large-scale projects, especially when files are shared between computers. We will look at tips for keeping file sizes low when using Materials later in this chapter.

Now that we've examined some potential reasons to use color Materials instead of texture Materials, let's jump right in and look at how to paint Materials onto Geometry using the Paint Bucket tool.

Paint Bucket Tool

The Paint Bucket tool is used to apply Materials to Faces. The Paint Bucket tool can be used to sample Materials that are already applied to Faces in the Model, and multiple modifier keys can be used to update Materials in the Model. The Paint Bucket tool is represented by a paint bucket spilling orange paint. When the Paint Bucket tool hovers over a Face or Object that can be painted, a circle will appear to show the cursor location:

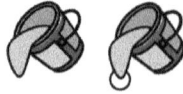

Figure 8.13 – The Paint Bucket Tool (Left) and the Paint Bucket Tool with a Cursor Circle (Right)

The Paint Bucket tool can be found on the **Large Tool Set** toolbar, the **Getting Started** toolbar, and the **Principal** toolbar:

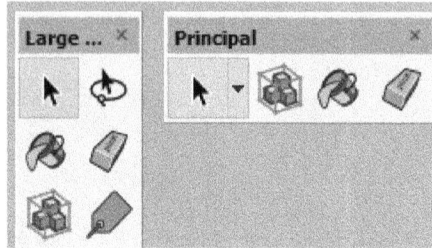

Figure 8.14 – The Paint Bucket Tool on the Large Tool Set Toolbar (Left) and Principal Toolbar (Right)

Additionally, the Paint Bucket tool can be activated by hitting *B* on the keyboard or by going to **Tools | Paint Bucket** in the **Main Menu** dropdown:

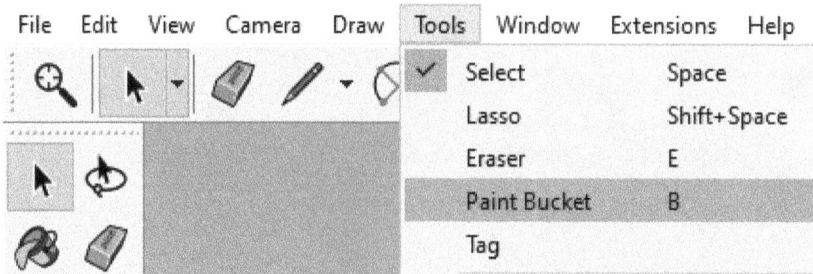

Figure 8.15 – The Tools | Paint Bucket Option in the Main Menu Dropdown

Applying Materials

When the Paint Bucket tool is activated, the **Materials** Panel will automatically open in the **Default Tray** window:

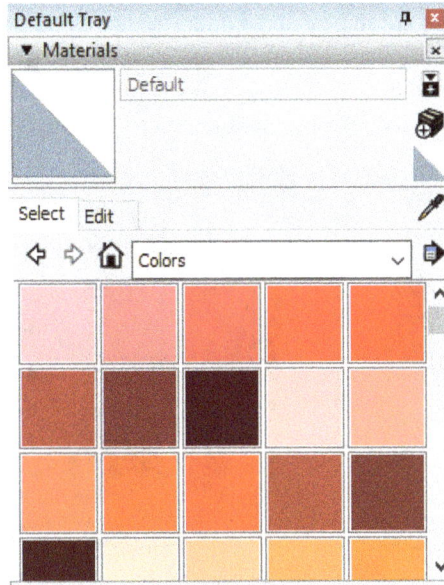

Figure 8.16 – Materials Panel in Default Tray

If the **Materials** Panel is already open in the Default Tray, it will become the active panel. The **Materials** Panel is where Materials can be selected. Once a Material is selected, the current Material icon and name will update at the top of the **Materials** Panel:

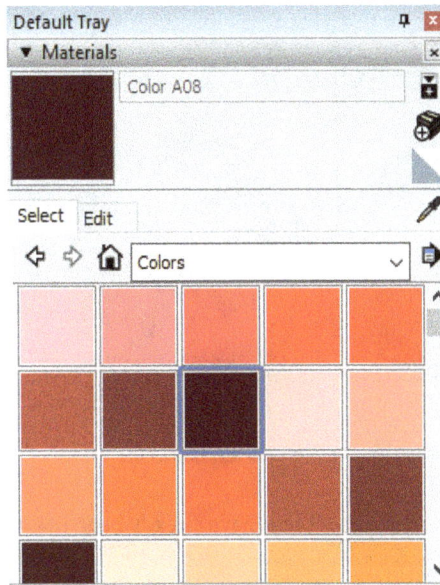

Figure 8.17 – The Material Selected and the Current Material Updated

Once a Material has been selected, it can be painted onto Faces or Objects in the Model. Materials are applied on a per-Face basis or a per-Object basis, whichever is first for that Geometry. In the next example, we will look at a simple ungrouped box. This box has six unique Faces that are not part of a Group or Component. When one Face is clicked with the Paint Bucket tool, it will be painted with the Material:

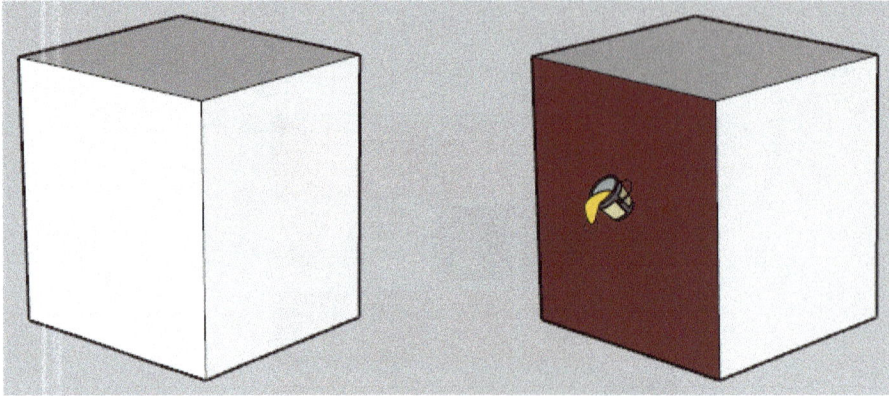

Figure 8.18 – A Simple Ungrouped Box (Left) and a Single Face Clicked
with the Paint Bucket Tool and the Material Painted (Right)

This process can be repeated with the same Material selected or a different Material selected. When clicking on a Face, the Material will always be overridden:

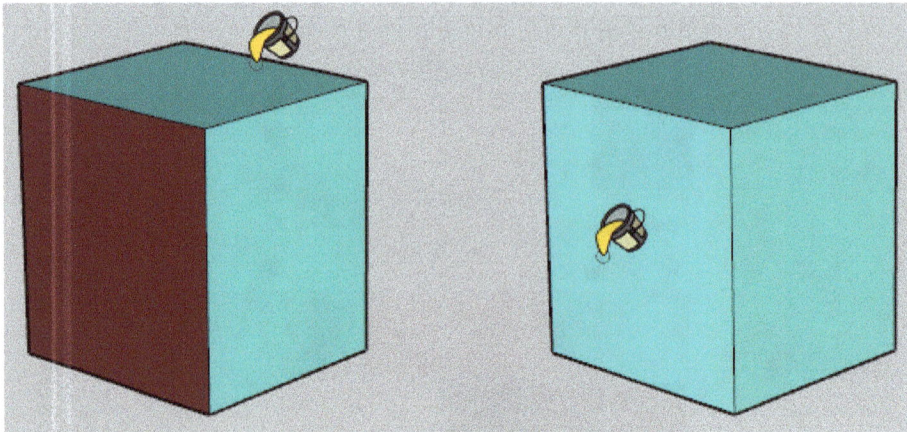

Figure 8.19 – The Remaining Faces Painted (Left) and the Material Overridden (Right)

The Paint Bucket tool will not override Faces that have already been painted when they are part of an Object. In this next example, we will look at the same simple box with one of the Faces painted. Now, we will convert the Geometry into a Group:

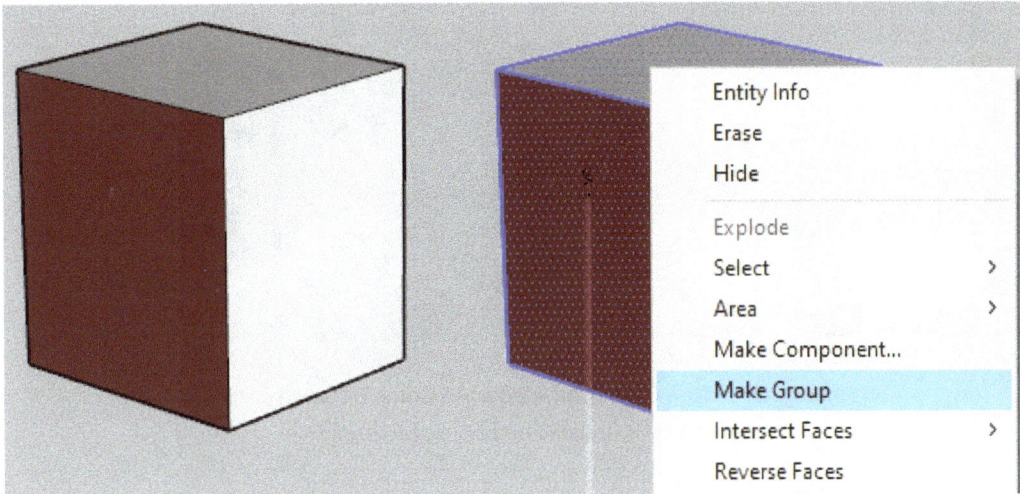

Figure 8.20 – A Simple Box with one Face Painted (Left) and a Box Converted to a Group (Right)

Now, the Paint Bucket tool will not be able to override the Material of any Face that does not have the Default Material applied. If the Paint Bucket tool is used to click on any Face in the Group, all Default Material Faces will be updated, and the remaining Faces will not be overridden:

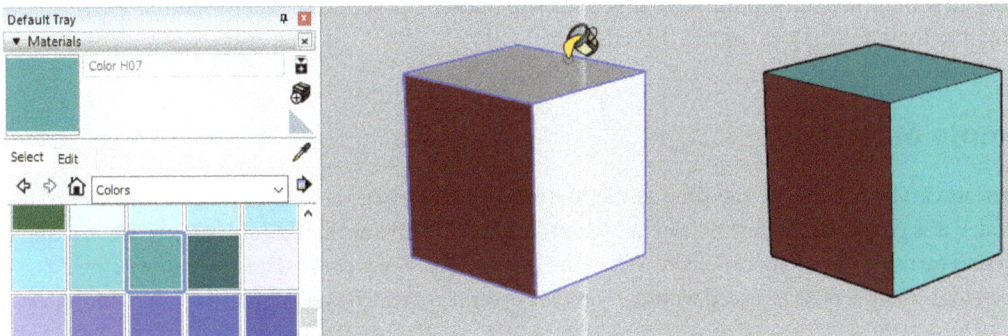

Figure 8.21 – A New Color is Selected and a Group is Clicked (Left) and Default
Faces are Painted and Other Faces are not Overridden (Right)

SketchUp will remember what Faces were painted on an individual basis and what Faces were painted as part of an Object. Using the same box from the previous example, a new Material is selected and the Group is clicked. The Faces that were painted as a Group will be overridden, and the remaining Faces will not be overridden:

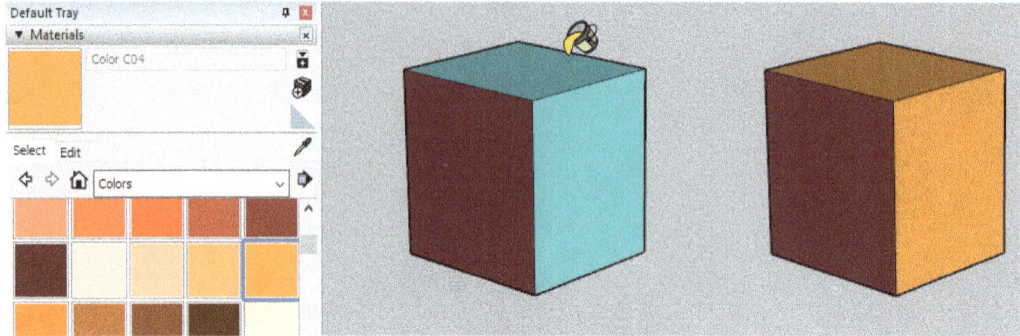

Figure 8.22 – A New Color is Selected and a Group is Clicked (Left)
and a Material is Updated on Specific Faces (Right)

In many cases, it is best to refrain from adding Materials until a Group or Component has been created that will receive the same Material. This way, individual Faces will not be forgotten when attempting to update a Material for an entire Object.

> **Note**
>
> Advanced texture Material editing techniques can only be performed at a per-Face level. Keep this in mind when attempting to change a texture Material position, scale, or rotation. We will discuss these techniques later in this chapter.

Sample Materials

Materials can be selected at any time in the **Materials** Panel when modeling. Once a Material is in a Model, it is common to want to use that Material again. The Paint Bucket tool can sample existing Materials by using the *Alt* modifier key. When the *Alt* modifier key is pressed and held down, the Paint Bucket tool cursor icon is replaced with the Sample cursor icon, which is represented by an eyedropper:

Figure 8.23 – A Sample Cursor Icon Shown with the ALT Modifier Key

> **Note**
>
> A user must hold down the *Alt* key until the material has been sampled.

The Sample option will activate the Material that is clicked, even if it has been used in a Group or Component:

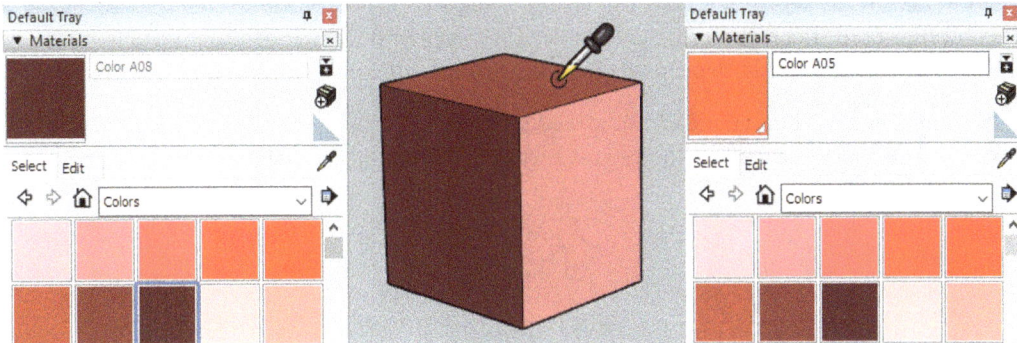

Figure 8.24 – An Active Material (Left), a Sample Option Clicked
(Middle), and an Updated Active Material (Right)

The Sample option can also be activated by clicking on the Sample button on the **Materials** Panel in the **Default Tray** window, and then selecting a Material in the Model to sample:

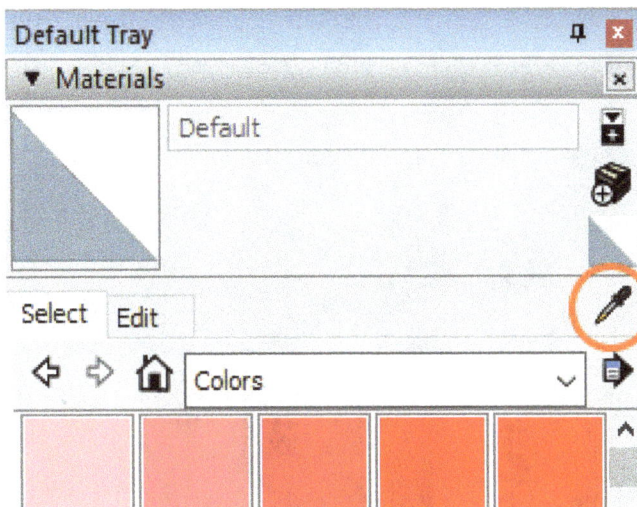

Figure 8.25 – The Sample Button in the Material Panel in the Default Tray

> **Note**
>
> This workflow cannot be used to create new materials. The **Sample** option in the Paint Bucket tool will only activate Materials that are in the Model. There are separate **Sample** options that can be used in the **Edit Material** workflow that will be discussed later in this chapter.

Overriding Existing Materials with Modifier Keys

We have already seen that a single click will update a single Face, or all Faces in an Object that have not yet been painted. The remaining modifier keys in the Paint Bucket Tool can be used to update multiple Faces simultaneously, depending on the desired outcome.

The *Shift* modifier key will override all Faces in a Model with matching Materials. The *Shift* modifier key is represented by three unconnected squares when the Paint Bucket tool hovers over a Face:

Figure 8.26 – The Paint Bucket Tool with the Shift Modifier Key

> **Note**
>
> Painting with the modifier keys will not override Faces that were individually painted within Groups or Components. However, all Faces that were painted on an entire Group or Component will be updated, as well as any ungrouped Faces.

In this next example, we have a stained-glass window. The rectangular window panes are ungrouped, while the top window pane is grouped:

Figure 8.27 – A Stained Glass Window (Left), Ungrouped Rectangular
Panes (Middle), and an Ungrouped Top Pane (Left)

We want the red shapes to be overridden with a pink Material. This is selected in the Materials Panel, and the *Shift* modifier key is held on the keyboard.

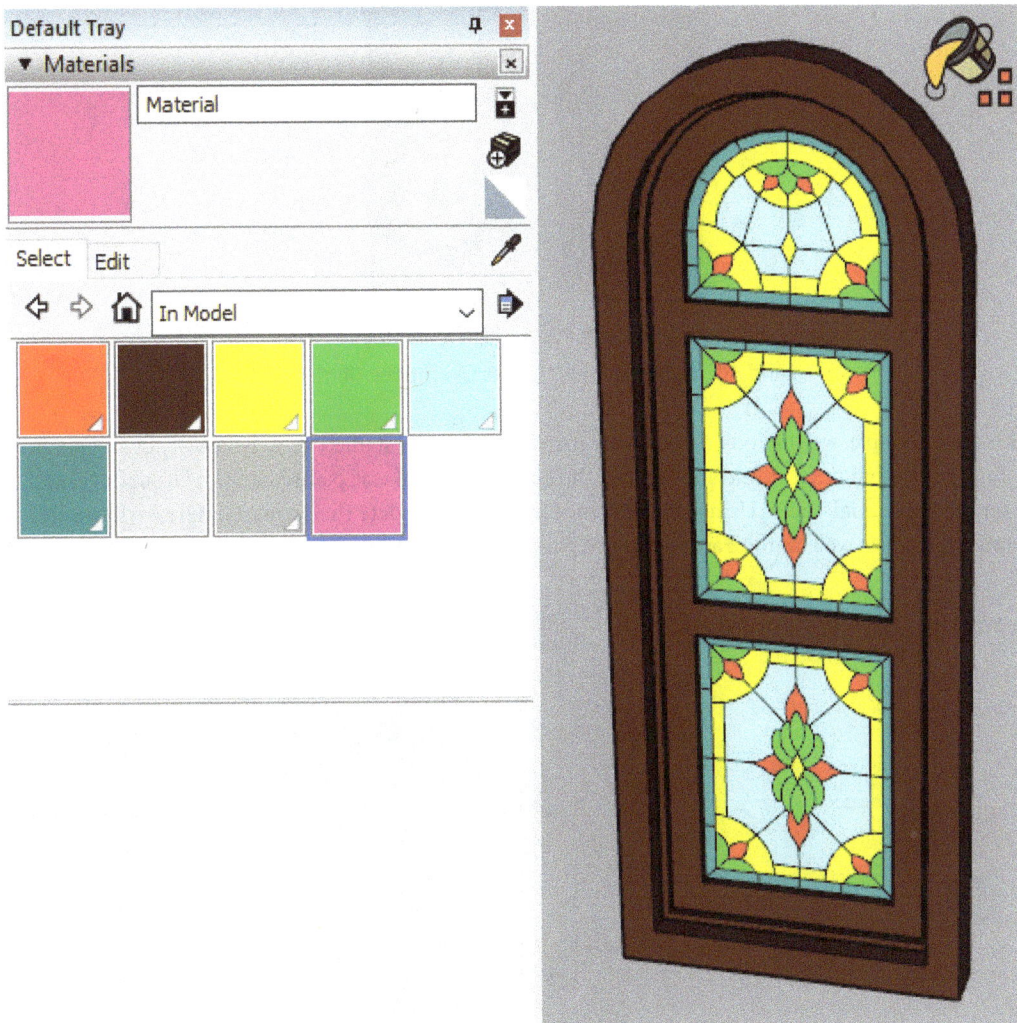

Figure 8.28 – A Pink Material Selected (Left) and the Shift Modifier Key Activated (Right)

When one of the red shapes is selected, all of the ungrouped red shapes will have their Materials overridden with the active pink Material. However, the red shapes in the top window pane will not be overridden, as they were individually painted while in a Group:

Figure 8.29 – A Red Shape Painted with a Shift Modifier Key (Left)
and All Ungrouped Red Shapes Overridden (Right)

In this next example, we will look at the frame that surrounds the stained-glass windows. Even though the inner frame and outer frame are Groups, the brown Material was painted on the Group level and not on the individual Faces. Using the *Shift* modifier key to update the brown Material will update the other Group as well, as the Faces were not painted individually:

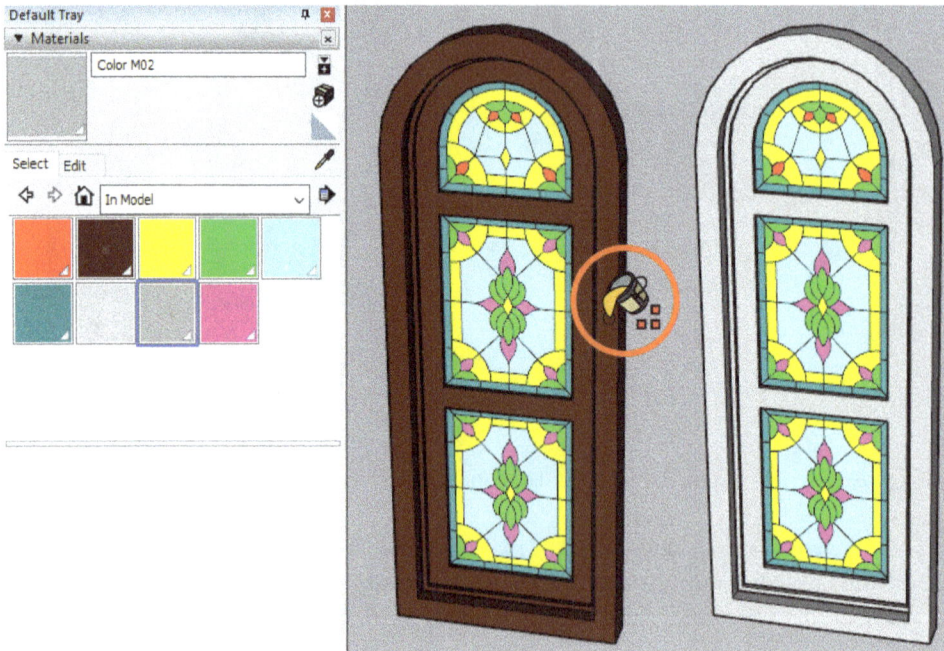

Figure 8.30 – The Gray Material Selected (Left), the Shift Modifier Key
Activated (Middle), and Group Materials Overridden (Right)

The Paint Bucket tool has two other modifier key options, *Ctrl* and *Shift + Ctrl*. These options will only update Faces within the same Object or connected, ungrouped Faces. Use these options when you do not want to update other Objects in your SketchUp Model.

The *Shift + Ctrl* modifier keys will override all Faces in the Object with matching Materials. This is similar to the *Shift* modifier key, although this workflow will only override Faces that are in the selected Object. The *Shift + Ctrl* modifier keys are represented by two red squares, separated by a black square when the Paint Bucket tool hovers over a Face:

Figure 8.31 – The Paint Bucket Tool with the Shift + Ctrl Modifier Keys

In this example, we want to update all of the green shapes in the bottom window pane, without updating the other window pane. We cannot use the *Shift* modifier key, as it will update the other window pane. A dark green Material is selected in the **Materials** panel and the *Shift + Ctrl* modifier keys are held down:

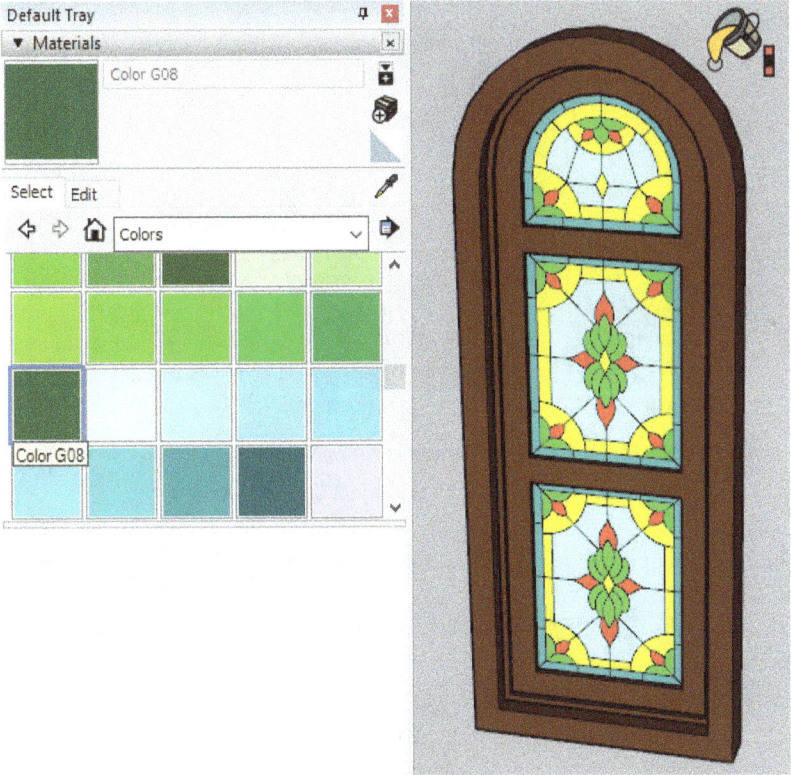

Figure 8.32 – A Dark Green Material Selected (Left) and the Shift + Ctrl Modifier Keys Activated (Right)

When one of the green shapes is clicked, the matching green Materials in the Object will be overridden. The other window panes are not included in the workflow.

Figure 8.33 – A Green Shape Painted with the Shift + Ctrl Modifier Keys
(Left) and Green Shapes on an Object Overridden (Right)

This process can be taken one step further by using the *Ctrl* modifier key. The *Ctrl* modifier key will override all connected Faces in the Object with matching Materials. Connected Faces in this workflow are Faces that share an Edge. The *Ctrl* modifier key is represented by three side-by-side squares when the Paint Bucket tool hovers over a Face:

Figure 8.34 – The Paint Bucket Tool with the Ctrl Modifier Key

In this last example, we will only want to update the green shapes in the middle of the window pane without updating the green shapes around the outside of the window pane. A dark green Material is selected in the **Materials** panel and the *Ctrl* modifier key is held down:

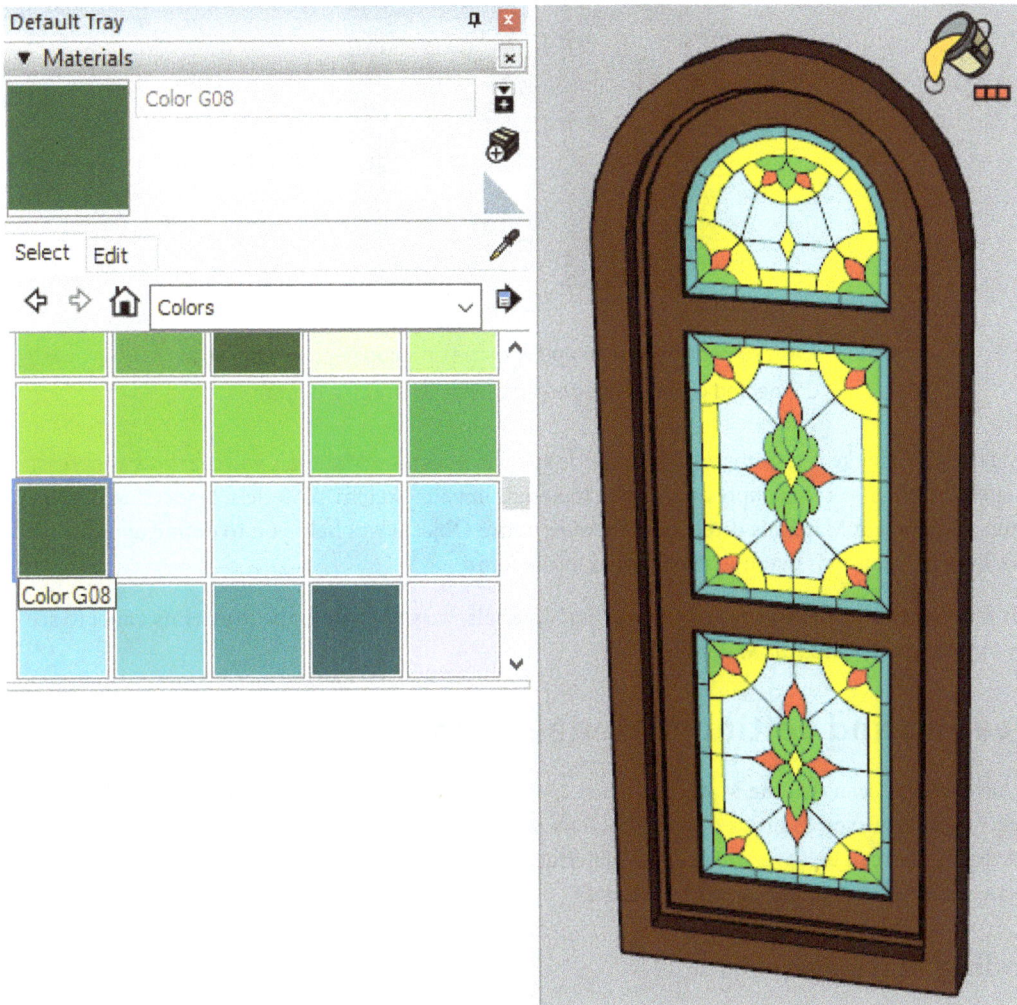

Figure 8.35 – The Dark Green Material Selected (Left) and the Ctrl Modifier Key Activated (Right)

When one of the green shapes is clicked in the middle of the window pane, only the matching Material Faces that share an Edge will be updated. This happens in a chain with all matching, connected Faces. The other green Faces in the Object are not included in the workflow.

Figure 8.36 – A Green Shape Painted with the Ctrl Modifier Key (Left) and
Connected Green Shapes on an Object Overridden (Right)

Using the modifier keys to Sample and Override specific matching Materials will drastically increase the speed at which you can apply and update materials in your SketchUp Models. Understanding the nuance of applying Materials directly to Faces or entire Objects will help you to define appropriate workflows for Materials that may be updated in the future.

Now that we understand the basics of applying Materials, we will explore the **Materials** Panel to see how to find, create, and edit Materials in it.

Creating and Editing Materials with the Materials Panel

We have already seen that the **Materials** panel can be used to select the active Material for the Paint Bucket tool. However, in order to use Materials in our Models, we must first understand how the **Materials** panel can be navigated to find Materials, and then how to create and edit Materials to be exactly what we need in our SketchUp Model.

Finding Materials

The **Materials** panel has two tabs underneath the current Material, the **Select** and **Edit** tabs:

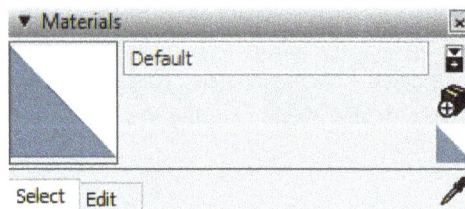

Figure 8.37 – The Materials Panel's Select and Edit Tabs Underneath the Current Material

The **Select** tab is used to find all Materials in SketchUp Pro. The Select tab has three navigation buttons, **Back**, **Forward**, and **Home**. The **Back** and **Forward** buttons allow the navigation to be moved backward and forward, and the **Home** button opens the **In Model** Collection. The **In Model** Collection is continuously updated as Materials are added and purged from the Model.

Figure 8.38 – The Select Tab's Navigation Buttons – Back, Forward, and Home

Next to the navigation buttons is the **Material** Collection dropdown, which includes the default SketchUp Pro Material Collections. These collections sort common Materials into groups that can be easily referenced when looking for Materials. The Collections can be viewed in a folder view by accessing the **Materials** option in this dropdown:

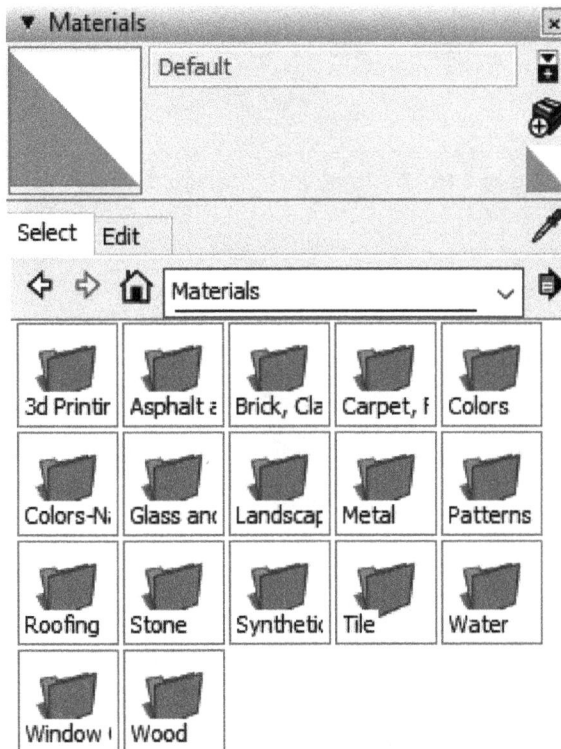

Figure 8.39 – The Material Collection Dropdown Open

To the right of the **Materials** Collection dropdown is the **Material Details** button, which includes view options for the size of the Material icons, as well as options to save and load Collections into your SketchUp Model:

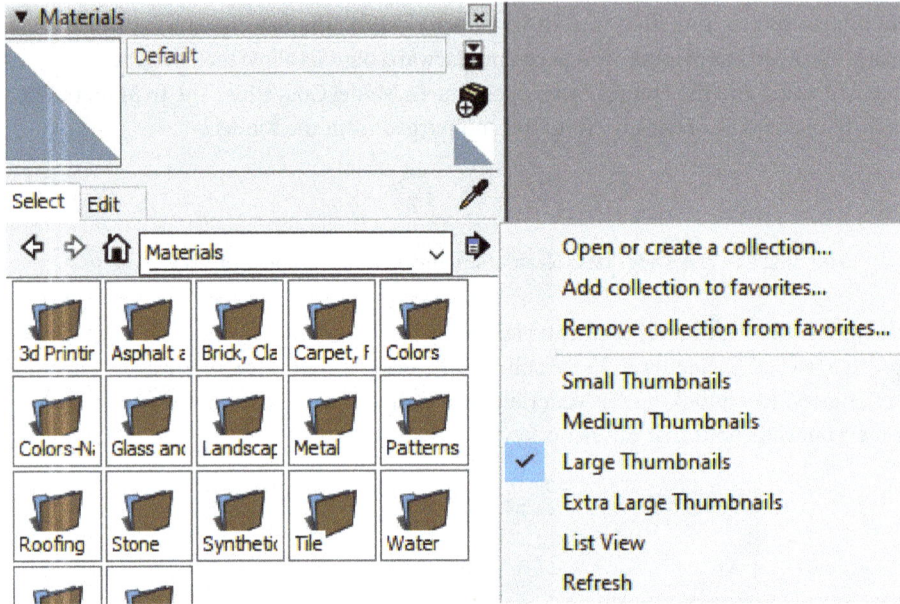

Figure 8.40 – The Material Details Button Flyout

In this example, we want to paint the Geometry in this scene to quickly represent real-world Geometry. As we have previously discussed, a great first step might be to assign color Materials to the Geometry and then refine it later with texture Materials if necessary. In this case, we will choose the **Colors** Collection from the dropdown:

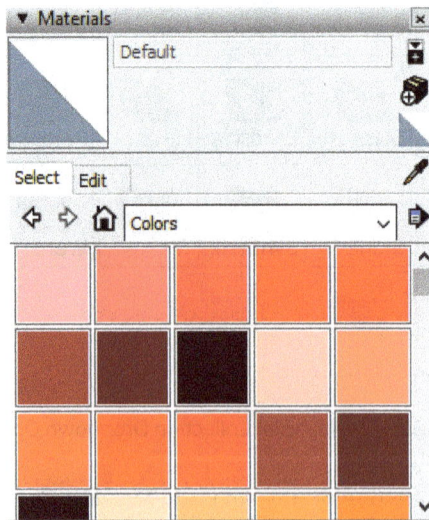

Figure 8.41 – The Colors Collection Selected in the Dropdown

With the **Colors** Collection open, we have the opportunity to scroll down to see the remaining colors, using the scroll bar to the right of the Material icons. The scroll wheel on the mouse can also be used to navigate the **Materials** Collections.

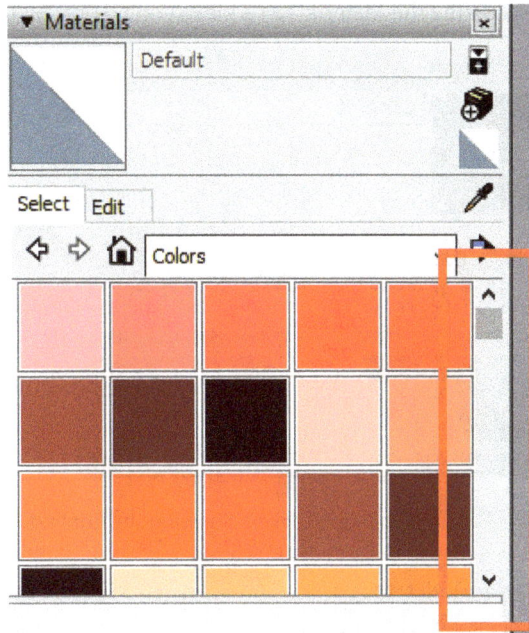

Figure 8.42 – The Scroll Bar in the Materials Panel

When a Material icon is clicked with a left-click, it will become the current Material. The current Material icon and name will update automatically in the **Materials** panel, and the Paint Bucket tool will also activate:

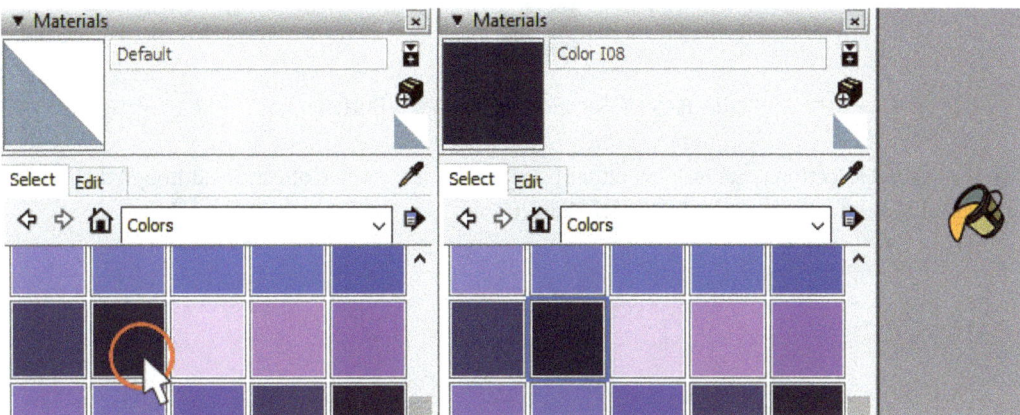

Figure 8.43 – A Material is Left-Clicked (Left), the Current Material is
Updated, and the Paint Bucket Tool is Activated (Right)

When a Material icon is clicked with a right-click, a context menu will appear with two options, **Delete** and **Add to Model**. The **Delete** option will delete the Material file from the SketchUp Pro folders on the computer, which is not recommended. The **Add to Model** option will add the Material to the **In Model** Collection, without requiring the Material to be placed on Geometry in the Model.

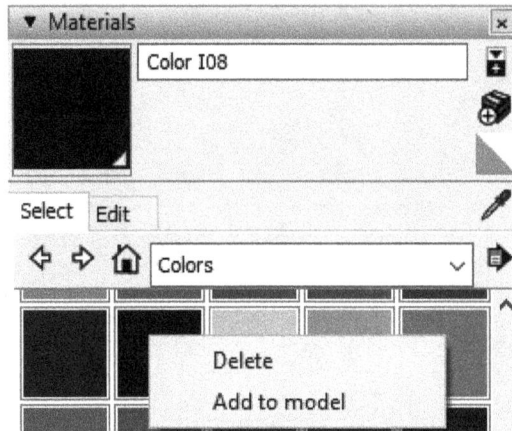

Figure 8.44 – A Material is Right-Clicked, Showing the Context Menu

When Materials are added to a Model, they are added to the **In Model** Collection. The **In Model** Collection can be quickly accessed by clicking the **Home** button. The **In Model** Collection is a great way to quickly find Materials that are already in use in the Model. If you want the **In Model** Collection open all the time, it can be beneficial to open a second selection pane. This can be opened by hitting the black and white button next to the current Material name:

Figure 8.45 – A Second Selection Pane Button

When the second selection pane is open, either pane can be set to any Collection, although it can be most helpful to keep one set to the **In Model** Collection:

Figure 8.46 – A Second Selection Pane Set to the In Model Collection

> **Note**
> A second selection pane can also be opened for the **Components** and **Styles** panels in the Default Tray.

With two selection panes open, we can continue to select Materials from the **Color** Collection and paint elements in the Model. In this next screenshot, multiple color Materials, texture Materials, and a translucent material from the **Glass and Mirrors** Collection have been used in the model:

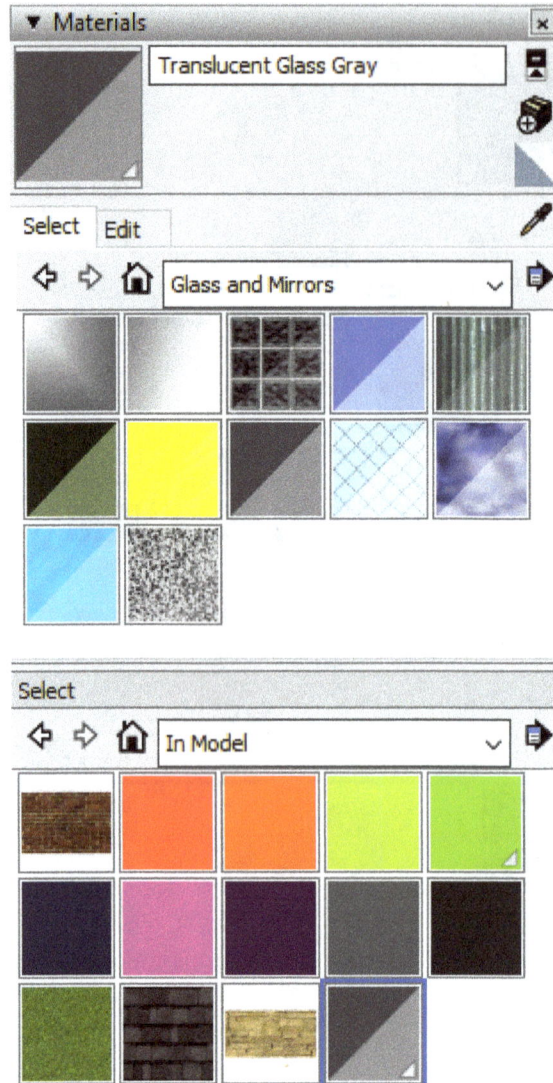

Figure 8.47 – The Materials Panel and the Model Painted with More Materials

In the **Glass and Mirrors** Collection, there are multiple color Materials and texture Materials that have a transparency to them, represented by a diagonal slash splitting the Material icon into darker and lighter triangles. The **Glass and Mirrors** Collection has a mix of color Materials, texture Materials, opaque Materials, and translucent Materials.

Now that we understand how to find Materials in SketchUp, let's take a look at two ways to create Materials in SketchUp Pro.

Creating a Material from an Existing Material

New Materials can be created in SketchUp Pro by clicking the **Create Material** button, which is underneath the second selection pane button. The **Create Material** button is represented by a striped cube and a plus sign:

Figure 8.48 – The Create Material Button

When the **Create Material** button is clicked, the **Create Material** dialog box will open. The Material options will be automatically filled out using the current Material settings, which essentially allows for a copy of an existing material to be made. The following screenshot has blank settings because the default Material was active when the **Create Material** button was clicked:

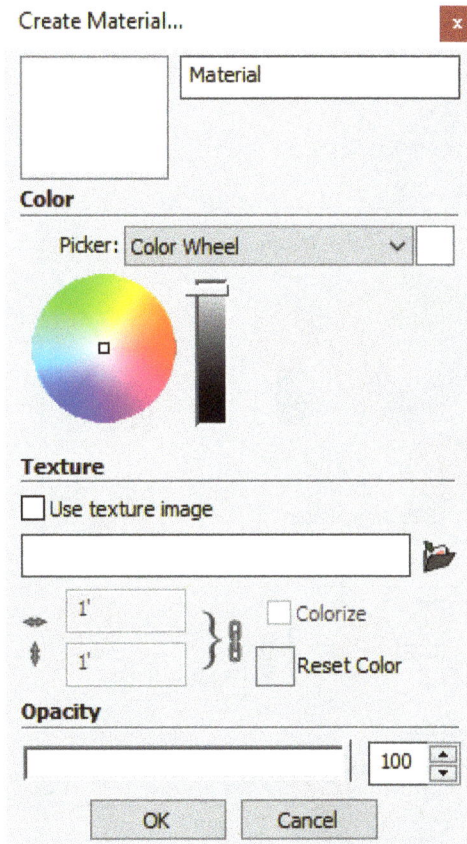

Figure 8.49 – The Create Material Dialog Box

The Create Material dialog box contains the following options:

- The **Material** Name
- The **Color** Picker (which can be set to **Color Wheel**, **HLS**, **HSB**, or **RGB**)
- The **Use texture image** checkbox
- The **Texture** File Path
- The Texture Dimensions (including a chain link button to lock in proportional scaling)
- The **Colorize** Texture
- The **Opacity** Slider and Percentage Value

We can see these options shown via numbers in the following screenshot:

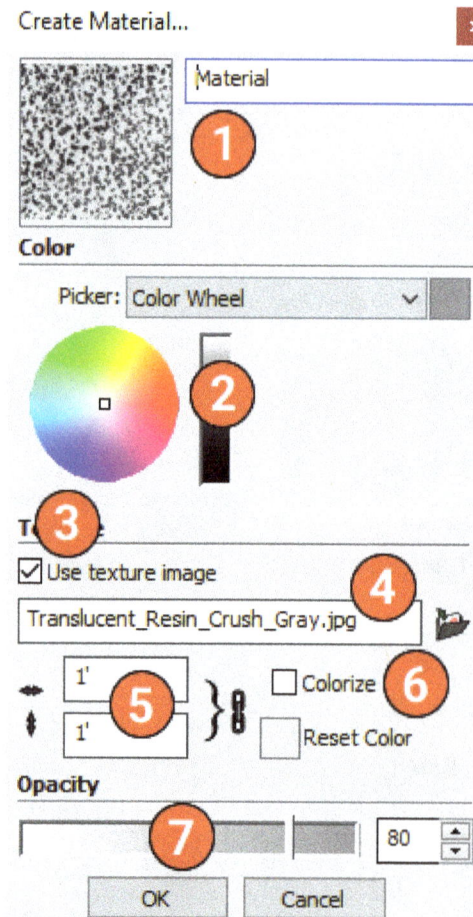

Figure 8.50 – Create Material Dialog Box Options

The Material name can be updated by typing into the name box at the top of the dialog box. The **Opacity** value can be changed with the slider or the percentage value box. The color can be set by using the **Color Wheel**, or one of the other color selection options.

Note

The color should be set manually if the Material is a color Material. If the Material uses a texture image, the color will be set automatically. This color will be the color used when using the **Shaded Face Style View** option.

The **Use texture image** checkbox determines whether the Material is a color Material or a texture Material:

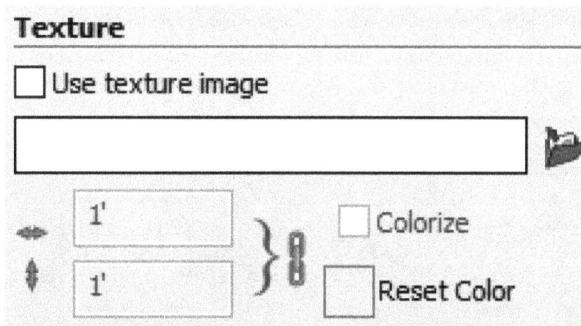

Figure 8.51 – The Texture Options in the Create Material Dialog Box

The **Texture** section can be ignored for color Materials. When the **Use texture image** checkbox is checked, the **Choose Image** dialog box automatically opens. This dialog box will accept many supported image types, including JPEG, PNG, PSD (a Photoshop document), TIF, TGA, and BMP:

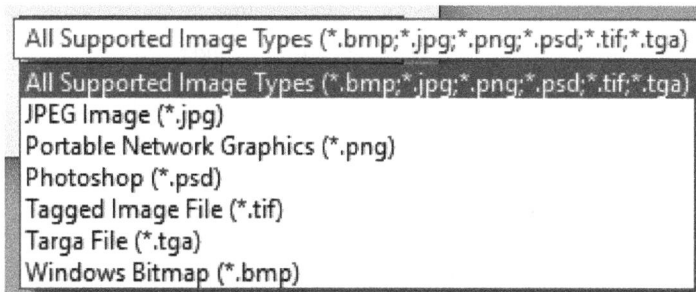

Figure 8.52 – The Supported Image Types Flyout

Once an image is selected in the **Choose Image** dialog box, the **Open** button can be clicked. Once the image is loaded, the Material icon will update. The **Create Material** dialog box will assign the default value of **1'** by **1'** for all new texture Materials, unless a texture image was already set to a different size by the copied Material:

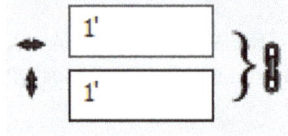

Figure 8.53 – The Texture Image Size set to 1' by 1'

The image dimensions should be updated to reflect the desired size for the image, which may need to be calculated separately. Texture images can also be colorized, which activates the Color Picker. The Color Picker can be used to shift the colors of the texture image. The **Reset Color** box can be clicked to return the color to its original hue:

Figure 8.54 – The Original Texture Image (Left) and the Texture
Image Colorized and the Color Shifted (Right)

When the Material options have been set, the **OK** button can be clicked. This will add the new Material to the **In Model** Collection. Texture Materials can also be created through the Import tool.

Creating a Material from an Image Import

We briefly saw that an image can be imported as a Matched Photo in the previous chapter. The same operation can be done to import an image as a Texture. Images can be imported by going to **File |
Import...** in the **Main Menu** dropdown:

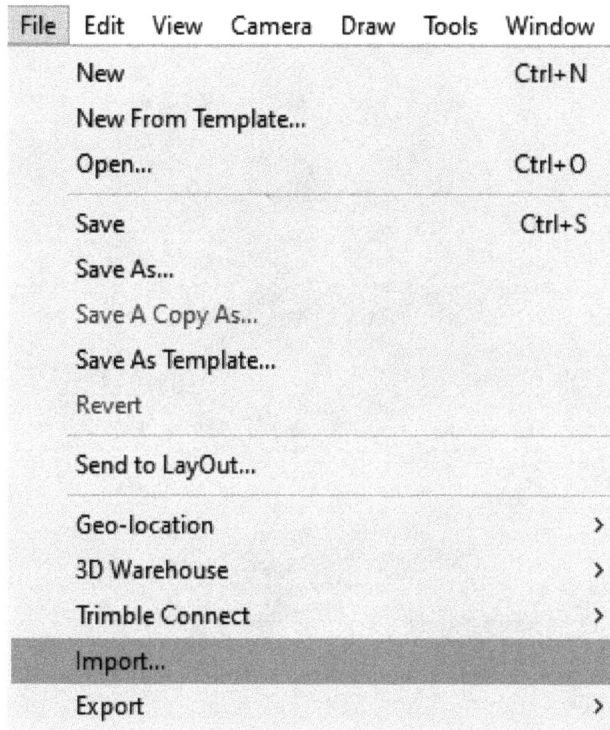

Figure 8.55 – File | Import in the Main Menu Dropdown

The **Import** window will allow for all supported file types to be selected, including other SketchUp Models, other 3D Model types, and images. When an image type is selected in the file type dropdown, or an image is highlighted in the File Explorer window, the **Import** dialog box will show the three **Use Image As** options – **Image**, **Texture**, and **New Matched Photo**:

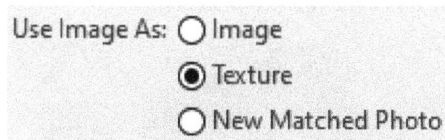

Figure 8.56 – Use Image As Options

If we want to make a new Material from the image, we should select the **Texture** option. When the image and the **Texture** option have been selected, the **Import** button can be clicked. The SketchUp cursor will show a preview of the image in the SketchUp Drawing Area:

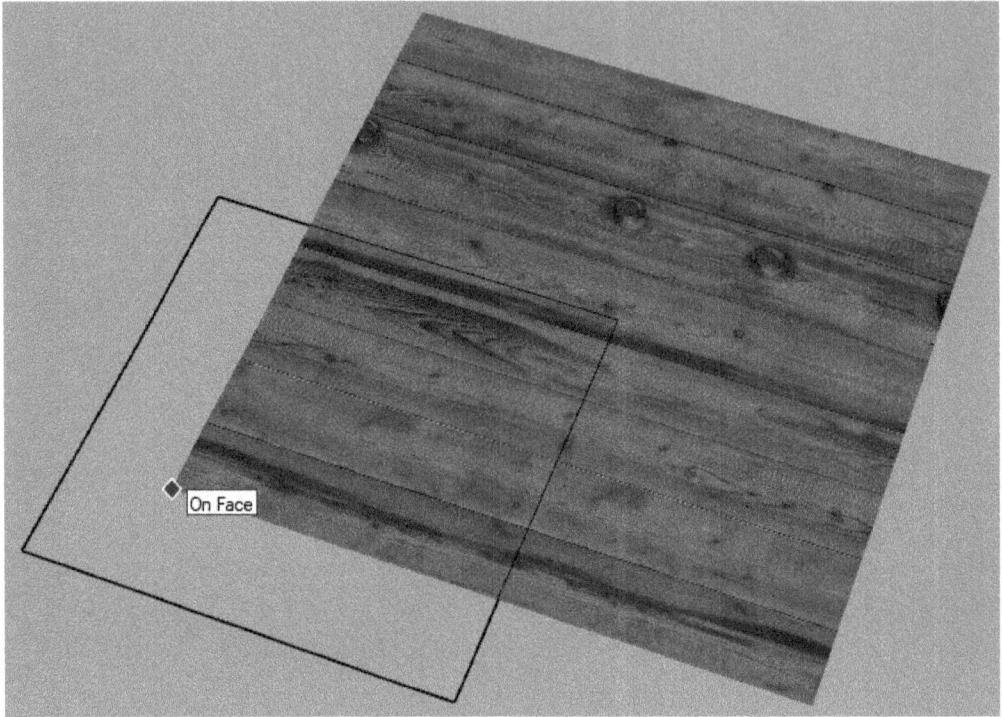

Figure 8.57 – The Import Button Clicked and the Image Preview in the Drawing Area

The image preview must be placed on a Face in the Model in order for the texture Material to be applied. The Face does not need to match the proportions of the image, but the Face will be used to scale the image texture. It can be helpful to match the image proportions and ensure that the Face is the exact size of the desired texture. With the image preview hovering over a Face, a point can be clicked to set the bottom-left corner of the texture Material:

Figure 8.58 – A Face Clicked to Set the First Point of the Texture Material

The texture Material will then stretch as the cursor is moved. The second point clicked will set the scale of the texture Material and will set the scale of the image in the Material. The texture Material will also be painted on the Face when the second point is clicked:

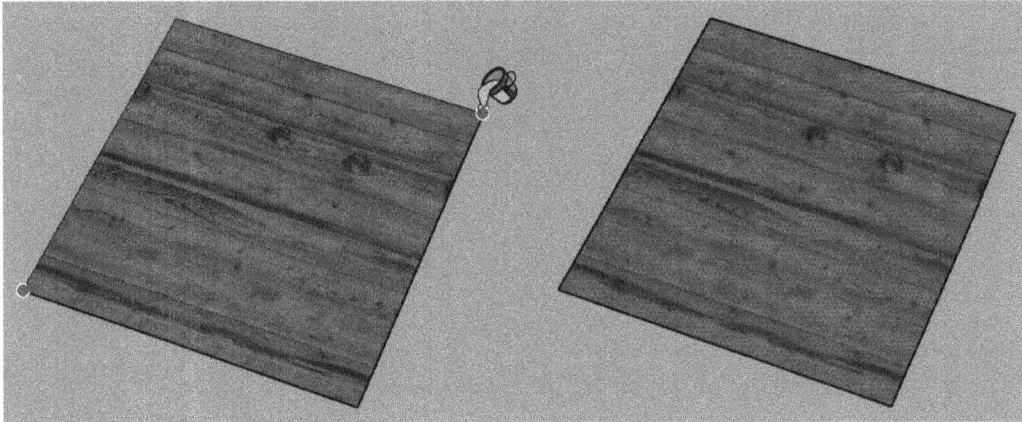

Figure 8.59 – The Material Stretched on a Face to Set the Image Scale (Left)
and the Second Point Clicked to Apply a Texture Material (Right)

Once this workflow is completed, a new Material will automatically be added to the **In Model** Collection in the **Materials** Panel:

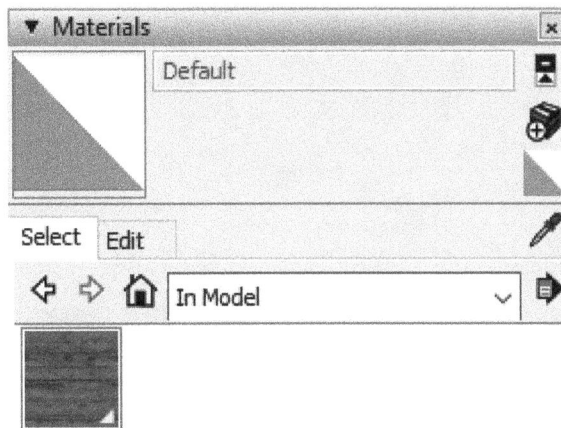

Figure 8.60 – A New Material Added to the Materials Panel

The new Material name will match the original image filename, and the Material icon will show the image without tiling. The new Material will not be activated as the current Material, but it can be selected and painted onto other Faces in the Model.

If the image is scaled so that it is larger than the Face, the Material will appear to be cropped. The image itself is not cropped, and the Material will show the full size if the Face is resized:

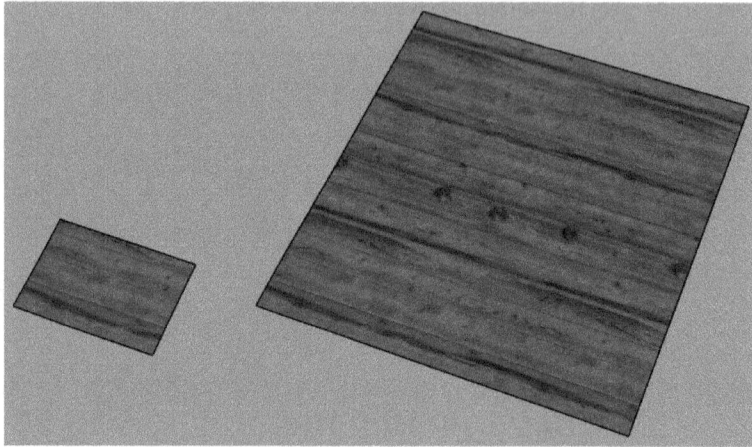

Figure 8.61 – The Image Appears to be Cropped (Left) and the Resized Face Reveals the Full Material (Right)

If the image is scaled so that it is smaller than the Face, the Material will automatically tile across the entire Face:

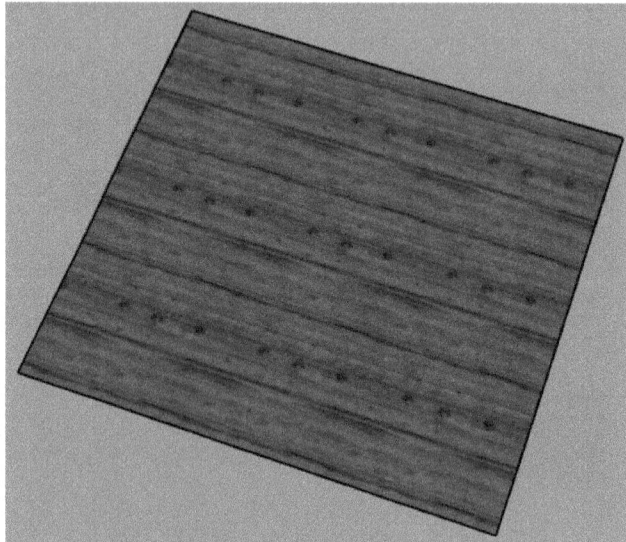

Figure 8.62 – The Image is Tiled Across the Face

Materials automatically tile in SketchUp Pro, so a good way to avoid tiling is to draw a new face around the texture Material, update the Material around the outside of the desired texture Material, and then hide the Edges so that the Face appears to be continuous:

Figure 8.63 – The Texture is Tiled (Left), New Edges are Drawn and Outside
Faces are Painted (Middle), and Edges are Hidden (Right)

We will look at more advanced techniques for editing texture Materials in the *Advanced Techniques* section later in this chapter. Next, let us look at how we can edit materials.

Editing Materials

Materials can be edited in the SketchUp Model on the **Edit** tab in the **Materials** panel:

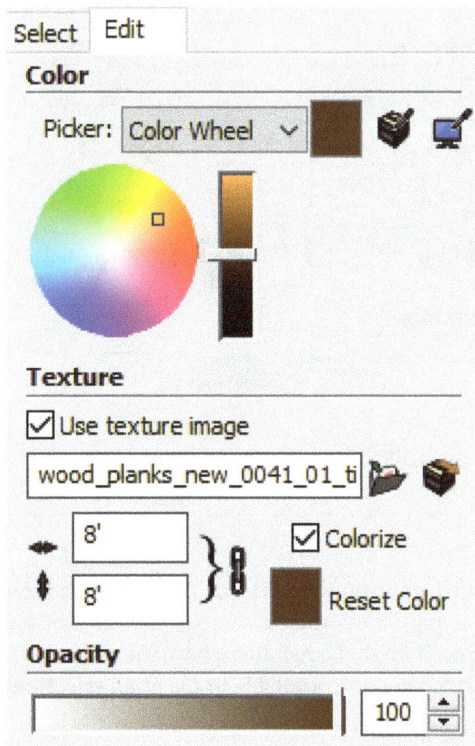

Figure 8.64 – The Edit Tab in the Materials Panel

The **Edit** tab and the **Create Material** dialog box have a similar interface, with the **Edit** tab offering a few additional options. These options are as follows:

- **Match Color of Object in Model**
- **Match Color on Screen**
- **Edit Texture Image in External Editor**

We can see these options shown via numbers in the following figure:

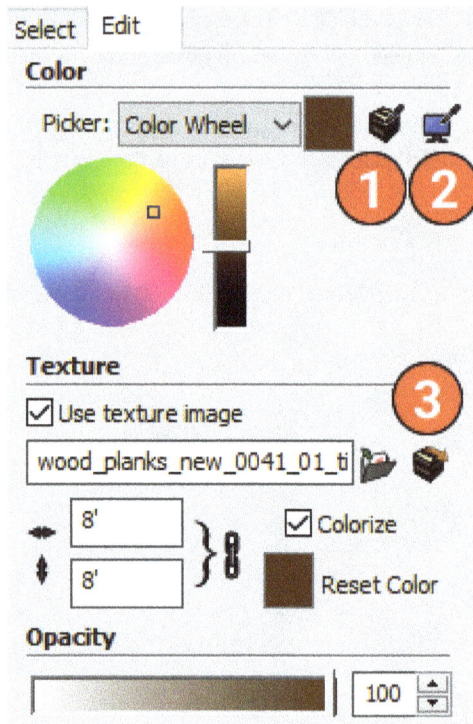

Figure 8.65 – The Edit Tab's Additional Options

> **Note**
>
> The **Sample** button is disabled while on the **Edit** tab. The **Sample** button is only used to select Materials in the current drawing, so in order to sample, you must be on the **Select** tab. If you are editing multiple Materials, it might be helpful to have the second selection pane open, but the **Sample** button and **Edit** tab are only available in the main selection pane.

The Material options can all be edited in the same way that the Materials were created. The color of the Material can be manually updated, including colorizing texture Materials. The scale of a texture Material can be updated and the scale ratio broken. Also, an opacity amount can be applied to any Material. Please refer to the *Creating a Material from an Existing Material* and *Creating a Material from an Image Import* sections for more details about these settings. Let's now dive deeper into the three additional options in the **Edit** tab.

The **Match Color of Object in Model** option will update the color of the current Material to the color of the picked object. The **Match Color of Object in Model** button is represented by a striped cube with a sample eyedropper:

Figure 8.66 – The Match Color of Object in Model Button

This option will update color Materials and colorize texture Materials. In this first example, we can see a texture Material and a color Material. If we want to colorize the texture Material to match the color Material, we can use the **Match Color of Object in Model** option. With the texture Material set as the current Material, the color Material can be sampled to colorize the current Material:

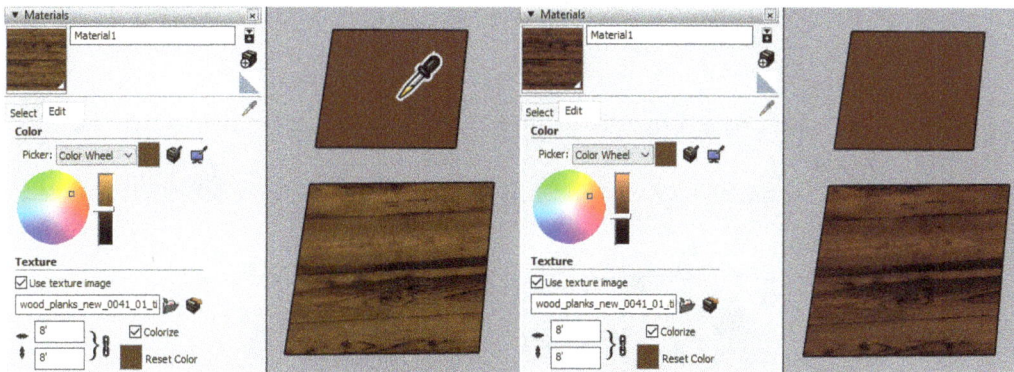

Figure 8.67 – The Current Texture Material and the Sampled Color
Material (Left) and the Updated Texture Material (Right)

The color may not match exactly, as the colorized texture Materials are not precise when viewed with the SketchUp renderer. However, this can provide a quick and easy way to colorize different options using the same transparent image, without going into any image editing software. The brick walls in the following figure were quickly updated using the **Match Color of Object in Model** option and SketchUp color Materials to colorize the texture Material:

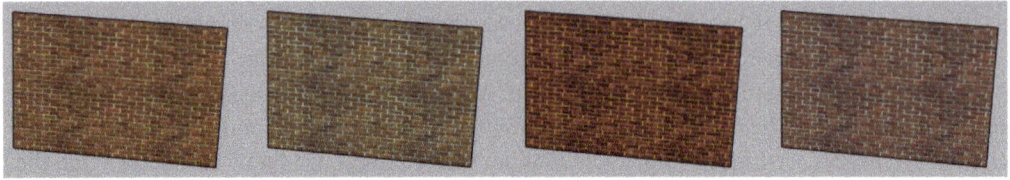

Figure 8.68 – The Brick Wall Options for a Color Study

In this second example, we can see a texture Material next to a color Material. If we want to highlight the logo color, we can update the color Material using the **Match Color Object in Model** button. With the color Material set as the current Material, we can sample the texture Material and update the color Material to match the average color of the texture Material:

Figure 8.69 – The Current Material Updated to Match the Average Color of the Sampled Material

The average color is the same color that is used when viewing the SketchUp Model with the Shaded Face Style:

Figure 8.70 – Shaded with Textures (Left) and Shaded (Right)

If we wanted to match a specific color from the texture Material instead of the average color, we would need to use the **Match Color on Screen** option instead. This option will identify the color of the specific pixel sampled and update the Material to match that color. This option can colorize texture Materials and update color Materials like the **Match Color of Object in Model** option. The **Match Color on Screen** button is represented by a computer monitor and a sample eyedropper:

Figure 8.71 – The Match Color on Screen Button

In this example, we can see the color Material updated to a dark brown color, which is sampled from a specific pixel in the texture Material when using the **Match Color on Screen** option:

Figure 8.72 – Match Color of Object in Model (Left) and Match Color on Screen (Right)

> **Note**
> Colors are best matched when the **Use Sun for Shading** option is active in the **Shadows** panel in the **Default Tray** window. SketchUp automatically adjusts the colors of Faces, depending on the camera angle, and **Use Sun for Shading** disables this SketchUp feature.

The third additional button on the **Edit** tab is **Edit Texture Image in External Editor**. This button is represented by a striped cube with a curved arrow pointing to the right:

Figure 8.73 – The Edit Texture Image in External Editor Button

The **Edit Texture Image in External Editor** option does just what it says; it allows an image used in the texture Material to be edited in external image editing software, and then it automatically updates the Material in SketchUp. The **Default Image Editor** option can be changed in **Window | Preferences | Applications**:

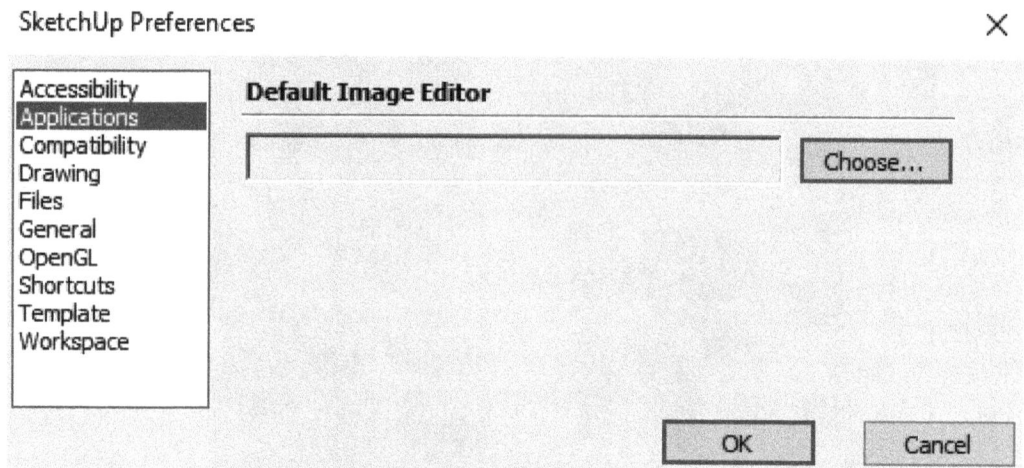

Figure 8.74 – The Preferences | Applications Dialog Box

This can be a useful tool when tweaking an image for a texture Material, removing the need to create new texture Materials each time the image is updated. Texture images can be exported and saved in their original image format. This can be done by right-clicking the Material icon in a selection pane and choosing **Export Texture Image**:

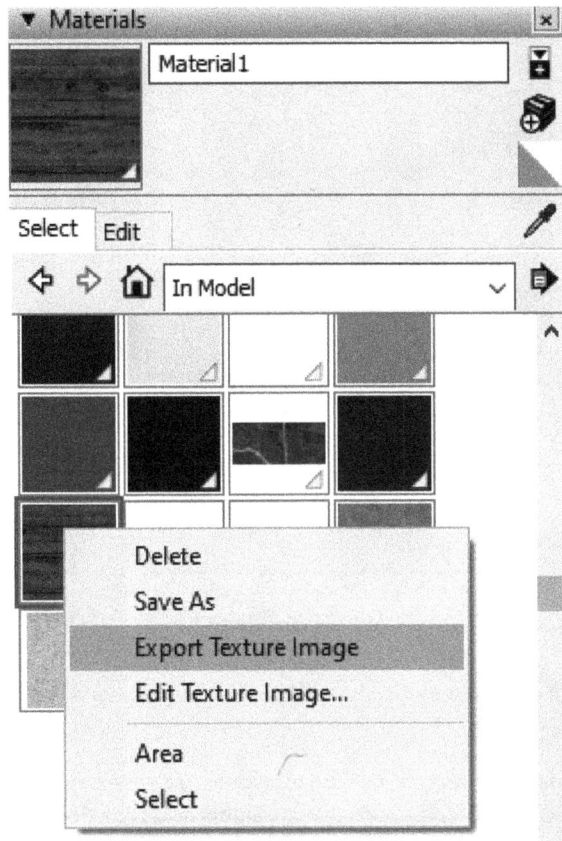

Figure 8.75 – The Material Icon Right-Clicked | Export Texture Image

This will open the **Export Raster Image** dialog box. The image can be saved to a location using the original image file extension.

Now that we understand how to create and edit Materials in SketchUp, let's look at how we can save our Materials for use in other SketchUp Models.

Saving Materials and Collections

Materials can be exported as individual files, called SketchUp Material Files (.skm). These files can be saved anywhere on your computer for future use in other SketchUp Models. SketchUp Materials do not need to be saved externally if they will only be used in your current Model; they will stay in the **In Model** Collection unless they are purged. We will discuss purging in the next section, *Managing Materials in Your Model*.

To save a SketchUp Material, simply right-click the Material icon in one of the selection panes and choose **Save As**:

Figure 8.76 – The Material Icon Right-Clicked | Save As

The **Save As** dialog box will open automatically, and the SketchUp Material can be saved to a location on your computer.

Collections are multiple Materials located in the same folder on your computer. Collections can be exported all at once. SketchUp saves Collections that are automatically loaded into the SketchUp Model when it is first created. These Collections include Colors, Patterns, Wood, and Tile. These Collections do not need to be exported, as SketchUp keeps them organized. However, it may be useful to export the **In Model** Collection, as this is updated by the user for each Model. A Collection can be saved by clicking the **Details** button and choosing the **Save collection as…** option:

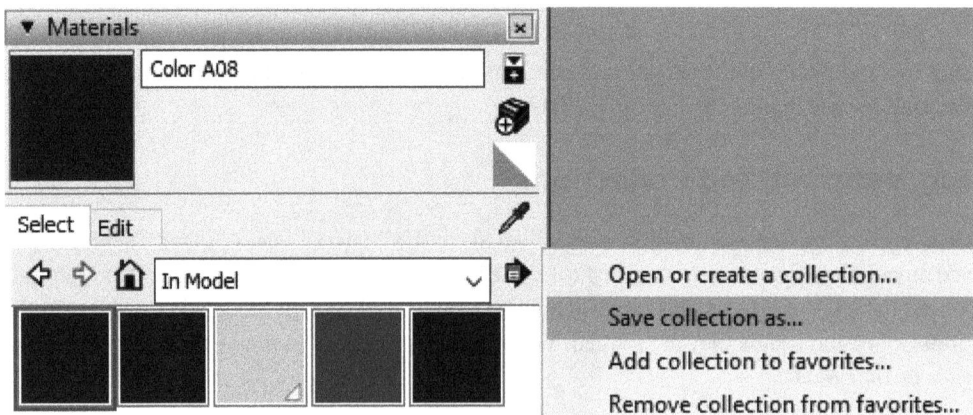

Figure 8.77 – The Materials Panel's Details Button | Save collection as… Option

This will open a dialog box, prompting the user to select a Collection folder or create a new one.

> **Note**
>
> It is recommended that a specific, empty folder is selected or created for Collections. If there are other files in the selected location, SketchUp Materials will fill up the folder and make it difficult to find the other files in the folder.

When the folder is selected, SketchUp will automatically save all of the included Materials to that folder. Then, the name of the folder on the computer will represent the Collection name.

Importing Collections into a SketchUp Model is a similar process. Collections can be loaded by clicking the **Details** button and choosing, the **Open or create a collection...** option:

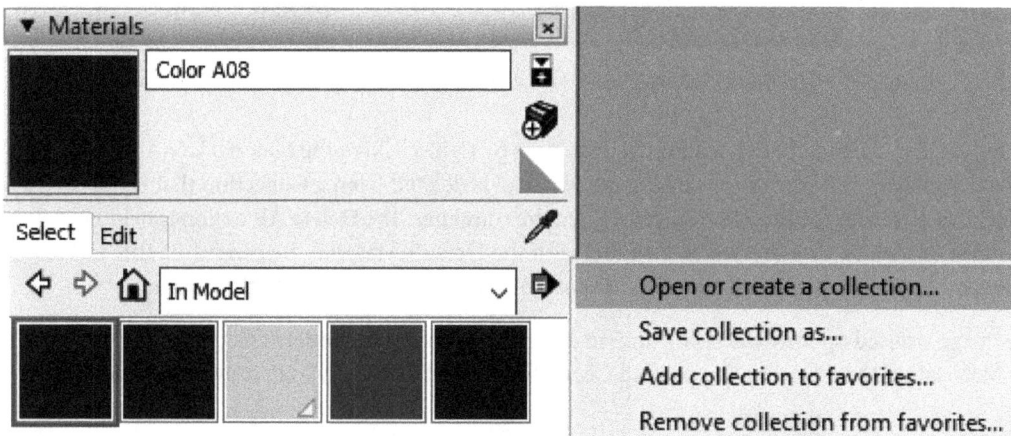

Figure 8.78 – The Materials Panel's Detail Button | Open or create a collection... Option

This option will open a similar dialog box, prompting the user to select a folder that represents a Collection. When the folder is selected, SketchUp will load all Materials in that folder into a Collection in the **Materials** panel dropdown, and the Materials can be selected for use in the Model.

> **Note**
>
> Collections can be saved to or loaded from the `Materials Favorites` folder, which saves the Collections in SketchUp folders on the local computer. This can be helpful to keep all of your favorite Collections in a safe place – just remember that you saved them there!

Let's take a look at a few options that can be helpful to manage the Materials in your SketchUp Models.

Managing Materials in Your Model

We have already looked at some of the options using the **Details** button in the **Materials** panel. In this section, we will look at the **Delete All**, **Purge Unused**, and **Refresh** options. The **Delete All** and **Purge Unused** options are only available in the **In Model** Collection.

The **Delete All** option will delete all Materials from the **In Model** Collection:

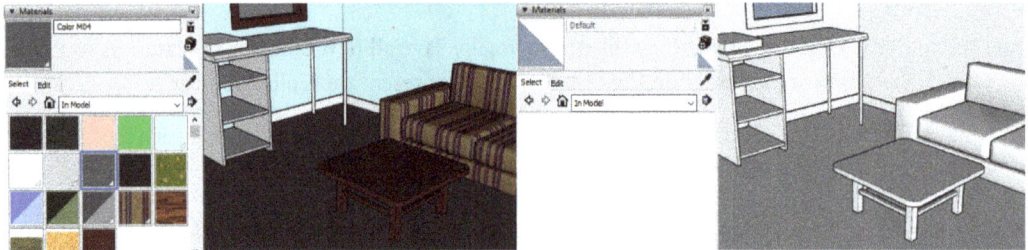

Figure 8.79 – The Model before Delete All (Left) and the Model after Delete All (Right)

Individual Materials can be deleted from a Collection by right-clicking the Material icon and choosing **Delete**. SketchUp will provide a warning if a Material is deleted from a Collection that is not the **In Model** Collection, as the file will be deleted from the computer. The **Delete All** option is helpful when you want to replace all Materials in a Model with the Default Material. Be careful, as this will also delete any Materials that were created in the current Model and not saved to the computer.

The **Purge Unused** option deletes any unused Materials from the **In Model** Collection. This will leave any Materials in the Model on the painted Geometry but will delete any Materials that are not in use:

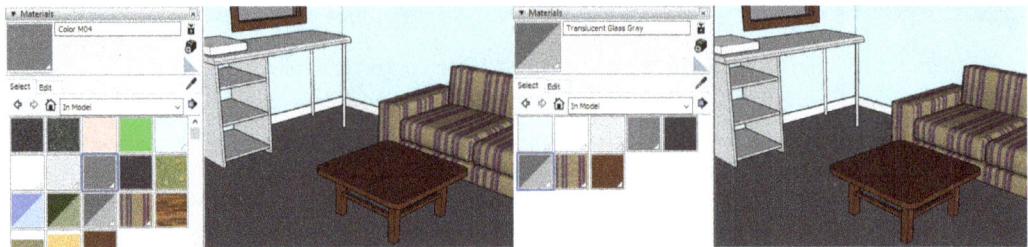

Figure 8.80 – A Model before Purge Unused (Left) and a Model after Purge Unused (Right)

This can be helpful when many Materials were overridden while attempting to find a specific Material, or a SketchUp Model was imported from the 3D Warehouse that came with many external Materials. The Materials must first be removed from the Geometry before they can be purged.

> **Note**
>
> The **In Model** Material Collection will include all Materials that are part of Components, even if the Components are not placed in the Model. It may be beneficial to Purge Components before Purging Materials; otherwise, you may have many additional Materials that will not Purge!

The **Refresh** Option will reload Materials in any loaded Collection. This can be helpful if the Material was edited and updated on another computer, or additional Materials were added to a Collection folder.

Now that we understand the basics of Materials and Collections, let's take a look at some advanced techniques to manage texture Materials in our SketchUp Models.

Advanced Techniques

Texture Materials can be manipulated in numerous ways in SketchUp Pro. Editing a Texture Position, including location, scale, rotation, skew, and perspective warping, or projecting the texture onto a curved surface can improve the realistic feel of a SketchUp Model. At the beginning of the chapter, we mentioned that it can be difficult to scale, position, and project texture Materials. While it can be difficult, it is not impossible. Let's jump in and take a look at ways to manipulate texture Materials in SketchUp Pro.

Texture Position

Individual Faces that have been painted with a texture Material can be edited with Texture Position. Texture Position can be activated by selecting the painted Face and right-clicking and selecting **Texture | Position**. Texture Position opens a contextual edit space that allows for the individual texture to be edited using the Texture Position Pins:

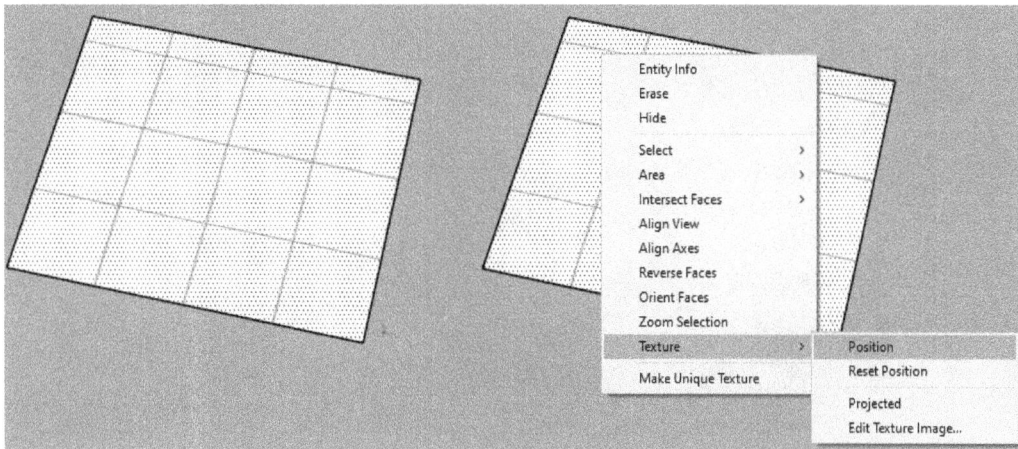

Figure 8.81 – A Selected Individual Face is Painted with a Texture Material
(Left) and Right-Clicking and Selecting Texture | Position (Right)

The four pins represent different ways to edit the texture. The pins are colored and include an icon to help remember their purpose:

- Red Pin: Move

- Green Pin: Rotate and Scale (Uniform Scale)

- Blue Pin: Shear and Scale (Scale from Red Pin to Blue Pin only)

- Yellow Pin: Distort (Perspective)

Figure 8.82 – Four Position Texture Pins

The texture will become transparent when it is edited. Additionally, the texture will tile to include at least one additional image on each side of the main texture, and as many texture images as needed to cover the Face. All texture Materials are in a SketchUp tile, so even if the texture is exactly the size of the Face, you will see at least one tiled copy on all sides of the texture:

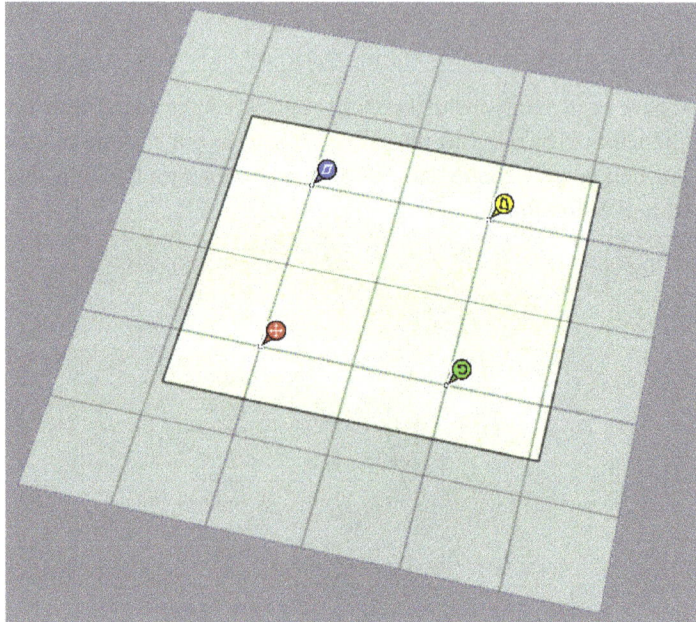

Figure 8.83 – A Transparent Texture with Tiled Copies

There are multiple ways to edit the texture once you are in the Position Texture contextual edit space. We will review each action next and include the ways to activate each workflow in the individual sections.

Move

A texture can be moved by clicking and dragging on the Red Pin or by clicking and dragging on the texture image itself. When dragging the image, the cursor will switch from an open-hand icon to show a closed-hand icon. The Red Pin acts as the snap point for moving the texture and is located in the bottom-left corner of the texture:

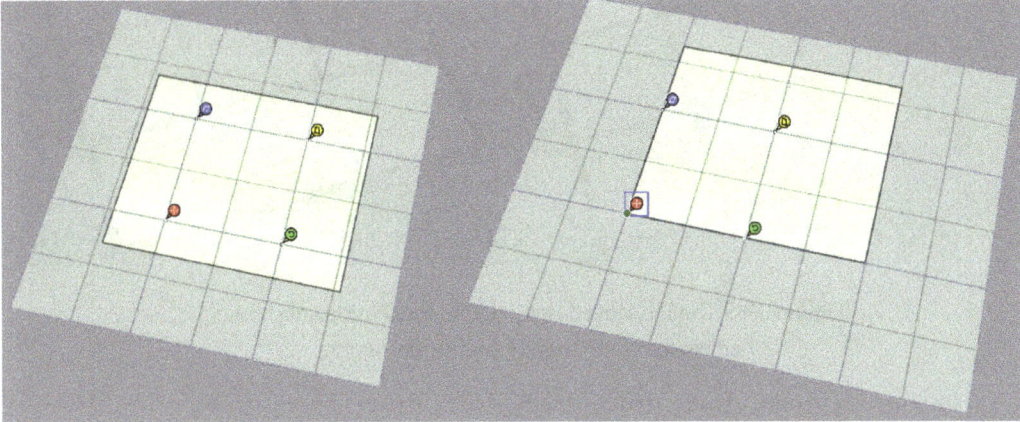

Figure 8.84 – The Texture in the Original Position (Left) and the
Texture Moved and Snapped with the Red Pin (Right)

Textures are often not aligned with the Geometry that they are painted on. In this example, we can see that there are a series of squares that represent tile. There is a small space between the sections of tile that should be ignored by the texture. We can see in the next figure that the texture is applied from the origin point in the Model and does not consider the individual Face origins (unless the Face is in a Component, which has different Axes and a different Origin Point):

Figure 8.85 – Squares Representing the Tile (Left) and the Tile Painted (Right)

The Texture Position move workflow can help align all of the tile textures so that the tile looks like how it would in the real world:

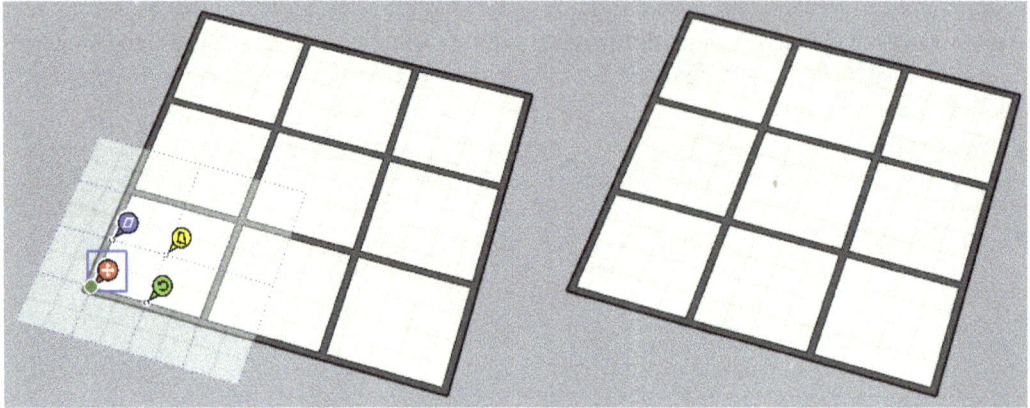

Figure 8.86 – The Tile Moved Using the Texture Position and the
Red Pin (Left) and the Finished Model (Right)

Rotate

A texture can be rotated by clicking and dragging on the Green Pin. The Green Pin acts as a snap point to set the rotation to exact angles on an existing Geometry or Guides created by the Protractor Tool:

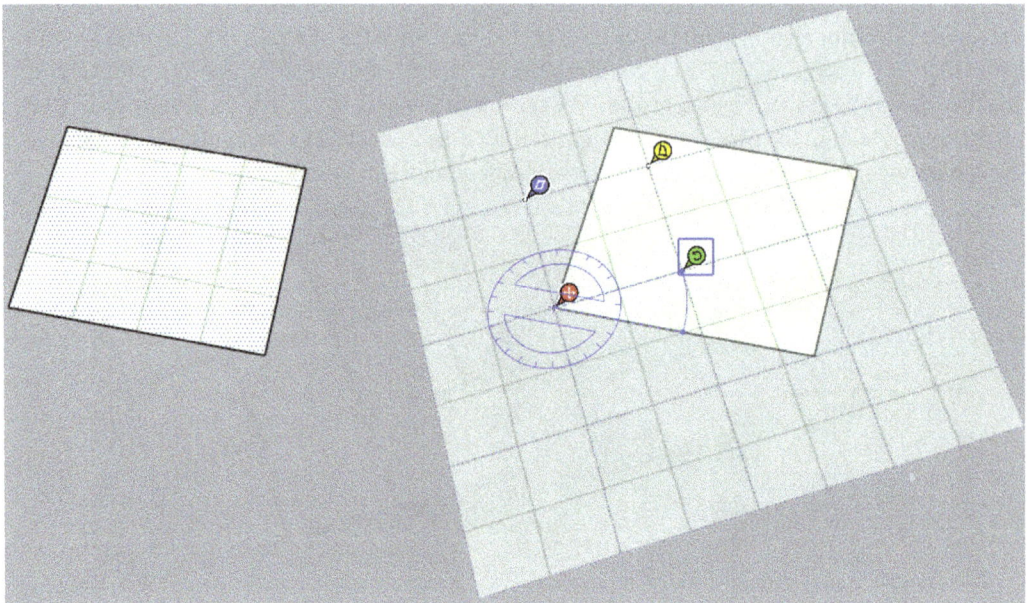

Figure 8.87 – The Texture in the Original Position (Left) and the Texture Rotated Using the Green Pin (Right)

> **Note**
> Be careful! The Green Pin can also scale the texture. Use the arc preview to ensure that the texture does not scale along the Red Pin/Green Pin direction.

In this next example, we can see that the Position Texture tool has been used to move all of the tile sections to line up with the Faces. However, the tile should be rotated in the center diamond to align with the 45-degree angle:

Figure 8.88 – The Geometry with the Tile Texture

The Texture Position rotate workflow can be used to rotate the texture to a specific angle. By clicking on the Green rotate grip and dragging it to snap to the SketchUp Geometry, the texture will be rotated 45 degrees:

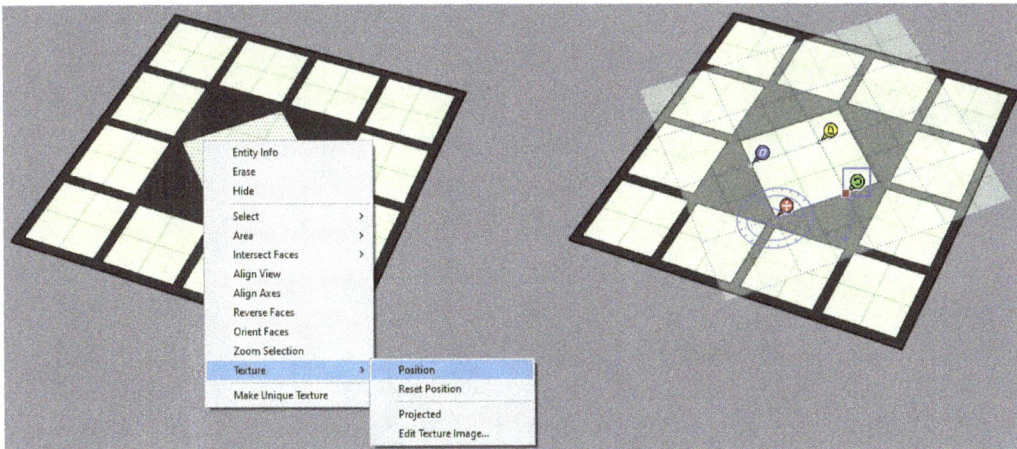

Figure 8.89 – The Selected Texture and the Position Texture (Left)
and the Green Pin Used to Rotate the Texture (Right)

The Texture Position edit workflow can be closed by clicking outside of the Texture or by hitting *Enter*. The finished texture now aligns with the specific angles of the Faces:

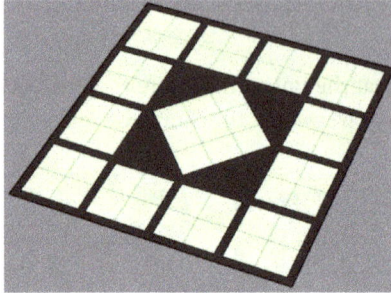

Figure 8.90 – The Finished Tile Textures with the Correct Rotation Angle

> **Note**
> When working in the Texture Position edit workflow, right-click on the Texture. Then, take a look at the **Flip and Rotate** section for rotation options of exactly 90, 180, or 270 degrees.

Scale

A texture can be scaled using either the Green Pin or the Blue Pin. The Green Pin scales the texture uniformly, and the Blue Pin scales the texture along the Red Pin/Blue Pin direction. This can be tricky because the Green Pin also rotates the texture, and the Blue Pin also shears the texture. Scaling must be done carefully to ensure that the texture does not get rotated or sheared accidentally.

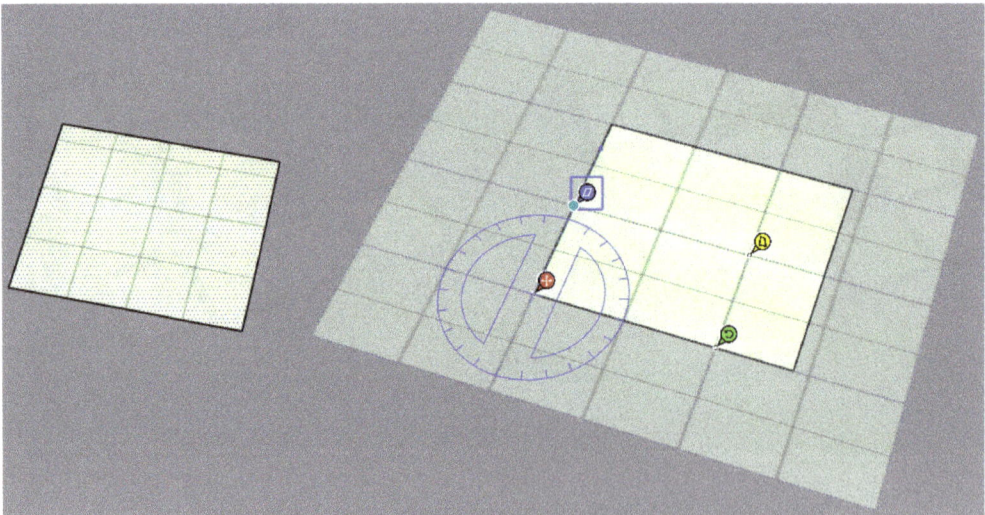

Figure 8.91 – The Texture in the Original Position (Left) and the Texture
Scaled in Two Directions Using the Green and Blue Pins (Right)

> **Note**
> Scaling textures using Position Texture will only update the scale for the active Face. Texture Materials can have their overall scale changed in the **Edit** tab in the **Materials** panel. It might be a good idea to update the overall scale of a texture Material instead of each Face one at a time.

In this example, we can see that the desired tile size does not match the square texture in the original texture image:

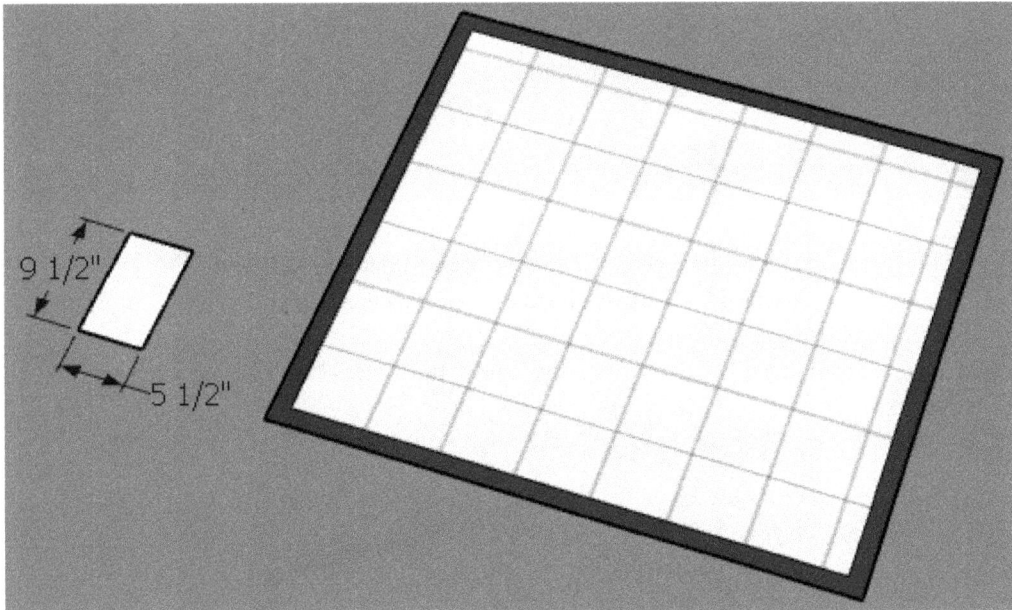

Figure 8.92 – The Desired Tile Size (Left) Texture on the Face with the Original Square Tiles (Right)

Guides can be created on the Face so that the texture can be snapped to an exact size. Because the Green Pin scales the texture uniformly, it should be used first to scale the texture, and then the Blue Pin can be used to set the scale in the other direction:

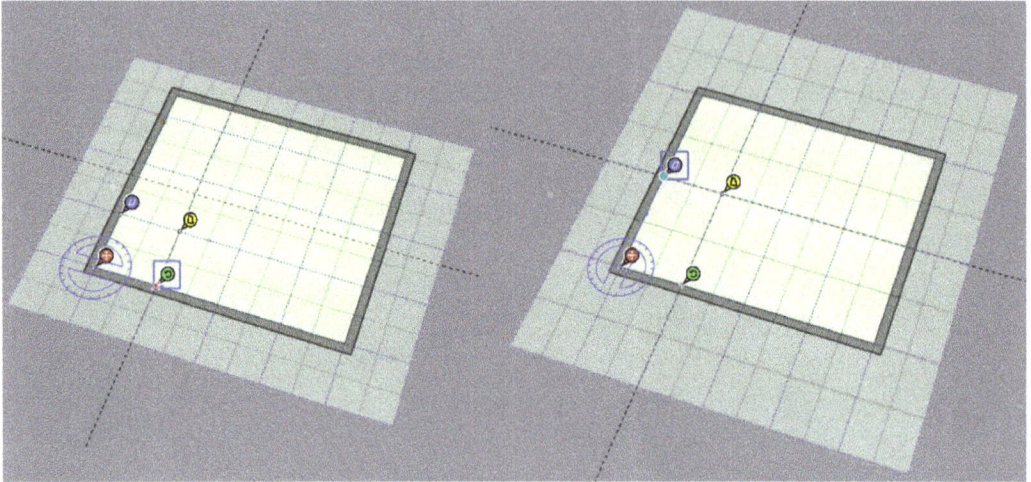

Figure 8.93 – The Green Pin Scales the Texture Uniformly (Left) and
the Blue Pin Scales the Texture in One Direction (Right)

The final Texture Position has not overwritten the scale of the Material but has only been applied to this one Face:

Figure 8.94 – The Finished Tile Texture with an Updated Non-Uniform Scale

Shear

All textures in SketchUp Pro are rectangles, and we can demonstrate shear using the rectangle in the next example. Shearing a rectangle means shifting the top Edge of it while keeping the overall height the same. This stretches the vertical Edges to make them longer while also changing their angle:

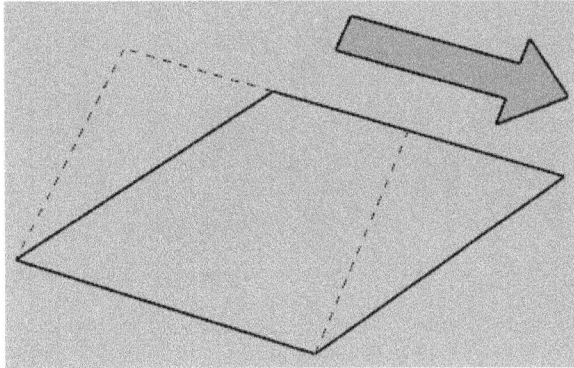

Figure 8.95 – A Rectangle is Sheared by Pushing the Top Edge and Slanting the Vertical Edges

When a texture is sheared, it updates the texture image in the same way; the horizontal lines are not adjusted, but the vertical lines are slanted. This is different from rotating, which turns the vertical and horizontal lines equally:

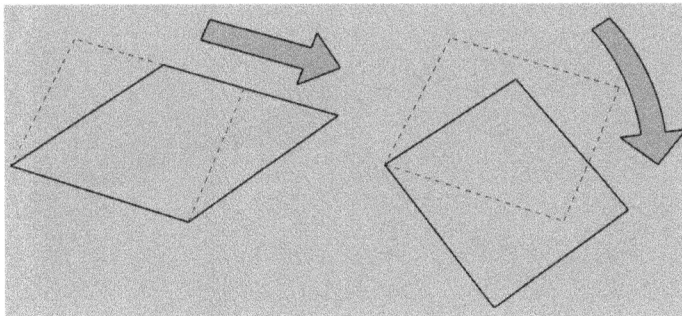

Figure 8.96 – A Rectangle Sheared (Left) and a Rectangle Rotated (Right)

A texture can be sheared by clicking and dragging on the Blue Pin. The Blue Pin must be moved across the contextual guideline in order to avoid scaling the texture when dragging the Blue Pin. The Blue Pin can be used to snap to existing Geometry that matches the angle of the desired texture shear:

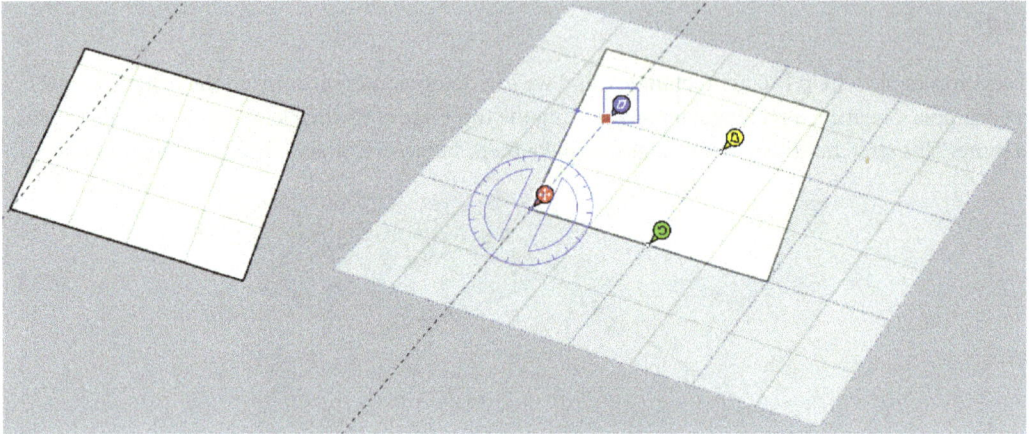

Figure 8.97 – The Texture in the Original Position (Left) and the Texture Sheared to a Specific Angle (Right)

In this next example, we can see a diamond-shaped tile and four diamond-shaped sections of the tile, but the tile texture is square in the original texture example:

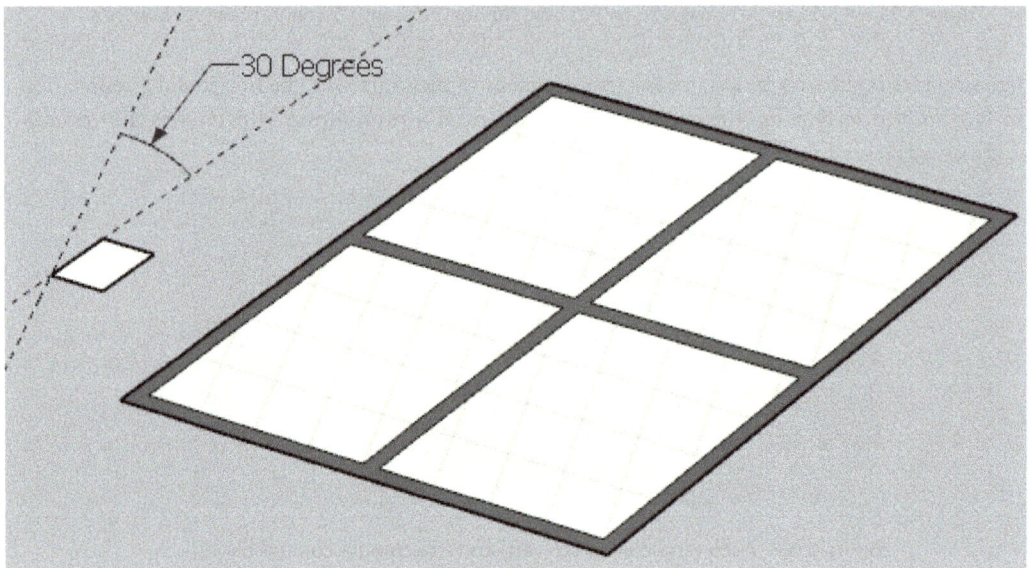

Figure 8.98 – The Desired Tile Shape (Left) and the Texture on the Face with the Original Square Tiles (Right)

The Blue Pin can be used to shear the tile texture so that the new angle of the texture matches the 30-degree angle of the Geometry.

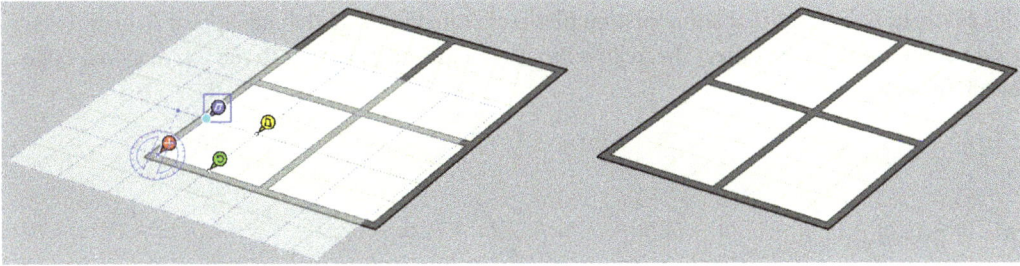

Figure 8.99 – The Blue Pin-Sheared Texture (Left) and the Updated Textures (Right)

> **Note**
>
> Shearing cannot be done within the **Edit** tab of the **Materials** Panel. Shearing can be done in some external image editors, so it may be beneficial to update a texture image before creating a texture Material if a shear angle must be defined for many Faces.

Distort

A texture can be distorted by clicking and dragging on the Yellow Pin. Distortion mimics a 3D perspective by applying a vanishing point to the texture image:

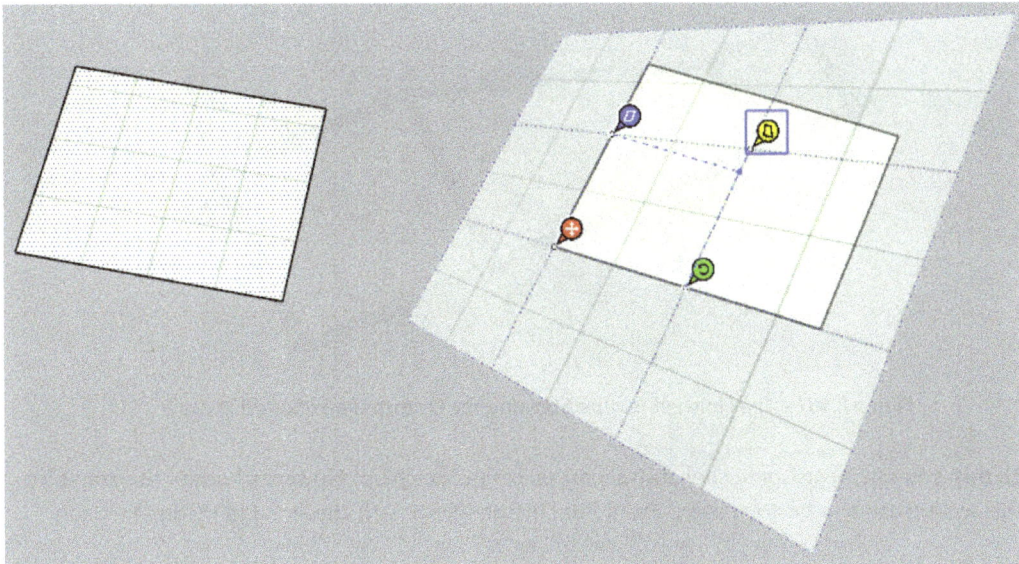

Figure 8.100 – The Texture in the Original Position (Left) and the
Texture Distorted to Create a Perspective View (Right)

In this example, we can see that some custom tiles are created to make the tiles appear to grow from one side of the shape to the other. The Yellow Pin can be moved to snap it to the top diagonal Edge to match the angle:

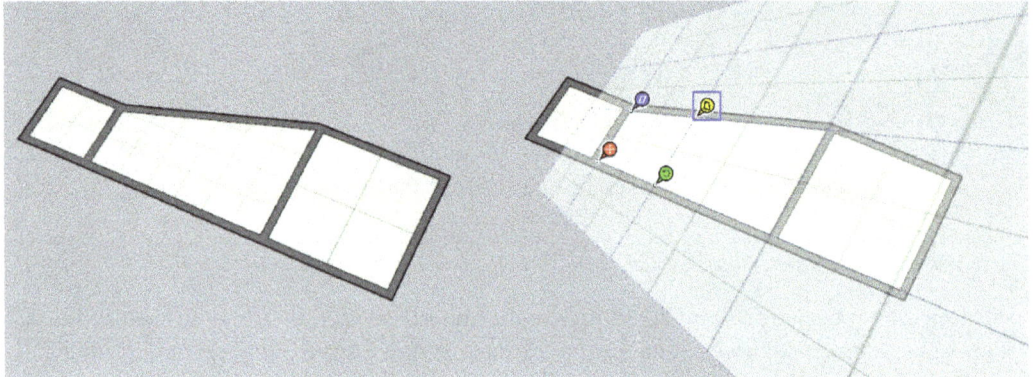

Figure 8.101 – The Custom Tile Shape (Left) and the Yellow Pin
Moved to Update the Distortion Angle (Right)

Distorting a texture creates a vanishing point, so the tiles that are closer to the vanishing point appear to be smaller vertically and horizontally than the tiles further away:

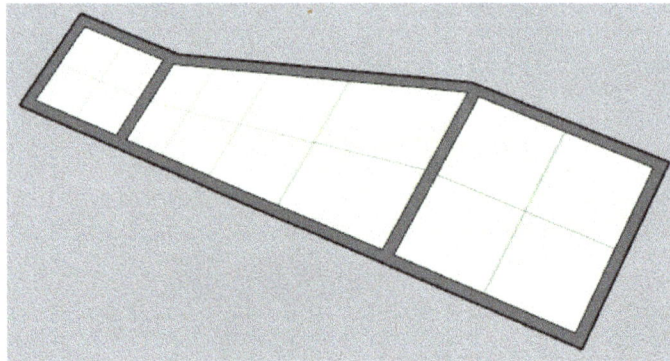

Figure 8.102 – The Finished Texture Showing the Distorted and Scaled Texture

With this, you can create some cool abstract art pieces in SketchUp! However, both of the examples in this section can also be done using Fixed Pin Distort, which we'll check out in the next section.

Fixed Pin Distort

The Freeform Distort option can be activated by holding the *Shift* modifier key to quickly toggle Fixed Pins, or by right-clicking and choosing the **Fixed Pins** option. All four Pins will turn white and show an icon representing a map:

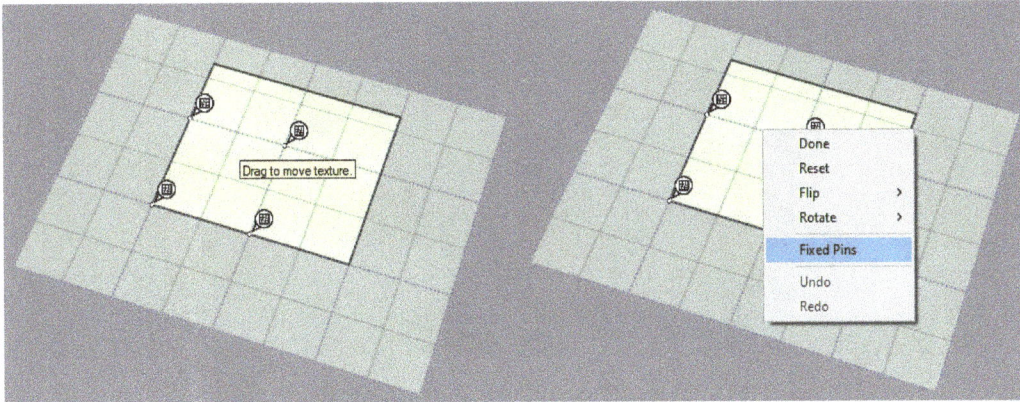

Figure 8.103 – The Texture Showing Fixed Pins (Left) and Right-Clicking Fixed Pins (Right)

Fixed Pin Distort is the best way to align a distorted photograph with a rectangular shape in SketchUp. In this next example, we can see that a photograph of a sign is meant to be placed on a rectangular face. The full photograph looks comical on the rectangle – Texture Position has been activated, and the **Fixed Pins** option has been toggled on by right-clicking and choosing **Fixed Pins**:

Figure 8.104 – Texture Position Active and the Fixed Pins Option Toggled On

It can be easier to turn this option on instead of using the *Shift* modifier key when all points need to be moved.

The Pins can now be clicked and dragged to match up with the corners of the sign in the image. Because there is no SketchUp Geometry to snap to, it can be helpful to zoom into the image to align the Pins accurately:

Figure 8.105 – The Pins Clicked and Moved to Match the Image
(Left) and the Pins in the Correct Positions (Right)

Once the Pins are aligned correctly, the Pins can be clicked and dragged to the desired location. The Pins will automatically snap to the Edges bounding the Face:

Figure 8.106 – The Pins Clicked and Dragged to Snap to the Face
(Left) and All the Pins Dragged to the Face (Right)

The completed Texture Position shows the image as if it were a rectangular texture without perspective distortion or the additional photograph around the sign:

Figure 8.107 – The Completed Texture Position

However, this does not update the overall texture Material, and this cannot be picked to place on other Geometry:

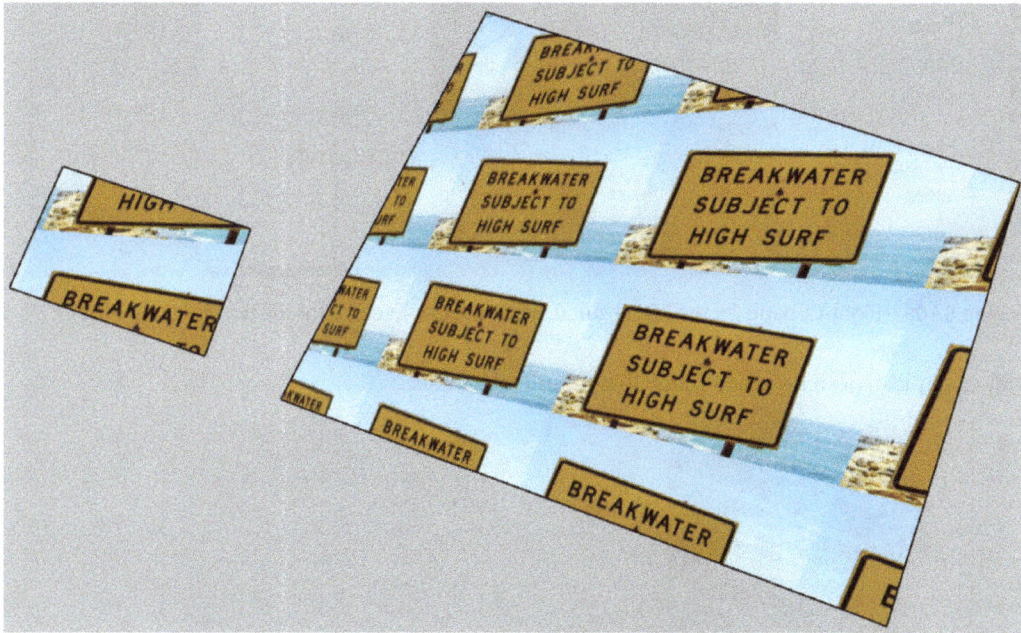

Figure 8.108 – The Texture Material Applied to the New Geometry (Left) and
the Existing Face Scaled to Show the Distorted Texture (Right)

This Geometry can be converted into a Component and multiple instances can be made, but this process does not create a material that tiles across a face as textures should.

No Snapping

It is possible to turn off snapping while changing the Texture Position. Snapping can be turned off by using the *Ctrl* modifier key.

Right-Click | Flip and Rotate

Many times, a texture only needs to be rotated or flipped in order to match the desired appearance. The right-click menu in the Texture Position workflow contains two menu flyouts, **Flip** and **Rotate**. The texture can be flipped **Left/Right** and **Up/Down** and rotated **90**, **180**, or **270** degrees:

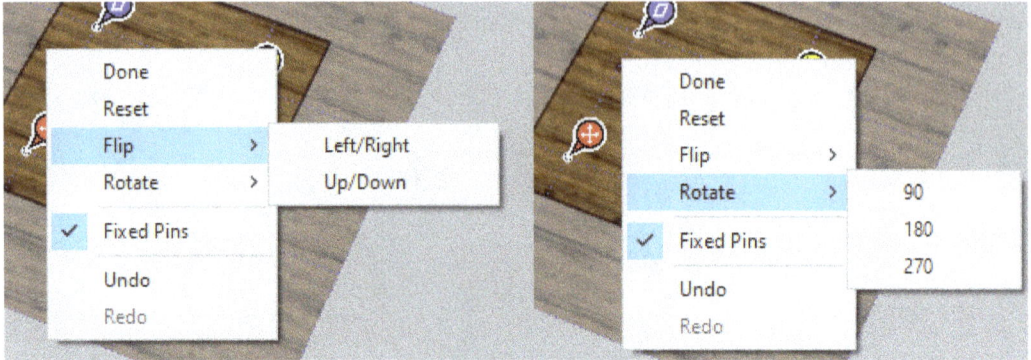

Figure 8.109 – Right-Clicking for the Flip Flyout (Left) and Right-Clicking for the Rotate Flyout (Right)

In this example, we can see a bed frame that is built using boxes:

Figure 8.110 – A Bed Frame with a Texture Material Applied to All Faces

The boxes are always at 90-degree increments, so it is easy to update the texture Position using the right-click options.

Remember, in order to update a texture with Position Texture, the Material must be painted onto the Face directly. These boxes are components that have been edited, and all Faces have been painted with the Paint Bucket's *Shift* modifier key:

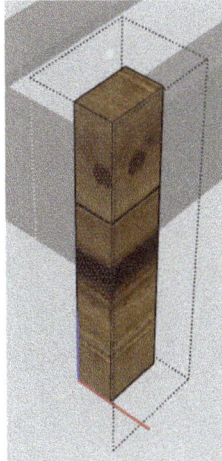

Figure 8.111 – All Faces Painted within Each Box Component

Each Face can be individually updated to Flip and Rotate the texture appropriately. The final product looks much more realistic, as each side of the boxes runs along the anticipated grain of the wood:

Figure 8.112 – The Bed Frame with the Default Texture Orientation (Left)
and the Bed Frame with the Updated Texture Positions (Right)

Reset

Textures can be reset to their original positions from the Texture Position right-click menu:

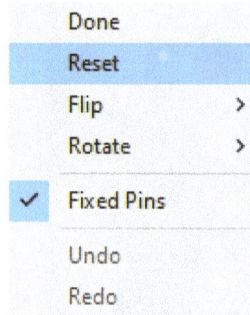

Figure 8.113 – The Texture Position Reset

This can save a lot of time if the editing process needs to be started over instead of using **Undo** multiple times.

Texture Position is complex and can be used to update individual Faces so that textures appear exactly as desired. In some cases, multiple Faces need to be spanned with a single texture image. This is the perfect use case for Projected Textures!

Projected Textures

Textures in SketchUp Pro automatically attempt to align with the individual painted Faces. By default, texture Materials do not attempt to connect from one Face to another, especially if the Faces are not coplanar. There are some extensions that can be downloaded in the Extension Warehouse to help with this, but we will not discuss individual Extensions in this chapter.

When a texture Material is applied to a curved object, the texture images can appear to scatter across the surface. This is especially noticeable in surfaces that curve in multiple directions:

Figure 8.114 – A Curved Object Painted with a Texture Material That Appears Scattered

The texture Material attempts to align with each individual Face, causing the image to fracture and appear scattered. Projected Textures can solve this problem by allowing a single texture image to span across multiple Faces. A Projected Texture can only be applied in one direction, which is perpendicular to the Face that has the Projected Texture:

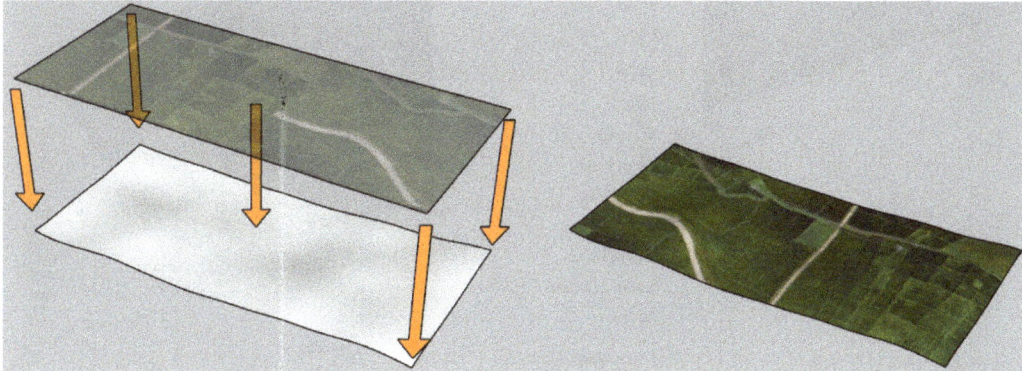

Figure 8.115 – The Texture Projected from Above (Left) and the Finished Model (Right)

Projected Textures can be used on all curved Geometry. Textures can be projected by first applying a texture Material to a Face in a Model. The orientation of the Face will determine the direction of the Projected Texture:

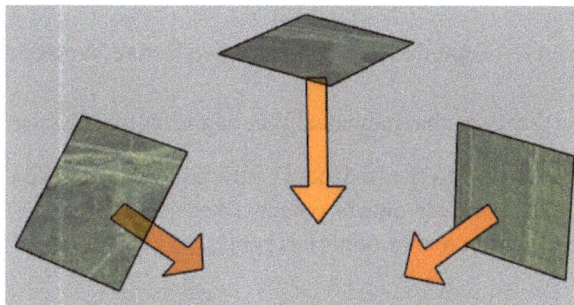

Figure 8.116 – Different Faces Project Textures in Different Directions

Once the texture Material has been painted to the individual Face, the texture can be converted to a Projected Texture. This can be done by right-clicking the Face and choosing **Texture** | **Projected**:

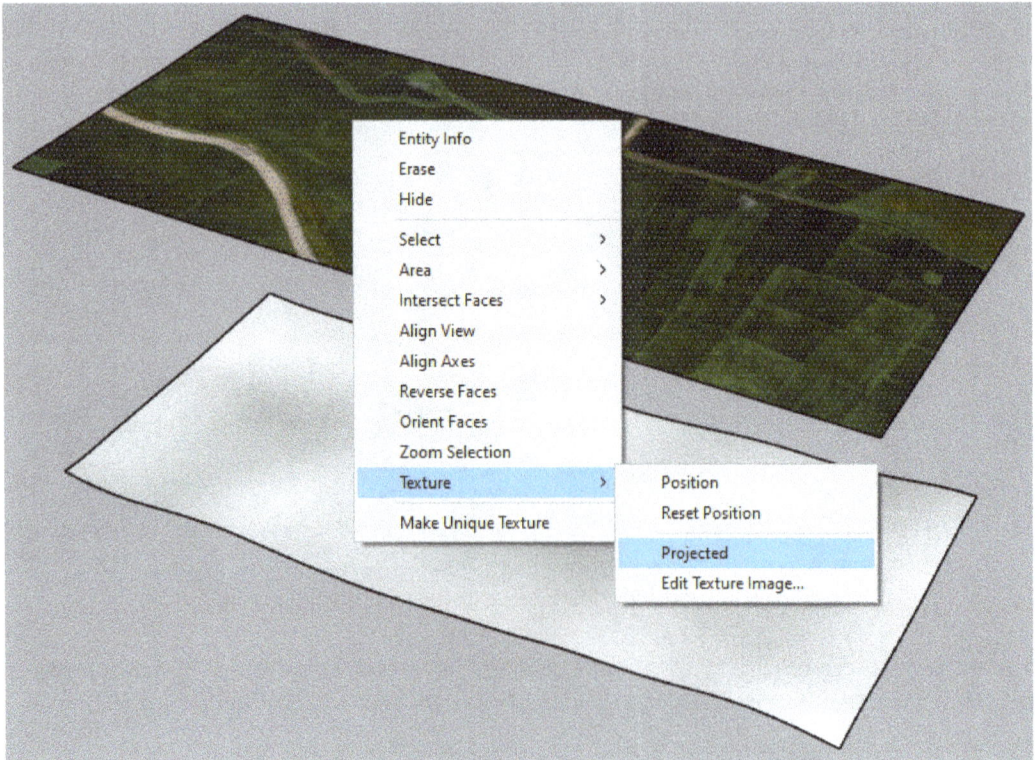

Figure 8.117 – Right-Clicking and Toggling on Texture | Projected

This will toggle the texture that is on that individual Face as a Projected Texture.

The Projected Texture is not added to the **In Model** Collection and cannot be saved as a Material for future use. The Projected Texture can only be referenced by sampling the Face, using the **Sample Material** option in the Paint Bucket tool or **Materials** panel:

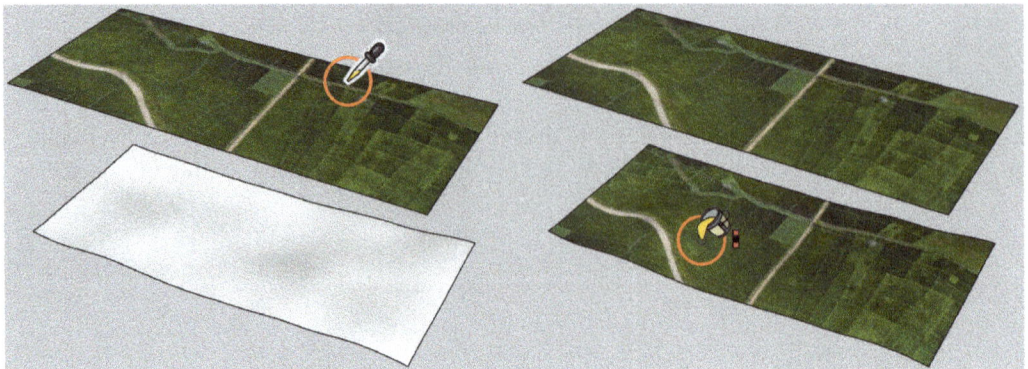

Figure 8.118 – The Projected Texture Sampled (Left) Projected Texture Painted on Curved Face (Right)

> **Note**
>
> SketchUp will still consider the Projected Texture to be a subset of the original texture Material. Projected Textures can be overridden with the Paint Bucket Tool, especially when using the modifier keys to paint multiple Faces at the same time. Be careful to not override Projected Textures in your Models!

In many cases, it is beneficial to use a combination of Projected Textures and regularly applied textures. Projected Textures can be applied to any Geometry, but if it is perpendicular to the original Face, the Projected Texture will only be represented by a single row of the pixels in the texture image:

Figure 8.119 – Projected Texture on a Cube, with the Vertical Sides Painted with a Single Row of Pixels

In this next example, we will look at a mattress for the bedframe we created previously. This is a complex Model, with some Faces along the standard Axes and multiple Faces that curve in two directions. Painting this Model with a standard texture Material will result in scattering on the top Faces:

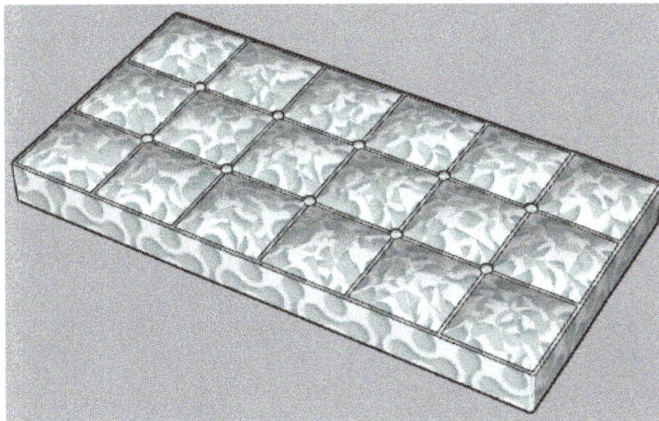

Figure 8.120 – A Mattress Painted with Texture Material Appearing Scattered

In this case, we will draw a simple Face next to the mattress. The Face does not need to be directly on top of the mattress or the same size as it. The Projected Texture will tile in all directions and project on any painted Faces. It can be helpful to align the Faces in some workflows, but it is not strictly necessary. The Face can then be painted with the desired texture Material:

Figure 8.121 – The Face Drawn Next to the Mattress Geometry and Painted with the Texture Material

The Face can then be selected and right-clicked on to activate the context menu, and the Projected Texture can be toggled on. Then, the Projected Texture can be sampled:

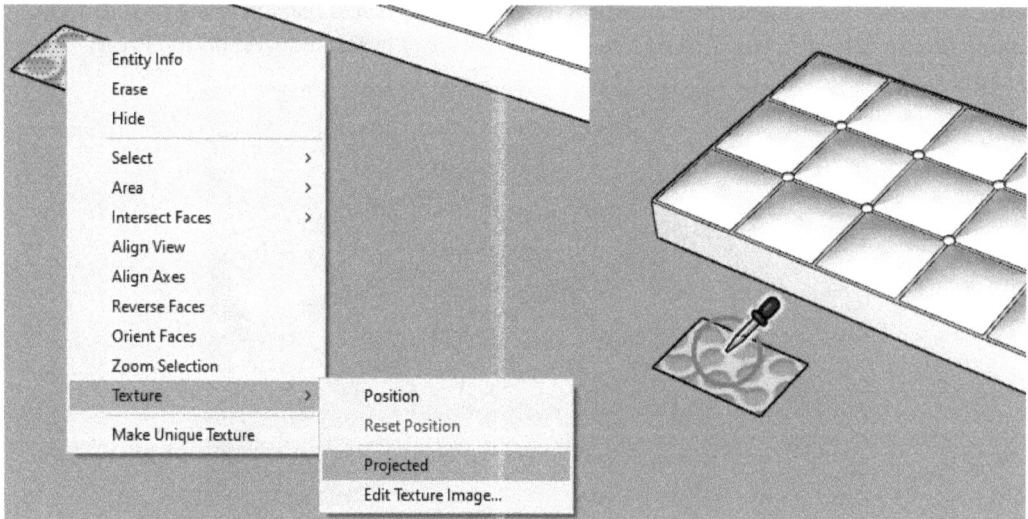

Figure 8.122 – The Projected Texture Toggled On (Left) and the Projected Texture Sampled (Right)

The mattress can then be painted using the Projected Texture. Note that the *Ctrl* modifier key is used in the Paint Bucket tool to paint all connected Faces on the mattress:

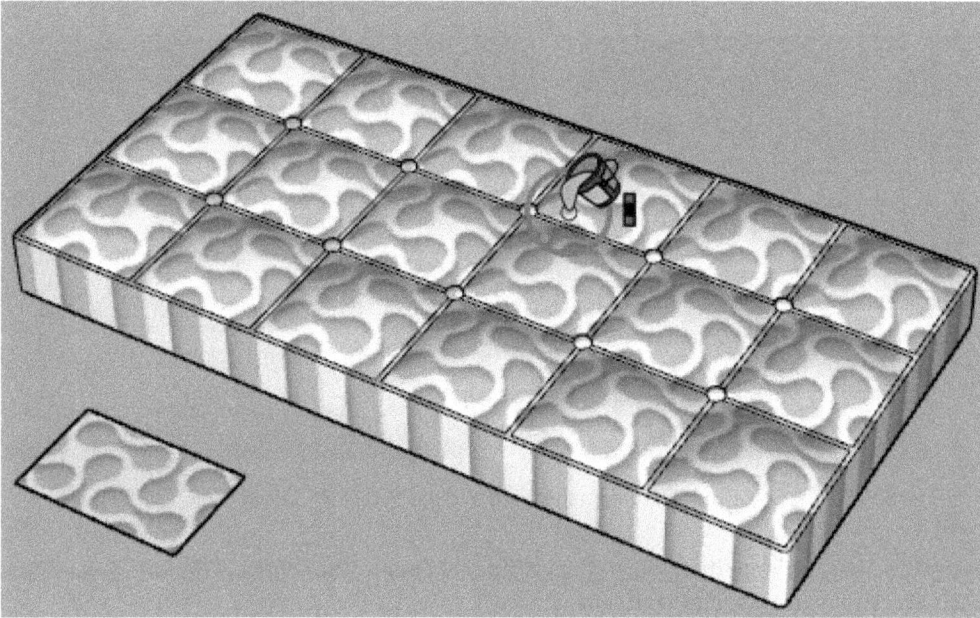

Figure 8.123 – The Mattress Painted with the Ctrl Modifier Key

The Projected Texture has painted the vertical sides of the mattress with a single row of pixels from the image, resulting in a striped appearance. This can be updated by selecting the main texture Material in the **In Model** Collection and painting the vertical sides of the Model with the standard Material. This removes the vertical striping from the Projected Texture:

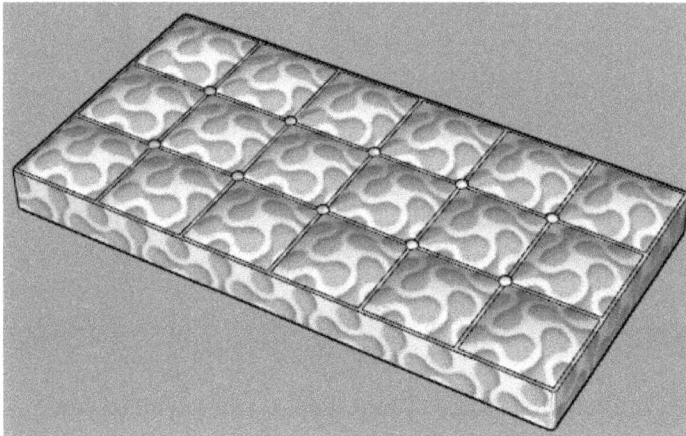

Figure 8.124 – The Sides of the Mattress Painted with the Standard Texture Material

SketchUp Pro can recognize when texture Materials are being applied to cylindrical connected Faces and will attempt to automatically wrap the texture Material around the object. In this next example, we can see a soft drink texture image and the soft drink can in SketchUp:

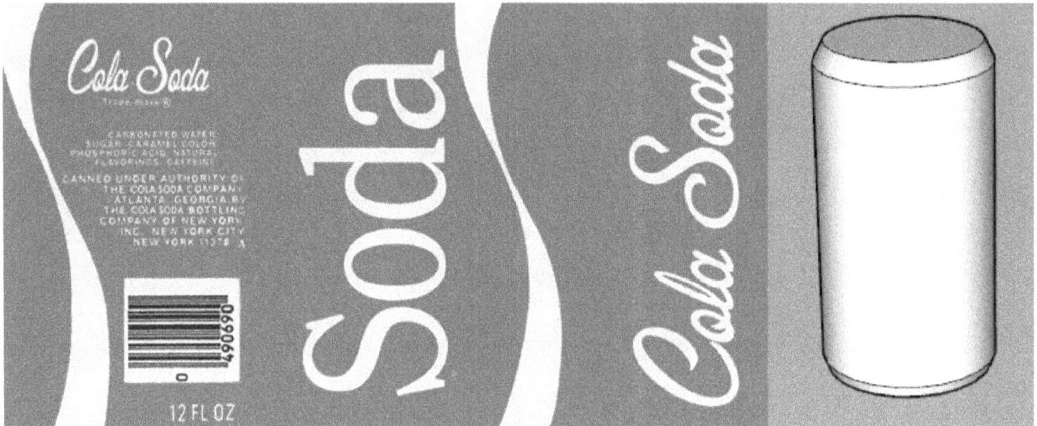

Figure 8.125 – The Soft Drink Texture Image (Left) and the Soft Drink Can in SketchUp (Right)

If this texture Material is applied only to the cylindrical Face in the center of the can, the texture Material successfully wraps around the can:

Figure 8.126 – The Texture Material Applied to the Cylindrical Face
(Left) and the Texture Wraps Correctly (Middle and Right)

If the texture Material is applied to all the connected Faces of the can Geometry, including the slanted Faces, then SketchUp will not be able to understand the texture wrapping, and the texture will appear as scrambled:

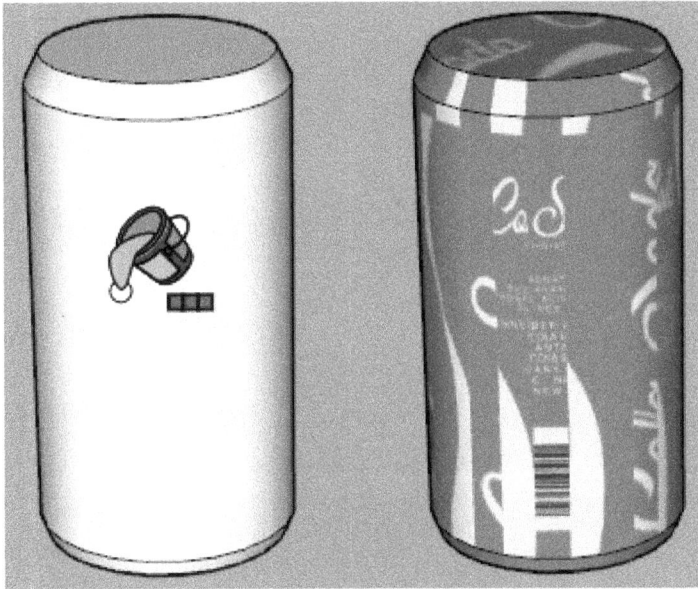

Figure 8.127 – The Texture Material Applied to All Faces (Left) and the Texture Appearing Scrambled (Right)

Projected Textures are very powerful in SketchUp Pro. Remember to correctly align the Face and Position of the texture Material before creating a Projected Texture. To wrap up this chapter, let's take a quick look at a couple of helpful Material options in SketchUp Pro.

Additional Material Options

SketchUp can be difficult to manage when it comes to Materials. To counter this, SketchUp offers the **Calculate Area** and **Select All** options to help with specific workflows using Materials.

Calculating the Area of a Material

It is often the case that a specific Material will represent a paint, tile, or other linear finish material in construction. SketchUp includes options to automatically calculate the area of individual Faces, multiple selected Faces, or even entire Tags. SketchUp has also included a quick way to find the area of all Faces painted with a specific Material.

The Material area calculation can be performed by right-clicking a Face with the desired Material and choosing **Area | Material**. A dialog box will appear, showing the calculated area of all Faces that are painted with that Material:

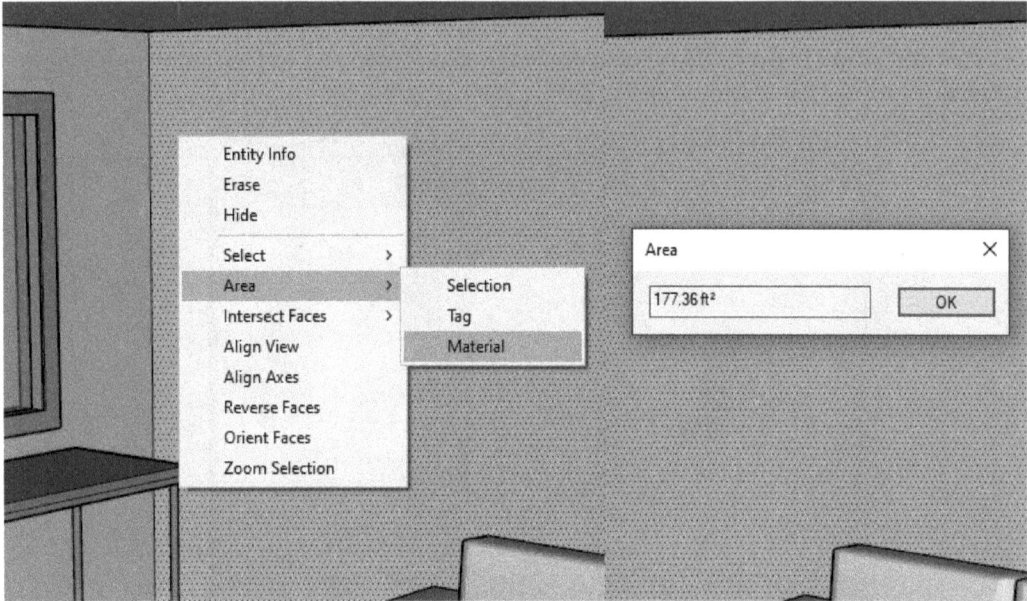

Figure 8.128 – Right-Clicking and selecting Area | Material (Left)
and the Area Calculation Dialog Box (Right)

> **Note**
> Make sure that you do not have any hidden Geometry that is painted with the Material, which should not be included in the calculation. All painted Faces – even hidden Faces – are included in the calculation.

Selecting All with Same Material

It can also be helpful to select all of the Faces painted with the same Material in a Model. Selecting all Faces painted with the same Material can be done by right-clicking a Face and choosing **Select | All with same Material**:

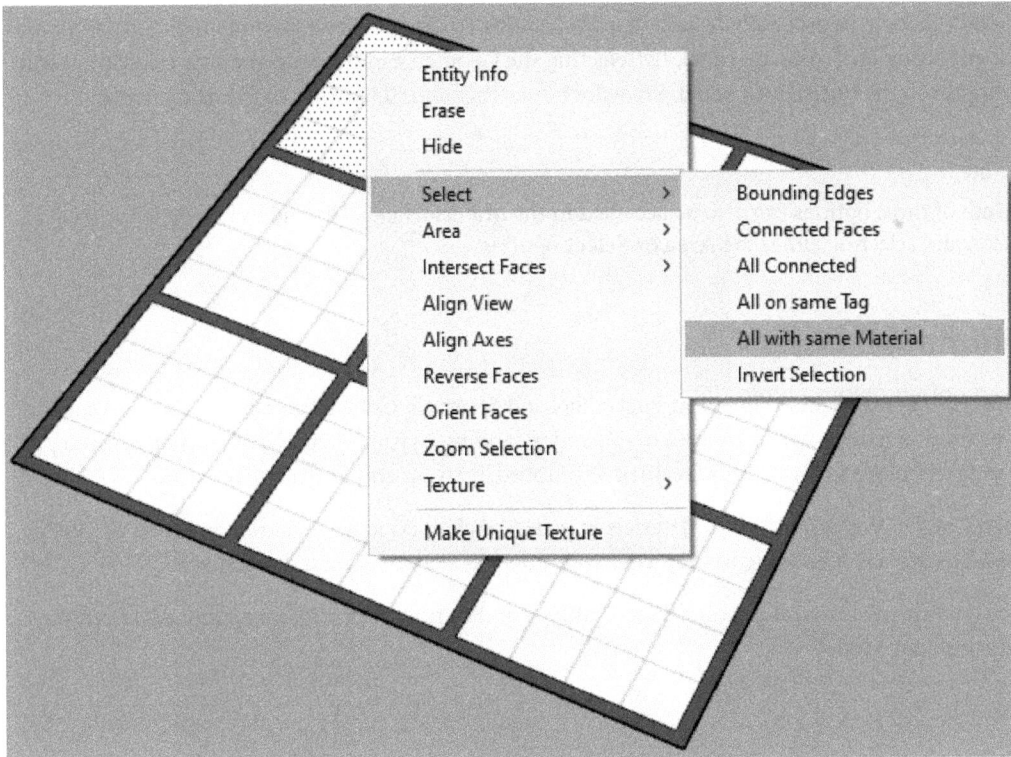

Figure 8.129 – Right-Clicking and selecting Select | All with same Material

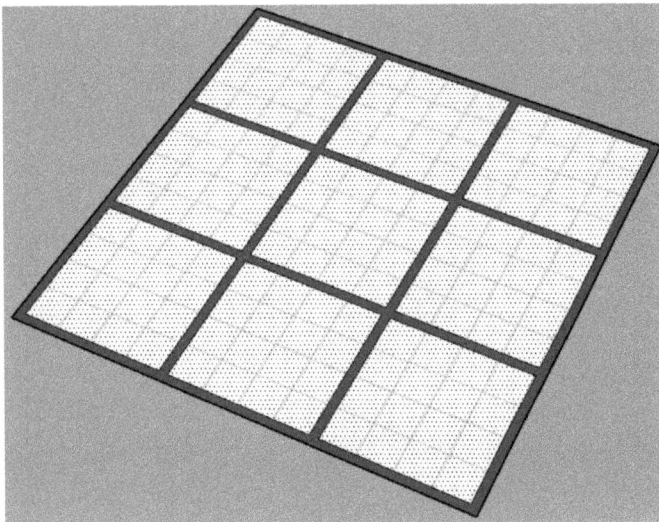

Figure 8.130 – All Painted Faces Selected

This workflow can be especially helpful when separating connected Geometry into individual Groups, or when setting up Tags in a Model. By selecting the Geometry in this way, the area calculation will also appear in the **Entity Info** panel. We will discuss Tags and Entity Info in the next chapter.

> **Note**
> Both of these options can also be accessed in the **Materials** Panel by right-clicking a Material icon and selecting either the **Area** or **Select** options.

Summary

In this chapter, we discussed the Paint Bucket tool and Materials. Color Materials and texture Materials can be used in many ways to create a quick understanding of your SketchUp Model or an in-depth mapping of images to create photorealistic Positioned Textures and Projected Textures.

I hope that you play around with Materials in your SketchUp Models – just remember to use **Purge Unused** every once in a while to keep your **In Model** Collection clean and organized.

In the next chapter, we will examine Tags, Outliner, and Entity Info as we learn more advanced ways to organize our Models.

Part 3 – Advanced Modeling and Model Organization

Part 3 will discuss different Default Tray panels, the Model Info and Preferences dialog boxes, and will include an advanced discussion about working with Components. Part 3 will conclude by looking at how to import and export models using various file types, the Extension Warehouse and Extensions – including Extension Overlays – and 3D Warehouse.

This part has the following chapters:

- Chapter 9, Entity Info, Outliner, and Tags Dynamically Organize Your Models
- Chapter 10, Model Info and Preferences
- Chapter 11, Working with Components
- Chapter 12, Import and Export, 3D Warehouse, and Extensions

9
Entity Info, Outliner, and Tags Dynamically Organize Your Models

In this chapter, we will look at three Default Tray panels: Entity Info, Outliner, and Tags. We will discuss how the Entity Info panel can be used to view and edit Geometry attributes. Then, we will discuss the Outliner panel and start to see our models in a new way. We will also look at the Tags panel (formerly called the Layers panel), which can be used to assign visibility options to sets of Geometry. These three panels can help to organize your models in new ways for modeling, editing, animation, and image export.

The following topics are covered in this chapter:

- Entity Info:
  - Edges
  - Faces
  - Groups
  - Components
  - Other objects
- Outliner
- Tags:
  - Tag Visualization Options
  - Tag folders
  - Cleaning Up Unwanted Tags
  - Using Tags with Scenes

Entity Info

The **Entity Info** panel displays attributes of any currently selected Geometry. These attributes change depending on the type of Geometry selected and can sometimes be edited within the Entity Info panel. The **Entity Info** panel can be activated via the **Window | Default Tray** dropdown. Additionally, **Entity Info** will always be the first option on the right-click contextual menu for any selection set:

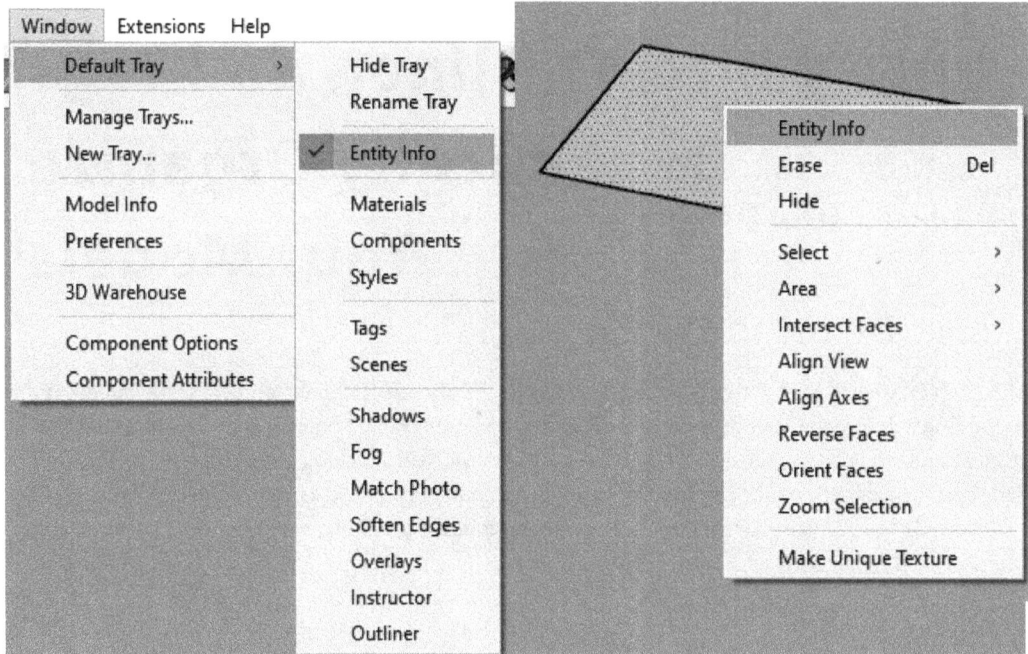

Figure 9.1 – Entity Info in Window | Default Tray Drop-down Menu
(Left); Entity Info in Right-Click Menu (Right)

The most common selection set while modeling is no selected Geometry. The **Entity Info** panel will read **No Selection** when no Geometry is selected:

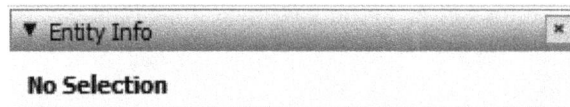

Figure 9.2 – Entity Info panel with No Selection

When Geometry is selected, **Entity Info** will always show the **Tag** attribute and the four main **Toggles**: Show/Hide, Lock/Unlock, Does Receive/Doesn't Receive Shadows, Does Cast/Doesn't Cast Shadows:

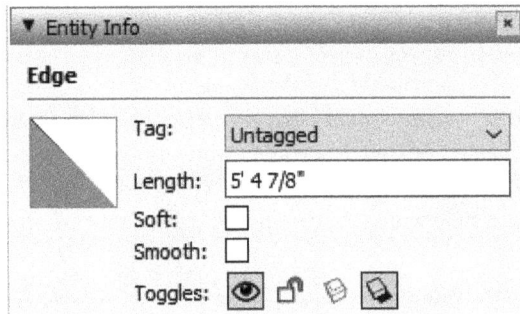

Figure 9.3 – Entity Info Showing Tag and Toggles of an Edge

These attributes will always be shown when any Geometry, Object, or Construction Geometry is selected, even if some of these toggles do not apply in certain cases. If the toggle does not apply to the current selection set, the buttons will appear grayed-out:

Figure 9.4 – Toggles Editable (Left); Toggles Grayed-Out (Right)

If multiple Objects, Edges, or Faces are selected that have different attribute values, the **Tag** dropdown will be blank, and the **Toggles** buttons will have their outlines removed:

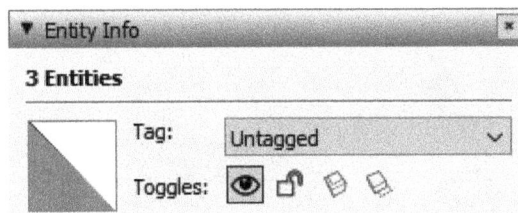

Figure 9.5 – Multiple Objects Selected and Entity Info Attributes Blank

> **Note**
>
> As previously mentioned, the **Tag** value for Edges and Faces should always remain **Untagged**. If you select only Edges and Faces (not Groups or Components) and Entity Info shows them with named Tags, use the dropdown to change the setting for Edges and Faces to **Untagged**.

Entity Info will also display additional attributes for Edges, Faces, Groups, and Components. There are also other types of SketchUp objects that will show unique attributes, including Welded Edges (Curves, Arcs, Polygons, and Circles) and Construction Geometry (Section Planes, Dimensions, Text, and Guides). We will quickly look at all the different attributes in the following sections.

Edges and Welded Edges

Entity Info will show different Edge attributes depending on how individual Edges are welded together. Individual Edges or selection sets of individual Edges will have a **Length** value and **Soft** and **Smooth** checkboxes. The **Length** value can be updated when a single Edge is selected that has at least one unconnected endpoint. When more than one Edge is selected, or when the Edge that is selected has two connected endpoints, SketchUp will show the approximate total **Length** value:

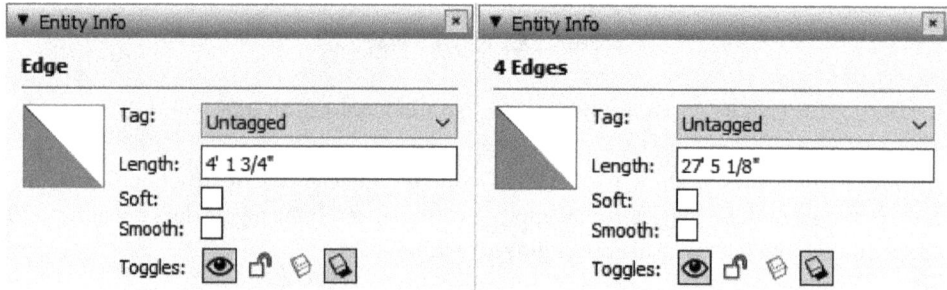

Figure 9.6 – Entity Info for One Selected Edge or Multiple Selected Edges

Edges can also be welded into Curves, which include Arcs and Circles. All Curves show a number of **Segments**, which can be edited when part of Arcs, Polygons, and Circles. Additionally, Arcs, Polygons, and Circles have editable **Radius** and **Circumference** attributes, and noneditable **Arc Length**, **Perimeter**, and **Circumference** attributes respectively.

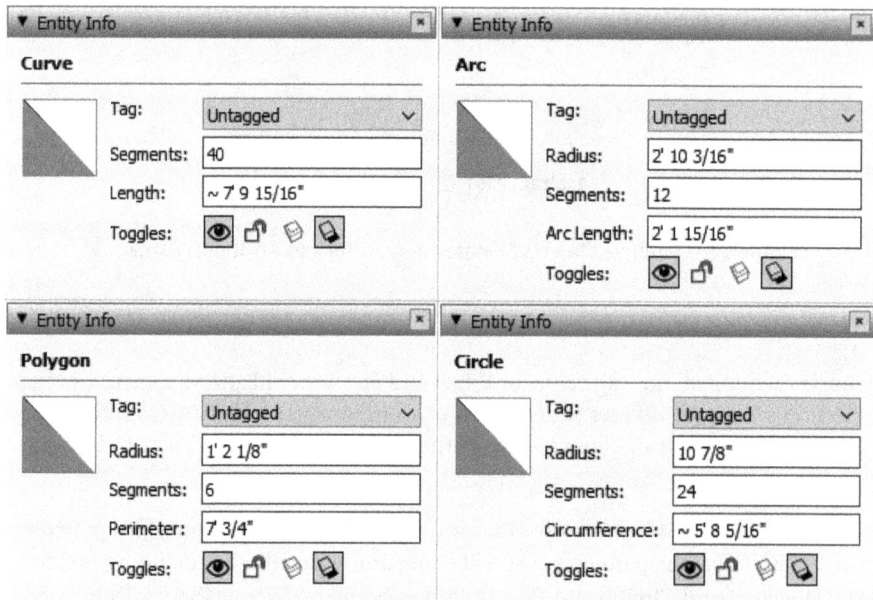

Figure 9.7 – Entity Info for Selected Curve (Top Left), Arc (Top Right),
Polygon (Bottom Left), and Circle (Bottom Right)

When more than one Curve is selected, or when Edges and Curves are selected at the same time, **Entity Info** will show the total number of Edges and Curve segments selected:

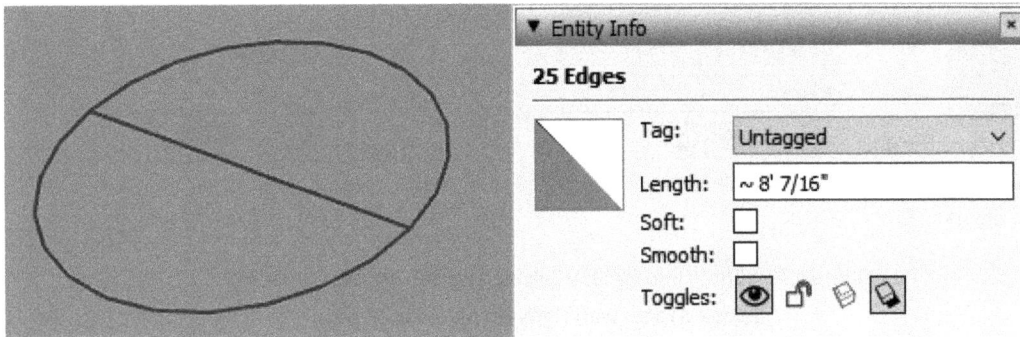

Figure 9.8 – Circle and Edge Selected (Left); Entity Info Shows 25 Edges Selected (Right)

In the preceding example, a Circle with 24 segments has been selected, as well as a single Edge. Entity Info shows that the selection contains **25 Edges**.

Faces

Faces have one additional attribute, which is the given **Area** of the Face. If multiple Faces are selected, then **Entity Info** will show the sum total area of all selected Faces:

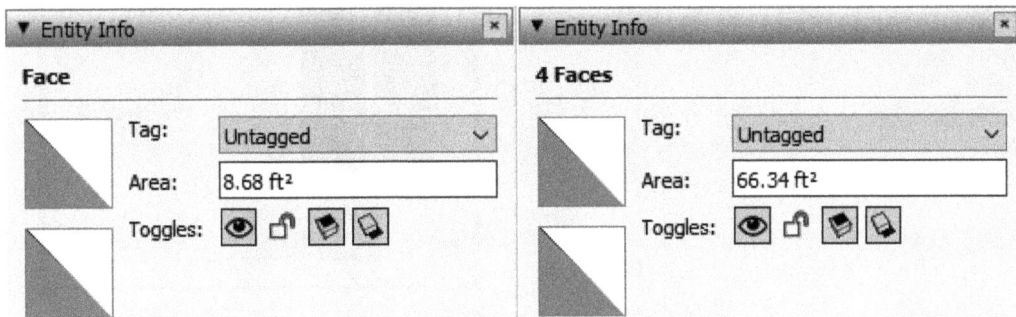

Figure 9.9 – Entity Info for One Selected Face (Left) and for Multiple Selected Faces (Right)

The Materials for the Face can also be edited in Entity Info. The top Material in Entity Info represents the positive normal of the Face, which is a white color in the Default Material. The bottom Material in Entity Info represents the negative normal of the Face, which is the blue/grey color in the Default Material. When multiple Faces that have been painted with different Materials are selected, the Material preview in Entity Info will show a question mark:

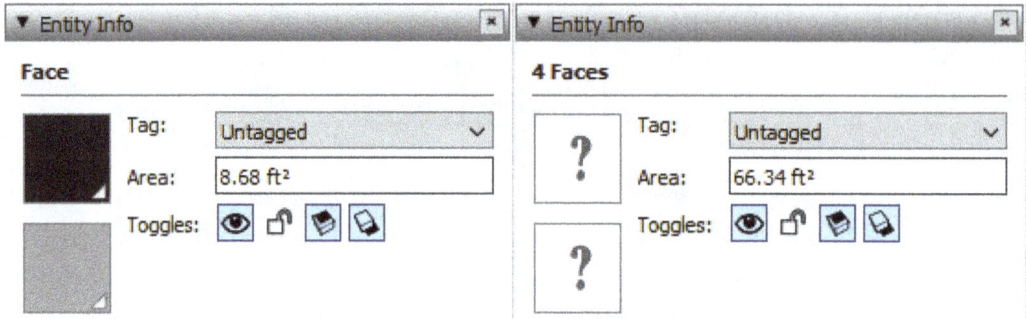

Figure 9.10 – Entity Info for One Selected Face (Left) and For Multiple
Selected Faces with Different Materials (Right)

Materials can be edited by clicking on the Material preview in Entity Info. This will open the **Choose Paint** dialog box, which will only show the In Model Collection:

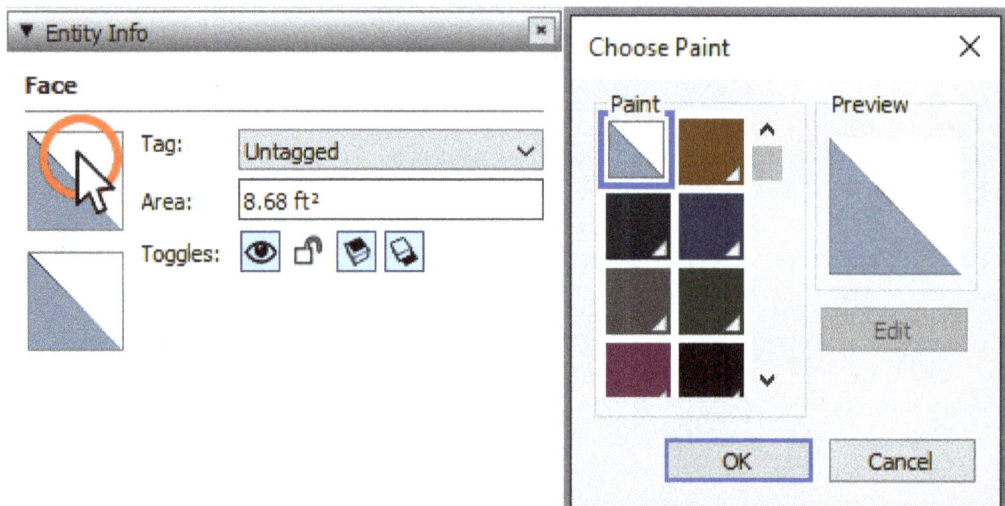

Figure 9.11 – Material Preview Clicked in Entity Info (Left); Choose Paint Dialog Box (Right)

Materials can be edited from the **Choose Paint** dialog box, but it is recommended to use the **Materials** panel in the **Default Tray** for any Material editing.

Groups

Groups have an **Instance** attribute, which is essentially the Group name. When a Group is initially created it does not have an Instance name, and the Group will be shown as **Group** in the **Outliner** panel. When an **Instance** value is added in **Entity Info**, the Group's name will be updated in the **Outliner** panel:

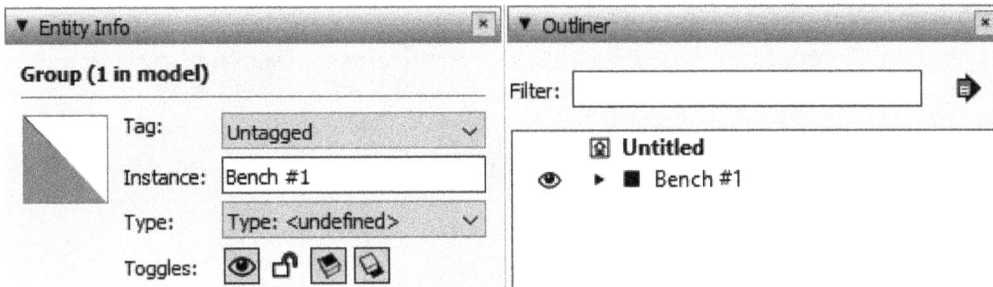

Figure 9.12 – Entity Info for Selected Group (Left); Outliner Showing Instance Name (Right)

We will look at the **Outliner** panel in a later section of this chapter.

Groups that contain watertight Geometry will be labeled as **Solid Groups** in Entity Info. A **Volume** attribute will also be shown for Solid Groups. When multiple Solid Groups are selected, SketchUp will list the number of Solid Groups, but SketchUp will only calculate the volume for one selected Group at a time:

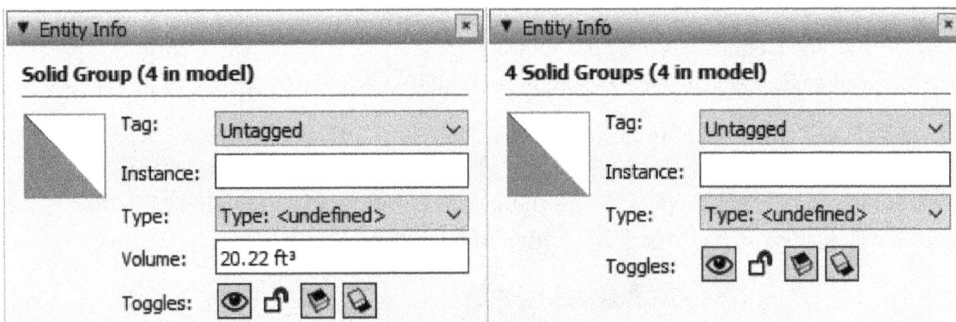

Figure 9.13 – Entity Info for One Selected Solid Group (Left) and for Multiple Selected Solid Groups (Right)

Groups also have a **Type** attribute. This attribute allows for Groups to have assigned Classifications. We will talk more about Classifications in *Chapter 10, Model Info and Preferences*.

Components

Components are similar to Groups in the Entity Info panel. Components show a Volume attribute when the Geometry forms a Solid Component. Components also have Instance Values, and can have Classifications assigned through the Type attribute.

> **Note**
>
> We will dive deep into Components in *Chapter 11, Working with Components*. This section is a brief overview of what you will see in the Entity Info Panel when a Component is selected.

Components contain more information than Groups and are also saved in the Model for easy editing and duplication. Components are differentiated in each model using the **Definition** attribute, which is the overall name for that Component. The **Instance** attribute is the name of individual Groups or Components, while the **Definition** attribute defines the Component as it is saved in the **Components** panel:

Figure 9.14 – Entity Info for Selected Component Showing Instance and Definition

The **Instance** value and **Definition** value both show up in the Outliner panel with **Definition** appearing in open and closed angle brackets. We will talk more about the Outliner panel later in this chapter.

Components are the only Object in SketchUp Pro that has an **Advanced Attributes** flyout in the **Entity Info** panel. The flyout can be toggled on by clicking the **Show Advanced Attributes** button. The Advanced Attributes for Components are rudimentary **Building Information Modeling** (**BIM**) data fields, which include **Price**, **Size**, **URL**, **Status**, and **Owner**:

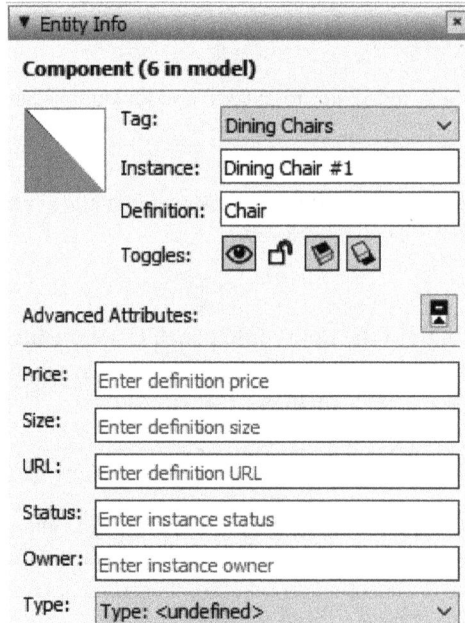

Figure 9.15 – Entity Info for Selected Component with Advanced Attributes Flyout Open

These attributes can be manually filled in to provide more information about each Component.

> **Note**
>
> Do you want to see how many Component Instances or copies of a Group are in your model? Entity Info will list the number of Component Instances or copies of a Group in your model when you select a single Object.

Figure 9.16 – One Component Selected and Entity Info Showing 10 Component Instances in the Model

Construction Geometry

The Entity Info panel will show information for all Geometry in SketchUp – including construction Geometry. These are elements that help during the modeling process but do not represent real-world Geometry. These include Section Planes, Dimensions, Text, and Guides.

Section Planes

Section Planes include two unique attributes in **Entity Info**: **Name** and **Symbol**:

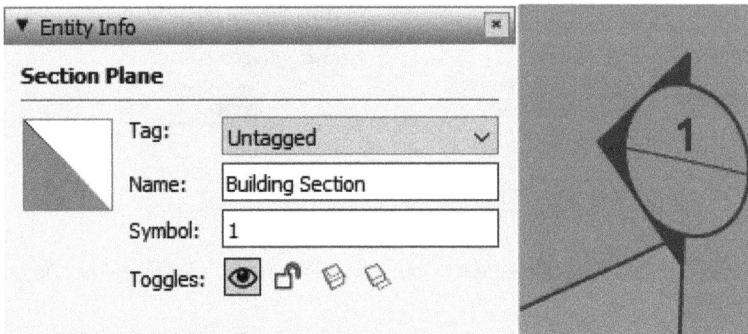

Figure 9.17 – Entity Info for Selected Section Plane (Left); Section Plane Symbol in the Model (Right)

The **Name** and **Symbol** attributes are both shown in the Outliner panel, but **Symbol** is also visible in the model.

Dimensions and Text

Dimensions and Text can be found in the Large Tool Set and the Construction Toolbar. These Tools are used for Construction Geometry and are not recommended to be used for final dimensioning of a drawing or model.

> **Note**
>
> For more detailed ways to use Dimensions and Text for final drawings or construction documents, check out the companion software to SketchUp Pro: LayOut. We will talk more about LayOut in *Chapter 12, Import and Export, 3D Warehouse, and Extensions.*

When selected, **Linear Dimension** and **Text** objects have many common attributes, including **Font**, font size, and a **Text** value:

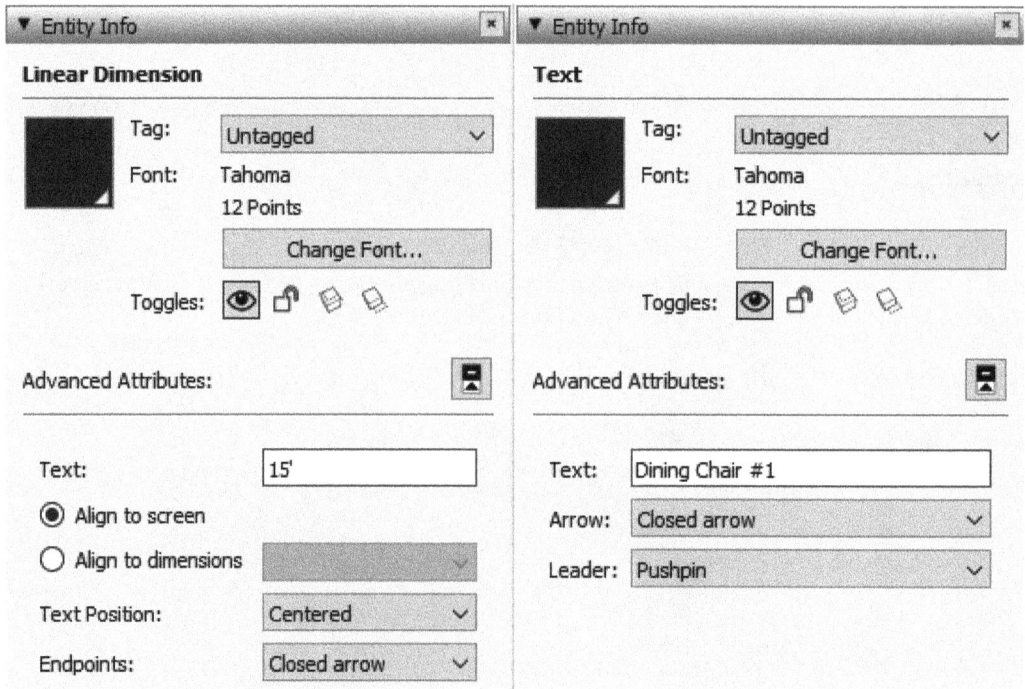

Figure 9.18 – Entity Info for a Selected Linear Dimension (Left) and for a Selected Text Object (Right)

The font and font size can be edited using the **Change Font** button. The **Font** dialog box will open when the **Change Font** button is clicked, and allows the user to select an installed font from the local computer as well as the **Font style** option and **Size** in **Points** or in-model **Height**:

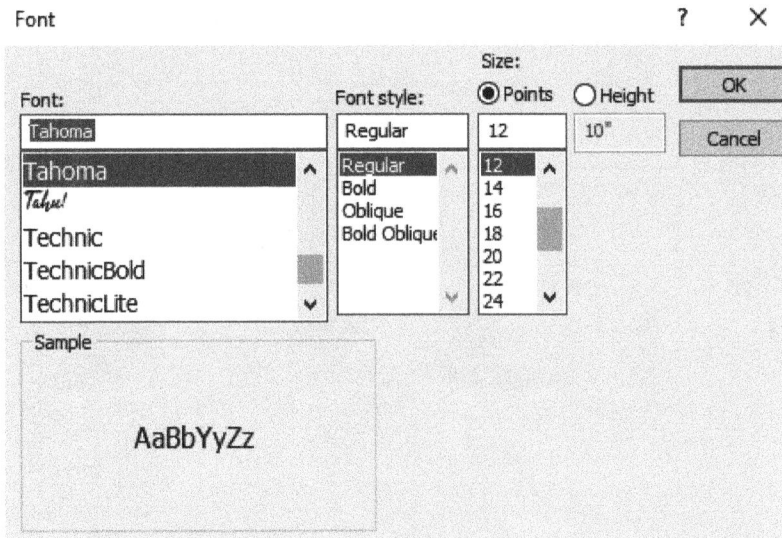

Figure 9.19 – Font Dialog Box

Dimensions and Text have slightly different Advanced Attributes. Dimensions define the distance between two points that are left-clicked in the model. The distance between these points is displayed between two extension lines, which are highlighted with two arrow heads. The Dimension Advanced Attributes panel allows for the Dimension text to be toggled between **Aligned to screen** or **Aligned to dimension**. When the dimension text is aligned to the screen it will always face the Camera. When the dimension text is aligned to the dimension, it will always be parallel to the dimension line:

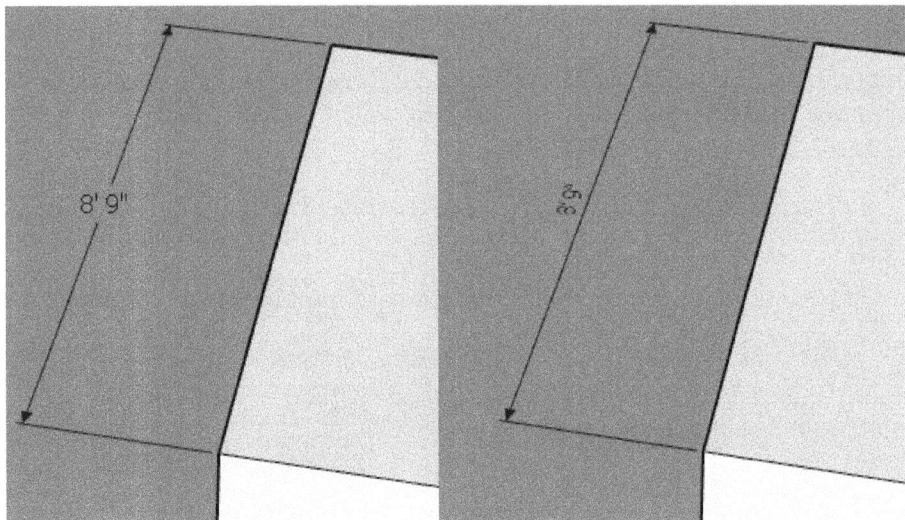

Figure 9.20 – Dimension with Text Aligned to Screen (Left); Text Aligned to Dimension (Right)

Dimension **Text Position** and **Endpoints** can also be changed, with each attribute restricted to a predetermined list:

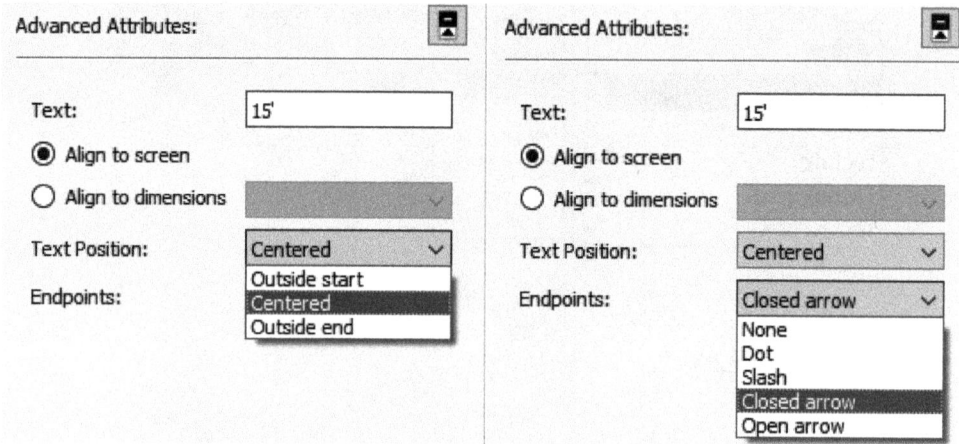

Figure 9.21 – Text Position Dropdown (Left); Endpoints Dropdown (Right)

Dimension **Text** cannot be edited in the **Entity Info** panel. To edit it, right-click the Dimension and choose **Edit Text**.

> **Note**
>
> Typically, Dimension Text should not be overridden with different distance values. However, it may be useful to change dimension text to the minimum and maximum value ranges if a part can be adjusted, or for descriptive terms such as "Equal" or "Same".

Text Objects in SketchUp will always face the Screen, so this attribute toggle is not shown in the Entity Info panel when a Text Object is selected. However, Text Objects have two other unique attributes, **Arrow** and **Leader**:

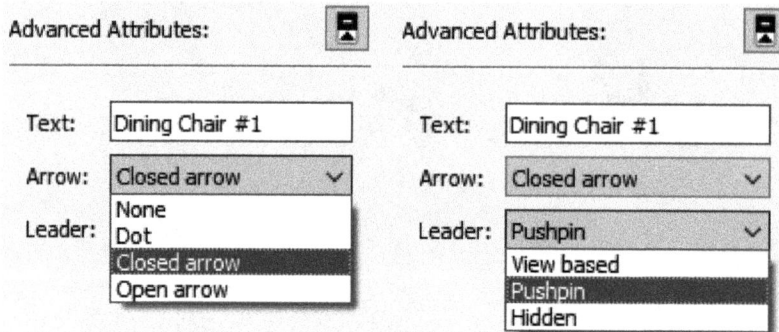

Figure 9.22 – Arrow Dropdown (Left); Leader Dropdown (Right)

The **Arrow** attribute is similar to the Endpoint attribute for the Dimension, but it only appears at one end of the leader. The **Leader** attribute defines whether the Leader itself faces the camera or is completely hidden.

Guides

Guides are unique in that they have no additional attributes:

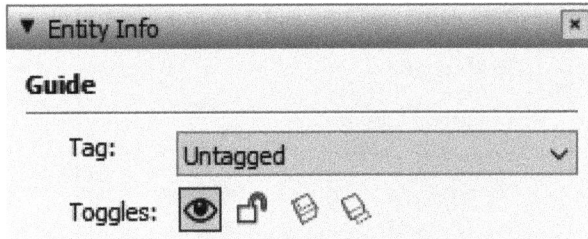

Figure 9.23 – Entity Info for One Selected Guide

> **Note**
>
> Guides can easily be selected when you are attempting to select Edges. If the Entity Info shows a number of Entities, then a Guide might have been selected alongside some Edges!

The Entity Info panel is the best way to update tags and attributes, especially when cleaning up a model. We have mentioned the Outliner panel multiple times in this section, so let's take a closer look at it now.

Outliner

The Outliner panel shows the Objects in the model in a hierarchical tree. This view shows only Objects (Groups, Components, and Section Planes) and their relationship to each other, specifically if they are nested. Object attributes and visibility can also be controlled in the Outliner panel. The **Outliner** panel can be activated in the **Window | Default Tray** dropdown:

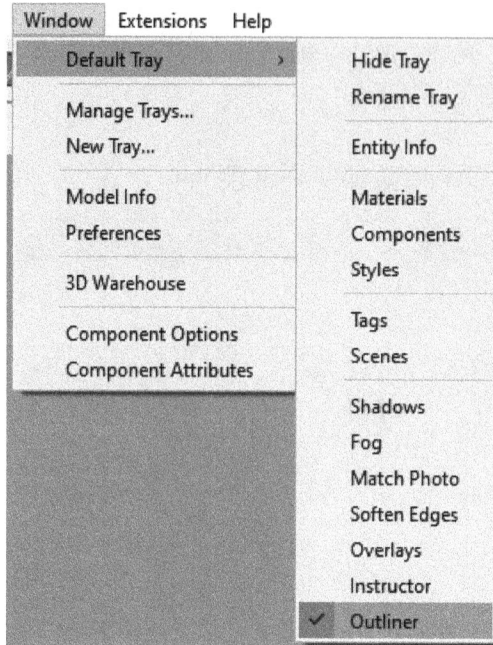

Figure 9.24 – Window | Default Tray Menu Dropdown

The Outliner panel contains the following tools:

1. A **Filter** search bar
2. The Details Flyout Button
3. The Model Title
4. The Selection pane

These tools are shown in the following screenshot:

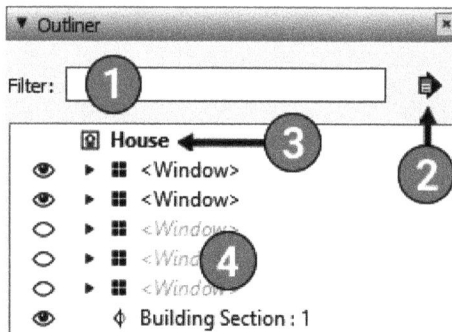

Figure 9.25 – The Outliner panel with the Filter Search Bar, Details, Model Title, and Selection Pane

The **Filter** search bar can be used to find specific Objects by searching for part of the Instance or Definition name. The search term must not contain typos, but it can use any letters of the Object's name. In the following example, CH was searched for, and SketchUp returned **Chris**, **Bench**, and **Chair**:

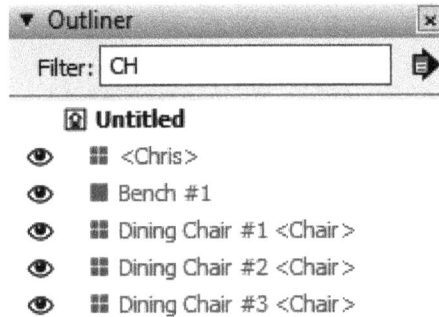

Figure 9.26 – Outliner Filtered to Show all Objects Containing the Letters "CH"

The Details Flyout button has three options that can help to sort the Outliner panel. Expand All and Collapse All will expand and collapse all nested Objects. In this next example, we can see multiple nested Objects in the model, and how different the views are with the Outliner fully expanded or collapsed:

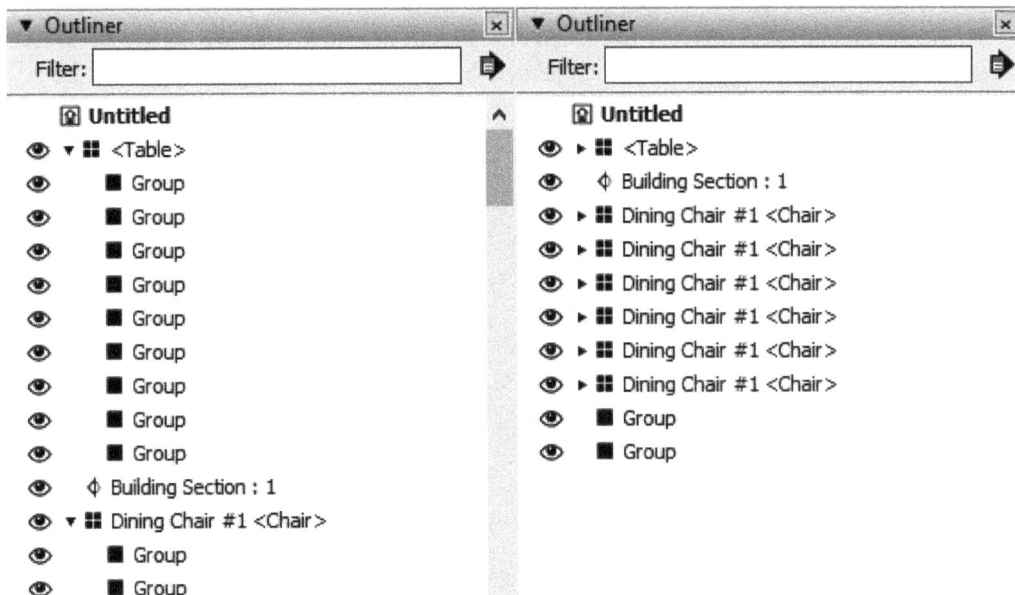

Figure 9.27 – Expand All Option Enabled in Outliner (Left); Collapse All Option Enabled in Outliner (Right)

The true power of the Outliner panel is in the Selection Pane. As previously mentioned, the Outliner panel shows a hierarchical view of the model and the relationships of the Objects in the model. In the following example, we can see a model named **House** with multiple Objects that are all at the same level, meaning that there are no nested Objects:

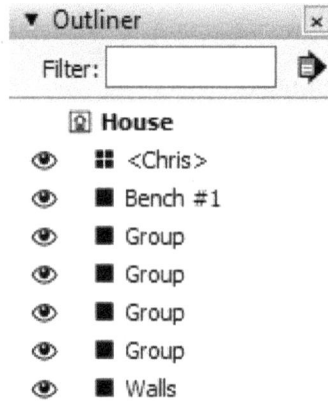

Figure 9.28 – Model With All Objects on the Same Level

The Outliner panel can be used to quickly identify the Objects in the model without having to find their actual position in the Drawing Area. This can be especially helpful in large and complex models with many Objects. Once an Object has been located in the Outliner panel, the Object can be double-clicked to Activate or Edit the Object. In this next example, the **Bench #1** Group is Double-Clicked in the Outliner panel to Edit the Group:

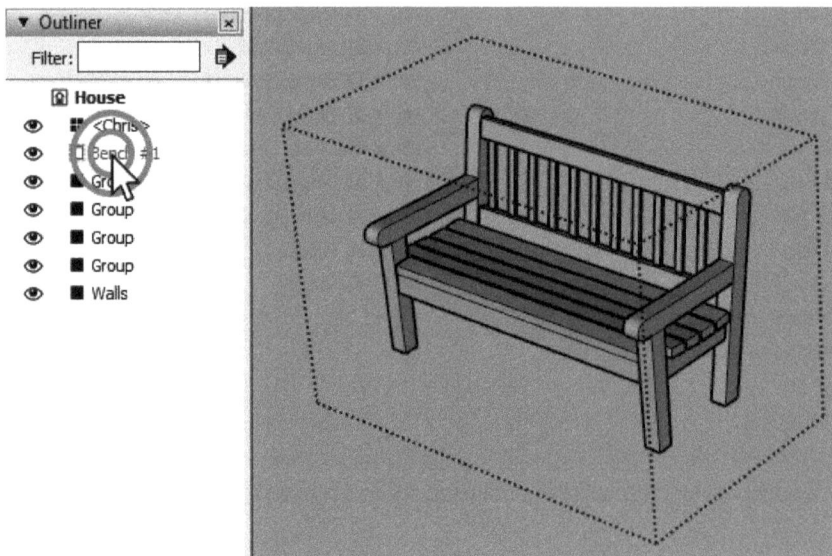

Figure 9.29 – Bench #1 Group Double-Clicked to Edit Group

When an Object is Open (currently being Edited or Active), the Outliner panel icon next to that Object will turn white. If the Object is hidden, the Eye icon will appear closed, while the eye will be open when the Object is visible. And, when an Object is Locked, the icon next to the Object will appear to be grayed-out and a small padlock will overlay the icon. In the following example, we can see the icons for **Groups** (Top), **Components** (Middle), and **Section Planes** (Bottom):

	Closed	Open	Locked	Hidden	Visible
Groups	■	□	■🔒	◇	👁
Components	⊞	⊞	⊞🔒	◇	👁
Section Planes	⬦	⬦	◁🔒	◇	👁

Figure 9.30 – Outliner Icons

One of the most common uses in the Outliner is Hiding Objects and Editing Hidden Objects. Objects can be hidden by clicking on the Eye icon next to the Object. Hidden Objects can still be selected and edited from the Outliner, as the Object is still in the model. In this next example, we can see that the Component instance, **Chris,** has been hidden by clicking the Eye icon, but then the Component instance can still be edited by selecting and right-clicking **Edit Component** in the **Outliner** panel:

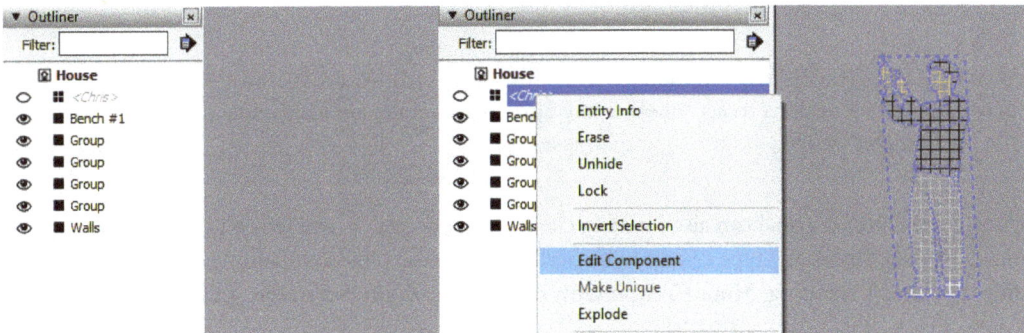

Figure 9.31 – Component Hidden (Left); Component Selected and Edited (Right)

When Objects are hidden using the Outliner panel, the Object itself is hidden, not the internal Geometry. This is a distinction in SketchUp Pro that is apparent in the **View** dropdown in the Main Menu. The **View** Dropdown shows two options: Hidden Geometry and Hidden Objects. We have seen Hidden Geometry in previous chapters in this book, and this workflow is the outcome when using right-click | **Hide** on any Geometry. When a Group or Component is hidden, it can be separately viewed in the Hidden Object set. In the following example, we can see multiple Objects hidden in the Outliner panel and Hidden Objects turned on in the Model:

Figure 9.32 – Objects Hidden in the Outliner panel (Left); Objects
Visible with Hidden Objects Toggled On (Right)

> **Note**
>
> SketchUp displays Hidden Geometry and Hidden Objects in the same way, so try toggling between the two options to see whether the Hidden elements are Geometry or Objects. Or use the Outliner panel!

Objects in the Outliner panel can also be right-clicked to show a very detailed contextual menu. This menu allows for almost any type of edit to be performed on the Object, including Editing, Erasing, Hiding, Locking, Exploding, **Make Component**, **Flip Along**, **Zoom Selection**, and **Rename**:

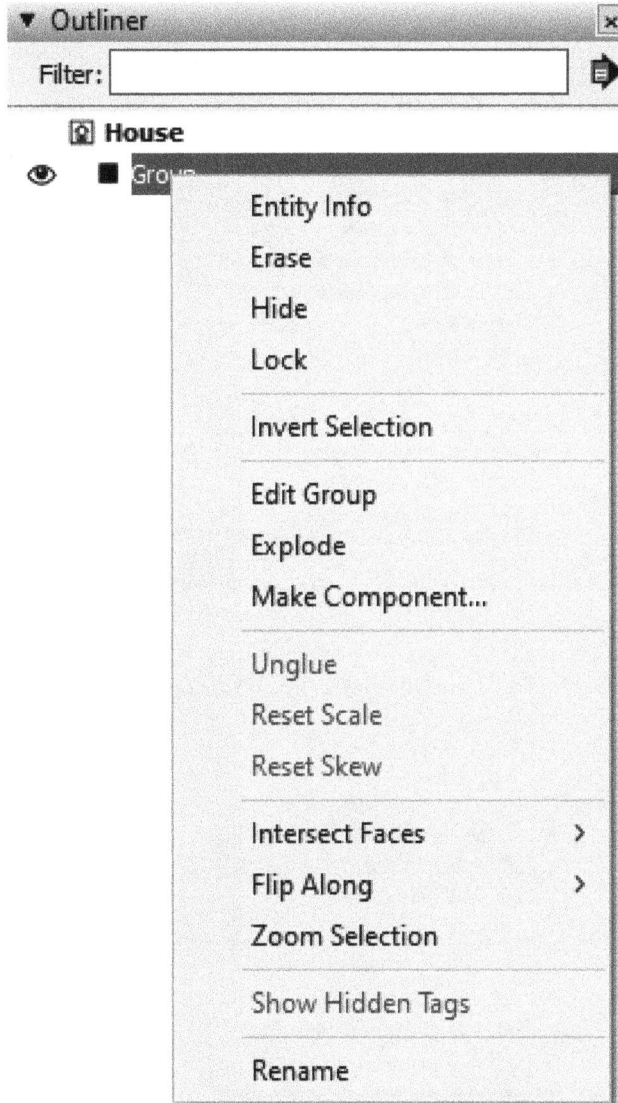

Figure 9.33 – Right-Click Contextual Menu for a Group in the Outliner panel

Components and Section Planes will have even more options, which are the same options found when right-clicking an Object in the Drawing Area.

There are many instances when a model needs to be restructured to arrange the Objects differently. This is especially the case when trying to add Groups to an existing Group or Component that is acting as a collection of Objects. In this next example, we can see the **House** model has a Group called **Window Group** and multiple **Window** Components nested inside:

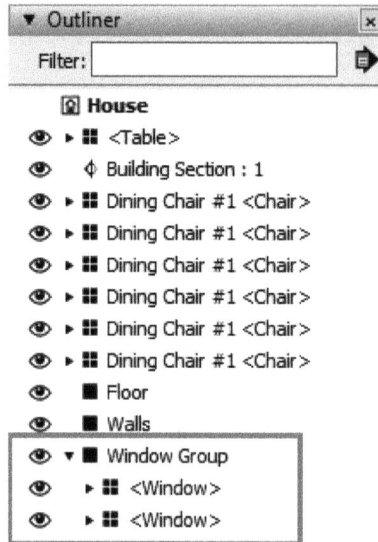

Figure 9.34 – The Outliner panel with Window Group and Nested Window Components

After further development on the model, three more **Window** Components have been added. However, these Components were not included in **Window Group** and are instead at the top level of the model.

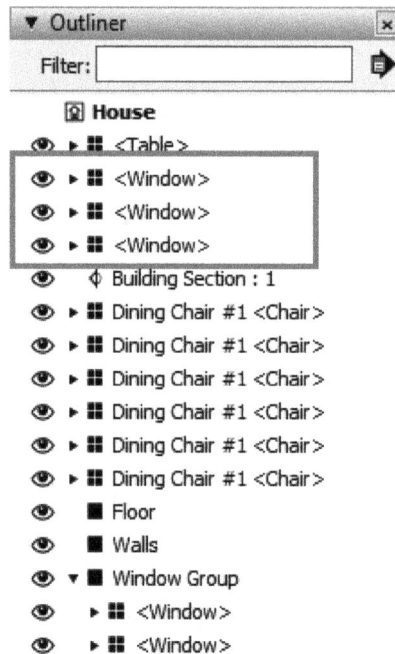

Figure 9.35 – Window Components Added at the Top Level of the Model

Groups and Components can be moved from their nested locations using the Cut Tool and Paste In Place Tool. This can help with quick fixes when part of a model has been created outside of a Group or Component, or in the wrong Object to begin with. However, the Outliner panel allows for Objects to be moved in the model hierarchy by simply clicking and dragging, while preserving the Object's location in 3D space. In this example, the **Window** Components were clicked and dragged on top of the **Window Group** entry:

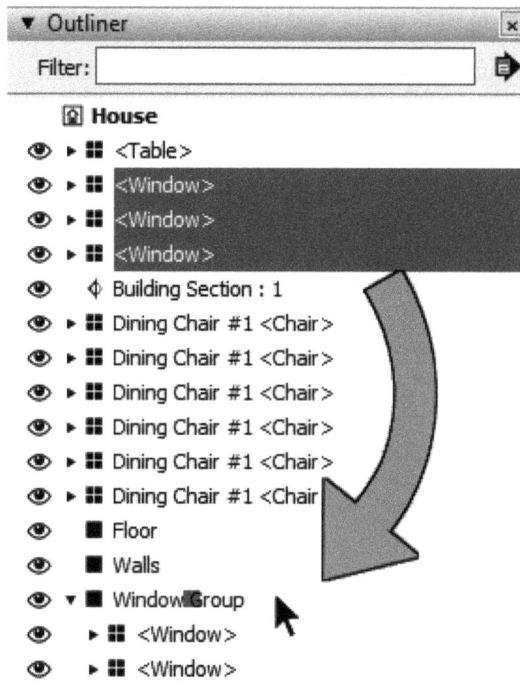

Figure 9.36 – Window Components Clicked and Dragged into Window Group

The **Window** Components remain in the same location in 3D space, but they are now inside the **Window Group**:

Figure 9.37 – Windows Nested in Window Group

> **Note**
> Multiple Objects can be selected at once in the Outliner panel by using the *Shift* and *Ctrl* Modifier Keys. This can help when moving multiple Objects in the model hierarchy.

The Outliner panel may not be useful for every SketchUp model, but it can be extremely useful when navigating large and complex models. Hiding and showing Objects while modeling and restructuring the model hierarchy will allow you to work quickly and efficiently when finalizing your SketchUp models.

Now that we understand the Entity Info and Outliner panels, let's take a look at a third way of controlling Geometry visibility, Tags.

Tags

The **Tags** panel (formerly called the Layers panel) organizes Geometry and Objects into sets that control visibility. Tags are used to change visibility settings for multiple Objects at once, including hiding and unhiding, as well as applying edge dashes (similar to linetypes in other software). The **Tags** panel can be activated in the **Window | Default Tray** dropdown:

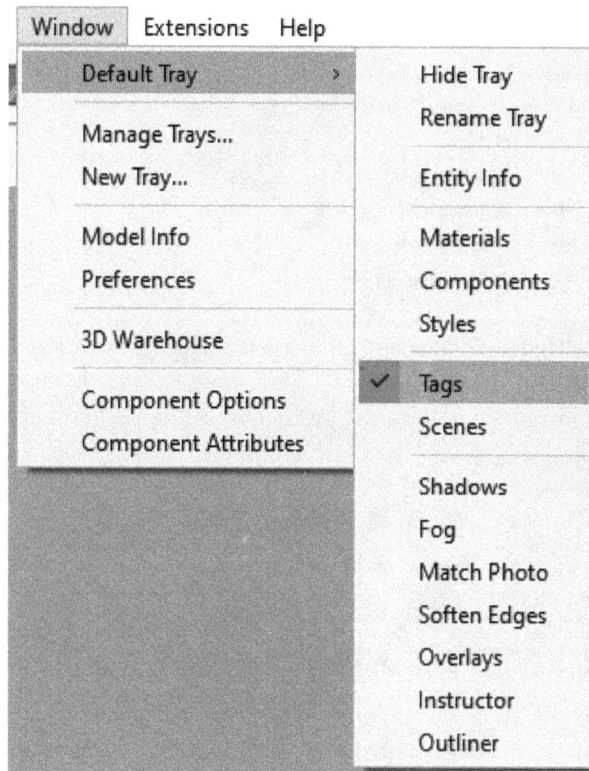

Figure 9.38 – Window | Default Tray Menu Dropdown

The **Tags** panel contains the following tools:

1. The Add Tag Button
2. The Add Tag Folder Button
3. A Search Box
4. The Tag Tool Button
5. The Color by tag Toggle Button
6. The Details Flyout Button
7. A Tags List including Visibility, Name, Color, Dashes, and Current Tag Icon

These tools are shown in the following figure:

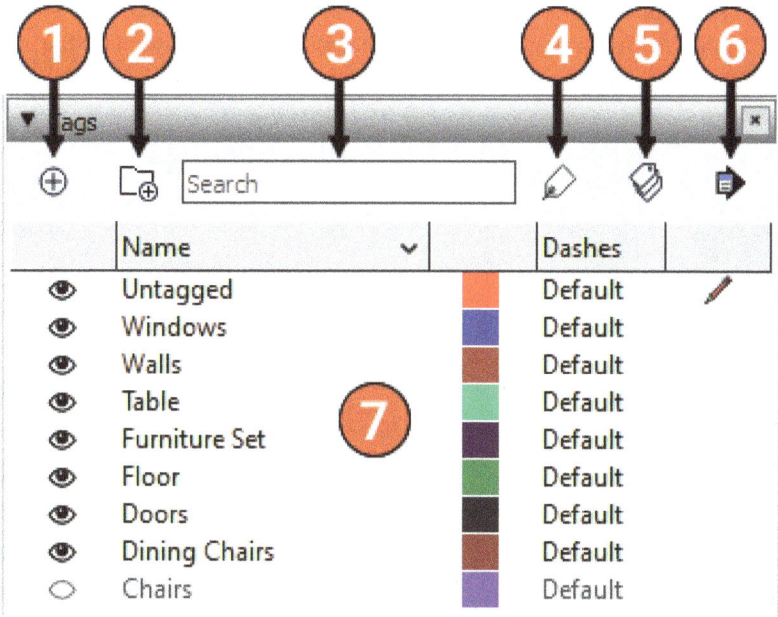

Figure 9.39 – Tags panel with Add and Delete Tags Buttons, Details, and a Tags List

There is also a Tags Toolbar in SketchUp Pro. The **Tags** Toolbar is simply a drop-down list of the Tags in the model. This list has the same functionality as the **Tags** dropdown in the **Entity Info** panel, but it's useful to quickly reference the current Tag. In this example, the **Tags** Toolbar is showing that the **Untagged** Tag is the current Tag:

Figure 9.40 – The Tags Toolbar with the Current Tag Set to Untagged

Unlike the Outliner panel, a Tag can be applied to any Object or Geometry. However, it is strongly recommended that all Geometry is drawn on the default **Untagged** Tag. Then, the Geometry can be converted into recognizable Groups and Components and assigned to any specified Tag in the model:

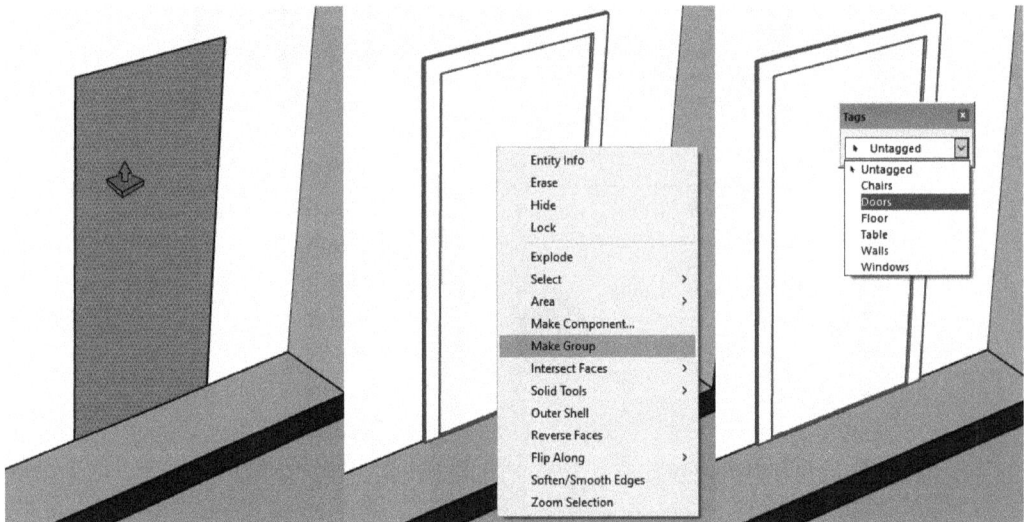

Figure 9.41 – Geometry is Drawn (Left); Group is Created (Middle); Group is Moved to Specified Tag (Right)

It is extremely important to always initially draw on the **Untagged** Tag and then move created Objects (Groups or Components) to a specified Tag after. When Objects and their included Geometry are on different Tags, or when the included Geometry is tagged and the Object is on the **Untagged** Tag, the visibility options of the Objects can lead to ghosted lines or Geometry refusing to be hidden. This can lead to frustrating modeling situations, as well as making it hard to work collaboratively in SketchUp Pro.

In this next example, we can see a window Group on the Window Tag with some internal Geometry on different Tags. When the Window Tag is hidden using the eye icon, some of the Geometry will be hidden, leaving behind a strange selection set of leftover Edges and Faces:

Figure 9.42 – Window Tag Hidden with Some Geometry Remaining Visible

> **Note**
> It is extremely important for model organization as a best practice workflow that Geometry is created on the **Untagged** Tag, and Objects are added to named Tags as the model progresses.

The **Tags** panel allows for any Tag to be set as the current Tag. Newly created Geometry, Imported Groups, and placed Components will automatically be assigned to the current Tag. Changing the current Tag can seem like a great workflow when drawing, but it can be easy to forget to switch back to the **Untagged** Tag when drawing individual Geometry for a different part of a model. It is recommended that **Untagged** should remain the current Tag in general, and that only once Groups and Components have been created should they be switched to a different Tag using the Entity Info panel or Tags Toolbar.

> **Note**
> One exception to this general rule could be while placing many components onto a specific Tag from the Components panel. The current Tag can be changed, Components Placed, then the Untagged Tag can once again be made current. Be careful to remember to activate the Untagged Tag at the end of the workflow!

If the current Tag needs to be switched, it can be done anytime a Tool workflow is not active. The current Tag can be changed by clicking on the white space to the right of the Tag, either above or below the Pencil Icon. The Pencil Icon represents the current Tag. When the white space next to a Tag is clicked, it will become the current Tag.

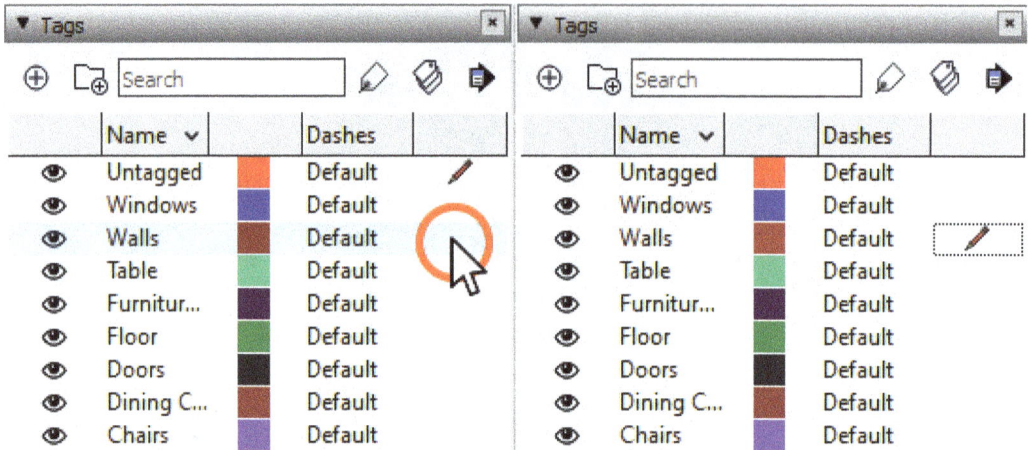

Figure 9.43 – White Space Next to Walls Tag is Clicked (Left); Walls Tag is New Current Tag (Right)

If a hidden Tag is selected for the new current Tag, SketchUp will automatically unhide (show) that Tag. If you attempt to hide the current Tag, SketchUp will show a dialog box stating that you cannot hide the current Tag:

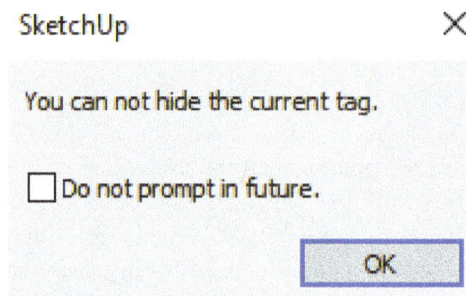

Figure 9.44 – Cannot Hide Current Tag Dialog Box

If the current Tag were hidden, any newly created Geometry would also be Hidden, so it would look like nothing was happening! This would be extremely confusing, so SketchUp Pro does not allow the current Tag to be hidden.

Now that we understand some of the rules and best practices for hiding Tags, let's take a closer look at some Tag visualization options.

Tag Visualization Options

Tags can be hidden similarly to how we hide Objects via the Outliner panel. The eye icon next to each Tag name can be toggled to hide or show all Objects that are on that Tag. In this example, we can see that all the Chair Groups in this model can be hidden by Hiding the Chair Tag, as long as those Objects have been manually assigned to that Tag:

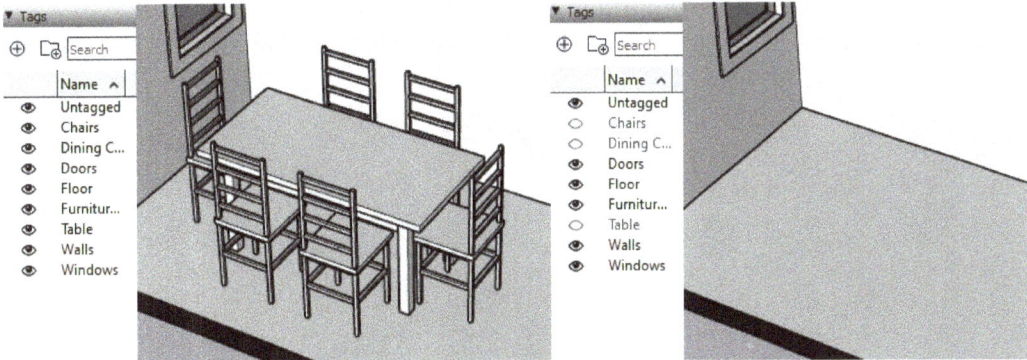

Figure 9.45 – Table and Chairs in the Model (Left); Tag Visibility Toggled and Chairs Hidden (Right)

We have already mentioned that it is a best practice workflow to have all basic Geometry on the **Untagged** Tag, and that Objects should be placed on other Tags. Occasionally, you may need to have nested Objects, and these Objects might be on separate Tags. If any Geometry, including nested Objects, is on a different Tag, it will still be hidden if the top-level Object is hidden using Tag visibility. In this example, we have the same table and chairs as previously, but all of the Objects are Grouped into a Dining Set Group:

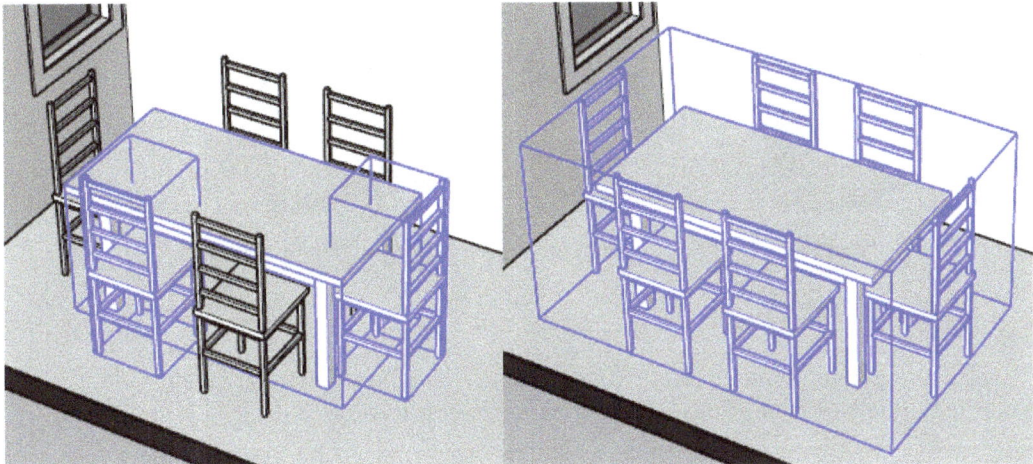

Figure 9.46 – Table and Chairs as Individual Groups (Left); Dining Set Group (Right)

The Table and Chairs Groups are on the corresponding Tags, but the Dining Set Group has been assigned to the new **Furniture Set** Tag:

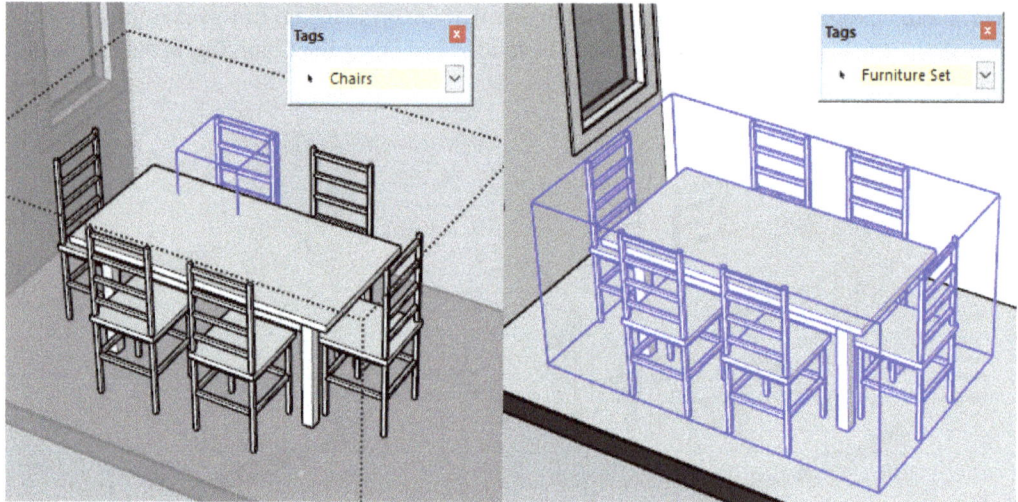

Figure 9.47 – Chair Group on Chair Tag (Left); Dining Set Group on Furniture Set Tag (Right)

When the **Chair** Tag is Hidden only the Chair Groups will be hidden. When the **Furniture Set** Tag is Hidden, all nested Objects in the Dining Set Group will also be Hidden:

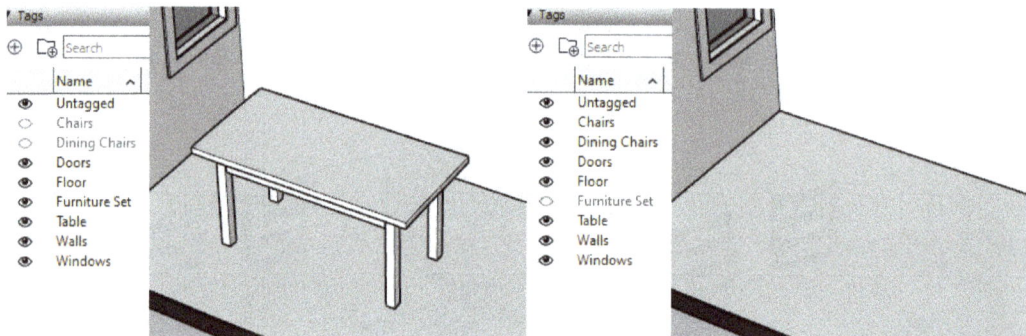

Figure 9.48 – Chair Tag Hidden (Left); Furniture Set Tag Hidden (Right)

> **Note**
>
> In newer versions of SketchUp, this workflow can also be achieved using Tag folders. Check out the *Tag folders* section of this chapter for more information.

Tags are unique in SketchUp Pro because they are the only way to assign different linetypes to Edges without using Extensions. Dashes in SketchUp Pro are only a visual effect, and do not cut or erase parts of the Edges. Dashes can be applied to Tags in the Tags panel. The **Default** linetype is a solid line, and it will be used by default when a new Tag is created:

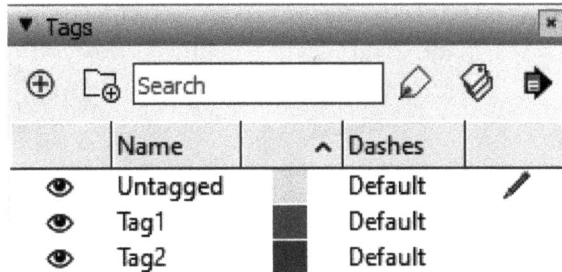

Figure 9.49 – New Tags Created with the Default Linetype in the Tags Panel

Dashes can be changed by clicking the current linetype (which may be **Default**) and choosing a different linetype from the contextual drop-down list:

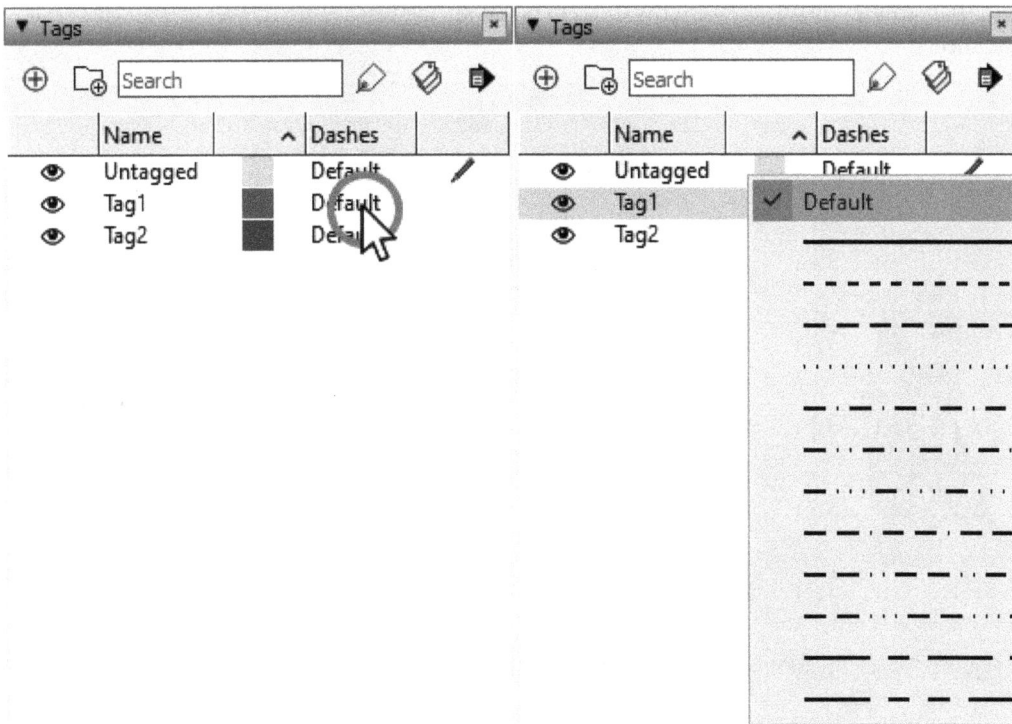

Figure 9.50 – Current Linetype Clicked (Left); Contextual Drop-down List (Right)

When a new **Dashes** option is selected, all Edges in the model that are associated with that Tag are updated. This includes Edges in Groups and Components, even if those Edges are assigned to a different Tag.

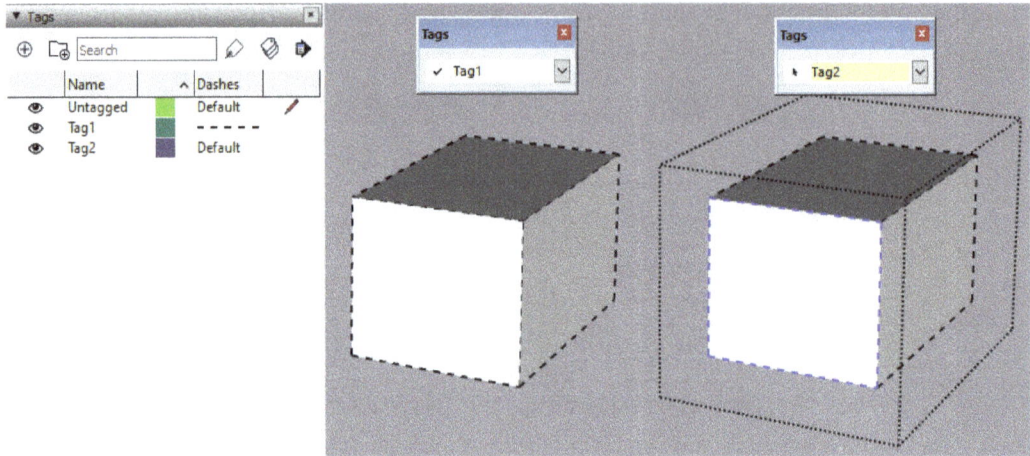

Figure 9.51 – All Associated Edges Show Dashes (Left); Nested Edges (Right)

Tag Dashes scale automatically with the Camera position. Tag Dashes are not set to a specific scale or distance:

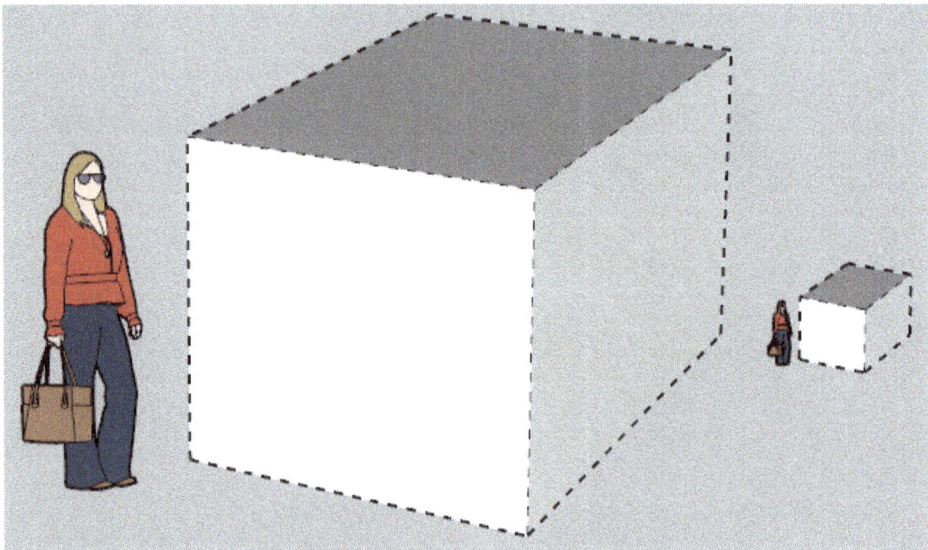

Figure 9.52 – Linetype Shown with a Close Camera Position (Left) and Far Camera Position (Right)

Typically, it is suggested to set exact line types and line weights when preparing final drawings in Layout, and not in SketchUp Pro while modeling.

The last helpful visualization that can be achieved with Tags is **Color by tag**. The **Color by tag** option can be toggled by clicking the Color by tag button in newer versions of SketchUp Pro, and in the Details flyout button in the **Tags** panel in older versions of SketchUp Pro:

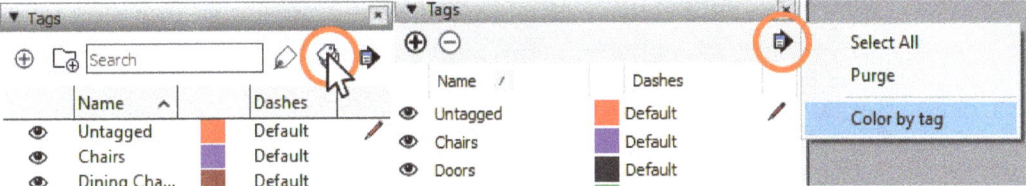

Figure 9.53 – The Color by tag Button (Left); Tag Panel | Details Flyout | Color by tag (Right)

Similar to how the Monochrome Face Style option will make it appear as though all Geometry in the model is set to the default Material, the Color by tag option will make it appear as though all Geometry and Objects on each tag are painted with a Material that matches the Tag color. The Tag colors are shown in the **Tags** panel next to the Tag name. In this example, we can see the Tag colors in the **Tags** panel and Color by tag toggled on to color the Geometry and Objects appropriately:

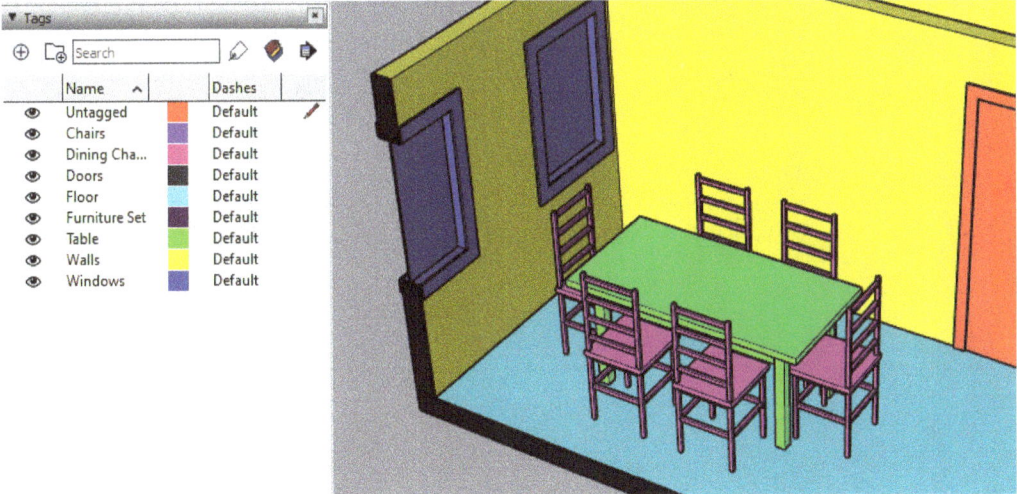

Figure 9.54 – Tag panel (Left); Color by tag in the Model (Right)

> **Note**
> This does not update the Material of the Geometry or Objects. The Entity Info panel will still show the correct Materials for a selected Face or Object even with Color by tag toggled on.

Tag colors can be updated by clicking on the Tag color box in the Tag panel. This will open the **Edit Material** dialog box, which will allow any Material to be set as the Tag color:

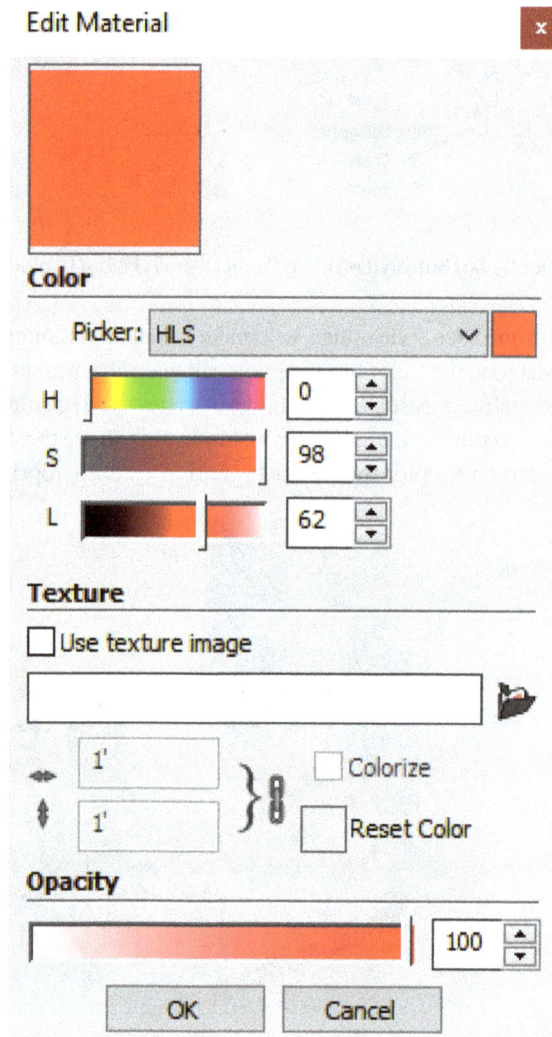

Figure 9.55 – The Edit Material Dialog Box to Update Tag Color

Typically, it is suggested that Tags be assigned representative colors, and individual Materials be applied to the Geometry and Objects in the Model. Any type of Material could be assigned to a Tag, including a Texture Material, so your model Materials could be managed entirely via Tags. However, this is not recommended, as Materials applied to Tags are not included in the Materials panel.

Color by tag can be a great tool to help quickly identify what Objects are on what Tags without selecting the Objects one by one. It can be helpful to update the Tag colors to be different from each other, instead of having similar colors. In this last example, we can see poorly chosen Tag colors that are hard to differentiate, and good Tag colors that can easily be identified in the model:

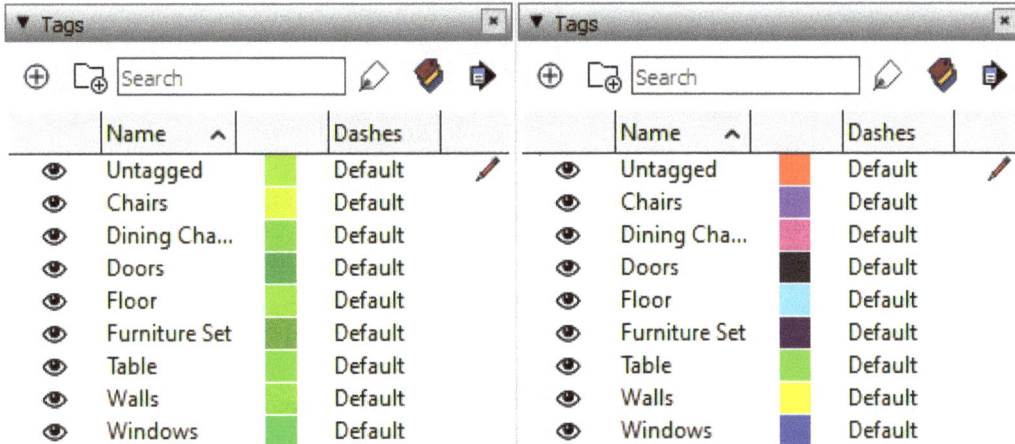

Figure 9.56 – Poor Tag Colors (Left); Good Tag Colors (Right)

Tags are extremely useful for organizing a model for quick visualization updates. SketchUp Pro has added even more functionality to Tags in more recent versions, and we will look at one of those features in the next section.

Tag folders

The Tags panel has gone through many changes and updates over the regular annual updates to SketchUp Pro. In previous versions of SketchUp Pro until the 2020 version was released, SketchUp Tags were known as **Layers**. Layers and Tags function the same way, so if you are using an older version of SketchUp Pro and you see the **Layers** panel, then you can still use the workflows in this chapter.

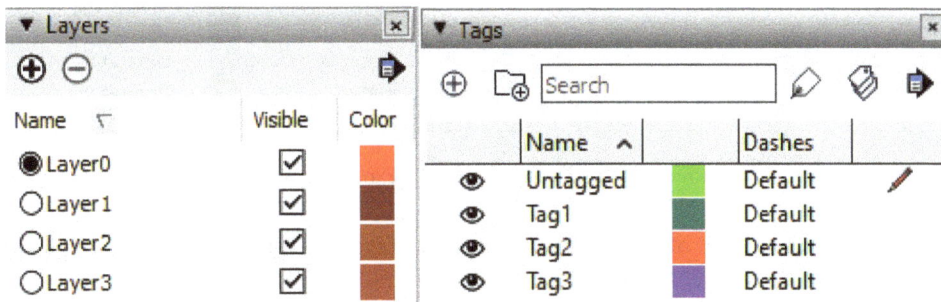

Figure 9.57 – SketchUp 2019 Layers panel (Left); SketchUp 2023 Tags panel (Right)

SketchUp Pro introduced new features to the **Tags** panel in the 2021 version of the software, including Tag folders. Tags are placed into folders in the model, and these folders are used to control the visibility of all included Tags. In versions where Tag folders can be used, the Create Tag folder button has replaced the Delete Tags button:

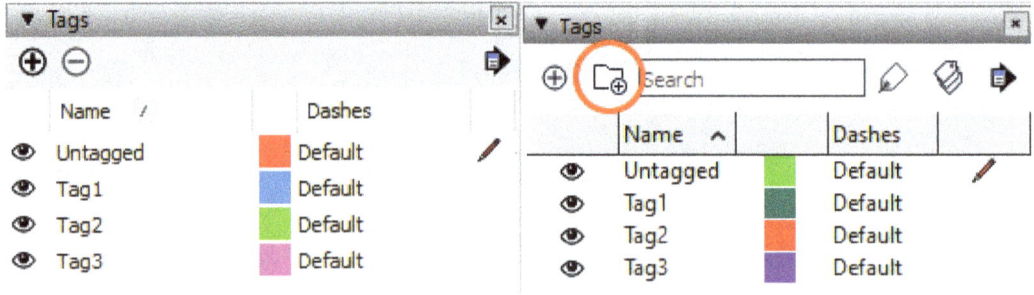

Figure 9.58 – SketchUp 2020 Tags panel (Left); SketchUp 2023 Tags panel (Right)

Tag folders can be identified by the black arrow next to the Tag Folder name. This allows the Folder to be collapsed and expanded to see the Tags included. The Tag Folder can be hidden or shown by clicking on the eye icon, and when this is clicked for a Tag Folder, the Tags in that folder will have their eye icons grayed-out:

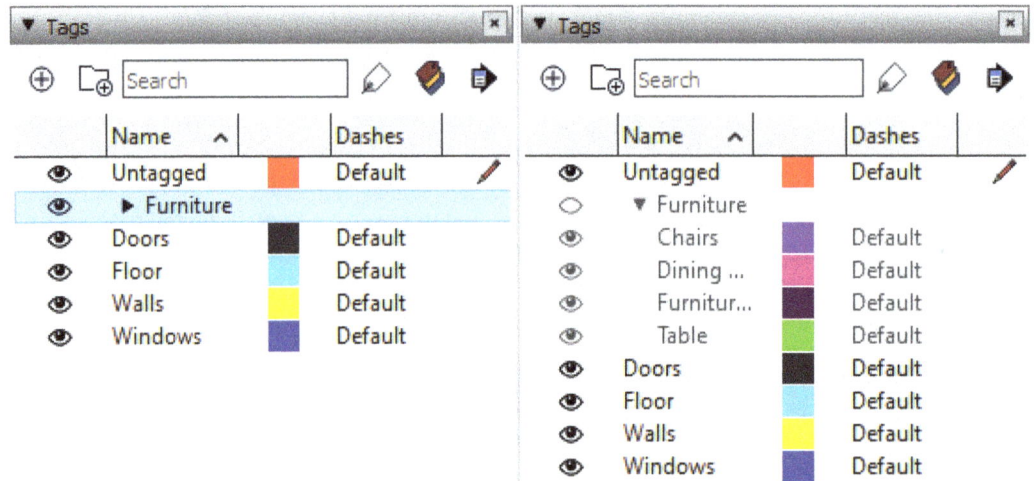

Figure 9.59 – Tag Folder Collapsed (Left); Tag Folder Expanded and Hidden (Right)

Tags can be moved in and out of folders by clicking and dragging. In the following example, the **Windows** and **Doors** Tags should be added to the Architecture Tag Folder. The Tags can be selected with the *CTRL* Modifier Key, then clicked and dragged with a left click. The Tags will now be added to the Architecture Tag Folder and can now be controlled from within that Folder:

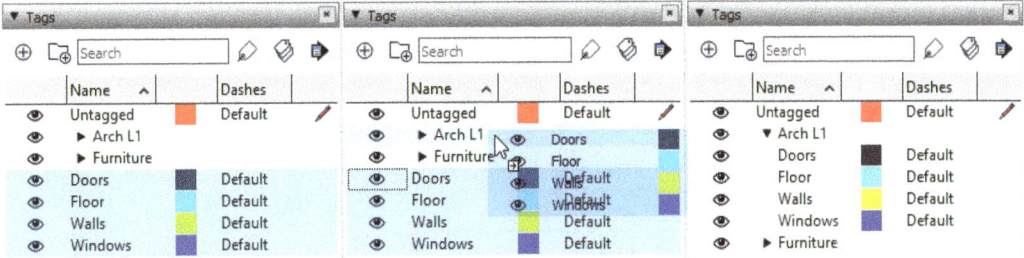

Figure 9.60 – Tags Selected (Left); Clicked and Dragged (Middle); Tags in Tag Folder (Right)

The Tags panel itself may need updating and cleaning up every once in a while. Let's take a look at when and how we should clean up the Tags panel.

Cleaning Up Unwanted Tags

There are many times when unwanted Tags end up in our SketchUp models. SketchUp allows for 2D and 3D drawings to be imported from many types of CAD software, and 3D models can be imported directly within SketchUp Pro from 3D Warehouse. We will discuss importing drawings and models in *Chapter 12, Import and Export, 3D Warehouse, and Extensions.*

SketchUp will convert Layers from software such as AutoCAD into Tags when the drawing is imported. SketchUp will also add Tags that are in Components imported from a local computer or via 3D Warehouse. In this next example, we can see an AutoCAD drawing has been imported and placed in the model as a Component. The **Tags** panel shows six layers imported from AutoCAD, showing the traditional CAD naming conventions:

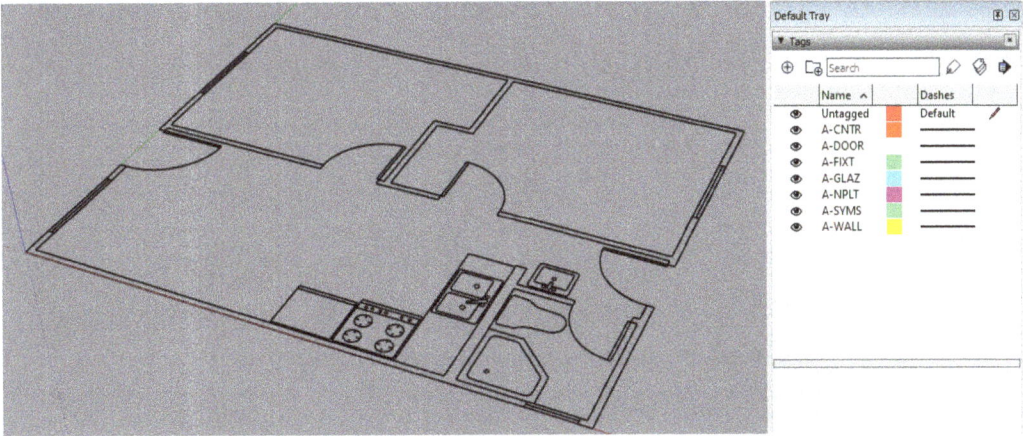

Figure 9.61 – AutoCAD Drawing Imported (Left); Tags panel with Additional Layers (Right)

These Tags are not necessary in SketchUp models and should all be replaced with a new Floor Plan Tag, merging all tags into one. To begin this process, the **Floor Plan** Tag can be created using the **Add Tag** button:

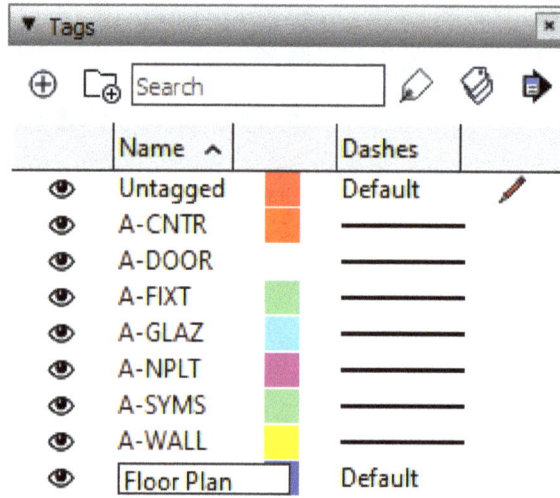

Figure 9.62 – Floor Plan Tag Added

Then, the imported Tags can be deleted by right-clicking on the selected Tags and choosing **Delete Tags**. When deleted Tags have Geometry or Objects assigned to them, SketchUp will prompt the user with the **Delete Tags Containing Entities** dialog box:

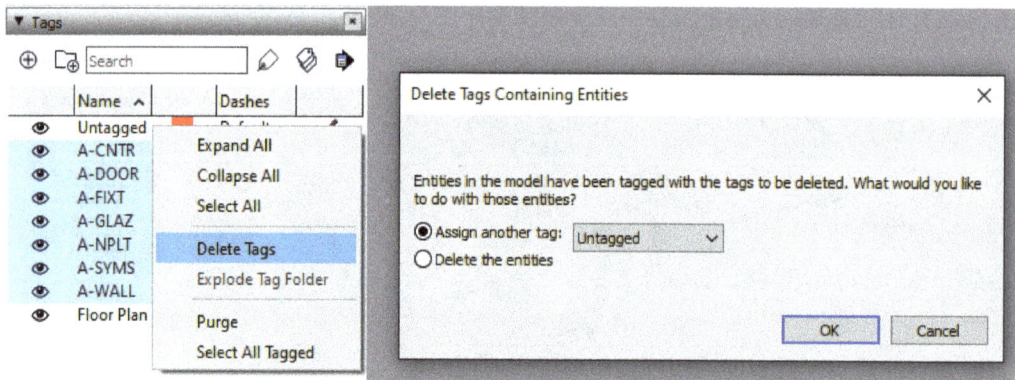

Figure 9.63 – The Delete Tags Containing Entities Dialog Box

This dialog box contains two options: to assign the Geometry to another Tag, or to delete the Entities. In this case, the **Floor Plan** Tag can be selected from the dropdown, and the Geometry will be assigned to the Floor Plan Tag:

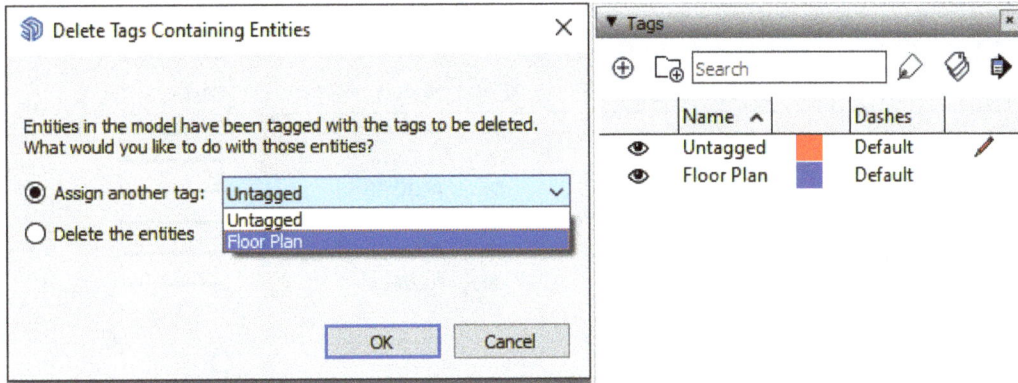

Figure 9.64 – Floor Plan Tag Selected From the Dropdown (Left); Floor Plan Tag in Tags Panel (Right)

In this following example, we can see the same drawing, but the CAD layers here should remain in the SketchUp model. However, it is unclear whether all of the drawing Tags contain Geometry. A fast way to quickly remove empty Tags is to use **Purge**. The **Purge** Unused Tags option can be found in the Tags Detail button at the top of the **Tags** panel:

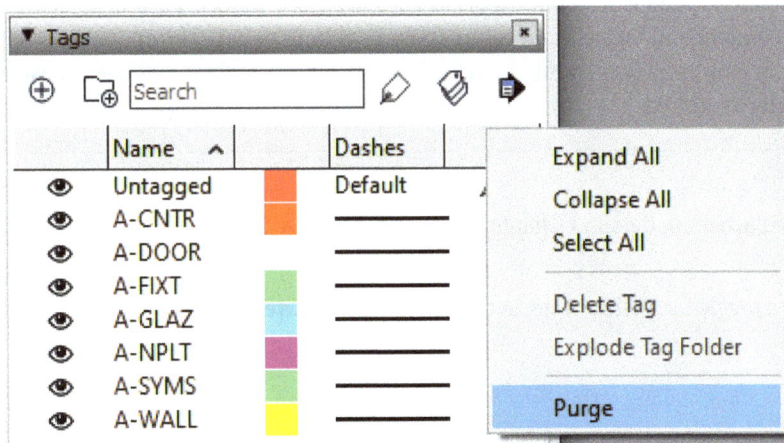

Figure 9.65 – Tag Details Button | Purge Unused Tags

Using **Purge** Unused Tags will only delete empty Tags in the model. We can see the drawing that was imported only had Geometry in five of the seven Tags present:

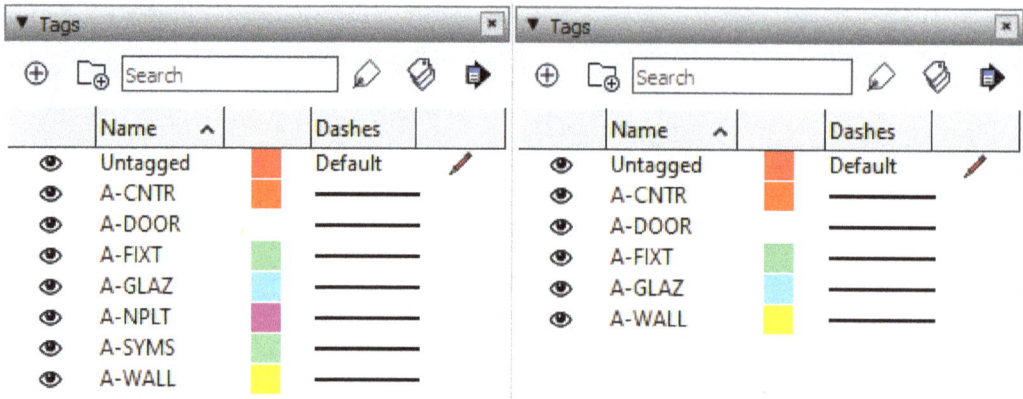

Figure 9.66 – Seven Imported Tags (Left); Five Used Tags (Right)

Purge and Delete Tags can be used to make your workflows more efficient by quickly cleaning up Components and drawings imported from external sources.

Using Tags with Scenes

Tags can be extremely useful when building Scenes in a SketchUp model. Tags can be used to Hide or Show sets of Geometry and Objects and can even be used to flash notes or dimensions onto frames in an Animation. However, Tags should be set up in a specific workflow in order to ensure that the Scenes are exactly as desired.

> **Note**
> We discussed Scenes in detail in *Chapter 7, View Options*.

Scenes include a toggle for **Visible Tags** in the **Properties to save** section:

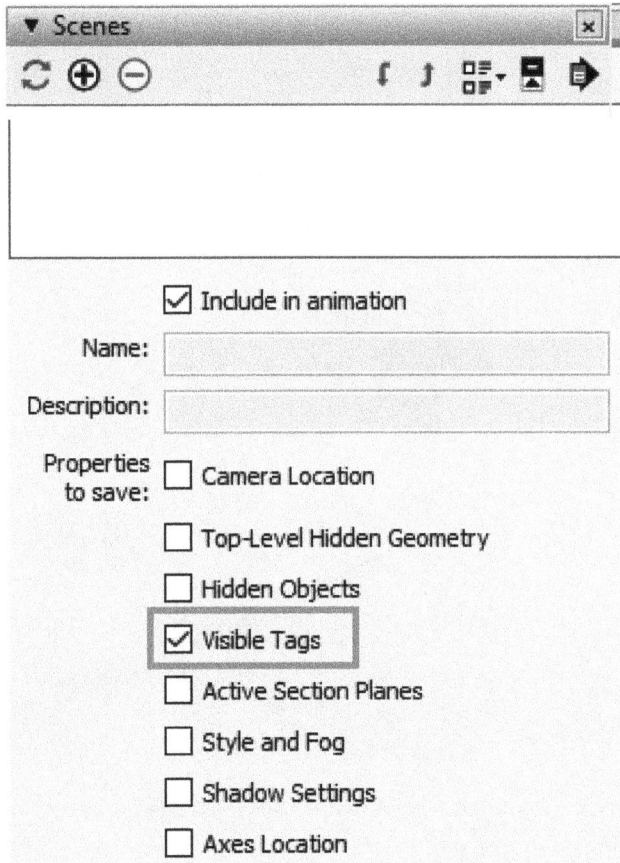

Figure 9.67 – The Scenes panel with the Visible Tags Property

This can quickly set visualization options for a model to be toggled between similar scenes or hide undesirable information in an Animation.

However, SketchUp will only save the Properties for currently existing Tags in the model. This means that any Tags created or imported after the Scene was created will be set to Visible in those Scenes! In this next example, we can see a Scene has been created, and the Chairs Tag has been Hidden:

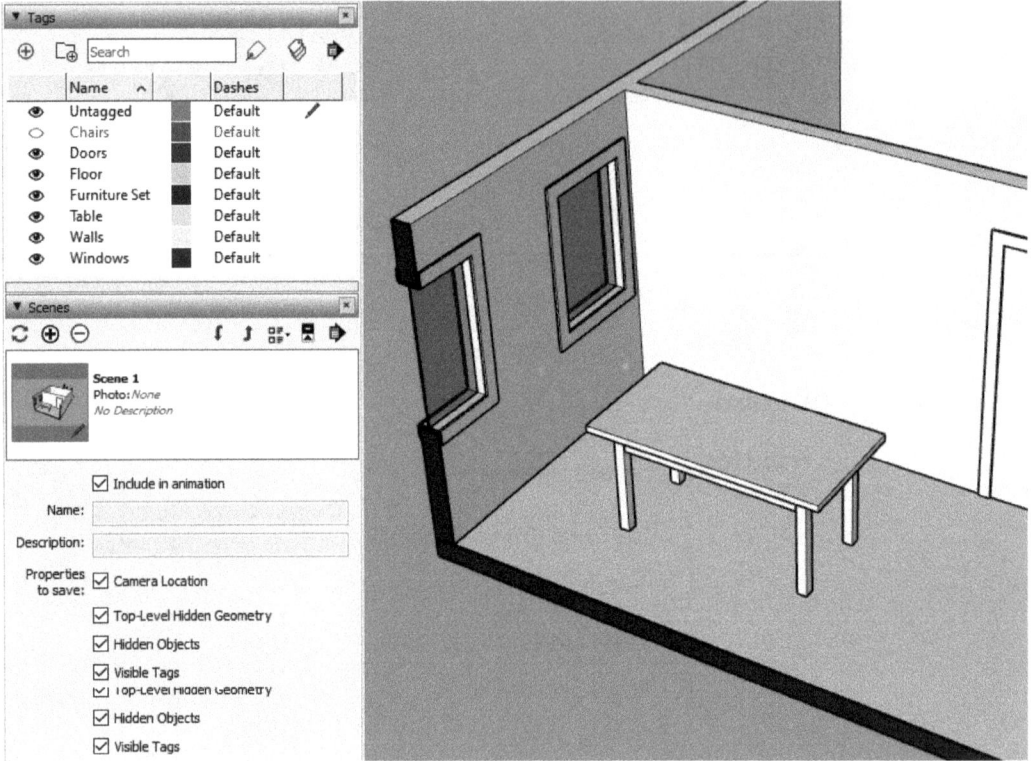

Figure 9.68 – Scene with Chairs Tag Hidden

While the model is developed, it is decided to split the Chairs Tag into a Dining Chairs Tag and Living Room Chairs Tag. The existing chair Components are moved from the Chairs Tag to the Dining Chairs Tag:

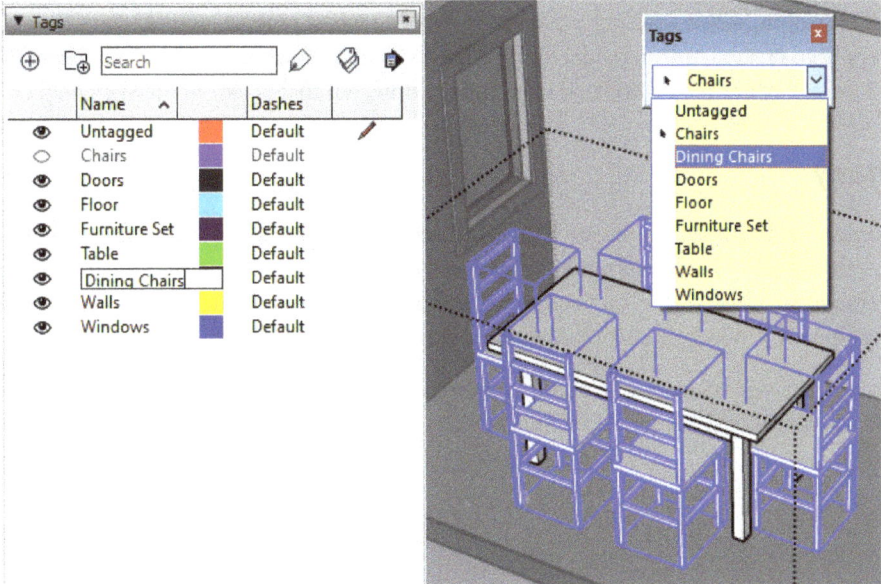

Figure 9.69 – New Tag Added (Left); Chair Components Moved to the Dining Chairs Tag (Right)

When the original Scene is activated, we can see that the chair Components are now visible. They were already in the model, but because the Chair Tag was hidden, and not the Components themselves, moving them to the new layer messed up the Scene visibility settings:

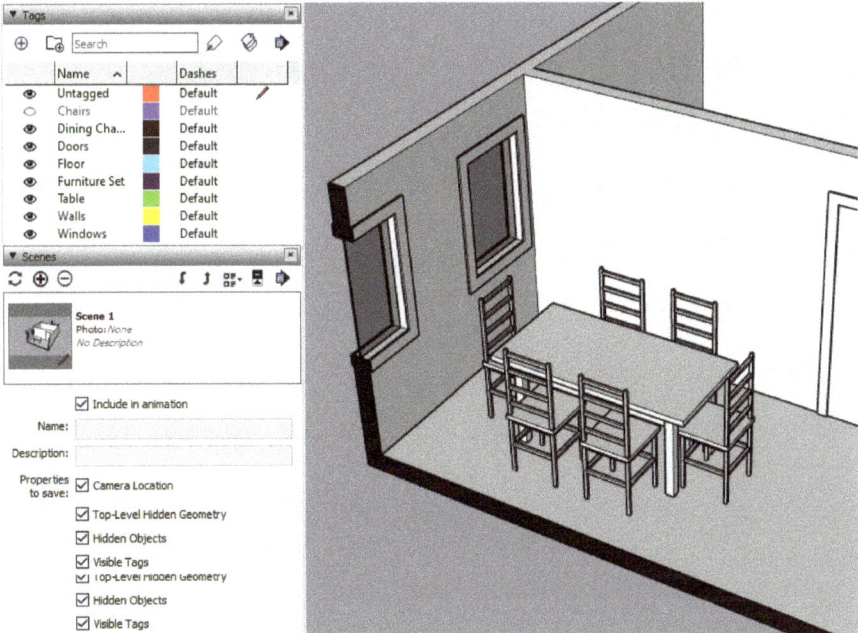

Figure 9.70 – Scene Showing Chairs Tag Hidden and New Tag Shown

This can be frustrating when working in SketchUp Pro. It is a best practice to make your Scenes after all Tags have been created for the model, or just make sure to keep an eye on your Scenes as you are working. Scenes may need to be updated from time to time, but that is part of the SketchUp process!

Summary

In this chapter, we discussed the **Entity Info**, **Outliner**, and **Tags** panels. These three Default Tray panels have the power to keep your models organized, and help by providing Geometry information, Object location, and large-scale visualization while modeling. Try using these panels in your project workflows, especially in large and complicated models.

In the next chapter, we will look at two more important parts of SketchUp Pro that cannot be found in the Default Tray: Model Info and Preferences.

10

Model Info and Preferences

In this chapter, we will look at two of the advanced dialog boxes in SketchUp Pro – **Model Info** and **Preferences**. These are not **Default Tray** panels but, instead, dialog boxes (Windows) that manage different elements of the current SketchUp model, or the overall SketchUp settings. We will discuss how to utilize the **Model Info** dialog box to update settings for the current file, including Objects, Statistics, and Units. We will also discuss how to utilize the **Preferences** dialog box to change overall SketchUp settings, including Accessibility, Drawing, and General.

The following topics are covered in this chapter:

- The **Model Info** dialog box:

  - **Animation**
  - **Classifications**
  - **Components**
  - **Credits**
  - **Dimensions**
  - **File**
  - **Geo-location**
  - **Rendering**
  - **Statistics**
  - **Text**
  - **Units**

- The **Preferences** dialog box:

  - **Accessibility**

  - **Applications**

  - **Compatibility**

  - **Drawing**

  - **Files**

  - **General**

  - **Graphics** (formerly **OpenGL**)

  - **Shortcuts**

  - **Template**

  - **Workspace**

Model Info

The **Model Info** dialog box contains 11 distinct tabs that contain settings to control the current model. These settings tabs address distinct Object types (**Components**, **Dimensions**, and **Text**), overall model settings (**Credits**, **File**, **Rendering**, **Statistics**, and **Units**), and specific aspects of unique workflows (**Animation**, **Classifications**, and **Geo-location**). The **Model Info** dialog box can be activated in the **Window** dropdown of the Main Menu:

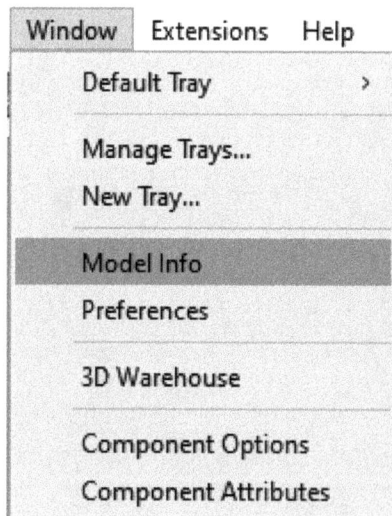

Figure 10.1 – Window | Model Info

Once open, the **Model Info** dialog box can be resized, but it has minimum height and width requirements. In earlier versions of SketchUp Pro, the **Model Info** dialog box could not be resized. On some tabs, additional information may need to be viewed by scrolling down in the dialog box. In newer versions of SketchUp Pro, there is no scroll bar to visually indicate that there is additional information. We can see this on the **Statistics** tab in the following example:

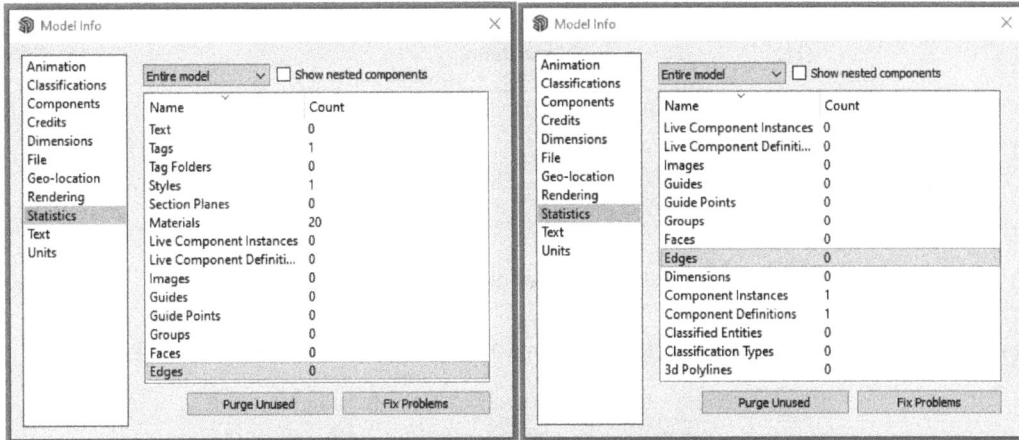

Figure 10.2 – Model Info | the Statistics Tab (Left) and the Statistics
Tab Scrolled to Show More Information (Right)

All settings on a tab can be viewed by resizing the **Model Info** dialog box. Different settings can be accessed by clicking a tab name on the left of the dialog box, and the settings for each tab will be visible on the right side of the dialog box.

Model Info settings will only be saved to a Template or SketchUp model file and are not meant for global SketchUp settings that continue from model to model. These global settings are located in the **Preferences** dialog box. We will review each **Model Info** tab in the next sections.

Animation

The **Animation** tab contains **Scene Transitions** and **Scene Delay** settings for Scene Animations in the current model. We discussed the **Animation** tab in detail in *Chapter 7, View Options*.

Model Info

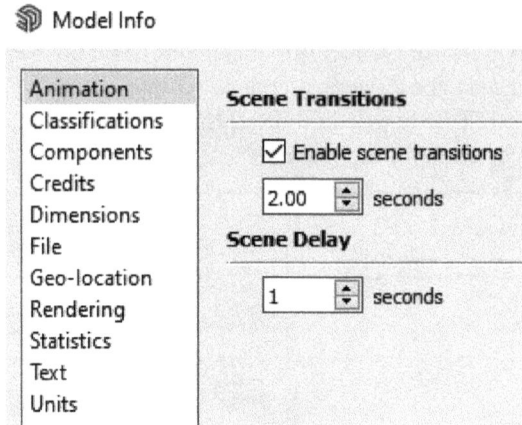

Figure 10.3 – The Model Info Animation tab

The **Animation** tab can be opened directly in the Main Menu by going to **View** | **Animation** | **Settings**:

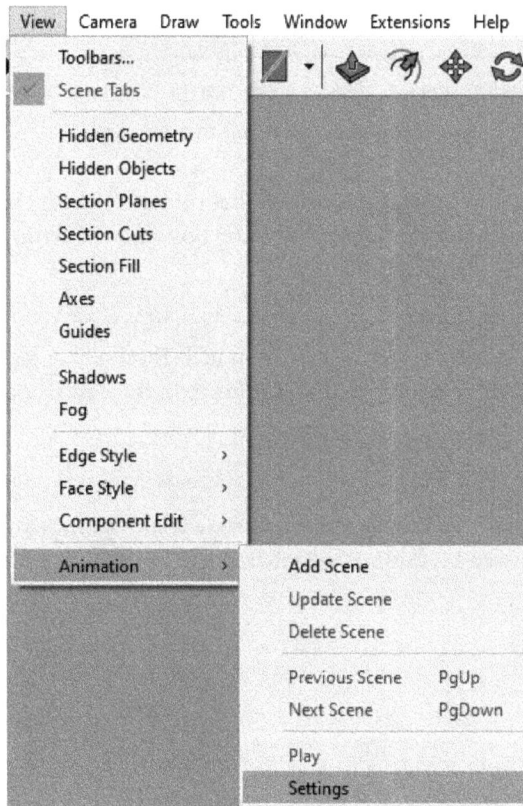

Figure 10.4 – View | Animation | Settings

Classifications

The **Classifications** tab contains and manages the classification libraries in the current model. Classifications are Component-specific definitions that embed data, including pricing, size, status, or manufacturer/owner. SketchUp classifications align with the **Industry Foundation Classes** (IFC). Models containing IFC data are considered **Building Information Models** (BIMs).

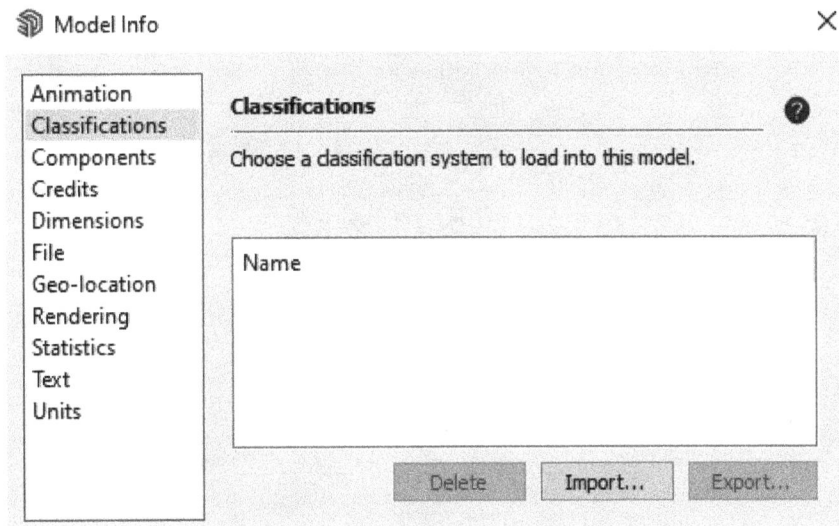

Figure 10.5 – The Model Info Classifications tab

Classifications and IFC data are an advanced topic. We will briefly discuss classifications in this section, but more detailed classification instructions, including how to make your own classification data files, should be researched online.

> **Note**
>
> Check out this SketchUp website for more information about classifications and IFC data: `https://help.sketchup.com/en/sketchup/classifying-objects`.

The **Classifications** tab can be opened directly by clicking the **Classification** button on the **Classifier** Toolbar if there is not a classification library imported into the current model:

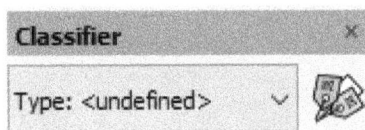

Figure 10.6 – The Classifier Toolbar

Classifications can be assigned to any Component or Group in the model, but not to other Geometry. The **Entity Info** panel will show a **Type** dropdown, which is the same as the **Classifier** Toolbar. Additionally, Components include **Advanced Attributes** options in the **Entity Info** panel, which are reserved specifically for classifications:

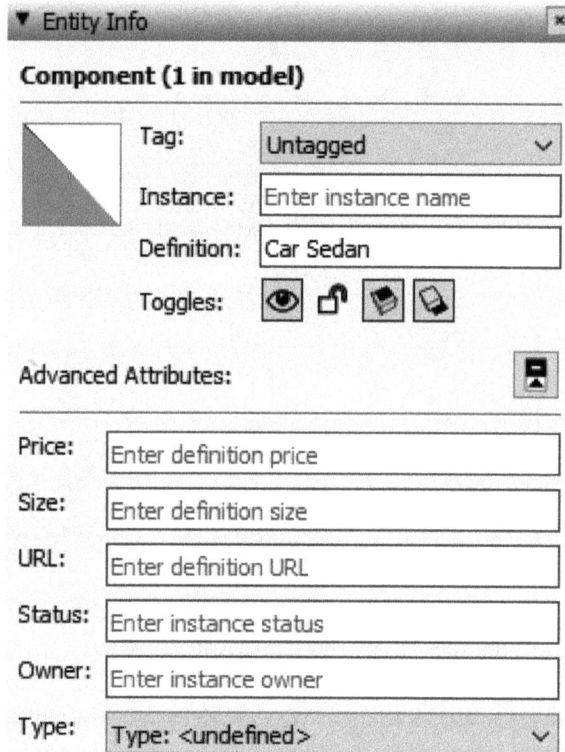

Figure 10.7 – The Entity Info Panel's Advanced Attributes

Before classifications can be used in the current model, an IFC library must be imported. When the **Import** button is clicked in the **Model Info** dialog box, an **Import** window opens. SketchUp Pro is installed with the standard IFC library included, and selecting this library file will import the library into the SketchUp model:

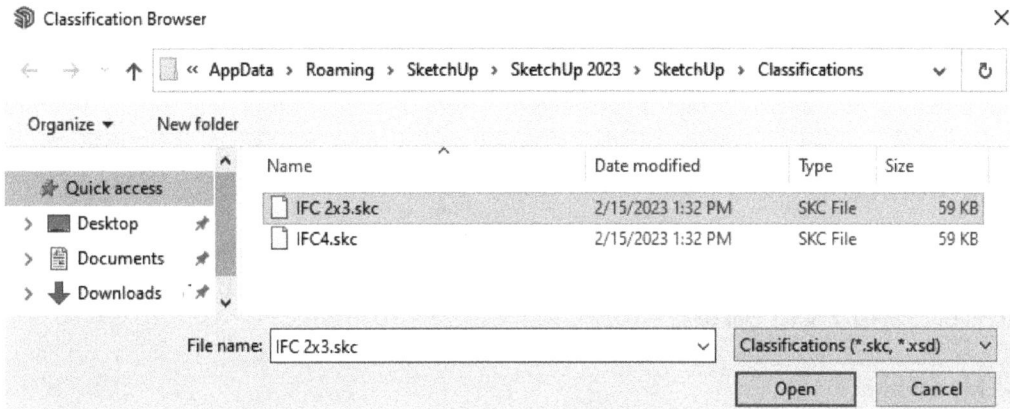

Figure 10.8 – The Import Button Clicked to Show the Import Window and the Included Library

The IFC library includes classifications for architectural, structural, fixture, and other construction disciplines. These classifications can now be applied to components in the model. Once applied, the classifications will appear in the **Entity Info** dialog box, which also includes any information typed into the **Additional Attributes** fields:

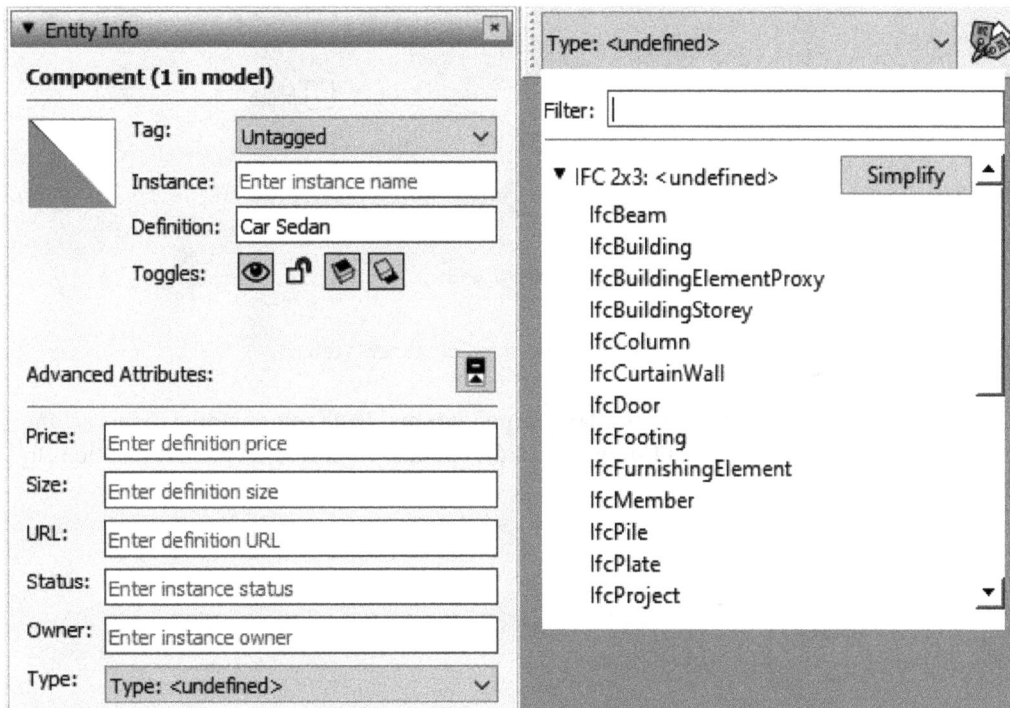

Figure 10.9 – Entity Info's Advanced Attributes and the Classification Dropdown

> **Note**
>
> Remember, the **Classifications** tab in the **Model Info** dialog box is not used to apply or manage the classifications themselves, only the classification libraries that are imported into the model.

Components

The **Components** tab contains Editing visualization and **Component Axes** settings for Groups and Components in the current model. We discussed the other Component Editing visualization options in detail in *Chapter 7, View Options*.

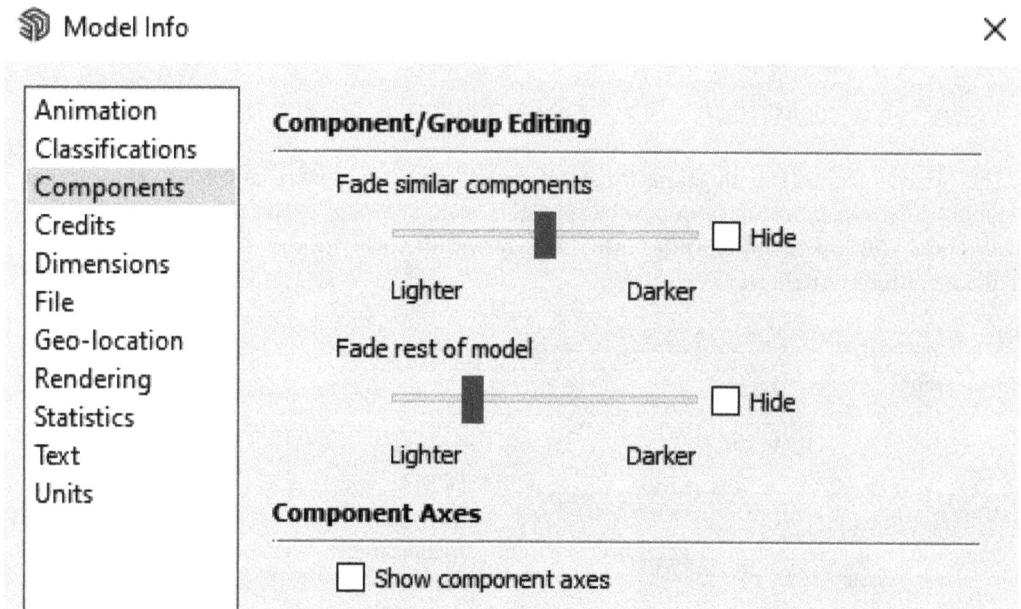

Figure 10.10 – The Model Info Components tab

The **Hide** checkboxes next to the **Fade similar components** and **Fade rest of model** sliders act the same way as the Main Menu | **View** | **Component Edit** options. We discussed these two options in detail in *Chapter 7, View Options*.

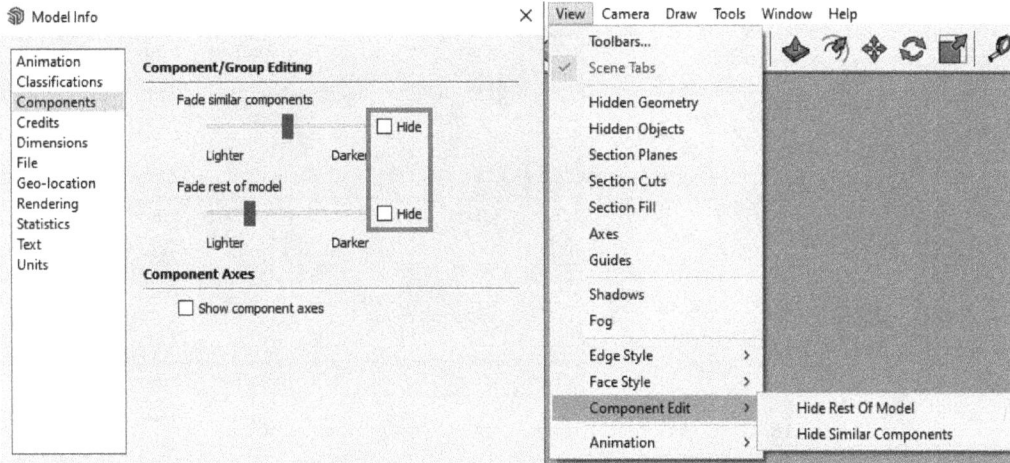

Figure 10.11 – The Model Info Components Tab Checkboxes (Left)
and the View | Component Edit Dropdown (Right)

The sliders themselves control the amount of fade applied to the Geometry when the **Hide** checkboxes are not selected. Sliding to the **Lighter** option fades the Geometry more, and sliding to the **Darker** option fades it less and less until no fade is applied when on the **Darker** setting.

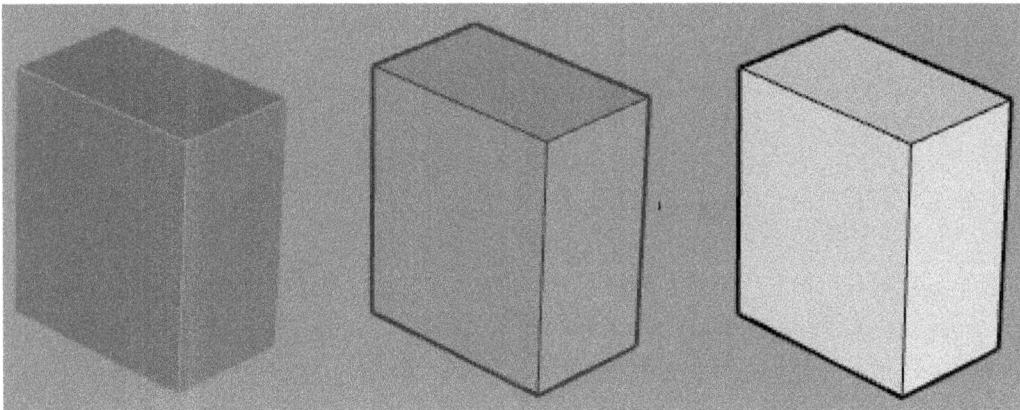

Figure 10.12 – The Slider Set to Lighter (Left), Default (Middle), and Darker (Right)

The **Component Axes** checkbox toggles the visibility of the Component Axes in the current model, which are different from the overall Model Axes:

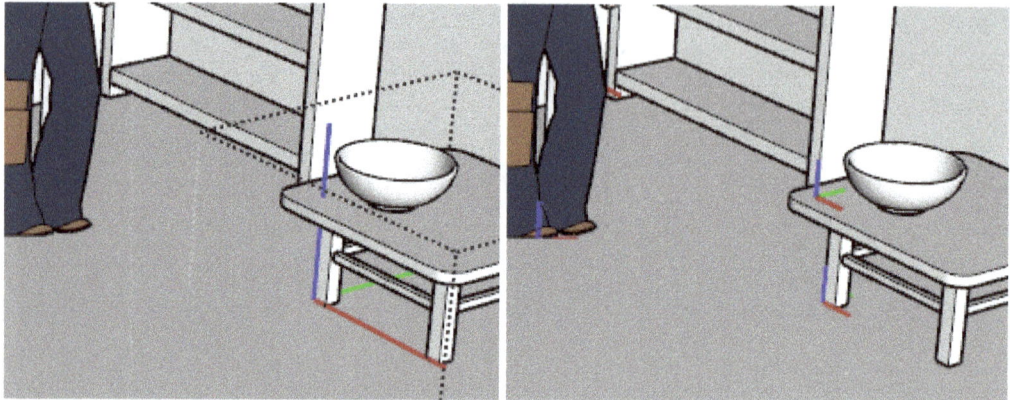

Figure 10.13 – Component Axes in the Active Component (Left) and
Component Axes Shown on All Components (Right)

The Component Axes are shown when editing a component but are typically hidden when the component is not active.

Credits

The **Credits** tab allows elements of a SketchUp model to be claimed by the original creator of the model. Credit can be claimed for the current model, and the **Component Authors** section will show credits from components loaded into the current model:

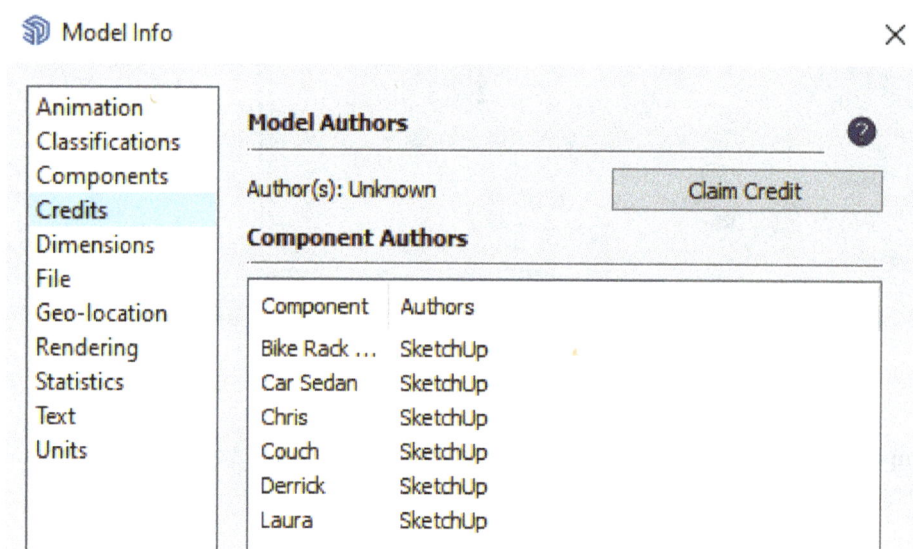

Figure 10.14 – The Model Info Credits tab

Claiming credit for a model is very simple, as SketchUp Pro now requires users to be logged in while working in SketchUp. Clicking the **Claim Credit** button will automatically assign credit to the account logged in to SketchUp Pro:

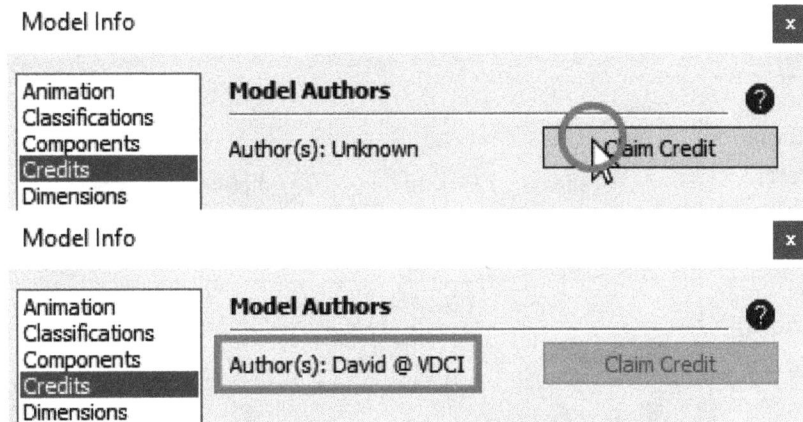

Figure 10.15 – The Claim Credit Button Clicked (Top) and the Author Credit (Bottom)

Component Authors will list the credits claimed from imported components:

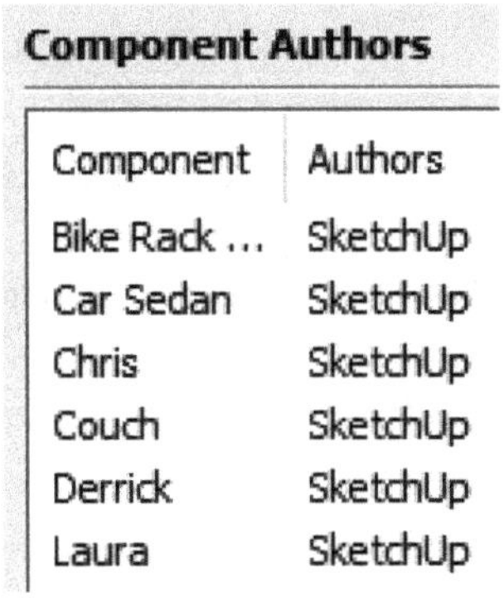

Figure 10.16 – Component Authors from Downloaded and Imported Components

This is important when downloading components from 3D Warehouse in a web browser or internally in SketchUp Pro. Components take a lot of work to create, so it is great to be able to acknowledge the original authors.

> **Note**
>
> To claim credit for individual components in your SketchUp models, it may be necessary to save them as individual SketchUp models first! You can use **Claim Credit** in the individual models, and then import them into your final model.

Dimensions

The **Dimensions** tab contains visualization options for **Text**, **Leader Lines**, and **Dimension**:

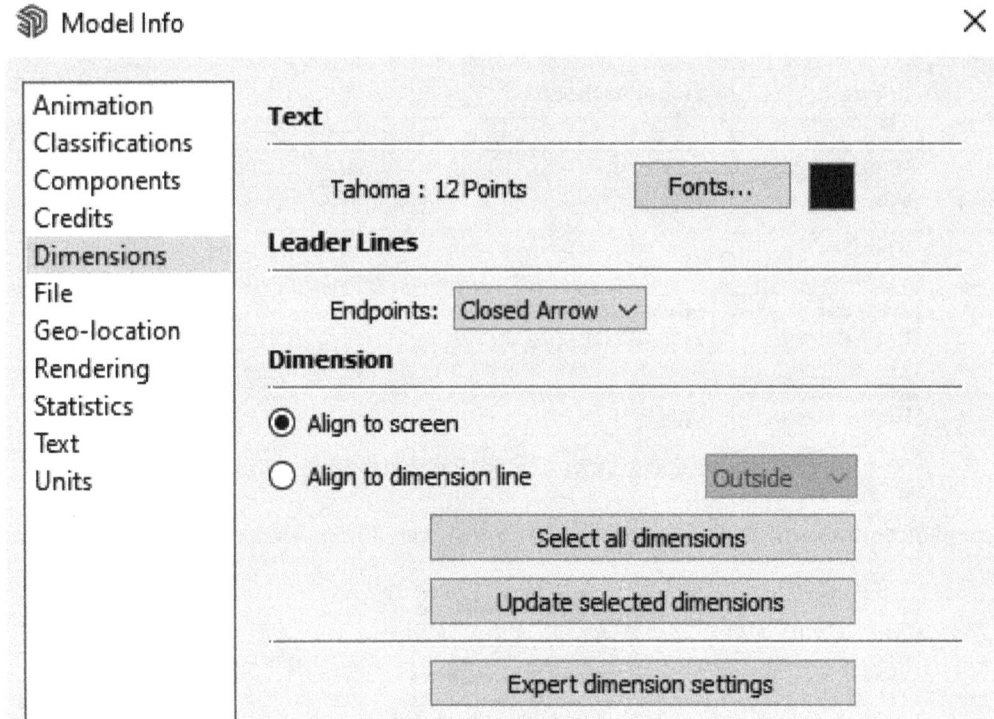

Figure 10.17 – The Model Info Dimensions tab

We introduced Text and Dimensions in *Chapter 9, Entity Info, Outliner, and Tags Dynamically Organize Your Models*.

> **Note**
> The options in this tab will change the appearance of any new Text, Leaders and Dimensions, but any Objects that have already been created will need to have their options updated in the **Entity Info** panel, or by clicking the **Update selected dimensions** button.

The **Text** options include a **Fonts** button and a color picker, represented by a colored rectangle. Clicking the colored rectangle will open the **Choose Color** dialog box, which is the same as that used for other color selection options. The **Fonts** button will open the **Font** dialog box, which includes many font options. All fonts that are installed on the local computer will be available in the drop-down list, including user-installed fonts. Font styles and size can also be set, including whether the size should be set by **Points** (Pixels) or in-model **Height**:

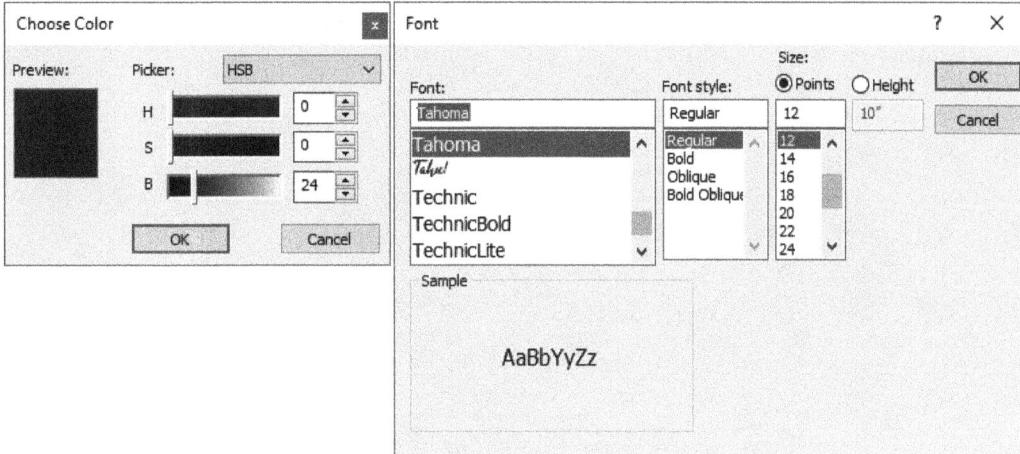

Figure 10.18 – The Choose Color Dialog Box (Left) and the Font dialog box (Right)

The only option for **Leader Lines** is to change the style of the Leader **Endpoints**:

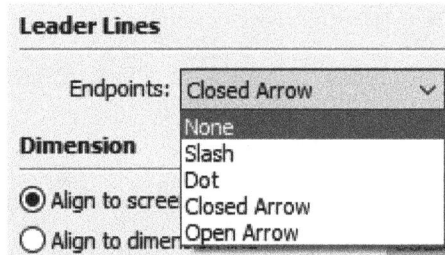

Figure 10.19 – The Model Info Dimensions Tab's Leader Lines Dropdown

The **Dimension** options include how to align the Dimension Text to the Dimension – either **Align to screen** or **Align to dimension line**. In the next example, we can see the dimensions that are **Aligned to screen** are horizontal relative to the screen, while the dimensions that are **Aligned to dimension line** are slanted at the same angle as the dimensions in the perspective view.

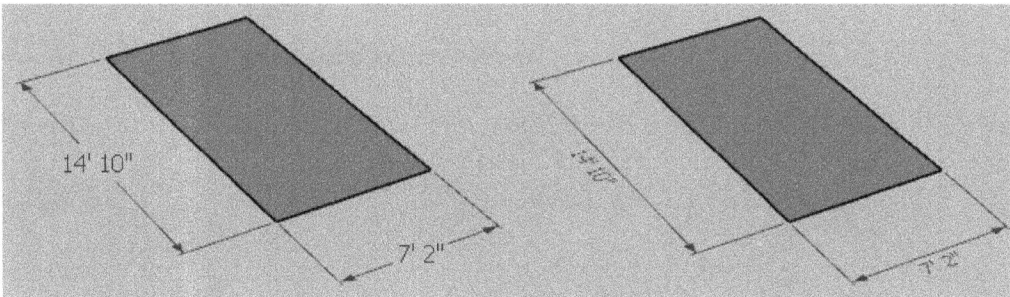

Figure 10.20 – The Dimension Aligned to the Screen (Left) and Aligned to the Dimension Line (Right)

Additionally, the **Model Info** window includes buttons such as **Select all dimensions** and **Update selected dimensions**:

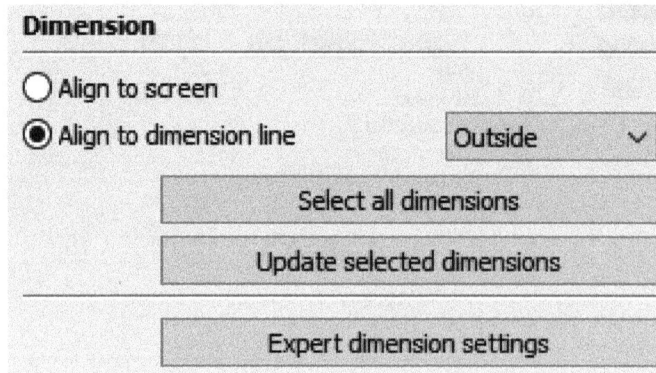

Figure 10.21 – The Model Info Dimension Tab's Dimension Options

This is a fast and easy way to update the settings of all the dimensions in the model, even after some have already been drawn!

The **Expert dimension settings** button will open a separate **Expert Dimension Settings** dialog box:

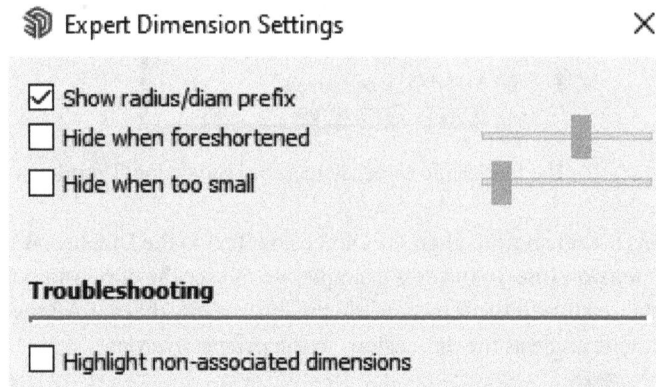

Figure 10.22 – The Expert Dimension Settings dialog box

This dialog box includes some high-level options to fine-tune the Dimension styles in the current model. These default options don't need to be changed for most models.

> **Note**
> The in-model dimensions are not always the best way to annotate a drawing for final production. Consider using different software such as Layout to create a presentation or construction documents!

File

The **File** tab contains **General** file information about the current model. This includes the **Location**, **Size**, **Name**, and **Description** options:

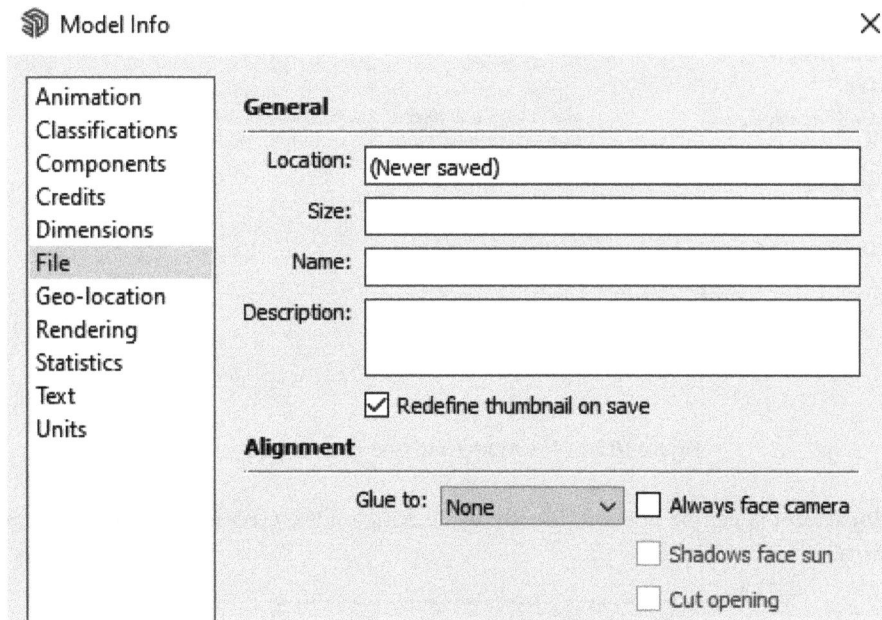

Figure 10.23 – The Model Info File tab

The **Name** and **Description** options are important if a model is ever imported as a component into other models. Additionally, the **Alignment** options are important if a model is imported as a component. We will talk more about advanced component settings in *Chapter 11, Working with Components*.

Geo-location

The **Geo-location** tab contains information about the real-world location of the SketchUp model. We discussed the **Geo-location** tab in detail in *Chapter 7, View Options*.

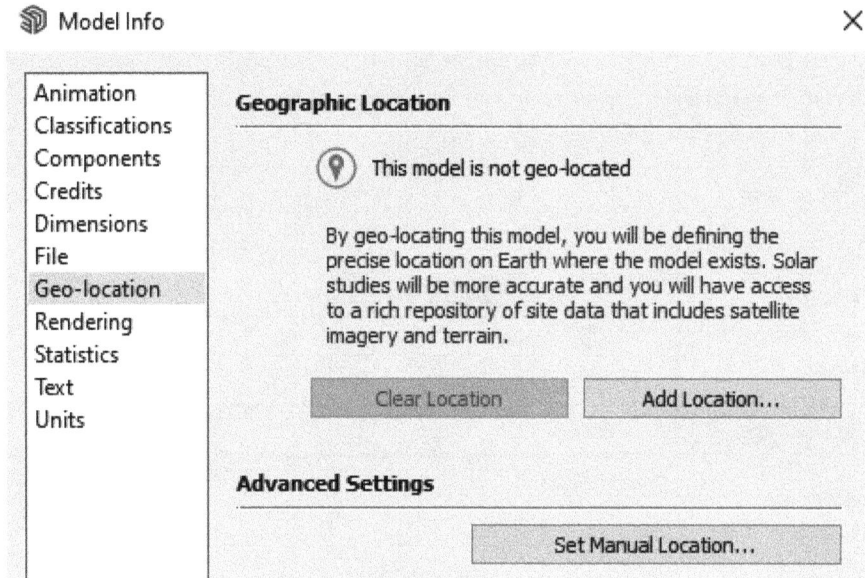

Figure 10.24 – The Model Info Geo-location tab

The **Geo-location** tab can be opened directly by clicking on the **Geo-location** Icon below the Drawing Area:

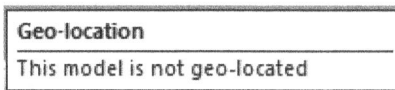

Figure 10.25 – The Geo-location Icon below the Drawing Area

Rendering

The **Rendering** tab is not a location to manage Rendering plugins or other Photorealistic Rendering options. The **Rendering** tab only contains one option – to toggle Anti-Aliased Textures:

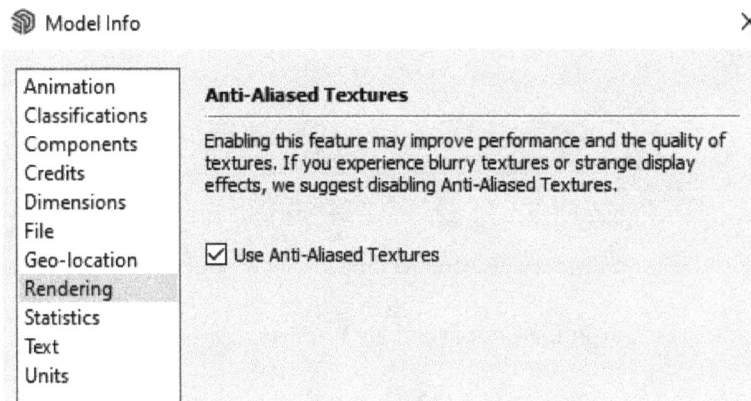

Figure 10.26 – The Model Info Rendering tab

Anti-Aliased Textures are images that SketchUp has "smoothed" to remove the jagged edges created by pixels in an image. This option can improve the performance of Texture Materials in the current Model but should be turned off when a computer is low on processing power.

Statistics

The **Statistics** tab contains a large amount of sorted data for the Geometry in the current model. There are too many Statistics categories to fit on one screen, and you may have to scroll down the **Statistics** tab to find the information you are looking for:

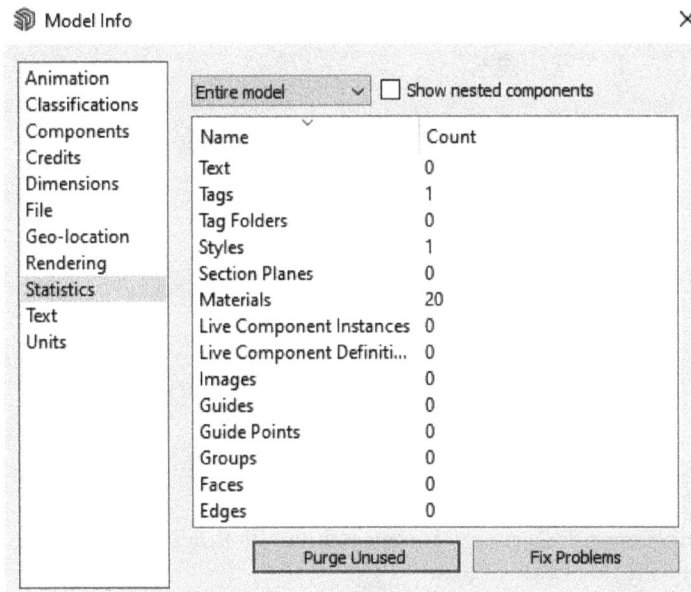

Figure 10.27 – The Model Info Statistics tab

The **Statistics** tab can also be sorted to view **Only components**, including an option to **Show nested components**:

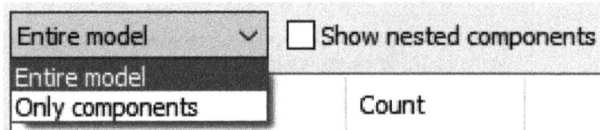

Figure 10.28 – The Statistics Tag Sorting Dropdown and the Show Nested Components Checkbox

The **Statistics** tab also contains **Purge Unused** and **Fix Problems** buttons:

Figure 10.29 – The Model Info Statistics Tab's Purge Unused and Fix Problems Buttons

The **Purge Unused** button will purge all unused Objects and Entities in the current model, including Components, Tags, and Materials. Individual categories can be Purged as well, but the **Purge Unused** button will Purge all unused Objects and Entities in the current model. Purging Unused Objects and Entities is a great way to clean up the **Components** and **Materials** panels, and to reduce the overall file size of the current model.

The **Fix Problems** button will run a high-level diagnostics check on the SketchUp file to see whether any file corruption has occurred, and it will attempt to fix any problems. The **Validity Check** dialog box will open to display the results of the test:

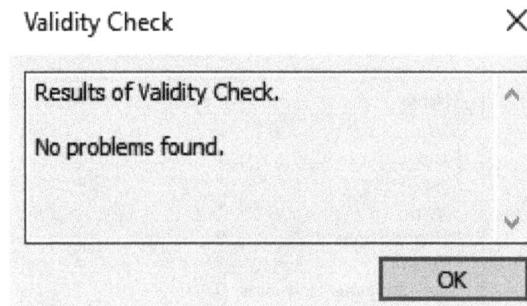

Figure 10.30 – The Validity Check dialog box

Text

The **Text** tab contains options similar to the **Dimensions** tab. The **Text** tab options are specifically for **Screen Text**, **Leader Text**, and **Leader Lines**:

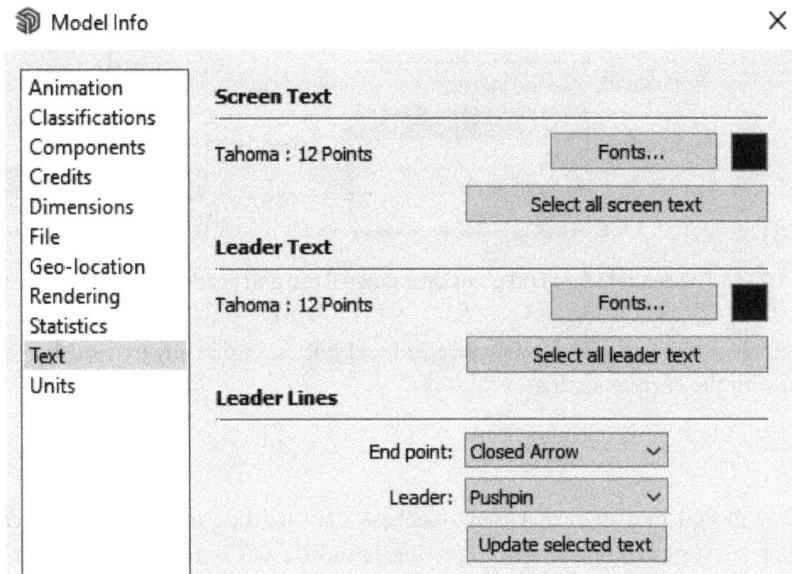

Figure 10.31 – The Model Info Text Tab

Screen Text is Text that is not connected to Geometry with a Leader. **Leader Text** is connected to Geometry with a Leader. The options for **Screen Text** and **Leader Text** are the same as the Dimension Text, including a **Fonts** button and color picker. Additionally, the **Text** tab contains buttons to either **Select all screen text** or **Select all leader text**:

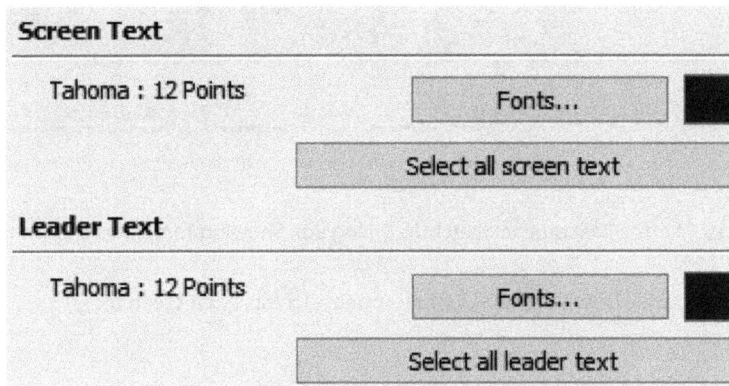

Figure 10.32 – The Screen Text and Leader Text Options

The **Leader Lines** options include **End point** options for the leaders, as well as options for the orientation of the Leader:

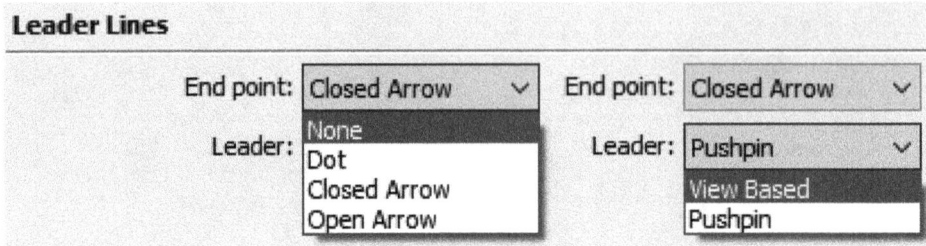

Figure 10.33 – The Leader Lines End point Dropdown (Left) and Leader dropdown (Right)

Finally, the **Text** tab also includes an **Update selected text** button so that any previously created Text can be updated with the current settings.

Units

The **Units** tab contains the measurements used to set base Units for the current model. Typically, Units are initially set by selecting a Template when creating a model. Each Template displays the base unit below the Template name, typically **Inches** (Imperial) or **Millimeters**, **Centimeters**, or **Meters** (Metric).

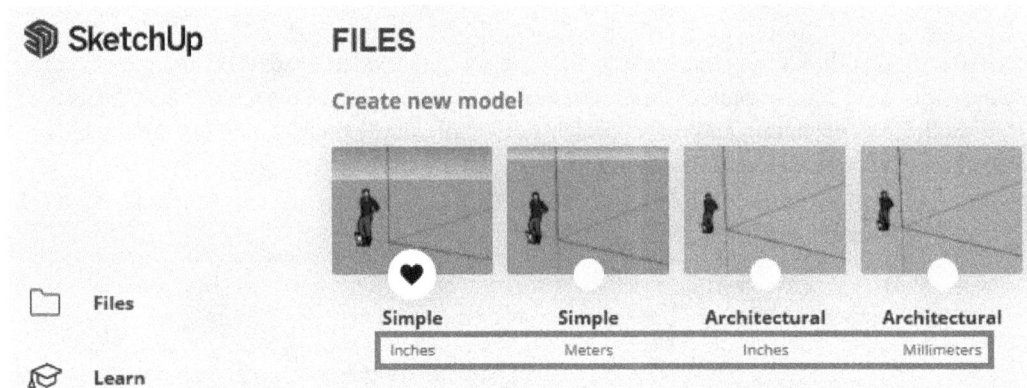

Figure 10.34 – The Welcome to SketchUp Dialog Box Showing Templates and Units

The **Units** tab allows these Units to be updated after a model has been created.

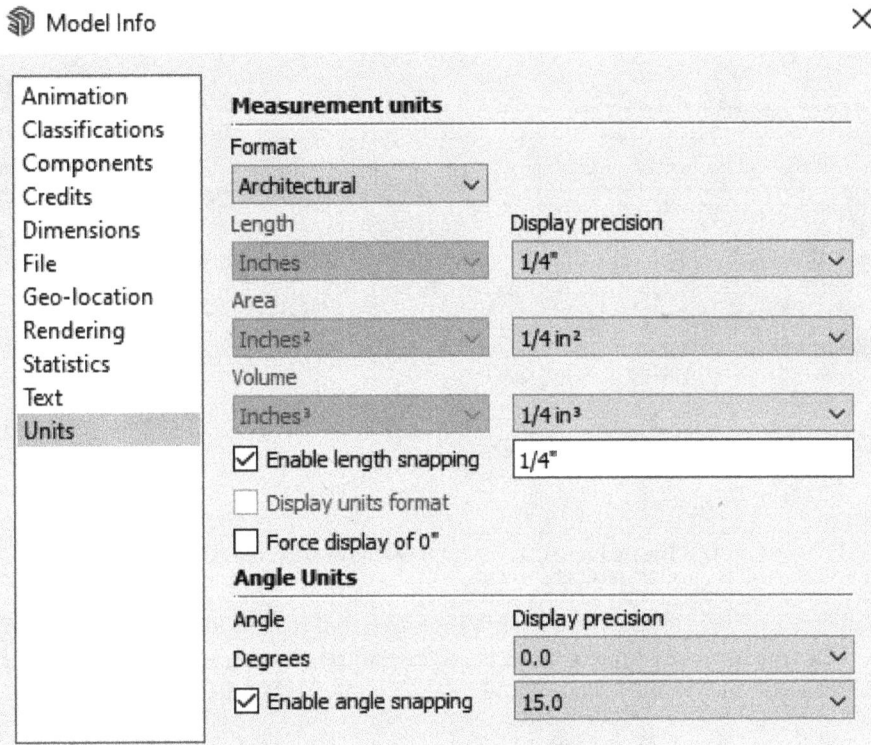

Figure 10.35 – The Model Info Units Tab

There are four **Format** options for Units – **Architectural**, **Decimal**, **Engineering**, and **Fractional**:

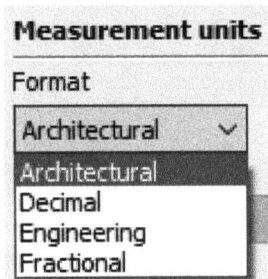

Figure 10.36 – The Model Info Units tab Format Dropdown

Architectural, **Engineering**, and **Fractional** Units are all exclusive to **Inches**. Choosing any of these options will gray out the additional options for **Length**, **Area**, and **Volume**:

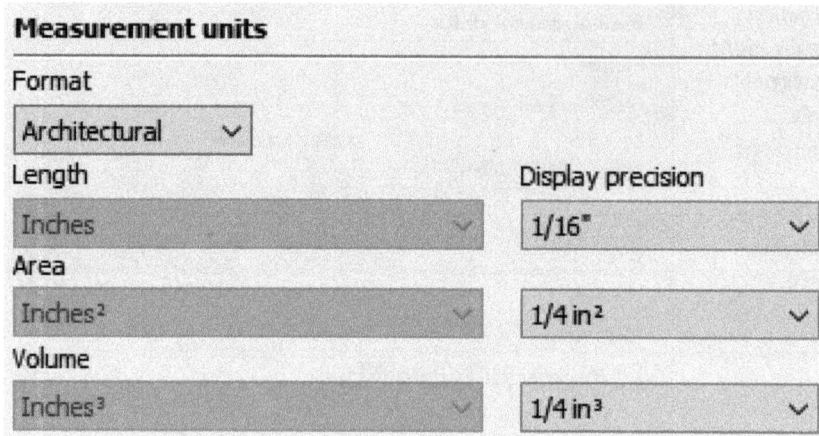

Figure 10.37 – The Architectural Format Selected and Options Grayed Out

The **Decimal** option allows for additional **Length** Units to be selected, including **Inches**, **Feet**, **Yards**, **Millimeters**, **Centimeters**, and **Meters**. **Area** Units are limited to the same options (squared), and **Volume** has the same units (cubed), with the addition of Imperial **Gallons** and Metric **Liters**:

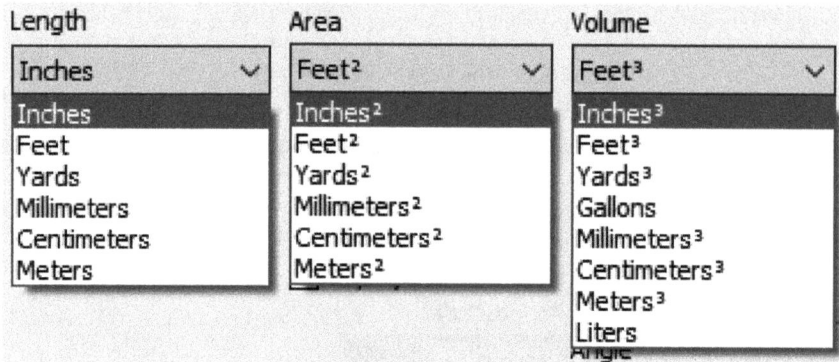

Figure 10.38 – The Length Dropdown (Left), the Area dropdown
(Middle), and the Volume dropdown (Right)

We previously mentioned that SketchUp Pro is an extremely precise program, calculating Geometry to many decimal places. The **Display precision** options do not change the actual measurements or locations of the Geometry but will limit the Dimensional displays of the Geometry for simplicity. The precision of a model will depend on the type of work being done and the scale of the overall project. If the model is a watch gear that is less than 1 centimeter wide, then the **Display precision** value should be set to a detailed level. If the model is a sports field layout, then the display precision can be set to 1/2" or even 0", meaning that there will be no fractional inches displayed:

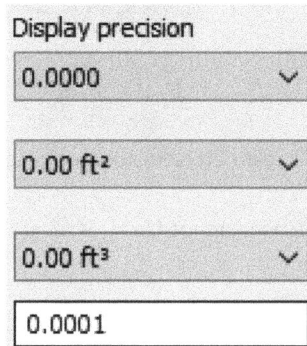

Figure 10.39 – Display Precision Options

The **Enable length snapping** checkbox and text field allow for a custom length to be set, which will be used incrementally when drawing new Geometry. With **Enable length snapping** toggled on, new Geometry will be forced in multiples of the specified distance.

The **Display units format** checkbox allows the Units marker for the **Decimal** and **Engineering** Units to be hidden. This option is not available for **Architectural** or **Fractional** Units. We can see that "cm" for Centimeters is hidden with the Units Format toggled off in the next example:

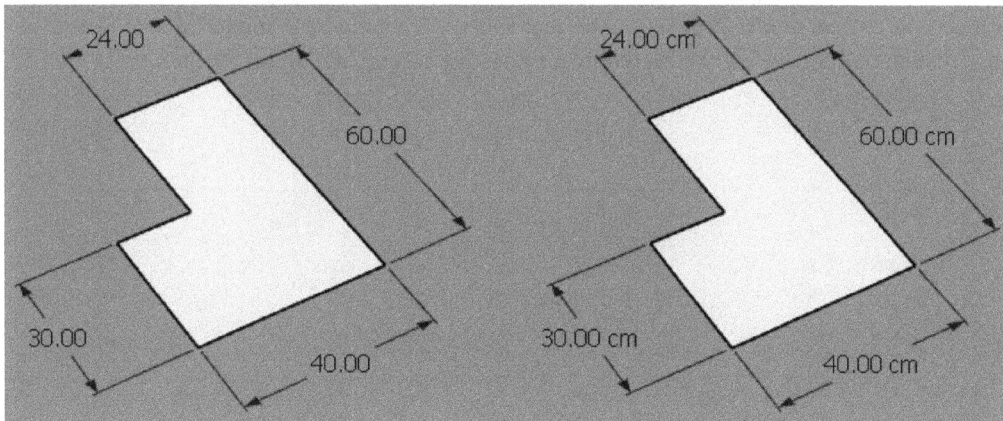

Figure 10.40 – The Units Format Toggled Off (Left) and the Units Format Toggled On (Right)

The **Force display of 0"** option will always display an Inches unit, even if the distance is an exact foot value. This is shown in the next example:

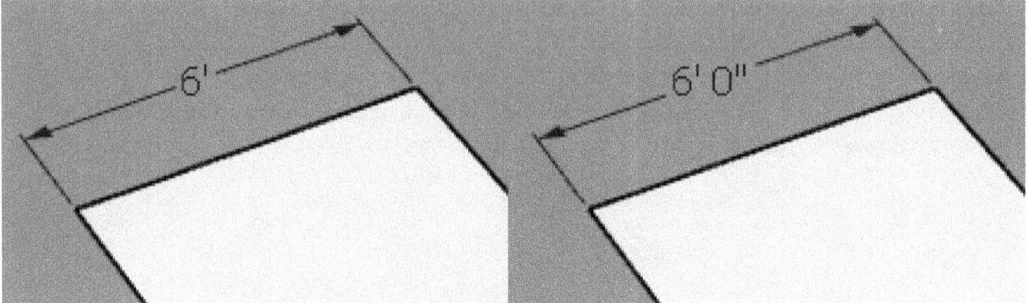

Figure 10.41 – The Force Display of 0" Toggled Off (Left) and the Force Display of 0" Toggled On (Right)

Angle Units is similar to the other Measurement Units, with specific options for Degree display precision and Angle Snapping:

Figure 10.42 – Angle Units Options

The **Model Info** options can be useful to update, depending on the type of project that you may be working on. Remember to stay on top of the **Statistics** tab in order to keep your models cleaned up and use the **Dimensions** and **Text** tabs to customize how your annotations appear. While **Model Info** only affects the current model, Preferences are remembered from model to model on each SketchUp Pro installation. We will review the **Preferences** tabs in the second half of this chapter.

Preferences

The **Preferences** dialog box contains 10 distinct tabs that allow for global preferences to be updated for each installation of SketchUp Pro. The **Preferences** tabs are not as easily sorted into groups as the **Model Info** tabs. The **Preferences** tabs range from drawing and accessibility options to default saving locations and keyboard shortcuts. The **Preferences** dialog box can be activated in the **Window** dropdown of the Main Menu:

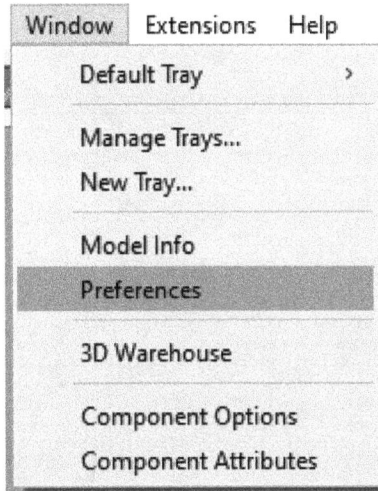

Figure 10.43 – Window | Preferences

Like the **Model Info** dialog box, the **Preferences** dialog box cannot be resized. Different settings can be accessed by clicking a tab name on the left of the dialog box, and the settings for each tab will be visible on the right side of the dialog box. Additional settings on the **Shortcuts** and **Template** tabs may need to be viewed by scrolling down in the dialog box. In the following sections, we will review all of the **Preferences** tabs and their unique options.

Accessibility

The **Accessibility** tab allows the default Axis/Directions and other inference Colors to be changed:

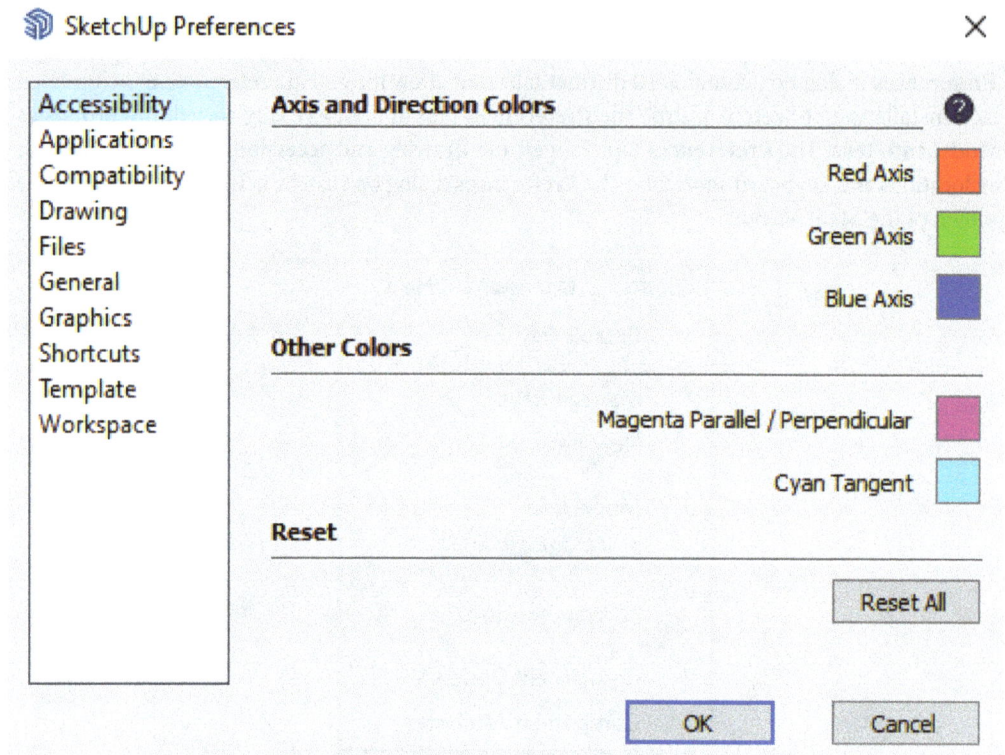

Figure 10.44 – The Preferences Accessibility tab

This is mainly for individuals who have a hard time distinguishing between the colors, but it can be used by anyone who would prefer different colors for the Axis and inferences.

The Colors can all be changed individually by clicking on the color picker, represented by a colored rectangle. The current colors for each option will be shown in the colored rectangle.

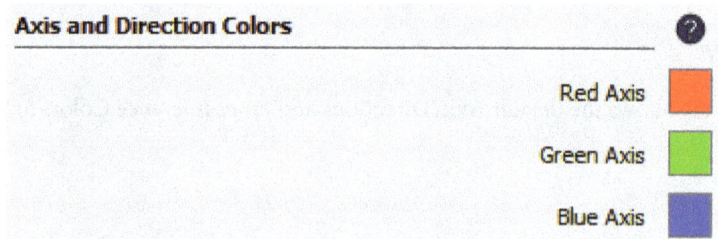

Figure 10.45 – The Current (Default) Colors for the Axes

There is also a **Reset All** button to return the colors to their default settings. So, feel free to experiment with the colors, and reset when you want!

Applications

The **Applications** tab is very simple in SketchUp Pro. The **Applications** tab allows the default image editor for Texture Materials to be specific to locally installed software. We discussed working with Texture Materials in detail in *Chapter 8*, *Materials*.

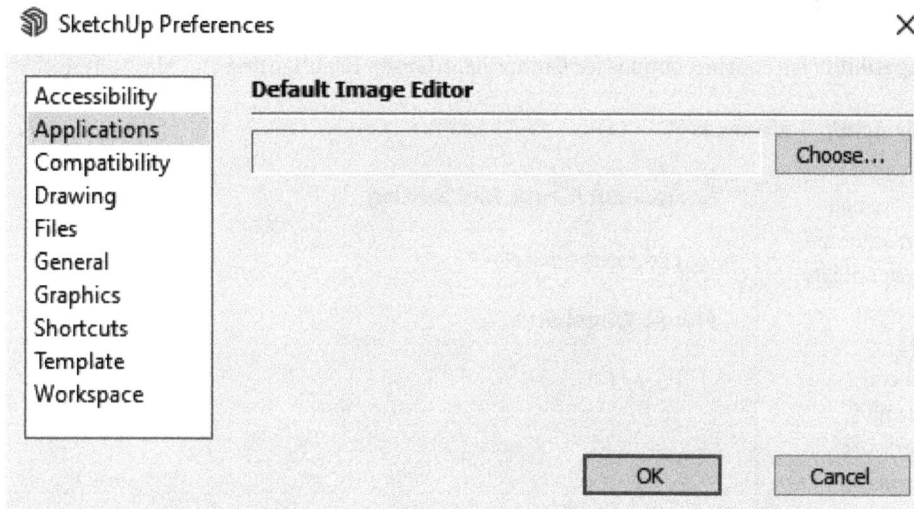

Figure 10.46 – The Preferences Applications Tab

When the **Choose** button is clicked, the **Image Editor Browser** dialog box will open:

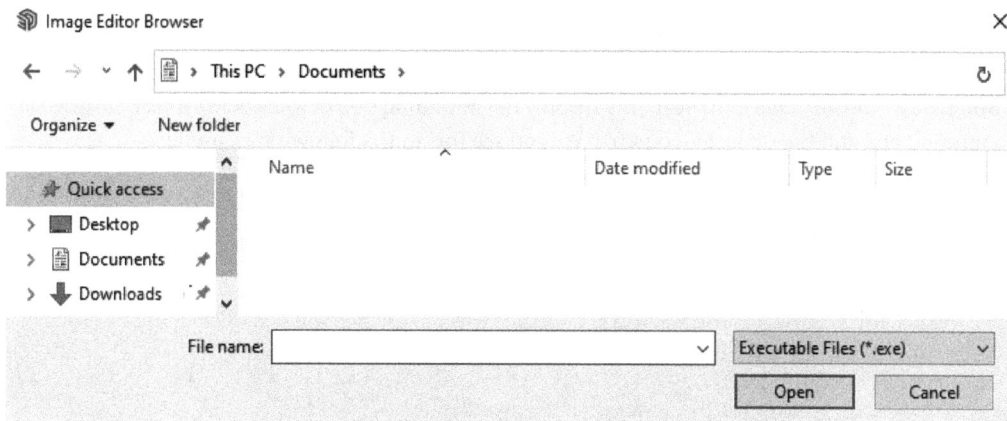

Figure 10.47 – The Image Editor Browser Dialog Box

This will allow any Executable File (commonly referred to as applications or apps) to be selected to edit images. Common options include the Photos App, MS Paint, or Photoshop.

> **Note**
>
> The image editor software must be installed on the local computer to be selected. You can view a list of applications installed on your computer by going to the `Program Files` folder.

Compatibility

The **Compatibility** tab contains options for **Component/Group Highlighting** and **Mouse Wheel Style**:

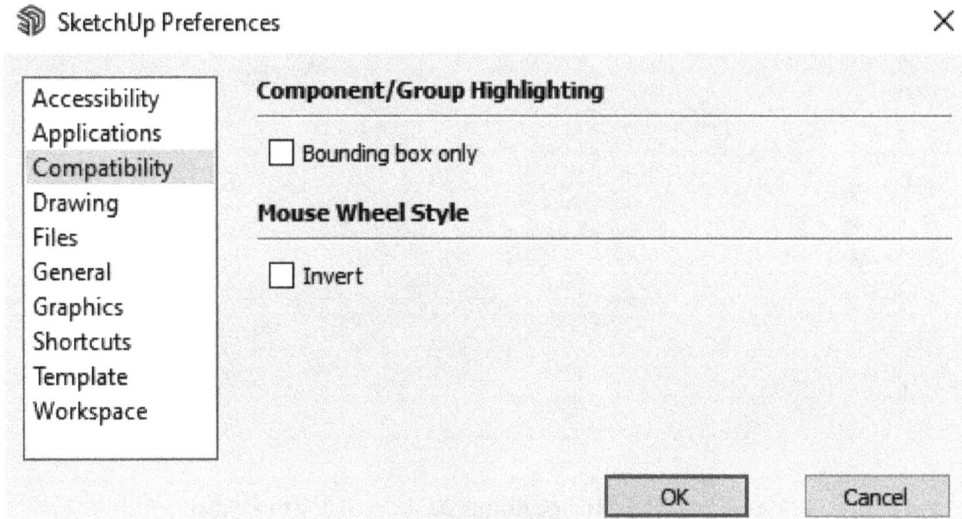

Figure 10.48 – The Preferences Compatibility tab

The **Bounding box only** checkbox will only display the bounding box of a selected Object, instead of the bounding box and the Object Geometry. We can see this in the following example:

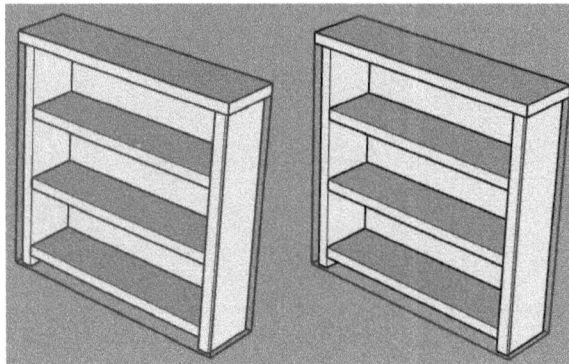

Figure 10.49 – Bounding Box Only Toggled Off (Left) and Bounding Box Only Toggled On (Right)

The **Invert** checkbox will change the direction of the Zoom Tool when scrolling with the scroll wheel. Many 3D modeling software programs have the default scroll direction set in the opposite direction to the SketchUp Pro default, and toggling this option may allow easier switching between these software programs.

Drawing

The **Drawing** tab allows the **Click Style** to be changed, as well as a couple of additional **Miscellaneous** specialized drawing options:

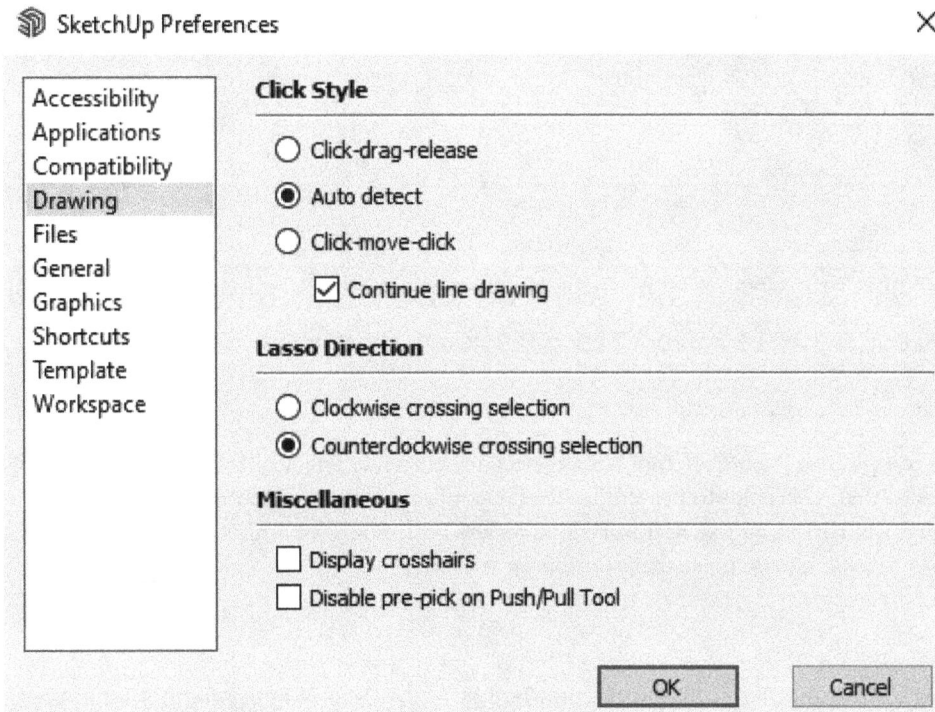

Figure 10.50 – The Preferences Drawing Tab

The **Click Style** options are exclusive, meaning that only one option can be active at a time. **Click-drag-release** will only allow one click per Drawing tool, such as Rectangles or Circles. **Click-move-click** will force two clicks, one for the first point and then a second for the second point. We discussed drawing with the Drawing Tools in *Chapter 4, Drawing Tools – We Begin Modeling!* **Auto detect** will attempt to recognize the drawing style used and allow for **Click-drag-release** and **Click-move-click** workflows to be used interchangeably.

The **Continue line drawing** checkbox is checked by default, and this allows a continuous set of connected lines to be drawn using the Line Tool.

The **Lasso Direction** options are also exclusive, and define the direction for the Crossing Selection, either clockwise or counterclockwise. Counterclockwise is selected by default for Crossing Selections. The opposite direction is used for Window Selection Lassos.

The **Miscellaneous** options include checkboxes for **Display crosshairs** and **Disable pre-pick on Push/Pull Tool**. With **Display crosshairs** toggled on, a 3D crosshair will appear when using a Drawing Tool or the Move Tool:

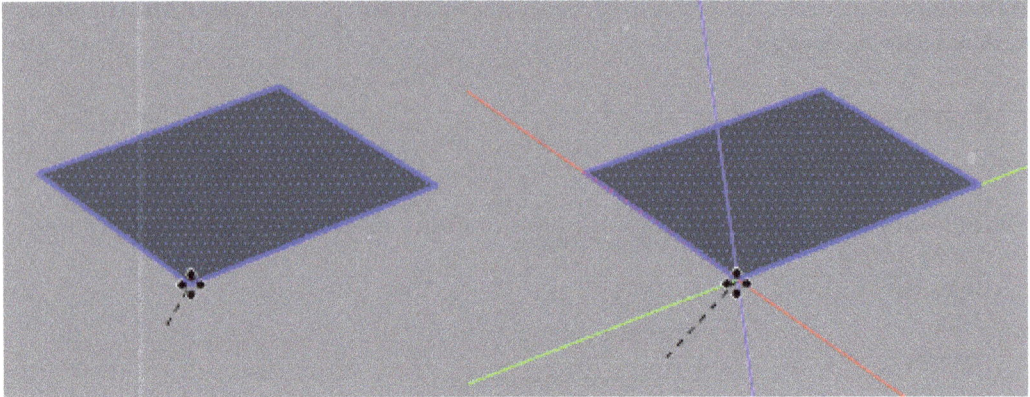

Figure 10.51 – Display Crosshairs Toggled Off (Left) and Display Crosshairs Toggled On (Right)

This is similar to some drafting programs, such as AutoCAD.

Disable pre-pick on Push/Pull Tool will deselect any currently selected Faces when the Push/Pull Tool is activated. This helps to prevent undesirable mistakes when a Face is accidentally selected, although it will remove all preselect workflows for the Push/Pull Tool, which may be your preferred workflow. You will have to make that decision for yourself!

Files

The **Files** tab lists the default file path locations of the different SketchUp Files, including Models, Components, and Materials. These folders are automatically created when SketchUp Pro is installed, or SketchUp selects the User's My Documents folder as the default:

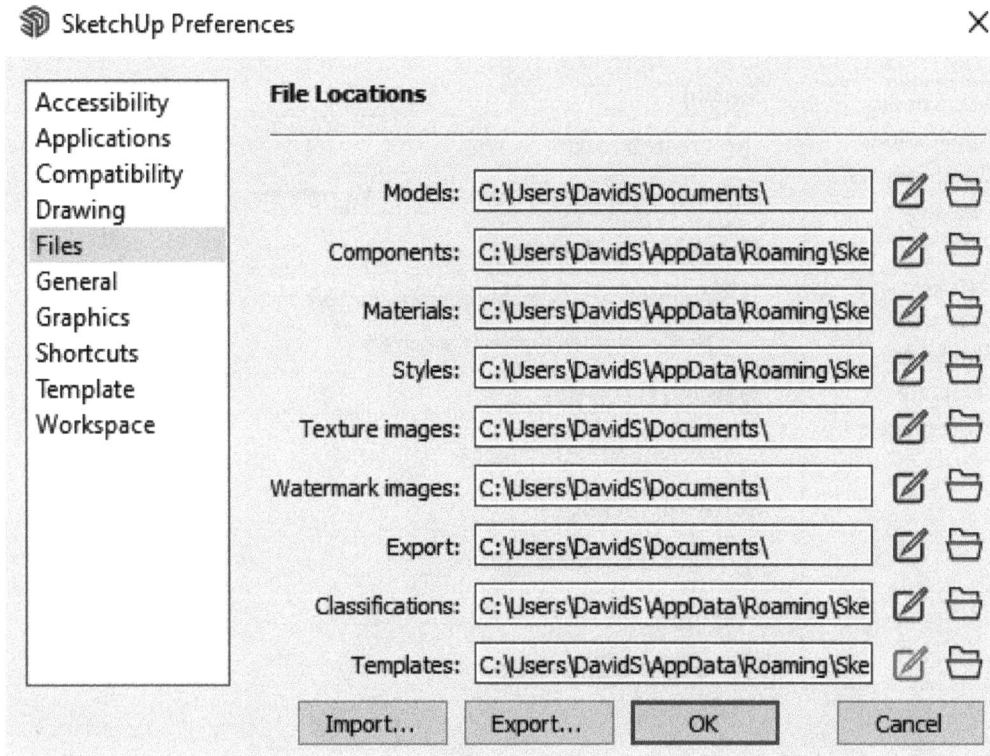

Figure 10.52 – The Preferences Files Tab

These File paths can be edited by clicking the **Edit** Icon or opened in a File Explorer window by clicking the Folder Icon:

Figure 10.53 – The Edit Icon (Left) and the Folder Icon (Right)

The File path locations can be Exported to a SketchUp Data File and Imported on a different computer or User.

General

The **General** tab has many options that do not fall under any of the other categories. The **General** tab includes **Saving**, **Check for Problems**, **Warning Messages**, **Software Updates**, and **Startup** options:

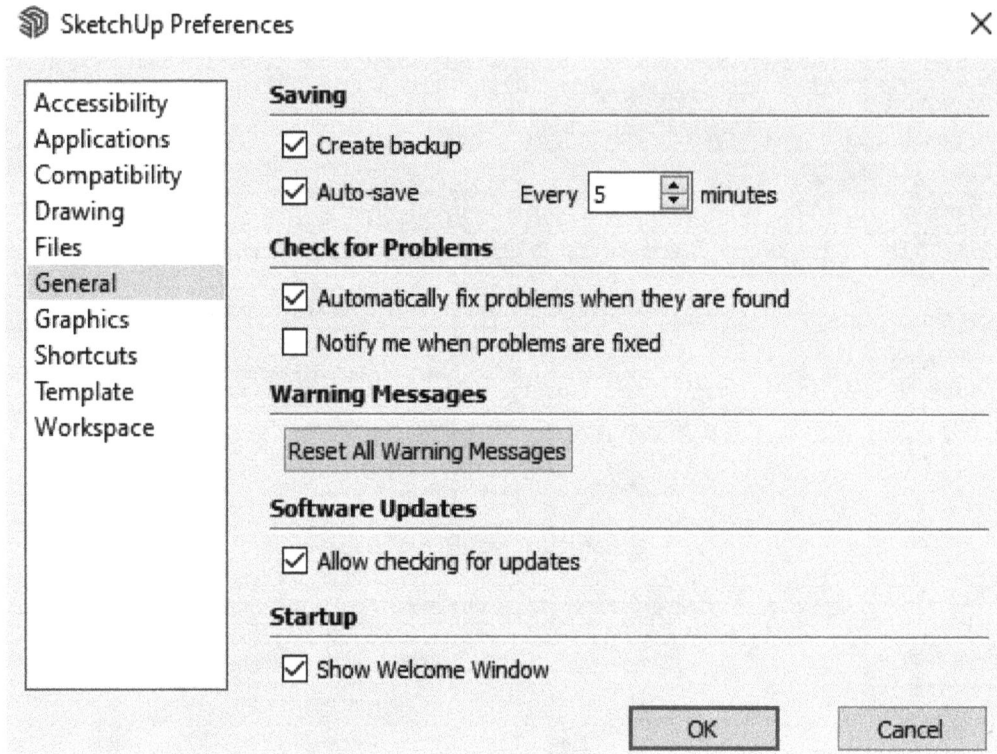

Figure 10.54 – The Preferences General tab

The **Saving** checkboxes are **Create backup** and **Auto-save**, which also includes a Text box for inputting the time increment for **Auto-save**. These checkboxes are checked by default.

The **Check for Problems** checkboxes help to automatically fix problems in SketchUp. The **Automatically fix problems when they are found** checkbox is checked by default, and notifications can be toggled on if problems are found and fixed automatically.

The **Warning Messages** section has one option, **Reset All Warning Messages**. This button will reset the Warning Messages for SketchUp errors or warnings.

The **Software Updates** checkbox allows SketchUp to check for updates for the installed version of the software. This will not automatically install the newest annual version of SketchUp, but it will check for bug fixes or incremental updates to the current version.

The **Startup** checkbox toggles the Welcome Window on or off. The Welcome Window can be accessed at any time under the **Help** dropdown in the Main Menu:

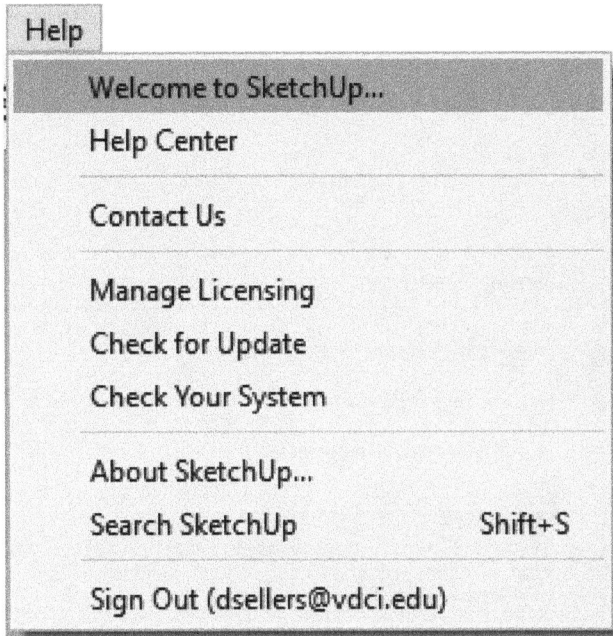

Figure 10.55 – Help | the Welcome to SketchUp Dropdown

The Welcome Window may not need to be opened every time for experienced modelers, but it is recommended to have it open when starting your SketchUp journey. The Welcome Window displays the SketchUp Templates for creating a new file, the **Open file...** button, and **Recent files**:

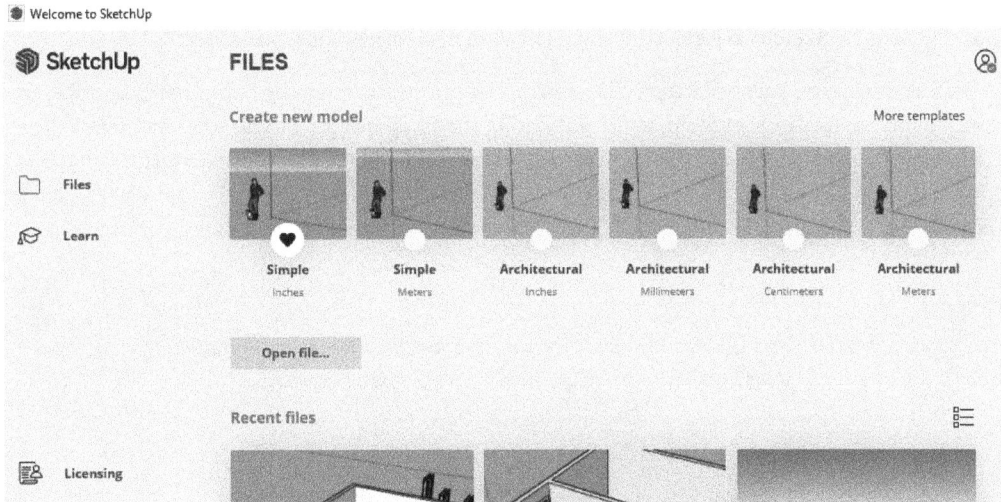

Figure 10.56 – The SketchUp Pro Welcome Window

Graphics (Formerly OpenGL)

The **Graphics** (formerly called **OpenGL**) tab contains and manages **Open Graphics Library** (**OpenGL**) settings, which improves the performance of hardware-accelerated rendering for anti-aliasing and Material Textures:

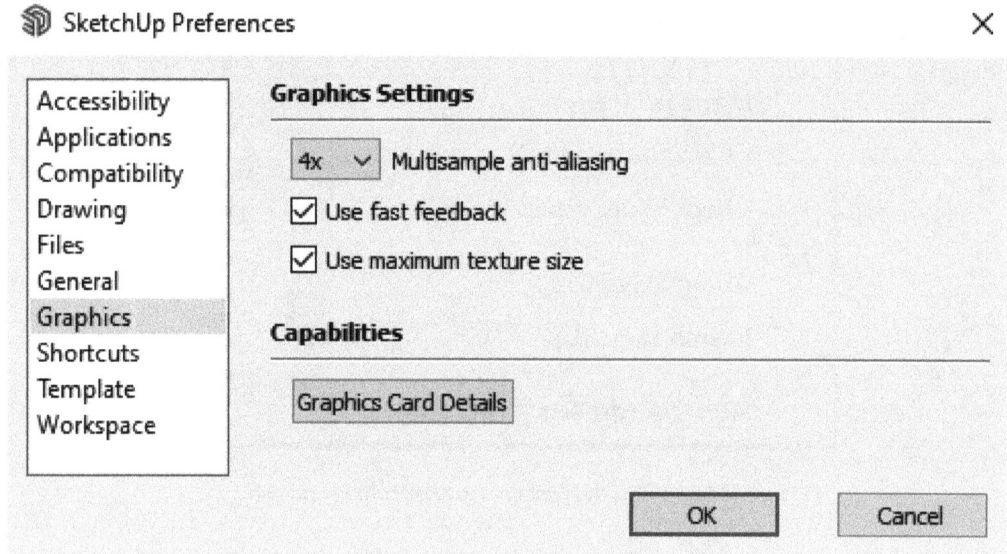

Figure 10.57 – The Preferences OpenGL tab

OpenGL is connected to SketchUp Pro through an **application programming interface** (**API**). *Anti-aliasing* is a term for SketchUp rendering jagged lines to make them look smoother.

OpenGL settings are an advanced topic. The OpenGL options focus on the computer being able to produce quality images in SketchUp, while not asking the Graphics card to do too much work. The default settings are set so that SketchUp can run quickly and efficiently, without overloading your computer. It is suggested that these settings are only changed if you understand the graphics card settings on your computer. More detailed OpenGL instructions should be researched online.

> **Note**
> Check out this SketchUp website for more information about SketchUp and OpenGl: `https://help.sketchup.com/en/sketchup/sketchup-and-opengl`.

Shortcuts

The **Shortcuts** tab lists all of the Tools and Functions in SketchUp Pro, and their assigned Keyboard Shortcut. Shortcuts can be added to Tools or Functions, as SketchUp Pro only assigns Keyboard Shortcuts to a small number of Tools:

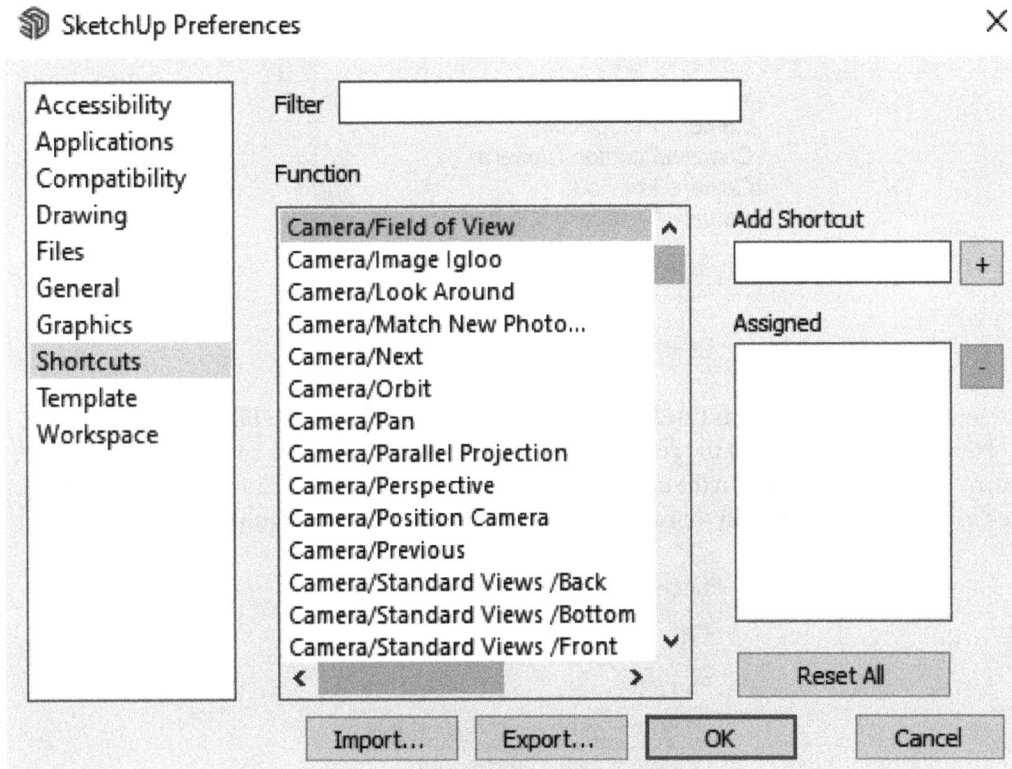

Figure 10.58 – The Preferences Shortcuts tab

Please reference the SketchUp Quick Reference Guides for a list of the Tools that are automatically assigned a Keyboard Shortcut.

The **Function** window shows a list of the SketchUp Functions and Tools listed alphabetically, first by their category and then the Tool name. The categories are **Camera**, **Draw**, **Edit**, **File**, **Help**, **SketchUp**, **Tools**, **View**, and **Window**:

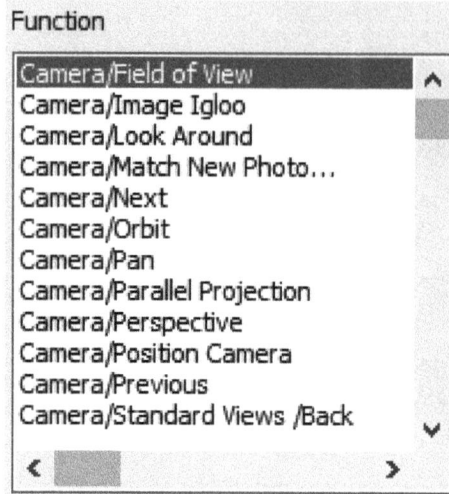

Figure 10.59 – The Function Window

There are many, many Functions listed in this window! SketchUp includes a **Filter** search bar at the top of the **Shortcuts** tab so that the Functions can be quickly filtered. Filters can be the name of the category, subcategory, or Tool. In the next example, we can see the phrase Animation has returned all tools in the **View/Animation** subcategory, as well as **File/Export/Animation…**:

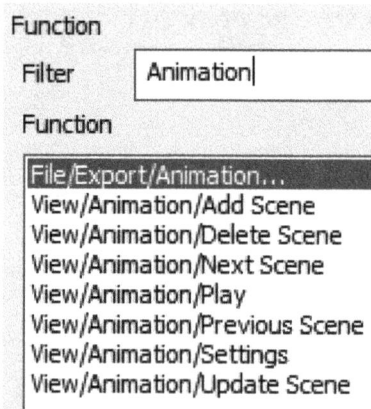

Figure 10.60 – The Animation Filter and Results

When a Tool is selected in the **Filter** window, the current Keyboard Shortcut will appear in the Assigned box if a Keyboard Shortcut already exists. In the following example, we can see that **View/Animation/Previous Scene** has the **PageUp** keyboard button assigned:

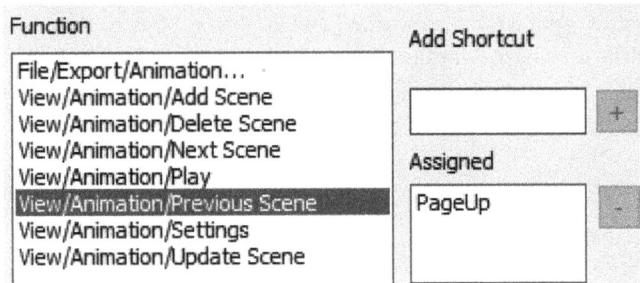

Figure 10.61 – View/Animation/Previous Scene with the PageUp Button Assigned

Assigned Keyboard Shortcuts can be removed by selecting the **Keyboard** button in the **Assigned** box and hitting the minus sign button:

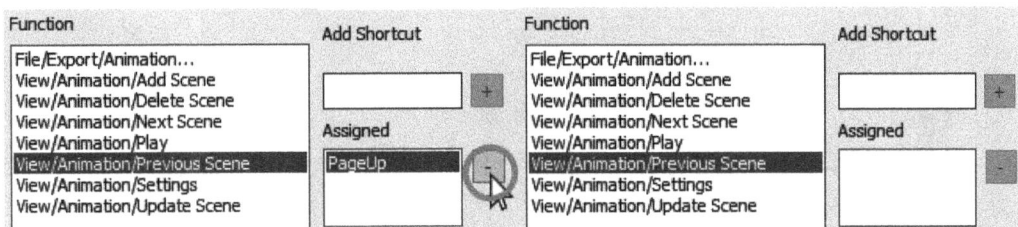

Figure 10.62 – PageUp Selected and the Minus Sign Button Clicked
(Left) and the Keyboard Shortcut Removed (Right)

Keyboard Shortcuts can be added to Tools in a similar way. With a Tool selected in the **Function** window, the **Add Shortcut** box can be selected. Then, a Keyboard Key (with or without Modifier Keys) can be typed, and the input will be recorded in the **Add Shortcut** Box:

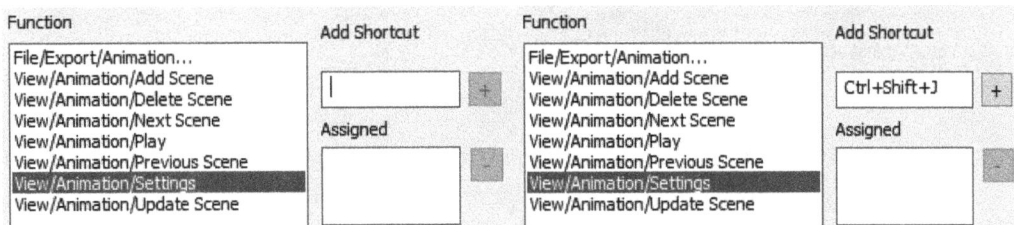

Figure 10.63 – The Tool Selected and the Add Shortcut Box
Activated (Left) and the Keyboard Key Typed (Right)

Once the desired keyboard input has been recorded, the Keyboard Shortcut can be added by clicking the plus sign button. If the Keyboard Shortcut is not already in use, the Shortcut will be automatically added. If the Keyboard Shortcut is already in use by another Tool, then SketchUp will recognize the conflict and prompt the user to confirm the request to reassign the Shortcut:

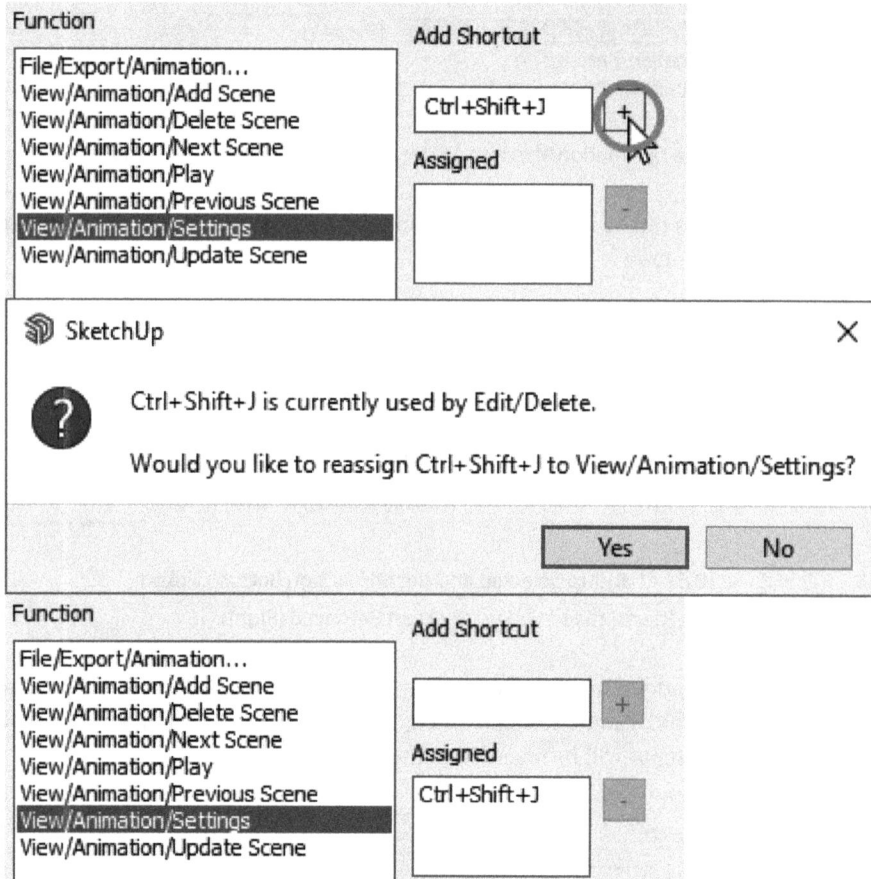

Figure 10.64 – The Plus Sign Button Clicked (Top), the Reassign Confirmation
Window (Middle), and the Assigned Shortcut (Bottom)

Multiple Keyboard Shortcuts can be added to a single Tool, although this is not common. Keyboard Shortcuts are an advanced technique when modeling, and not all Tools need a Shortcut to be added. However, if you find yourself using a Tool many times while modeling, it may be helpful to add a new Shortcut.

Keyboard Shortcut settings can be Exported and Imported using the buttons at the bottom of the tab, and all Shortcuts can be reset using the **Reset All** button.

Template

The **Template** tab allows **Default Drawing Template** to be selected:

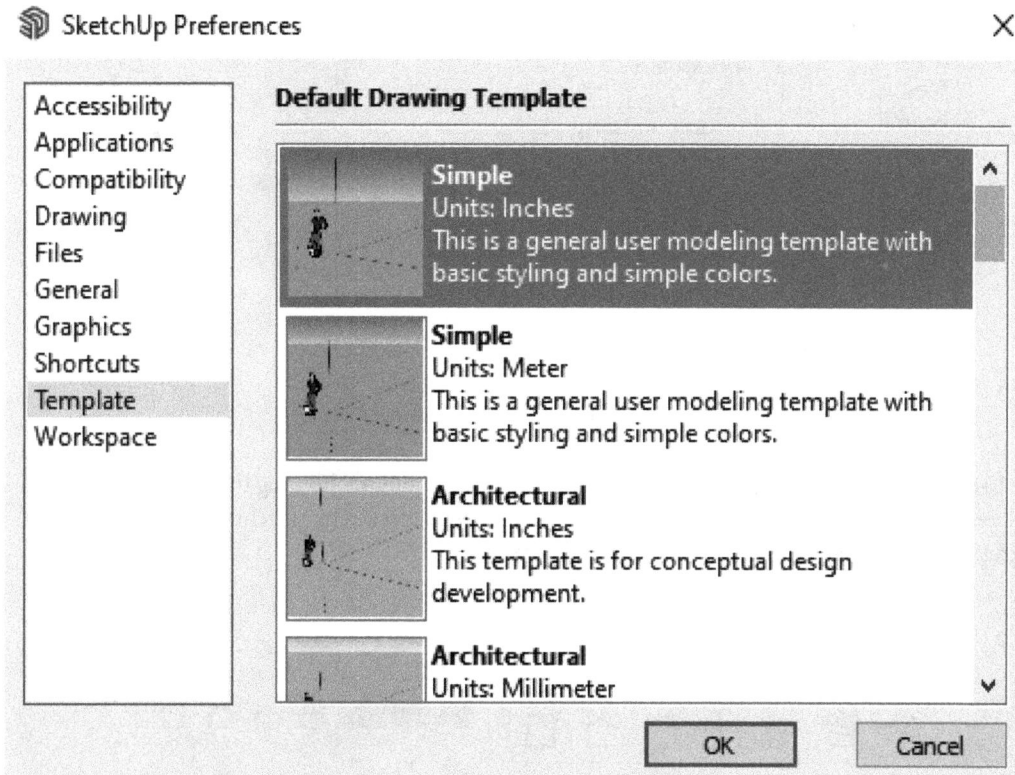

Figure 10.65 – The Preferences Template tab

Templates cannot be edited in this tab.

Default Drawing Template is used automatically if the **Welcome to SketchUp** Window is toggled off.

Workspace

The **Workspace** tab contains options to manage the Toolbars in the SketchUp Workspace:

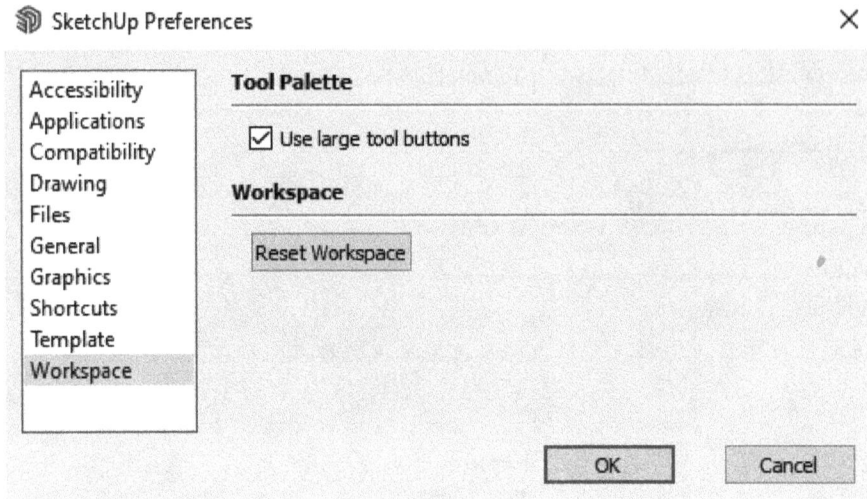

Figure 10.66 – The Preferences Workspace Tab

The **Use large tool buttons** checkbox is toggled on by default, and unchecking this option will shrink the size of the Toolbar buttons. This option is not commonly unchecked by SketchUp users, as computer display sizes have increased dramatically in recent years.

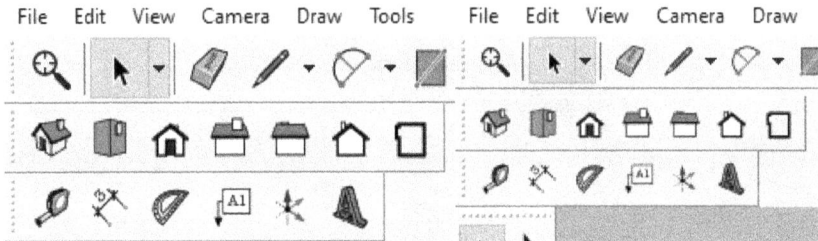

Figure 10.67 – Use Large Tool Buttons Toggled On (Left) and Use Large Tool Buttons Toggled Off (Right)

The **Reset Workspace** button will reset any changes made to the current Workspace.

Summary

In this chapter, we discussed the **Model Info** and **Preferences** dialog boxes. These dialog boxes contain helpful and powerful options for the current model and overall SketchUp settings. The **Model Info** tabs allow the current model to customize View settings, Object settings, and custom workflow settings. The **Preferences** tabs allow a user to update overall SketchUp settings, such as accessibility options and performance optimization. Once you have been modeling for a while, it may be worth poking around in these tabs to see how you can customize SketchUp Pro to fit your preferences! In the next chapter, we will take a deep-dive into working with Components.

11
Working with Components

In this chapter, we will explore Components and what makes them special in SketchUp Pro. Components allow us to keep our files more organized, easily use Objects in multiple files, and automatically align Geometry in our models. Component attributes can be set and edited to achieve many different functions in SketchUp Pro. We will wrap up this chapter by briefly discussing Dynamic and Live Components.

The following topics are covered in this chapter:

- Using the Components Panel
- Creating Components:
  - Creating New Components
  - Nested Components
- Updating and Editing Components
- Component Statistics
- Component Instances
  - Instance Options
- Component Attributes
- Dynamic and Live Components

Using the Components Panel

Components are fundamental to elevating SketchUp Pro from a simple 3D software working with Edges and Faces into a scalable architecture, landscape architecture, interior design, or woodworking tool. Components allow SketchUp models to be protected, imported, edited, and referenced throughout a model.

Components are managed in the **Components** panel, which can be toggled on in the default tray by going to **Window | Default Tray | Components** in the Main Menu:

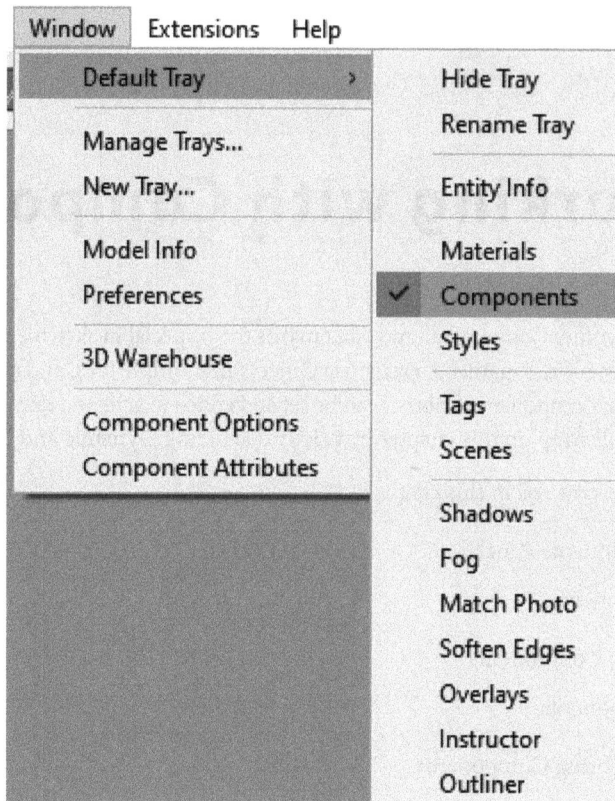

Figure 11.1 – Menu Dropdown for Components Panel

The **Components** panel is broken into two parts, with active Component information at the top and the **Select**, **Edit**, and **Statistics** tabs at the bottom. Additionally, the black-and-white button at the top right of the panel will display or hide the secondary selection pane, which works the same way as the **Select** tab:

Figure 11.2 – Display the Secondary Selection Pane Button and Tooltip

The **Components** panel works very similarly to the **Materials** panel, which we discussed in *Chapter 8, Materials*. SketchUp has some example Components available in the **Select** panel, and some of these are included in the default Component collection, **Component Sampler**:

Figure 11.3 – Component Sampler Collection Showing Example Components

The view options for the **Select** pane can be changed by clicking the button directly below the **Select** tab and choosing one of the view options:

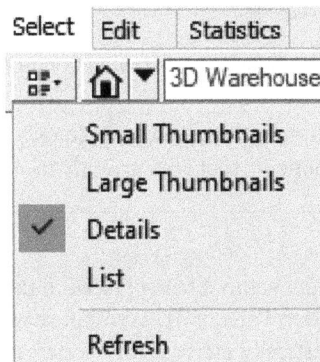

Figure 11.4 – View Options Dropdown

Additional Components can be found by changing the active Component collection, and this can be done by accessing the drop-down arrow next to the house icon – which represents the **In Model** collection:

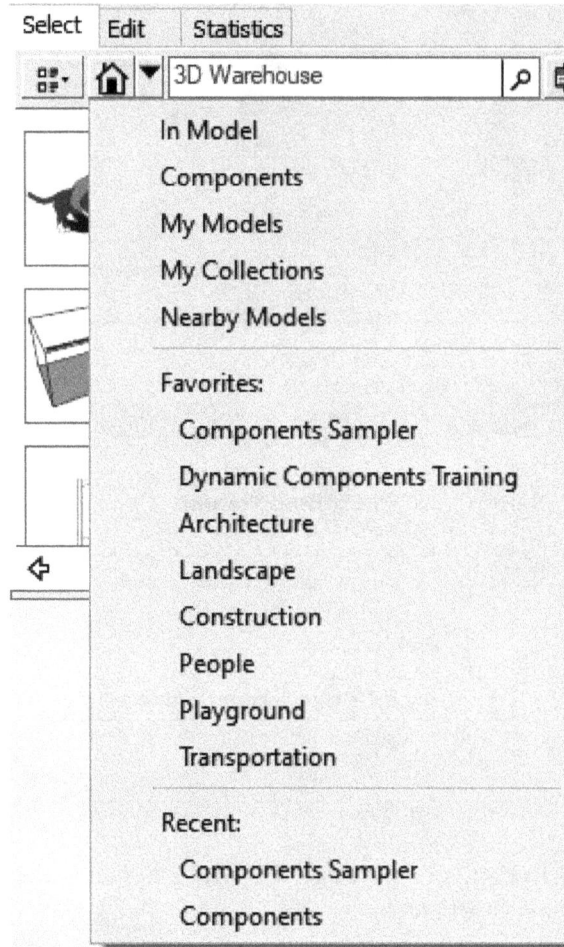

Figure 11.5 – Navigation Dropdown Showing Component Collections

The **In Model** collection is the top option in the list and can quickly be accessed by clicking the house icon. This will show a list of all Components that are currently in the model.

> **Note**
>
> Components can be saved in a model without being placed in the Drawing Area. Components that are brought in from the Selection Pane or 3D Warehouse will be "remembered" in the In Model collection, and this can drastically increase the model's file size! It is recommended to regularly delete unused Components manually or by using Purge.

In this next example, we can see that a model has four **Components** in the **In Model** collection:

Figure 11.6 – Components in the Drawing Area (Left) and the In Model Collection (Right)

Even if one of the Components is deleted from the Drawing Area, we can see that this does not remove the Component from the model:

Figure 11.7 – Component Deleted in the Drawing Area (Left),
Component Remains in the In Model Collection (Right)

This can be extremely helpful when modeling, as we can save Objects in our In Model collection and use them later!

Components can be brought into the model by clicking on the Component image or description in the **Components** panel. This will preview the Component on the cursor in the Drawing Area, similar to how Geometry might appear when using a Copy/Paste workflow:

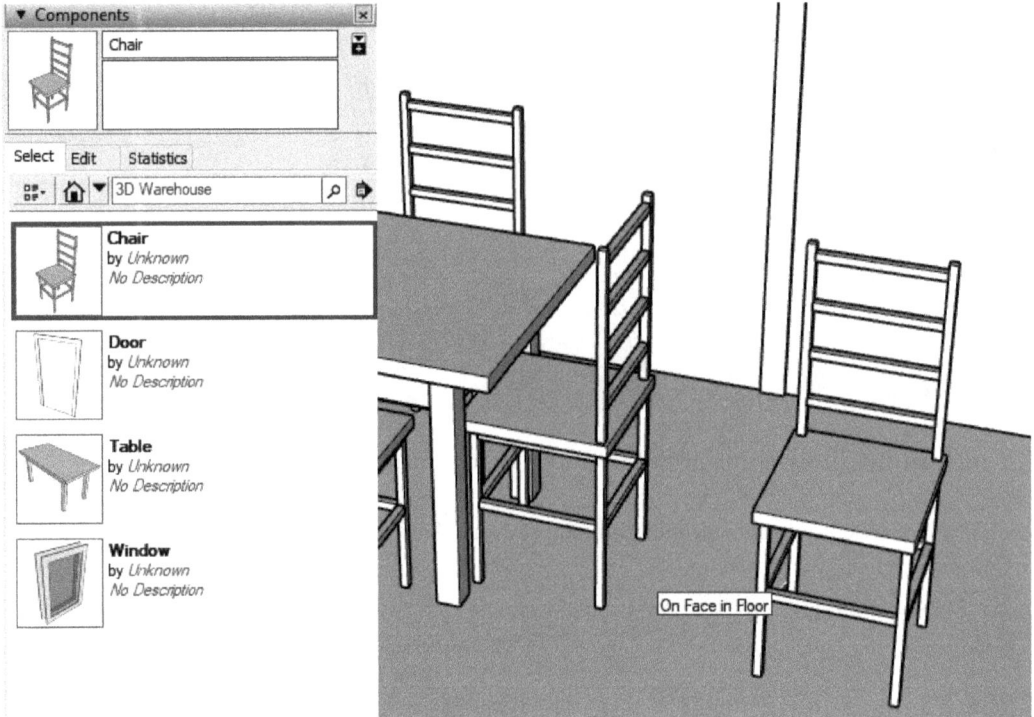

Figure 11.8 – Component Clicked in the Components Panel (Left), Component Previewed on Cursor (Right)

The Component can then be placed in the model by left-clicking in the Drawing Area. Components may have specific requirements as to how they can be placed; we will discuss these options in the *Creating Components* section of this chapter.

> **Note**
>
> Components can also be clicked and dragged from the **Components** panel to be brought into the model, but it is recommended that a two-click workflow be used.

Once a Component is placed in the model, it behaves similarly to a Group. It is protected Geometry and can be edited as discussed in *Chapter 3, Working with Groups and Components*. We will discuss the **Edit** tab and advanced editing techniques for Components in the *Updating and Editing Components* section of this chapter.

The top section of the **Components** panel previews the currently selected Component in the **Components** panel, not what is currently selected in the Drawing Area. The preview section is a fast way to update the Component name and description. Additionally, even after the Component has been deselected in the **Components** panel, the preview section will remember the Component information. Clicking the image of the Component will re-activate the Component and allow you to place more Component Instances in the model. We will discuss Component Instances in the *Component Instances* section of this chapter.

The 3D Warehouse search bar can be used to search for Components that have been uploaded to SketchUp's 3D Warehouse. 3D Warehouse is a great tool for finding and downloading Components or even full SketchUp models. We will take a closer look at 3D Warehouse in *Chapter 12, Import and Export, 3D Warehouse, and Extensions*.

Creating Components

Components are so fundamental to the SketchUp Pro workflow that the Make Component tool is one of the Principal tools. The Principal toolbar was first introduced in *Chapter 2*, but we will cover the Make Component tool in this section. The Make Component tool has two designs, with a new design being introduced in SketchUp Pro 2021. The old Make Component button looked like the SketchUp logo inside a red bounding box. The new Make Component button is three boxes inside of a blue bounding box:

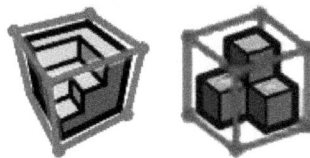

Figure 11.9 – Old Make Component Button up to 2020 (Left), New
Make Component Button 2021 onward (Right)

Components can also be made by hitting the *G* keyboard shortcut or by right-clicking and selecting **Make Component**.

Creating New Components

Clicking the Make Component button with no selection will begin the Make Component process, and SketchUp will prompt for Component Axes to be placed. We will discuss Component Axes later in this section. It is recommended that the Make Component button be clicked when there is an active selection of Geometry:

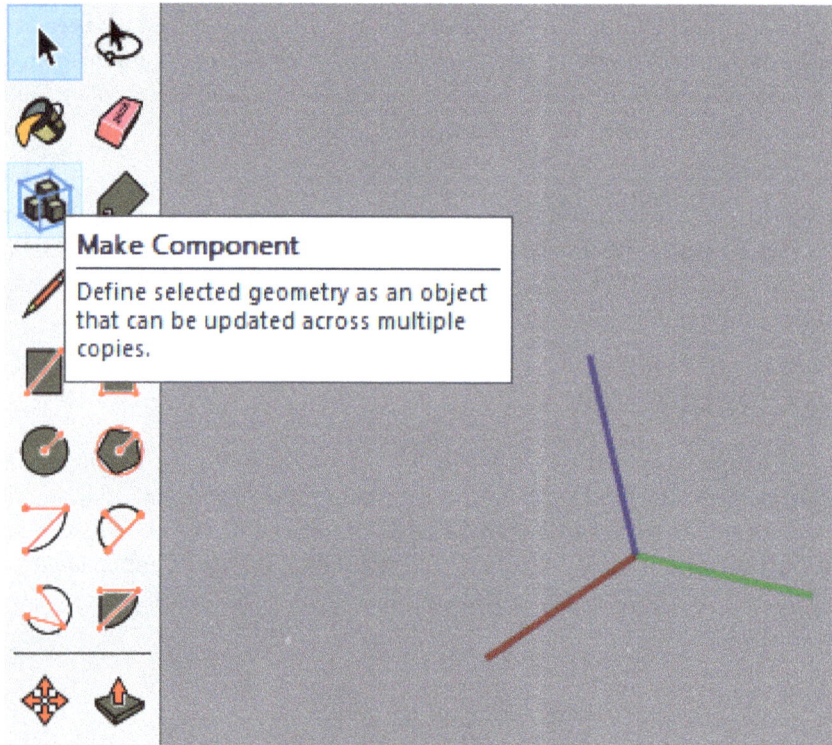

Figure 11.10 – Component Axes Appear when Make Component Is Clicked with No Selection

If there is an active selection set, or after the Component Axes have been placed in the drawing area, the Create Component dialog box will appear. The **Create Component** dialog box contains **General**, **Alignment**, and **Advanced Attributes** sections for the new Component:

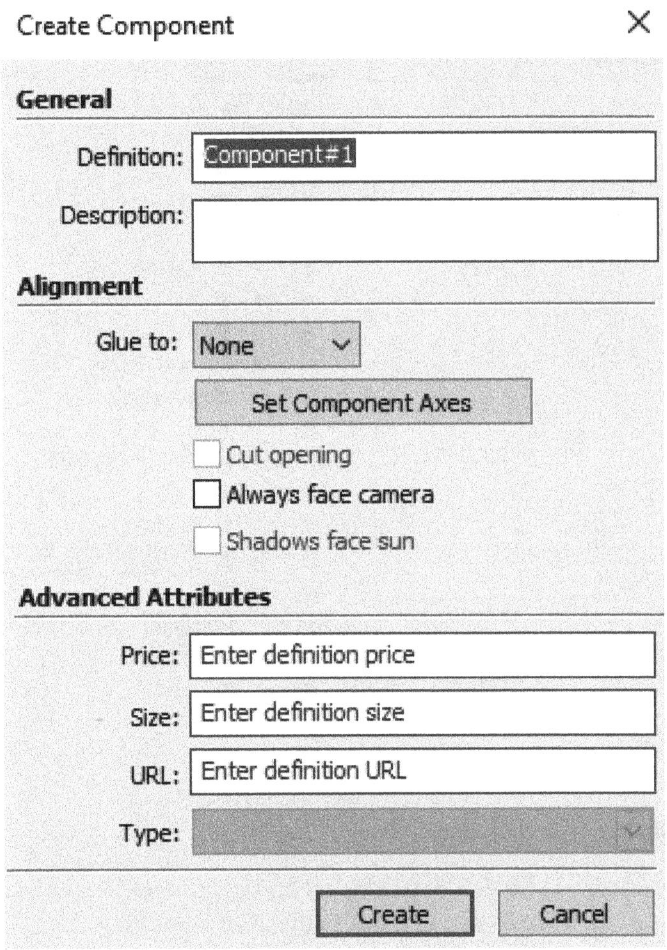

Create Component   ✕

General

Definition: Component#1

Description:

Alignment

Glue to: None ⌄

Set Component Axes

☐ Cut opening
☐ Always face camera
☐ Shadows face sun

Advanced Attributes

Price: Enter definition price

Size: Enter definition size

URL: Enter definition URL

Type:

Create   Cancel

Figure 11.11 – Create Component Dialog Box

General Options

The **General** options include **Definition** and **Description**. **Definition** is the name of the Component, although individual Component Instances can also receive unique Instance names once placed. The default Definition will be the word Component and a unique number starting at #1. The Definition name and Instance names appear in the **Outliner** and **Entity Info** dialog boxes:

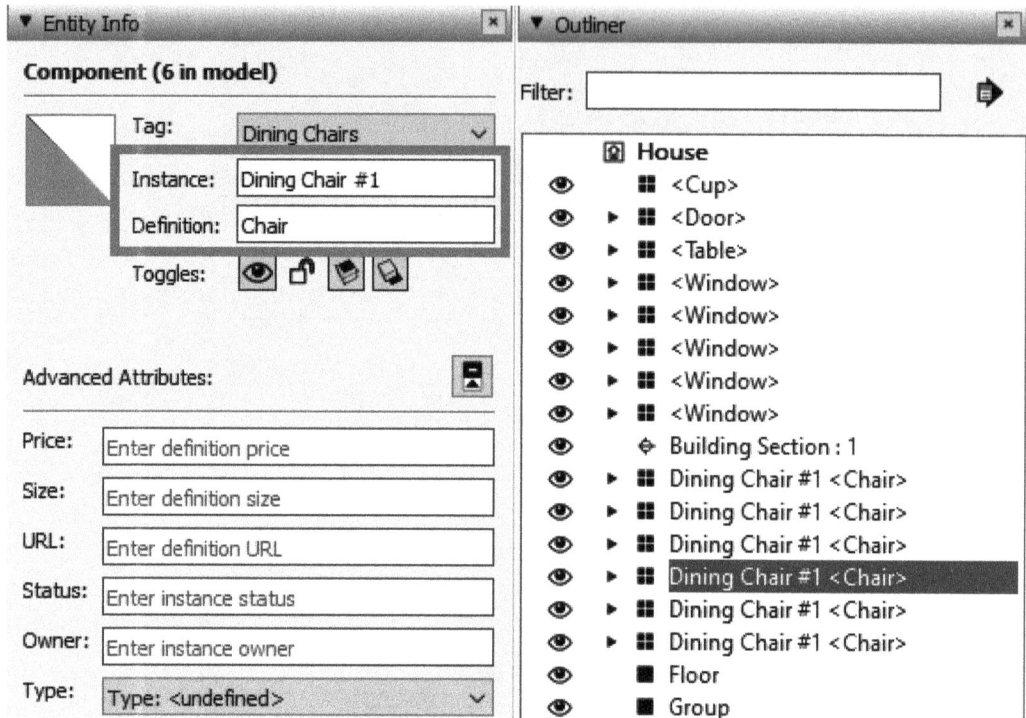

Figure 11.12 – Definition and Instance in Entity Info (Left) and Outliner (Right)

Alignment Options

The **Alignment** options include **Glue to**, **Set Component Axes**, **Cut opening**, **Always face camera**, and **Shadows face sun**. The **Glue to** option is one of the most powerful and forgotten settings for SketchUp Components. The **Glue to** dropdown contains **None**, **Any**, **Horizontal**, **Vertical**, and **Sloped**. Selecting a **Glue to** option will align the Component with one of these options when placing the Component in the model. In this next example, we can see a wall light Component that has **Glue to** set to **Vertical**. The wall light Component cannot be placed on any horizontal or sloped surface and can only be placed on existing vertical Geometry.

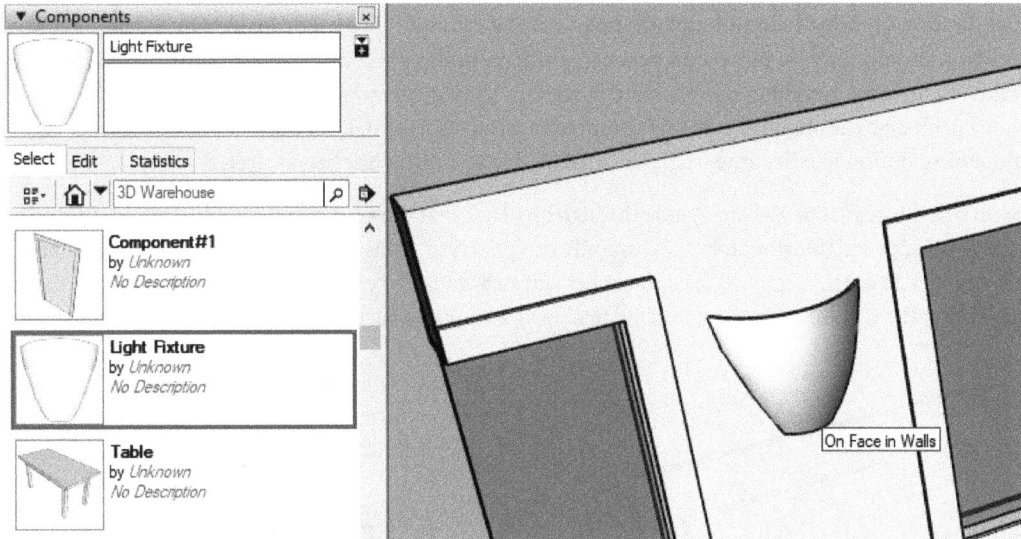

Figure 11.13 – Wall Light Component with Vertical Glue to Setting (Left) Placed on Wall (Right)

> **Note**
> SketchUp Pro will allow Components to be placed in the standard direction (positive blue direction up) when placing the Component in space. Only Components placed on other Geometry will be forced to align with the **Glue to** options.

If we attempted to place this Component on a horizontal surface, **SketchUp would show a Cannot place component there! warning**:

Figure 11.14 – Glue to Placement Warning

Setting appropriate **Glue to** options greatly helps when defining what a user can do with a Component. Wall lights should only be placed on vertical surfaces, tables and chairs should only be placed on Horizontal surfaces, skylights may only be placed on sloped surfaces, and so on. Selecting the **Any** option will change the alignment of the Component to any horizontal, vertical, or sloped surface. The **Cut opening** option will become available once a **Glue to** option has been selected.

The **Cut opening** function will cut a hole through any Face that the Component is placed on so that an opening will appear. Unfortunately, this can only be applied to a single Face, so a window Component will not be able to cut through two Faces on either side of a wall. There are some Extensions available that expand upon the **Cut opening** functionality to allow for multiple Faces to be cut with one Component:

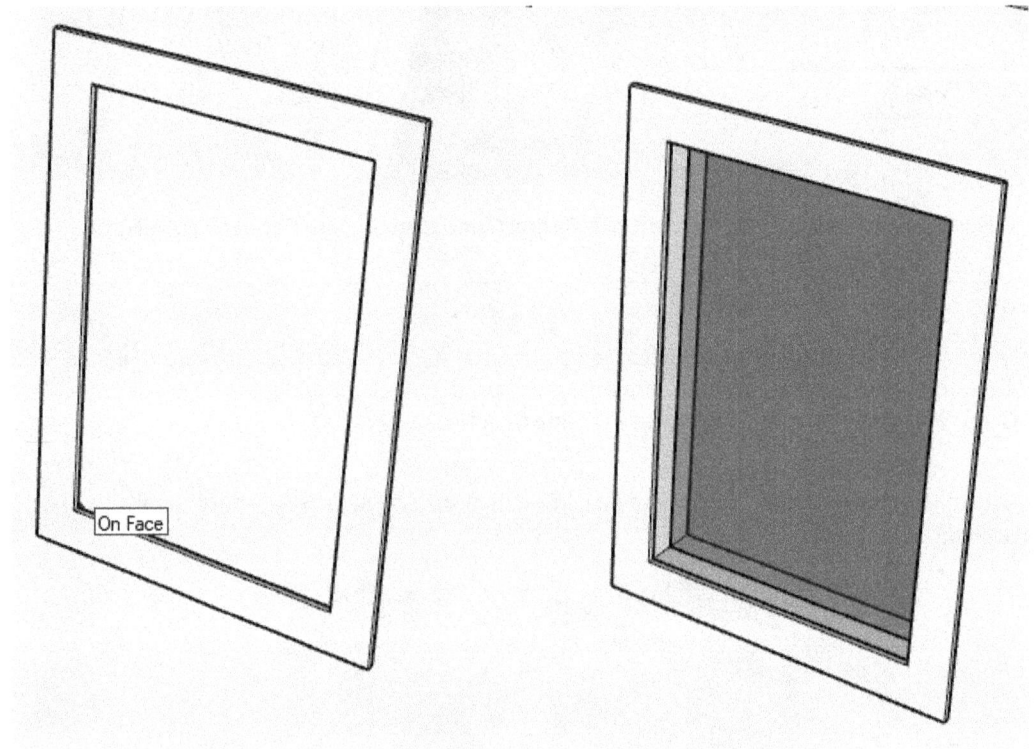

Figure 11.15 – Component Placed on Wall (Left), Hole Automatically Cut in Outer Face (Right)

The **Always face camera** and **Shadows face sun** options are unique to SketchUp, and a Component with these attributes will be immediately recognizable! We have all seen a Face Camera Component before – the Person Component that is automatically included every time a new drawing is created. This Component is a 2D set of Edges and planar Faces, and it appears to rotate when the Camera Orbits around the model:

Figure 11.16 – Person Component Appears to Rotate when the Camera Orbits

SketchUp promotes the use of 2D Components to keep the model file size light, and they can help the model to run more efficiently. Using the **Always face camera** option will remove the need for complex fully 3D models and is a much better solution than using old-school arcade-style graphics, which represent 3D with multiple 2D planes. In this next example, we can see a 3D model with 30,000 Edges and 20,000 Faces, an old-school arcade-style model, and a Face Me model:

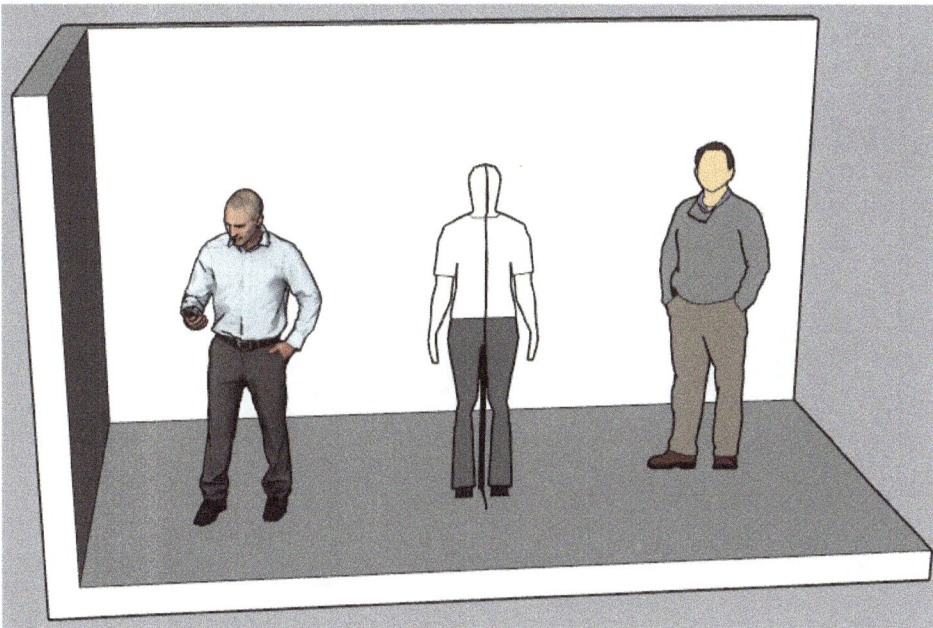

Figure 11.17 – 3D Model (Left), 2D Model (Middle), and 2D Face Me Model (Right)

When the Camera is Orbited around the model, we can see how each model updates.

Figure 11.18 – Updated View when the Camera Is Orbited

The downside of a Face Me model is that it will always give the exact same view, but that will also mean that there will be no awkward angles. Also, it provides a full view of an entourage without increasing the file size and slowing down the model!

When **Always face camera** is turned on, the **Shadows face sun** option is also available. This will allow the Component Geometry to face the camera, as well as casting the full profile shadow facing toward the direction of the sun. That way, if the Camera angle is perpendicular to the sun angle, a full shadow will still be cast. We can see this example clearly with the following Face Me tree Component. When the Camera angle is perpendicular to the sun angle, the tree only casts a narrow shadow. However, with **Shadows face sun** toggled on, the full profile of the Face Me Component is cast on the ground, which adds to the model representing a 3D tree:

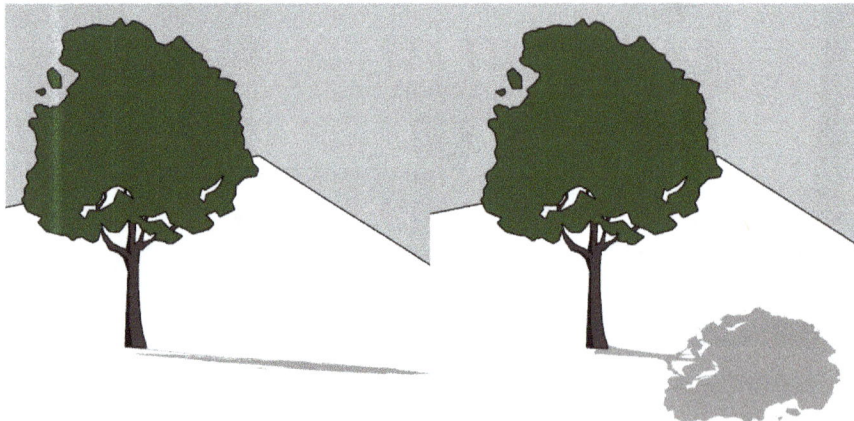

Figure 11.19 – Shadows Face Sun Off (Left), Shadows Face Sun On (Right)

The last option in the **Alignment** section is the **Set Component Axes** button. Clicking this button will temporarily hide the **Create Component** dialog box. Component Axes will appear on the cursor, and by clicking new Axes can be set for the new Component. This is the same workflow as when creating a Component with nothing pre-selected.

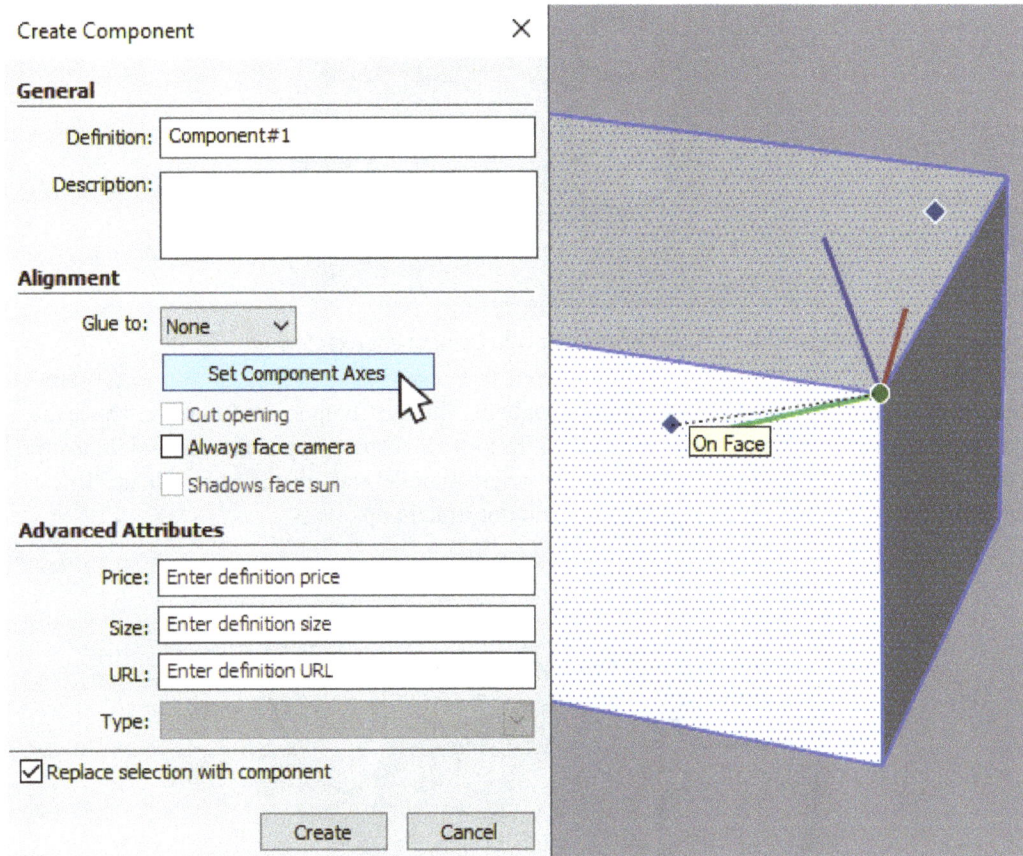

Figure 11.20 – Set Component Axes Button Clicked (Left), Component Axes Cursor (Right)

Setting the Component Axes is important as this location and orientation will determine where a Component will be placed when selecting a new Instance from the Components. Additionally, the orientation of the Component Axes will be updated based on the **Glue to** setting. In this next example, we can see a cup Component that has Component Axes set away from the Geometry. Additionally, the Component Axes show that the Blue direction (Positive Z) is pointing sideways. This Component has the **Glue to** option set to **Horizontal**:

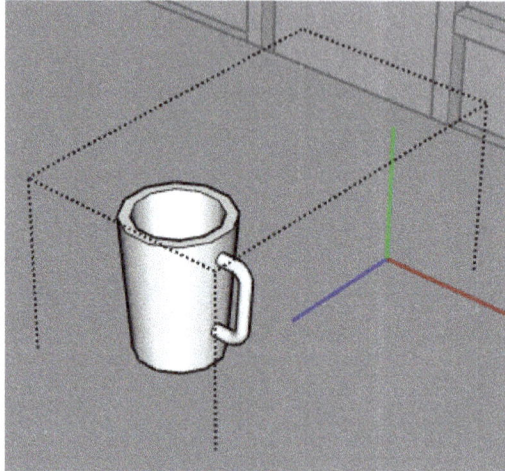

Figure 11.21 – Cup Component with Component Axes Shown

When a new Instance of the cup Component is brought in from the Components Panel, we can see the Move cursor is away from the cup Geometry. This is the location of the Component Axes. Additionally, the cup appears to be on its side, even though the original model was created right side up. This is because the Blue direction is pointing sideways, and not straight up.

Figure 11.22 – Component Instance Created from Components Panel

This Component will be able to be placed on Horizontal surfaces, but the cup will not rest on the surfaces directly. In this case, a more appropriate location for the Component Axes would be the base of the cup, either on one of the corners of the bounding box or the center of the cup. Additionally, the Component Axes should be shown with the blue direction (Positive Z) pointing up, so that the cup sits "the right way up" when placed:

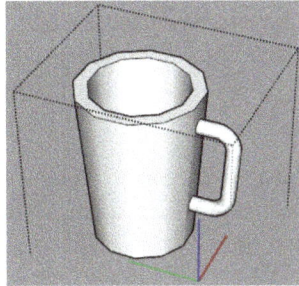

Figure 11.23 – Component Axes Set at Base of Cup with Blue Direction Up

When this Component is brought in from the Components Panel, we can see that the cup sits on the horizontal surface in the correct orientation:

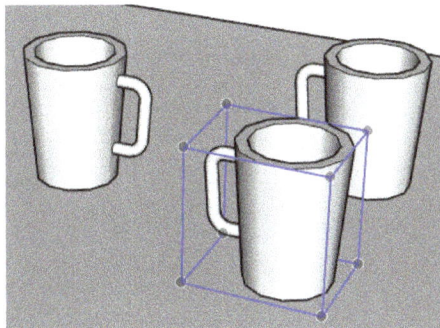

Figure 11.24 – Multiple Cup Components Placed on a Horizontal Surface

Advanced Attributes

The **Advanced Attributes** section contains options attaching information to the Component, such as **Price**, **Size**, **URL**, and **Type**. The **Type** option is directly linked to the IFC Classifications, which should be loaded into the project before being assigned to a new Component. We discussed IFC Classifications in *Chapter 10, Model Info and Preferences*.

Figure 11.25 – Advanced Attributes Options in the Create Component Dialog Box

The **Advanced Attributes** options can be manually filled out for a basic Component and will be the same for each instance of the Component in the model. **Advanced Attributes** can also be used inside of Dynamic Components, which we will briefly discuss later in this chapter in the *Dynamic and Live Components* section.

While the **Advanced Attributes** options prompt for specific values, the text boxes can be filled in with any text or numerical value. It is suggested that a standard format is used from Component to Component if these values will be used to run reports for a project, but this is an advanced workflow.

The **Entity Info** panel will display two additional **Advanced Attributes** options once the Component has been created and is selected in the model:

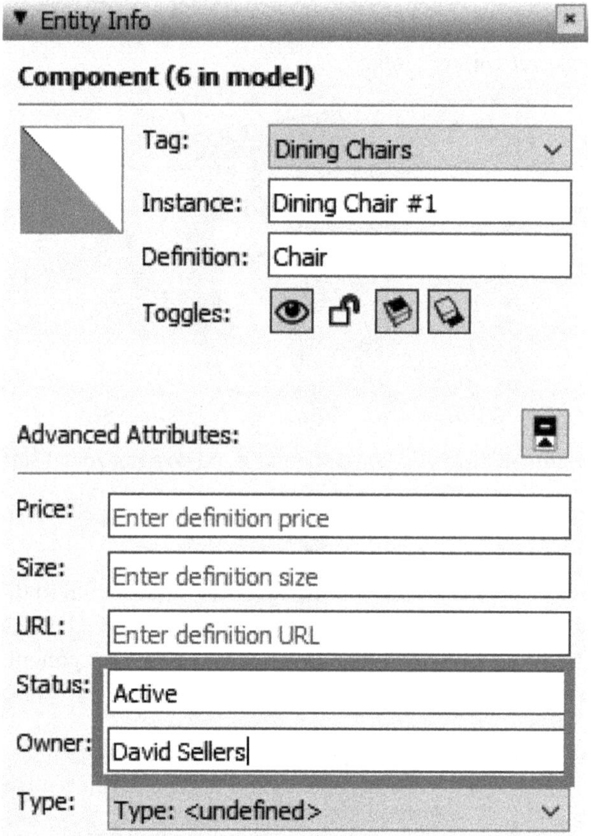

Figure 11.26 – Status and Owner Shown in Entity Info Dialog Box

The **Status** and **Owner** options are **Instance** options, while the **Price**, **Size**, **URL**, and **Type** options remain as **Definition** options. If the same Component should have different Definition values, then Unique Components should be created, or Dynamic Components should be used. We will discuss both workflows later in this chapter in the *Dynamic and Live Components* section.

The final option in the **Create Component** dialog box is a checkbox that by default replaces the selection with an Instance of the new Component:

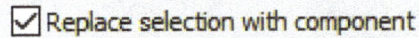

Figure 11.27 – Replace selection with component Checkbox

This is a common workflow, where the Geometry should be replaced by/converted into the new Component. However, in some workflows, it can be useful to add the Component Definition to the In Model tab of the Components panel but keep the loose Geometry while working through the model. If this is desired, the replace selection checkbox should be unchecked.

Once all options have been reviewed, the **Create** button can be clicked to create the new Component.

Nested Components

Nested Components are simply Components that are included in the selection when creating another Component. Another workflow is Components can be created while editing an existing Component, which adds the nested Component when updating the top-level Component. In this next example, we can see a should be - **Table and Chairs** dining set defined as a Component, and the table and chairs Objects are nested Components:

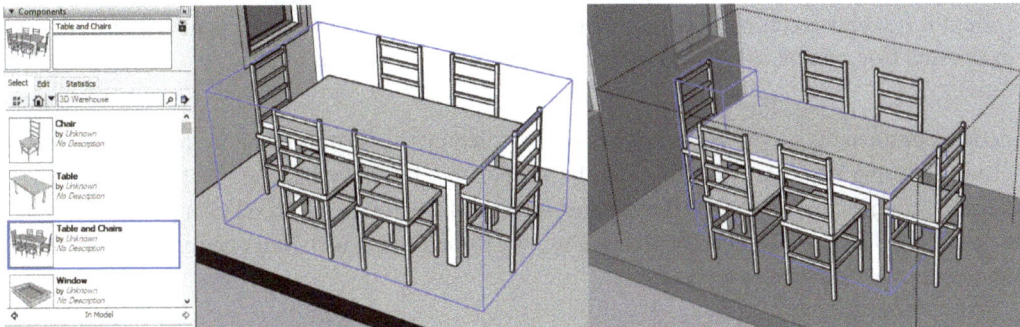

Figure 11.28 – Table and Chairs Component (Left), Nested Individual Components (Right)

Nested Components do not need to be created in any special workflow or using a different tool; they are simply Components! But using nested Components can provide some great advantages to how models are updated over time. In the dining set example, we can see how changing the design of one of the chairs would update the other chairs at the table, and that makes sense – we want all of these chairs to match!

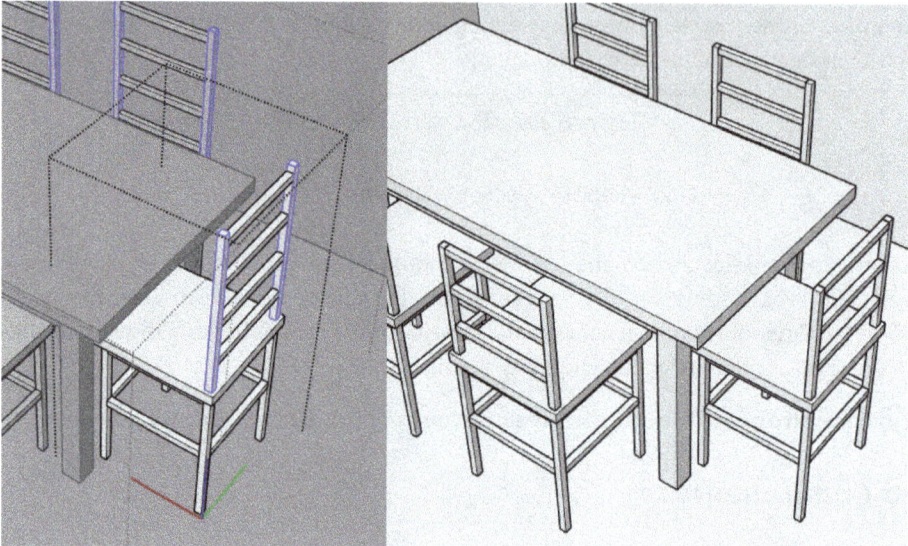

Figure 11.29 – Chair Component Is Edited (Left), All Chair Components Update (Right)

> **Note**
> When using nested Components, make sure to examine what should be a Component or a Group! It is highly recommended that all Geometry in the top-level Component is included in a nested Component or Group and that no loose Geometry is mixed in!

All nested Components can be viewed in a model by clicking the **Details** flyout in the Components panel and choosing **Expand**. This will show all Components and nested Components in the model. In this next example, we see the **Outliner** panel for a model with one top-level Component, **South Wall**. Inside the **South Wall** Component, there are two **Door** Components, but once one is expanded, it shows four other nested Components!

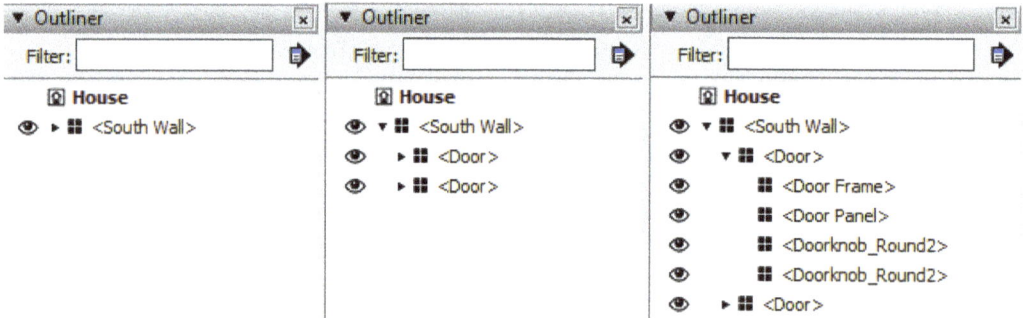

Figure 11.30 – One Top-Level Component (Left), Expand Selected
(Middle), Four Nested Components (Right)

We will look at updating Components, swapping Components, and information about Component Instances in the next sections. These workflows are essential for updating nested Components in our models! But first, we will quickly take a look at the **Components** panel's **Statistics** tab.

Component Statistics

The Component **Statistics** provides information about the selected Component at a glance. The **Statistics** tab shows **Count** information for **Edges**, **Faces**, (nested) **Component Instances**, and more:

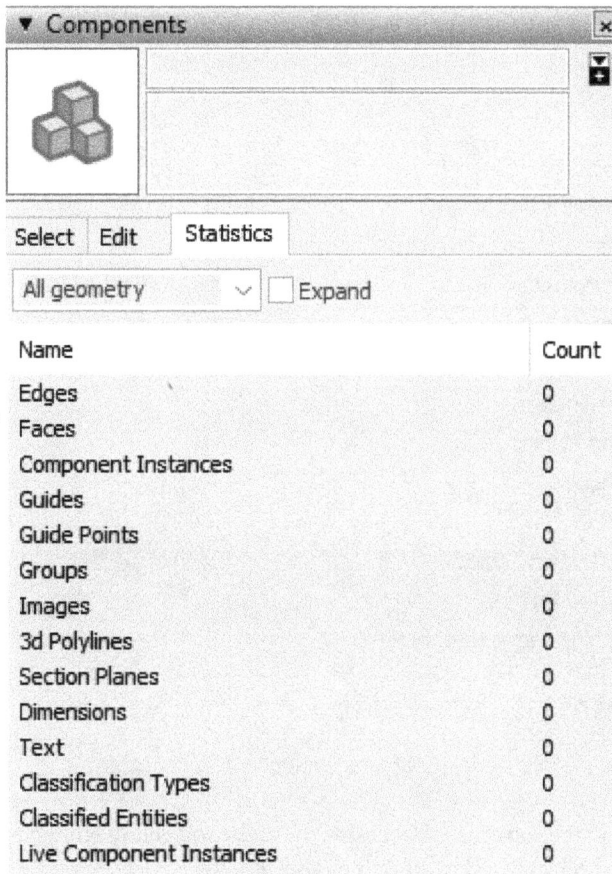

Figure 11.31 – Components Panel Statistics Tab

The **Statistics** tab provides information for the first level down in the selected Component, meaning what can be selected when the top-level Component has been activated. In this example, we can see a **Bedframe** Component is selected, and the Statistics tab show that there are **0** Edges and **0** Faces in this Component. That is because there is no loose Geometry one level down in the Component; everything is located in a different Component. There are 23 Component Instances – these are nested Components.

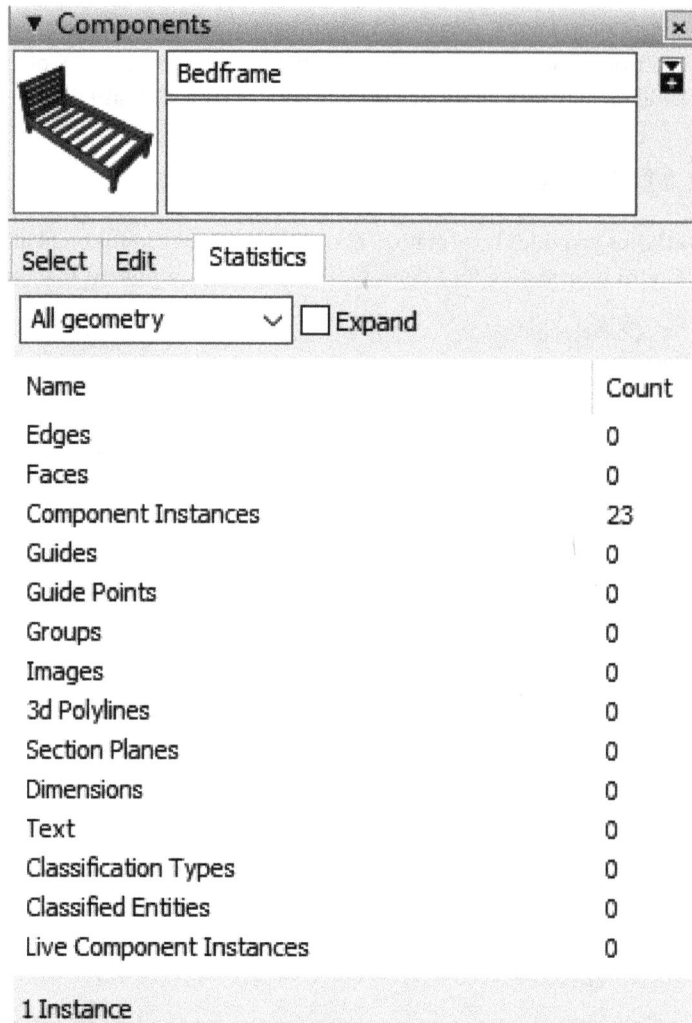

Figure 11.32 – Bedframe Component Statistics Tab

The **Expand** checkbox at the top of the **Statistics** tab can be toggled to show the grand total of all nested Geometry, no matter how many times it has been nested in a Group or Component. In this example, the **Expand** checkbox can be toggled to show a total of **276** Edges and **138** Faces:

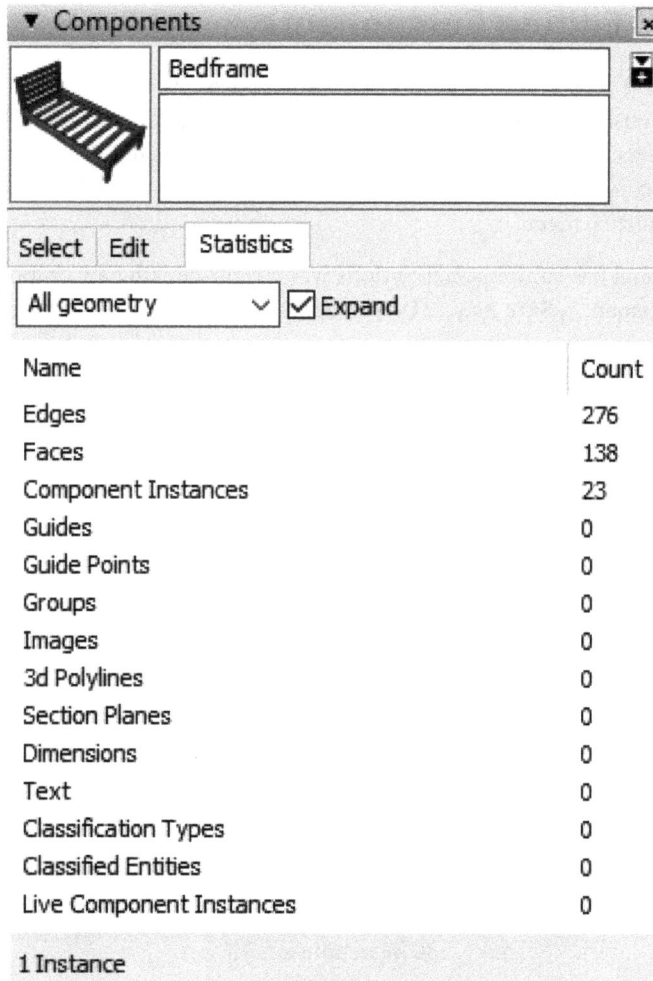

Figure 11.33 – Bedframe Component Statistics Tab with Expand Toggled On

Similar to the **Outliner** panel, the **Statistics** tab can show a quick tally of the nested Components at a glance. Switching the drop-down option from **All geometry** to **Components** will show the names of all nested Components and their total counts.

The **Statistics** tab can be used to quickly understand what is going on in a Component, especially if that Component was created by another user or downloaded from 3D Warehouse. Often, these Components need to be updated to work well in our models, so let's now take a look at updating and editing Components.

Updating and Editing Components

Once Components have been created, they are not static models! We change, update, and swap out Components continuously as we work through our SketchUp models. In *Chapter 3, Modeling with Groups and Components*, we looked at how to activate a Component and Edit the Geometry using the Edit tools. This section will focus on the advanced editing options in the Components panel and in the right-click contextual menu.

The **Components** panel has some special options when right-clicking a Component. These are **Properties**, **Delete**, **Reload…**, **Save As…**, **3D Warehouse**, **Select Instances**, and **Replace Selected**:

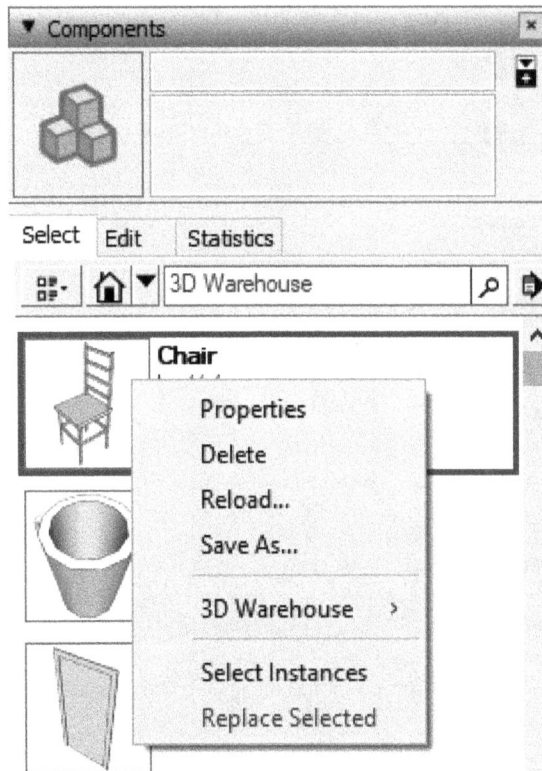

Figure 11.34 – Components Panel Right-Click Menu

Additionally, when a Component is right-clicked in the Drawing Area, there are additional options that are not found when clicking on Groups or Geometry, including **Make Unique**, **Change Axes**, **Scale Definition**, and **Dynamic Components**. We will discuss these options in the *Component Instances* section of this chapter.

Entity Info

Erase   Del

Hide

Lock

Edit Component

Make Unique

Explode

Select   >

Unglue

Reload...

Save As...

3D Warehouse   >

Change Axes

Reset Scale

Reset Skew

Scale Definition

Intersect Faces   >

Outer Shell

Zoom Selection

V-Ray Object ID   >

Camera Focus Tool

V-Ray UV Tools   >

Dynamic Components   >

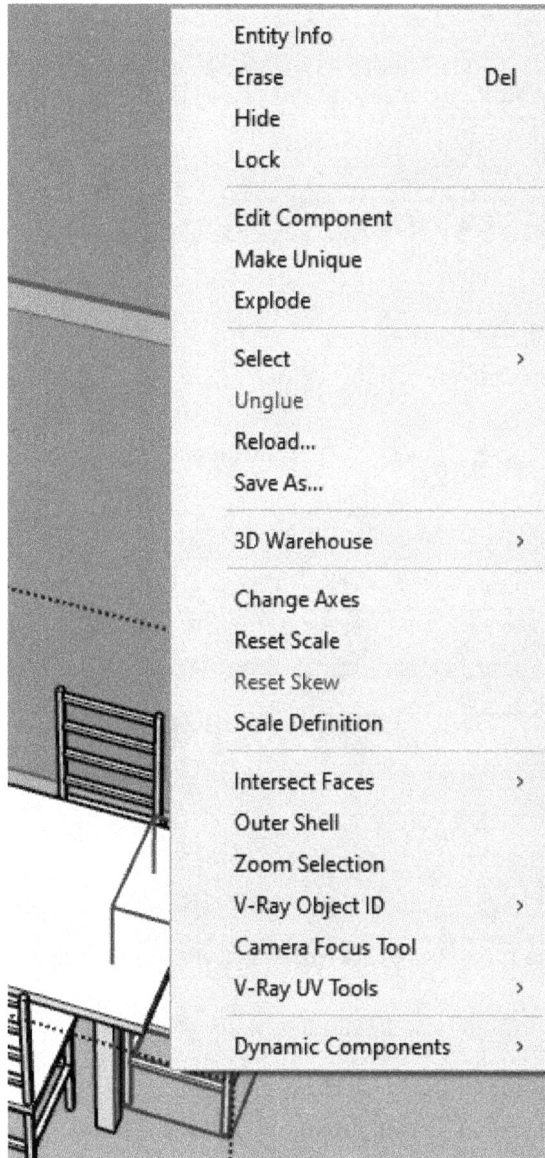

Figure 11.35 – Group Right-Click Menu

These options are what make Components essential to working in SketchUp Pro, as they allow for flexibility and additional control while working on your SketchUp models.

The **Properties** option opens the **Edit** tab in the **Components** panel. The **Edit** tab contains the same options as the **Alignment** section of the **Create Component**.

> **Note**
>
> The Preview section of the Components panel and the Entity Info panel contain the remaining options found in the Create Component. All three panel locations can be used to update an existing Component.

Figure 11.36 – Right-Click Properties (Left), Edit Tab (Right)

As a reminder, the **Edit** tab will only edit the currently selected Component in the Components panel **Select** pane, not in the Drawing Area. If there is no Component selected, it will need to first be picked in the **Select** pane before it can be edited in the **Edit** tab. The right-click **Edit Component** option in the Drawing Area activates the selected Component Instance and does not activate the **Edit** tab.

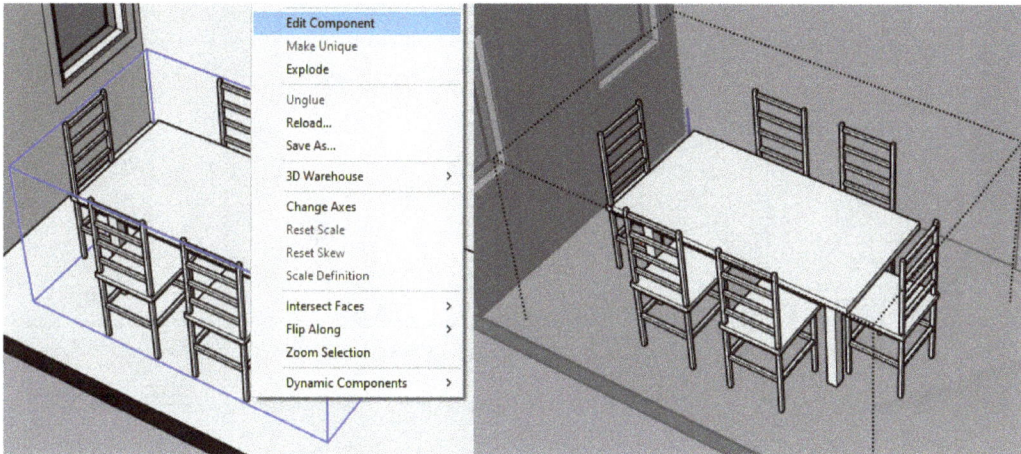

Figure 11.37 – Right-Click Edit Component in Drawing Area (Left), Component Activated (Right)

The **Edit** tab also contains the file path for a loaded Component if it was imported or downloaded from 3D Warehouse. In this example, the Component was installed with SketchUp Pro, so the file path shows the C: Drive path for the SketchUp Pro installation.

Figure 11.38 – File Path for Selected Component

Clicking the Open Folder button to the right of the file path will open the folder location in Windows Explorer.

If a Component was created in the current model, the Component has not been saved to a specific location. This will show in the file path as **Internal Component**:

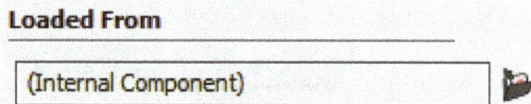

Figure 11.39 – Local Component File Path

If you would like an internal Component to be saved to a location on your computer, you can use the right-click **Save As…** option:

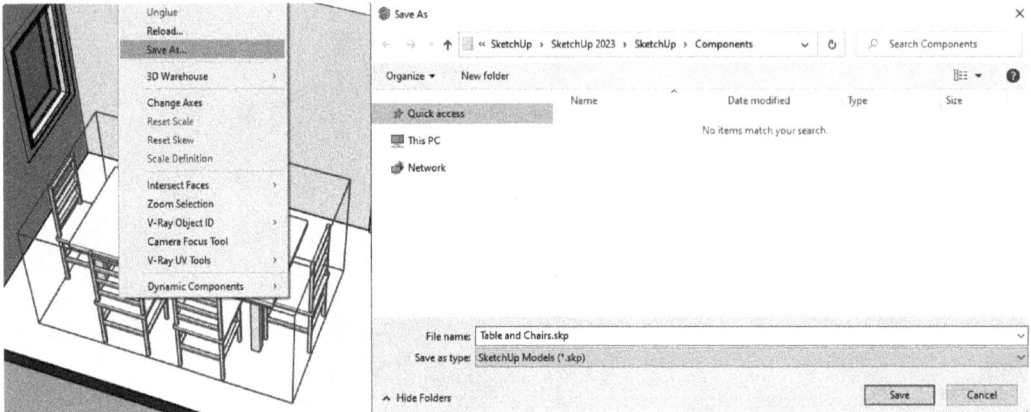

Figure 11.40 – Right-Click | Save As... (Left), Save As Dialog Box (Right)

> **Note**
>
> The **Save As…** option in the Drawing Area right-click menu and the Components panel right-click menu will perform the same workflow. Just make sure you are not accidentally selecting a unique Component in the Drawing Area!

The **Delete** option is fairly straightforward, as it will delete the Component from the model, including all Instances in the Drawing Area. If there are Component Instances in the Drawing Area, SketchUp will show a warning dialog box:

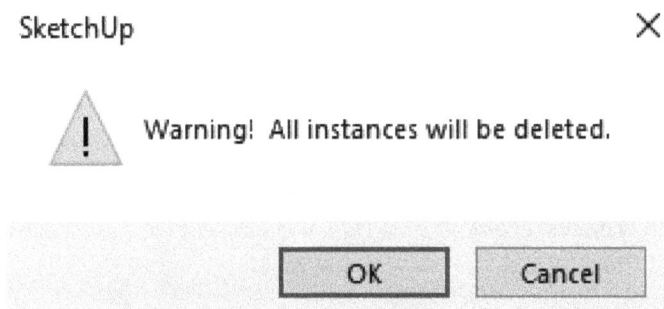

Figure 11.41 – Warning for Deleting Component Instances

Typically, deleting Component Instances is not the best workflow for working with Components, especially if those Components are meant to be replaced with something else. In this next example, the Green Tree Components in the model should be replaced with the Red Tree Component:

Figure 11.42 – Tree Components in a SketchUp Model

Right-clicking and selecting **Delete** would remove the trees entirely, but we want to reuse the tree locations. In this case, we first want to select all of the Green Tree Component Instances in the model so that we can see all of their locations. The Green Tree Component can be right-clicked and **Select Instances** clicked:

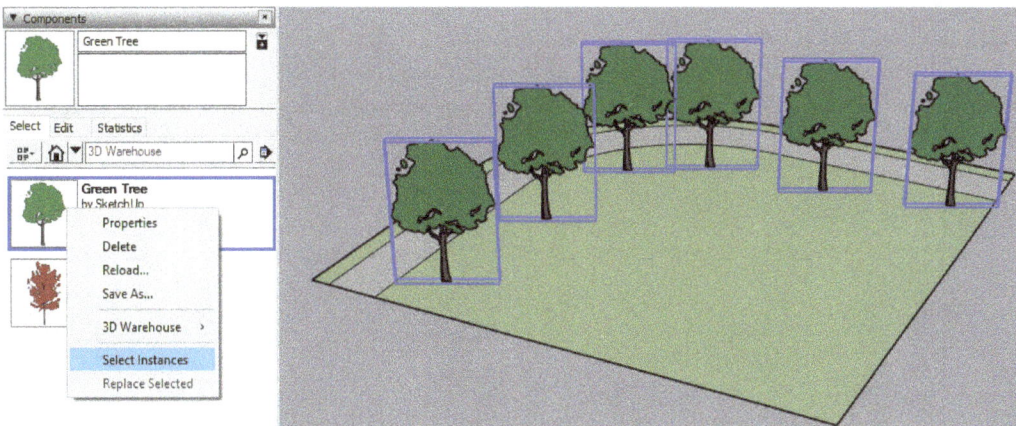

Figure 11.43 – Right-Click | Select Instances (Left), All Component Instances Selected (Right)

Now, the Red Tree Component can be selected and right-clicked in the **Components** panel, and **Replace Selected** can then be clicked. This will replace the selected Components with the Red Tree Component. This will use the same coordinates and rotation settings for the Components, so it is important to know where the Component Axes are located for each Component!

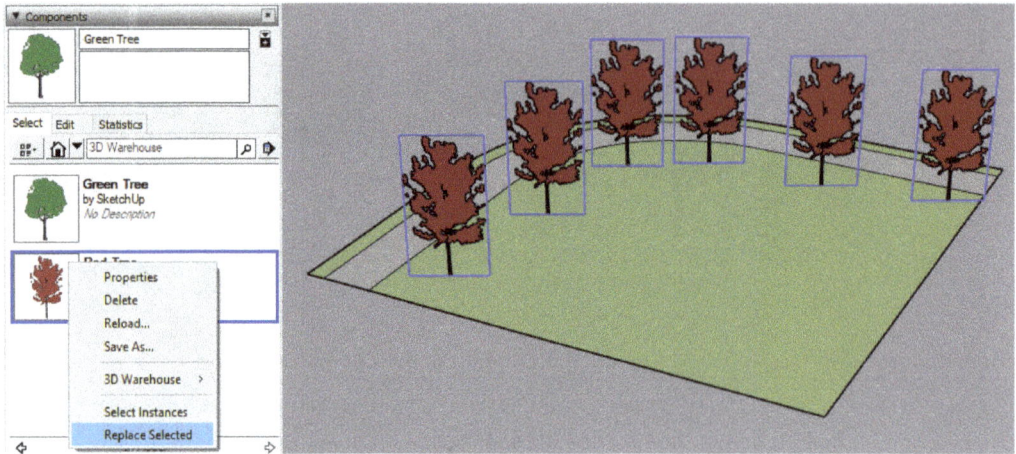

Figure 11.44 – Right-Click | Replace Selected (Left), Selected Components Are Replaced (Right)

> **Note**
>
> Individual Components or any selection set of Components can also be replaced with this workflow. Once the selection set is ready, simply find the new Component in the Components panel and choose **Replace Selected**.

The **Reload…** and **3D Warehouse** options relate to the original location of the loaded Component. The **Reload…** option will allow the Component to be reloaded from a different file path. In many professional workflows, teams of modelers will be sharing updated Components, and the updated Components should be loaded into the working model. This workflow is identical to clicking the Open Folder button in the Components Panel **Edit** tab.

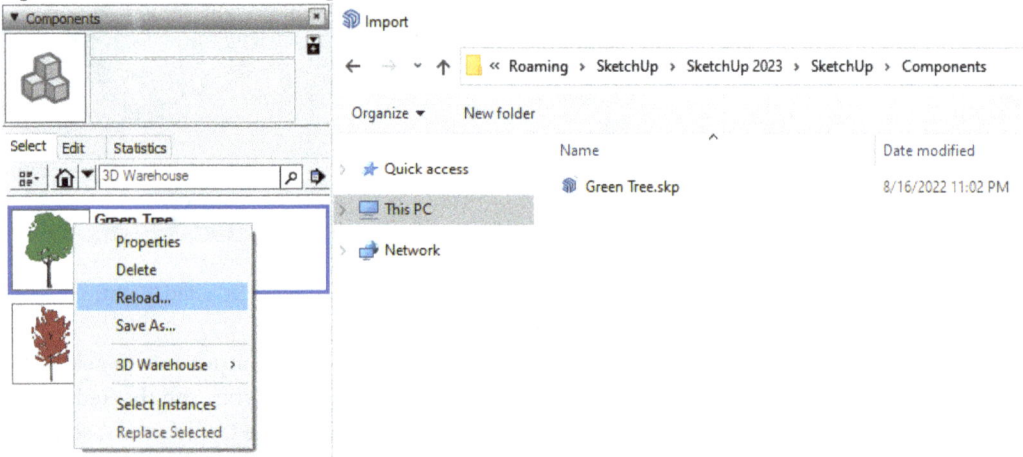

Figure 11.45 – Right-Click | Reload… (Left), File Explorer Window Opened (Right)

> **Note**
> This could also be a workflow to completely swap out a Component in your Model – simply choose a different Component file after clicking the **Reload…** option!

The **3D Warehouse** option has three flyout options, **Share Component**, **Reload**, and **View Details**. This **Reload** option is the same as the local **Reload…** option, except it will open a 3D Warehouse. The **Share Component** option allows you to upload the selected Component to 3D Warehouse, and **View Details** will show the details of the Component if it was originally downloaded from 3D Warehouse. We will discuss these options in more detail in *Chapter 12, Import and Export, 3D Warehouse, and Extensions*.

The options in the Drawing Area right-click menu are considered to be Instance options. These options allow for special workflows when working with single instances in SketchUp Pro. Let's take a look at them in the next section!

Component Instances

We have already discussed the difference between Groups and Components. Groups are single instances of Geometry that can be copied but are not linked to other copies of the same Group. One Group can be updated or changed and the other Groups will not be impacted at all. Component Instances are different from Group instances because Component changes update across all Instances. In this next example, we can see this in action. There are four desk Groups and four chair Components:

Figure 11.46 – Four Desk Groups and Four Chair Components

When a change is made to one of the desk Groups, nothing changes in the other Groups. The Groups are not linked together once the initial copy has been made:

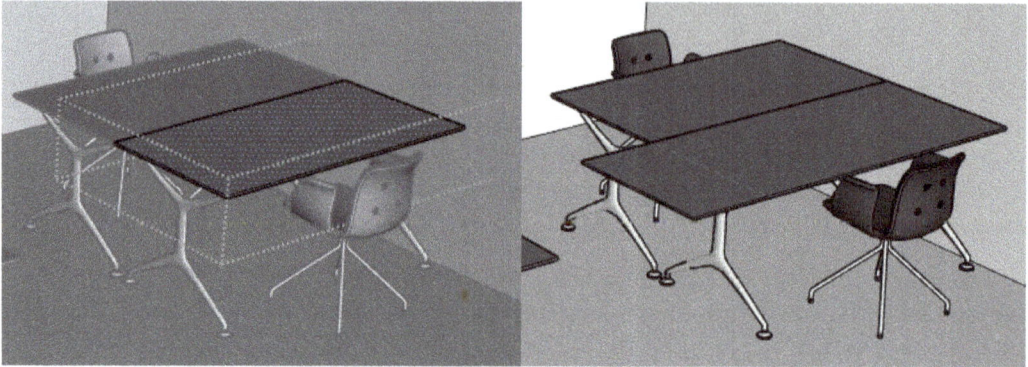

Figure 11.47 – Desk Is Edited (Left), Other Desk Groups Not Updated (Right)

But when one of the chair Components is edited, the other chairs in the model also update to reflect those changes. This is because the four chair Objects are Component Instances of the same Component and are linked to internal changes:

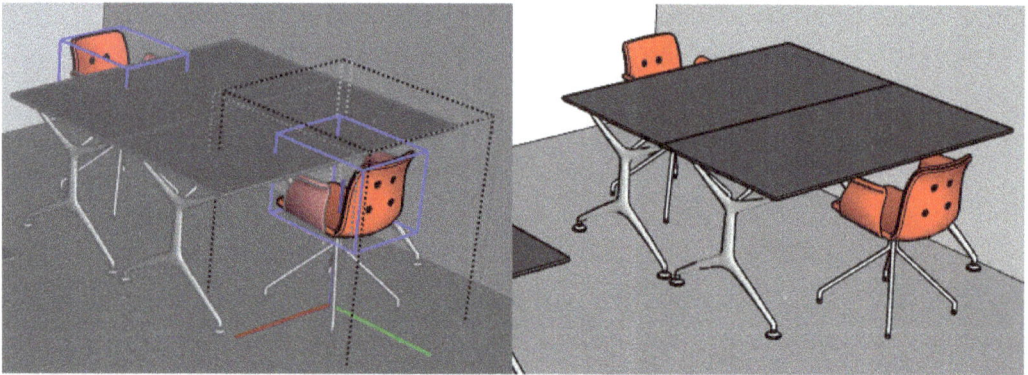

Figure 11.48 – Chair Is Edited (Left), Other Chair Component Instances Are Updated (Right)

Component Instances will change based on changes made when the Component is active, meaning the contained Geometry or nested Objects are changed. External edits made to individual Component Instances will not update the other Component Instances, including Move, Rotate, or Scale. But if these edit workflows are done to all of the Geometry within the active Component, then this will update all Component Instances in the model. In this next example, we can see that in action. We have the same four chair Components, and we will Rotate and Scale one of the chair Components:

Figure 11.49 – Four Chair Components (Left), Chair Rotated and Scaled (Right)

In this case, because these were external edits, the other Component Instances are unchanged. If the changes should be made to all Components, we can activate one of the Component Instances, then make the change to the Geometry:

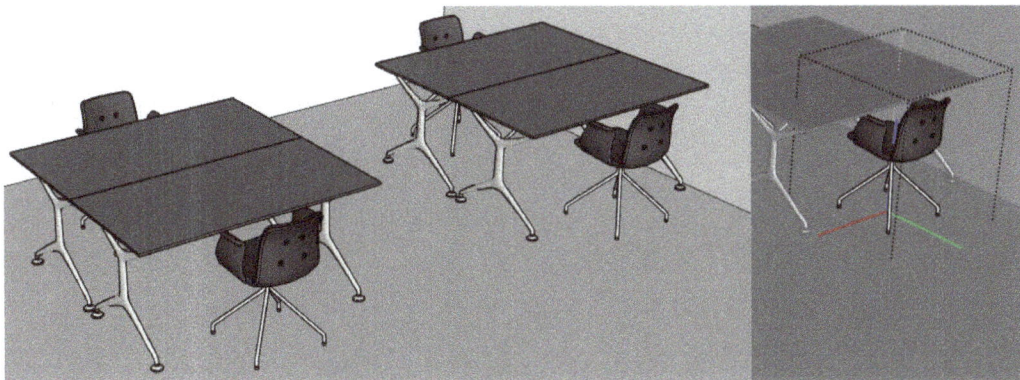

Figure 11.50 – Four Chair Components (Left), Chair Component Activated (Right)

Once the Component is active, the Geometry and Objects can be selected and rotated, then scaled:

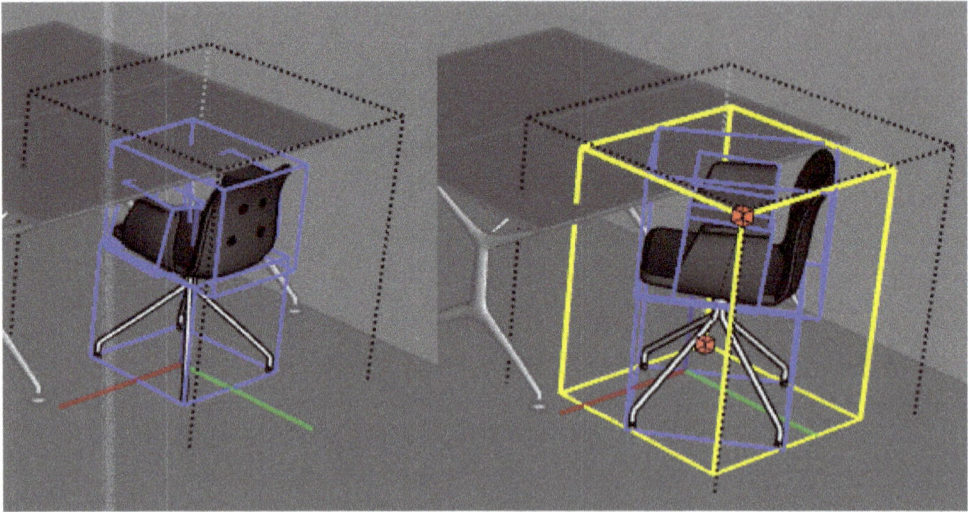

Figure 11.51 – Geometry and Objects Selected (Left), Geometry and Objects Rotated and Scaled

We can see that when a Component is active, the other Component Instances will preview the changes being made, even previewing the active selection set. Once the Component is closed, we can see all of the Components have been updated accordingly.

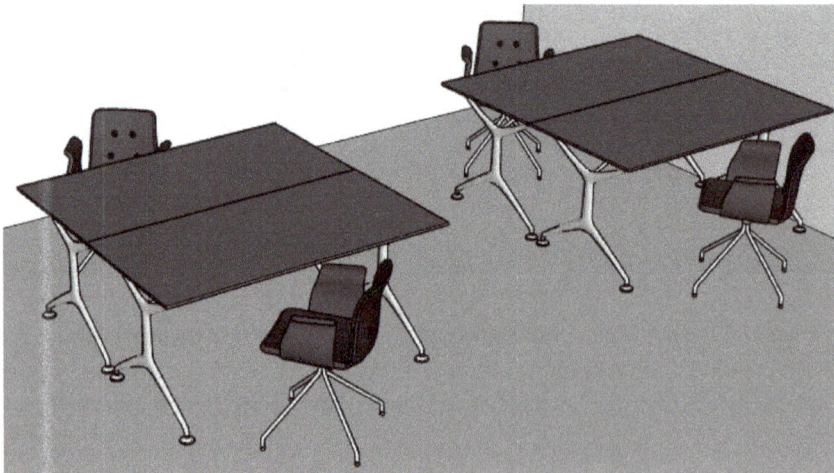

Figure 11.52 – All Chair Components Updated

In many cases, editing the contained Geometry is not the best workflow for working with Components. There are Component Instance options that make working with Components very powerful, and we were introduced to some of these options in the right-click contextual menu earlier in this chapter. Let's take a look at some of these options now!

Instance Options

The Components panel does not have a Copy option, which you may have noticed when we reviewed the right-click options earlier in the chapter. Components can be copied in the Drawing Area, but this workflow will only create more Component Instances. In order to make a copy of the Component Definition, the **Make Unique** option can be used. In this example, we have four chair Components:

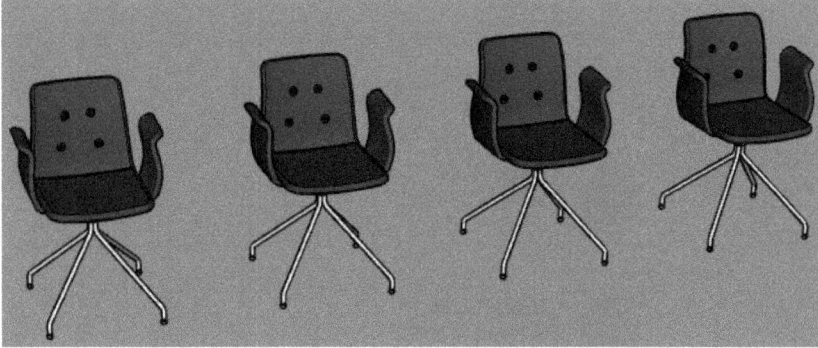

Figure 11.53 – Four Chair Components

If we wanted to swap one of the chair Components, we could use the **Replace Selected** option, but we do not have a Component in our library that we want to replace it with. Instead, we need to model this Component using the existing chair Component as a starting point. We could explode one of the Components and make a brand-new Component, but this would cause us to lose the data that has been saved in the original Component. In this case, we can click on one of the chairs, right-click, and choose **Make Unique**:

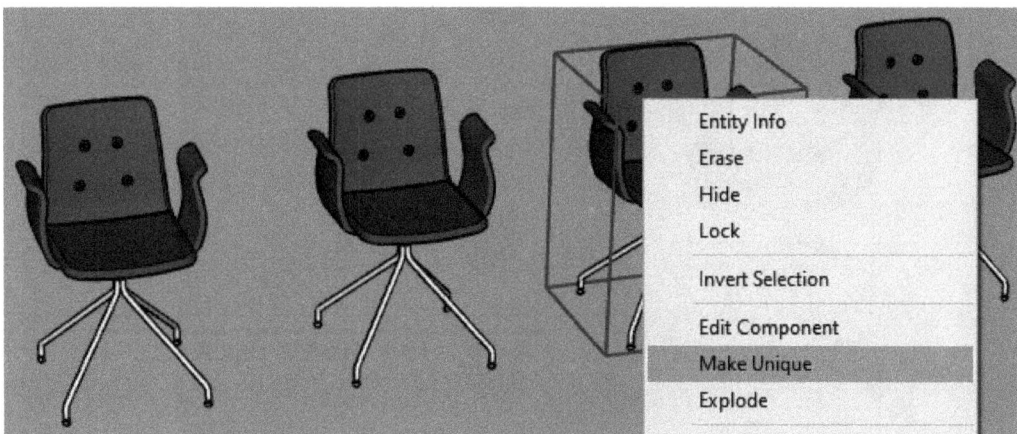

Figure 11.54 – Chair Component Selected | Right-Click | Make Unique

This will add a new Component to the Components Panel and add a number to the Component Definition Name. Typically, this will be **#1**:

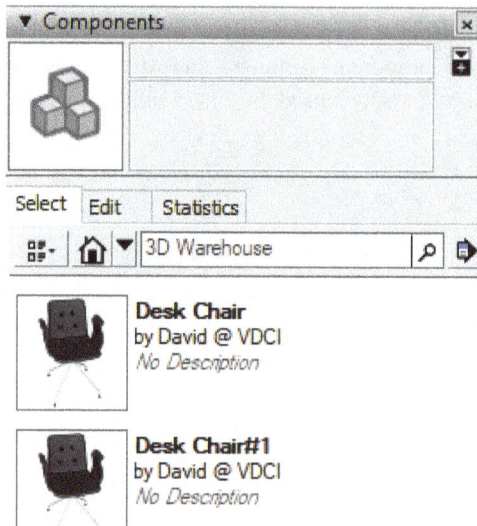

Figure 11.55 – Chair Component Added to Components Panel

This new Component is a copy of the Component Definition, and now represents the single chair that we selected earlier. We can then make edits to this Component, and we will have a new, unique Component in the model:

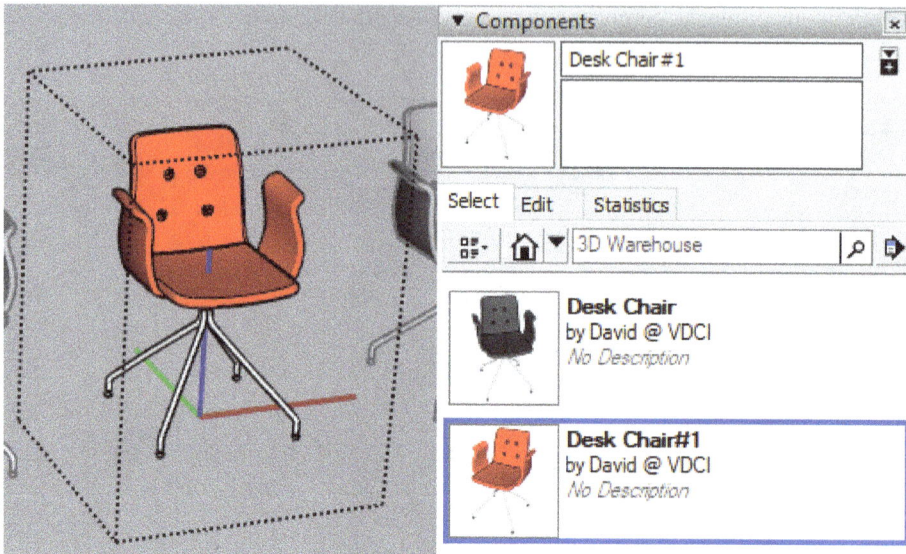

Figure 11.56 – Unique Component Edited (Left), Component
Preview Updated in Components Panel (Right)

In some cases, we want to adjust a Component without making a Component active and working on the contained Geometry. In some cases, this is because there is hidden Geometry inside of the Component, or it could be for other reasons! We have already discussed why setting appropriate Component Axes is important in the *Alignment Options* section of this chapter. If we find a Component from 3D Warehouse that has poor Axes for alignment, it may be appropriate to use Change Axes. In this next example, we can see that we have found a floor lamp Component in 3D Warehouse:

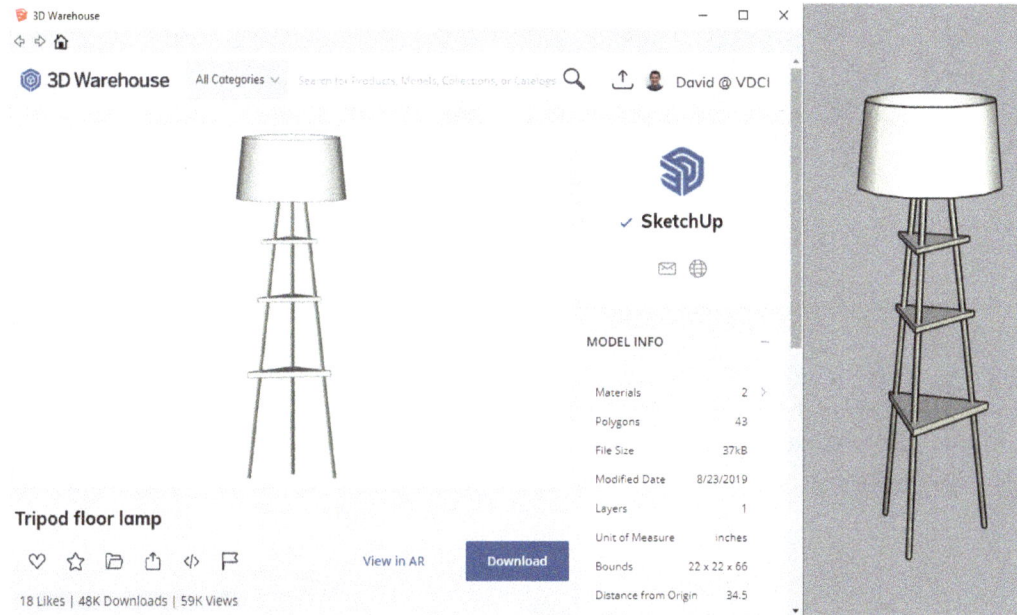

Figure 11.57 – Component in 3D Warehouse (Left), Component Loaded in Model (Right)

When the Component is loaded into the model, we can see that it appears to be lying flat on the floor.

The Component should clearly be standing up. We can Rotate the Component Instance using the Rotate tool. However, when the next Component is brought in from the Components panel, we can see that it has the original orientation as when it was downloaded.

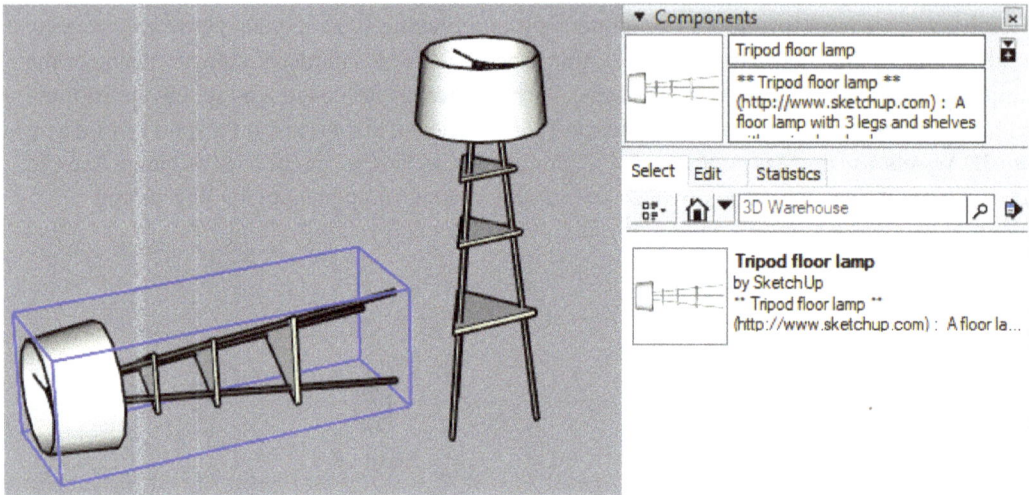

Figure 11.58 – Component Rotated (Left), Component Instance In Components Panel (Right)

In this case, before placing any Components in the model, we should use the Change Axes tool. This tool works similarly to the Axes tool on the Construction toolbar, but it will only change the Axes for the selected Component. In this example, we will select the lamp Component, right-click, and choose **Change Axes**:

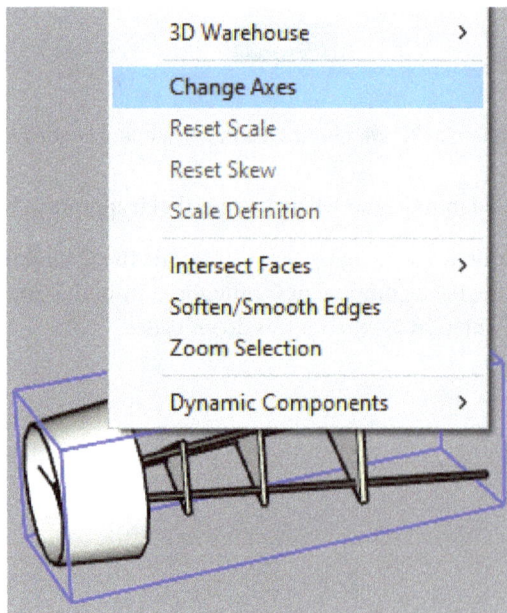

Figure 11.59 – Lamp Component Selected and Right-Click | Change Axes

The Axes gizmo will be shown, and a new Origin point can be selected for the Component. Once the Origin point is set, then a new Positive Red direction can be selected, followed by a new Positive Green direction. Once the directions are chosen, the Component Axes will be updated.

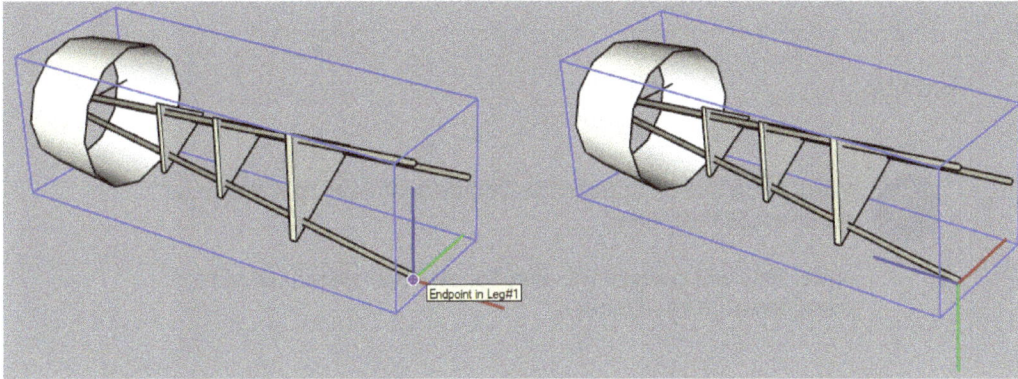

Figure 11.60 – New Origin Point Set (Left), New Component Axes Set (Right)

This will not move the Components in the model, but new Components can be placed from the Components panel in the correct orientation.

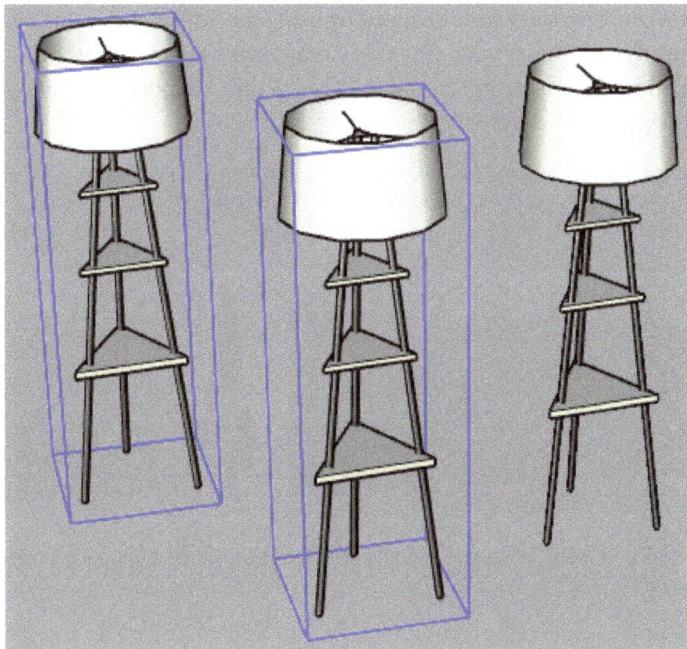

Figure 11.61 – New Component Instances Placed from Components Panel

> **Note**
>
> The Axes tool can be used while in an active Component to change the Component Axes. If the Sketch Axes are changed in an Active Component, SketchUp will ask whether the Component Axes should be updated when the Component is closed.

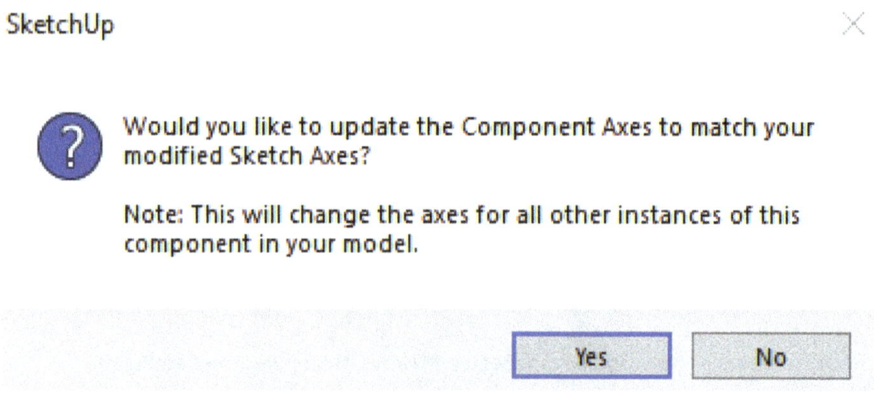

SketchUp   ✕

(?) Would you like to update the Component Axes to match your
     modified Sketch Axes?

     Note: This will change the axes for all other instances of this
     component in your model.

                                              [Yes]   [No]

Figure 11.62 – SketchUp Warning for Changing Component Axes

The **Scale Definition** option allows for the scale of the Component Definition to be changed. Individual Component Instances can have their scale changed by using the Scale tool. In this next example, we have a wall and a row of tree Components. The tree Components are meant to show over the top of the wall, but the loaded tree Component is too short. One of the Component Instances can be scaled to show the correct height for all the trees:

Figure 11.63 – Wall and Tree Components (Left), One Tree Component Instance Scaled (Right)

If all of the Component Instances are meant to be at this scale, the tree Component Instance with the correct scale can be selected, right-clicked, and **Scale Definition** can be clicked. Similar to the **Change Axes** option, any other existing Component Instances will not be updated. However, any new Components placed from the Components panel will have the newly set scale:

Figure 11.64 – Existing Component Instances Unchanged (Left),
Newly Placed Components Updated (Right)

The final unique options in the Component right-click menu have to do with Dynamic Components. Dynamic Components are complex, and we will discuss them in the next section!

Dynamic and Live Components

Dynamic Components and Live Components are special types of Components that rely on Extensions to give them additional features. Dynamic Components were released in 2008 in SketchUp Version 7. Since then, users have been able to create and modify their own Dynamic Components, and use the Dynamic Components created by the SketchUp team. The Components panel even has a Dynamic Components Training Collection in the **Select** pane to help users get started with Dynamic Components.

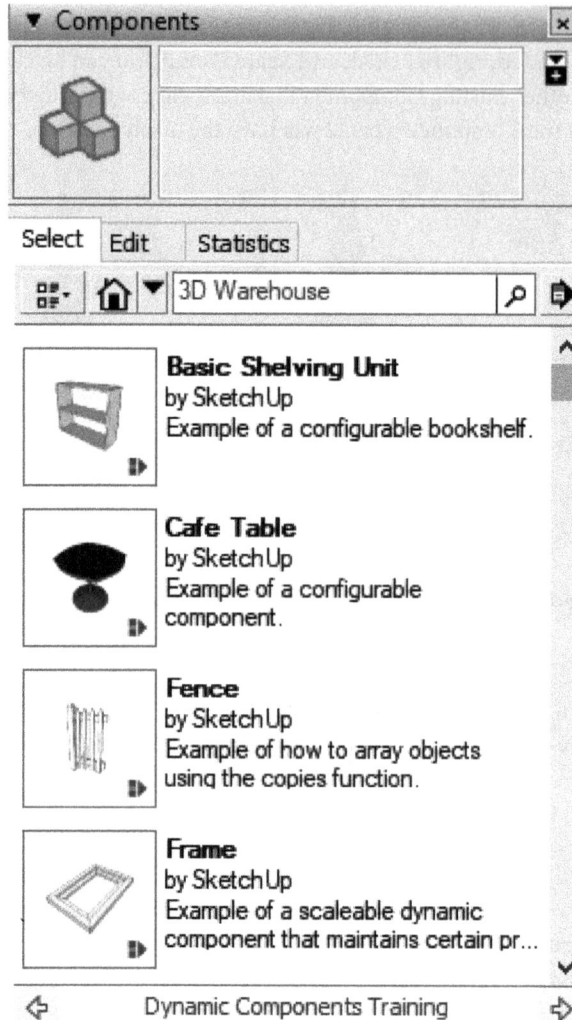

Figure 11.65 – Dynamic Components Training Collection in Components Panel

Live Components are the more recent addition, as they were introduced in 2020. Live Components are web-generated Components that can be loaded into a SketchUp model but rely on a connection with the SketchUp cloud server to be updated. Live Components cannot be created by users but can be edited from a collection in 3D Warehouse, and then used in local SketchUp models.

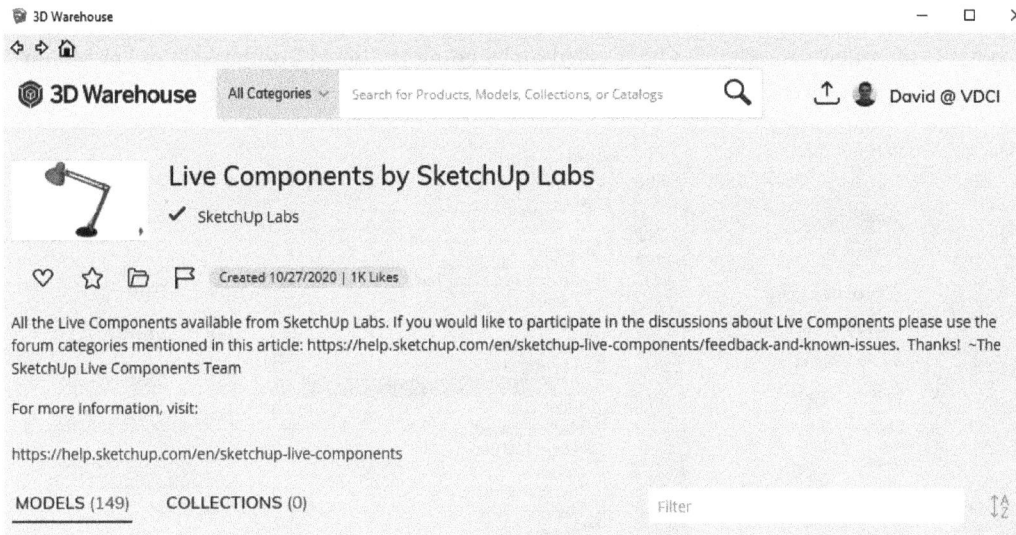

Figure 11.66 – Live Components Collections in 3D Warehouse

Dynamic Components may be phased out of SketchUp in the future, but currently still have support in SketchUp Pro. Dynamic Components are still the more widely used configurable Component, as they provide more customization and unique designs.

Dynamic Components and Live Components are complex Objects, and we will cover them briefly in this chapter. For a deep-dive into Dynamic Components and Live Components, check out the online help articles provided by SketchUp – there is a lot of great (and complex) information about both Component types!

```
https://help.sketchup.com/en/sketchup/making-dynamic-component
```

```
https://help.sketchup.com/en/sketchup-live-components
```

Dynamic Components

Dynamic Components are configurable Components that can be created, edited, and used in SketchUp Pro models. The Dynamic Components Extension is one of the default Extensions included when SketchUp Pro is installed.

Dynamic Components can have parameters applied to them, which can add a dynamic – or changing – attribute to the Component. These changing attributes include Materials/colors, dimensions, rotation, OnClick behaviors, number of nested Components, and more!

Dynamic Components are controlled by the tools on the Dynamic Components toolbar. The Dynamic Components toolbar can be toggled on by going to **View | Toolbars…** and checking **Dynamic Components**:

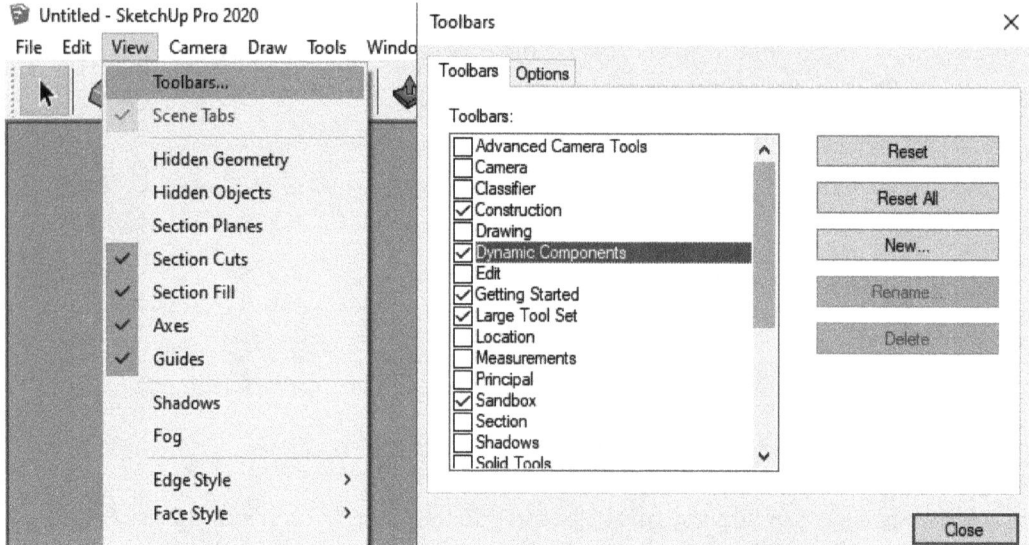

Figure 11.67 – View | Toolbars (Left) Dynamic Components Checkbox (Right)

The Dynamic Components toolbar contains three tools: Interact, Component Options, and Component Attributes. The Component Attributes button opens the Component Attributes dialog box, which can be used to create and edit parametric and dynamic attributes. The Component Attributes button looks like two stacked, green squares with a green triangle on the right. This icon will also be at the bottom right-hand corner of any Dynamic Component in the Components panel:

Figure 11.68 – Component Attributes Button (Left), Dynamic
Component Preview in Components Panel (Right)

Component attributes can be applied to any Component in the model, although it can be very challenging to set parametric values using formulas. It is suggested that only advanced modelers attempt to create complex Dynamic Components. Dynamic Components include many customizable attributes in the **Component Attributes** dialog box, and these attributes can be applied to any Component or nested Component.

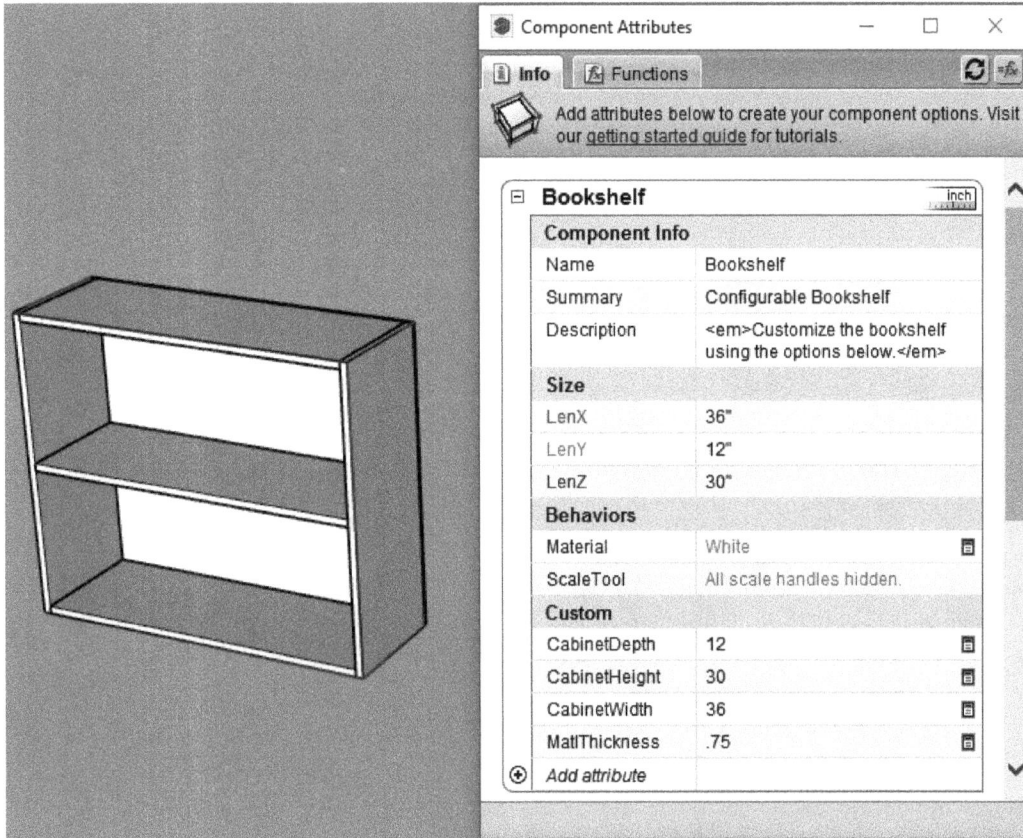

Figure 11.69 – Component Attributes Options

In this example, we will look at one of the simpler Dynamic Component attributes, adding Color options. In this example, we have a Bookshelf Component that we want to show with Color options. Instead of loading multiple Components into the model, we can use Dynamic Components to add a range of selectable Colors. At the moment, the Bookshelf Component has nested Groups, but no nested Components. With the Component Attributes Dialog Box open, we can select the Bookshelf in the Drawing Area. This will update the dialog box to show that there are currently no assigned attributes for this Component.

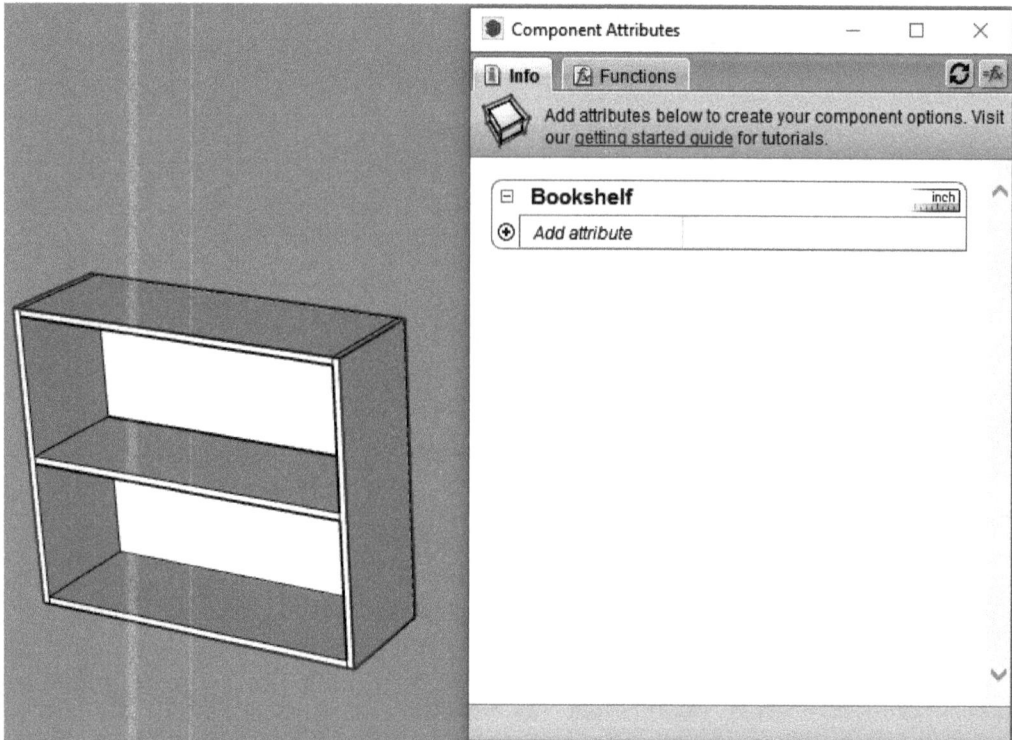

Figure 11.70 – Component Attributes Dialog Box Showing No Assigned Attributes

The **Add Attribute** text can be clicked to add a new attribute, and in this case, we will select **Material** under the **Behaviors** section:

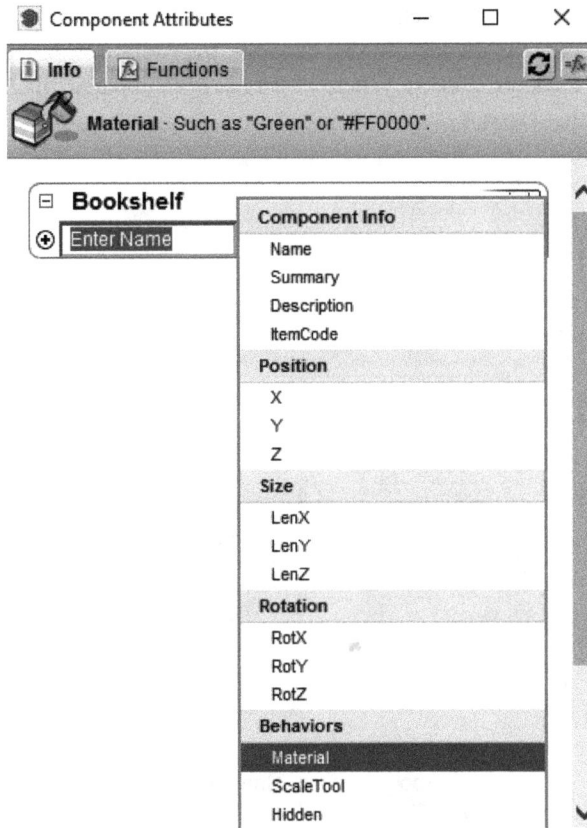

Figure 11.71 – Add Attribute Is Clicked and Material Is Selected

This will add the Materials attribute, and it can now be configured using the details arrow at the right side of the text box. The configuration window allows the user to change **Display rule**, which defines how users can interact with the attribute. In this case, we will set the option to **Users can select from a list**. This will generate a blank list.

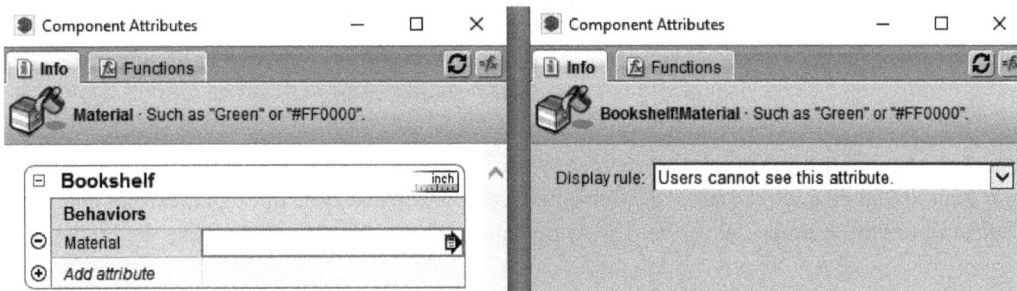

Figure 11.72 – Details Arrow Clicked (Left), Configure Window with Display Rule Option Set (Right)

Material options can now be added. SketchUp can support some named Colors, such as Red or Blue, and will also support hexadecimal codes. In this example, the Color options are set to **Tan** (**Tan**), **Grey** (**#AAAAAA**), and **Pink** (**#CC0055**). When the list has been updated, the **Apply** button can be clicked at the bottom of the screen:

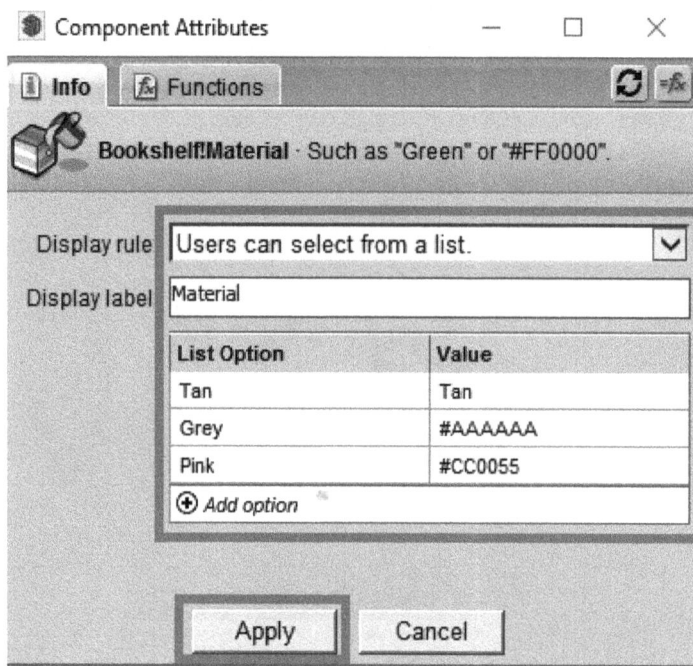

Figure 11.73 – Color Options Set and Apply Button Clicked

Now that an attribute has been set for this Component, we want to be able to interact with the Dynamic Component. This is achieved using the **Component Options** dialog box. The **Component Options** button looks like a dialog box with a red title bar:

Figure 11.74 – Component Options Button

The **Components Options** dialog box will show the attribute options for the selected Dynamic Component. In this case, we can select the **Bookshelf** Component, and cycle through the Color options in the **Component Options** dialog box. The **Apply** button should be clicked to update the model:

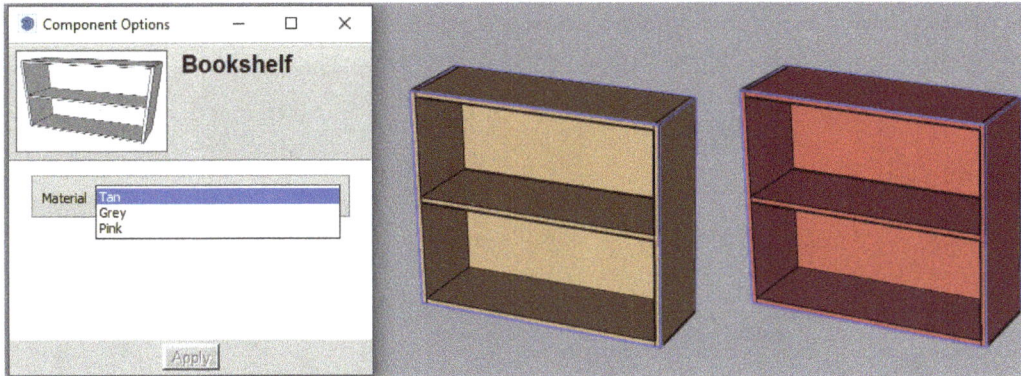

Figure 11.75 – Component Options for Bookshelf Component
(Left), Tan Option (Middle), Pink Option (Right)

The final button is the Interact with Dynamic Components tool, and it is represented by a hand with a pointing finger.

Figure 11.76 – Interact Button

This tool will Interact with dynamic components that have OnClick attributes defined, which are done with formulas. These formulas are very complex and are too detailed for the scope of this book. In the next figure, we can see an example of a SketchUp Dynamic Component and the defined Component Attributes.

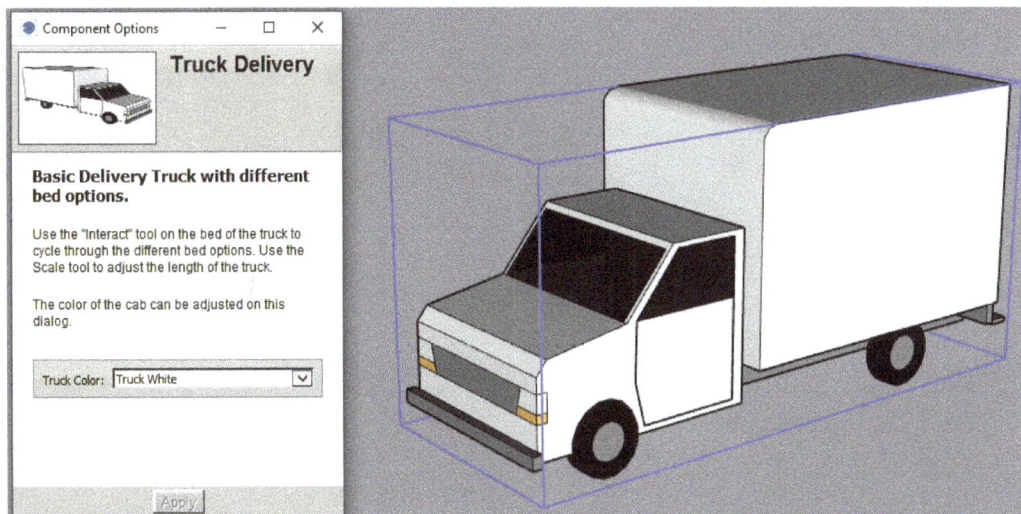

Figure 11.77 – SketchUp Dynamic Component and Defined Component Attributes

Using the Interact tool on this Dynamic Component will cycle through the Component Options every time the Component is clicked. This can be done for any of the Component Attributes, not just Materials!

Figure 11.78 – Component Clicked with Interact (Left), Component Options Are Cycled (Right)

Dynamic Components offer exciting ways to create and interact with SketchUp models. They are not for everyone, as they can be difficult to create and edit, but they might find a place in your models!

Live Components

Live Components are configurable Objects in SketchUp Pro, but they cannot be created from scratch by SketchUp users. Instead, Live Components can be found in Live Components Collections in 3D Warehouse and can be configured to specific settings. This could also be achieved using certain Dynamic Components, but this relies on the local computing power. Live Components utilize a cloud-based platform to configure the models before they are loaded into a SketchUp model, and the cloud-based platform is also used to update the Live Components.

Live Components do not have a toolbar in SketchUp Pro. Instead, they can be found by using advanced search tools in 3D Warehouse. The SketchUp Labs account creates all of the available Live Components, so searching for the SketchUp Labs account will allow you to find all of the Live Component Collections.

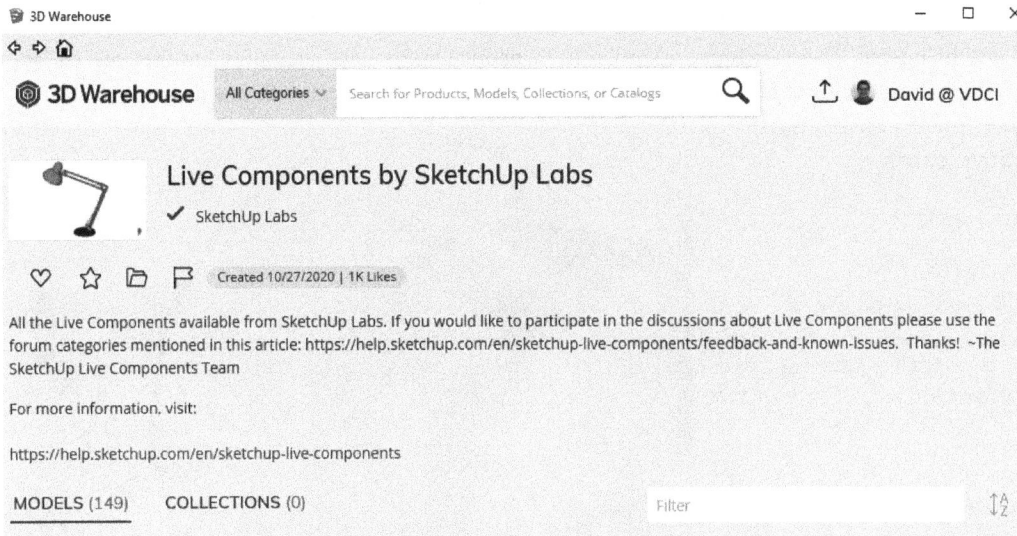

Figure 11.79 – SketchUp Labs Account in the 3D Warehouse

> **Note**
> Clicking the **COLLECTIONS** option and toggling on **Verified Collections** will help ensure that you only see the Live Component Collections while searching in 3D Warehouse!

Live Components can be used even if working on an older version of SketchUp Pro. When accessing 3D Warehouse in an older version of SketchUp Pro, Live Components can only be configured in an internet browser and then downloaded as static Components. In this example, we can see a Live Component that prompts the user to open the page in an internet browser:

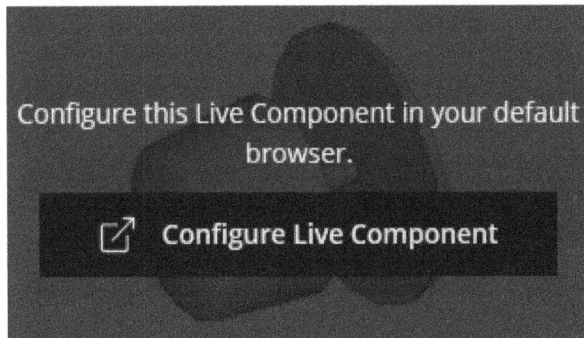

Figure 11.80 – Configure in Browser Button

Once in the internet browser 3D Warehouse, the Component can be Configured in the browser using the Configure button. This will be the default setting when loading this page.

The Configure workflow allows the user to change specific settings that have been predetermined by the SketchUp team. These settings are similar to Dynamic Component attributes, but they can be far more complex in terms of how the parts of the SketchUp model interact with one another.

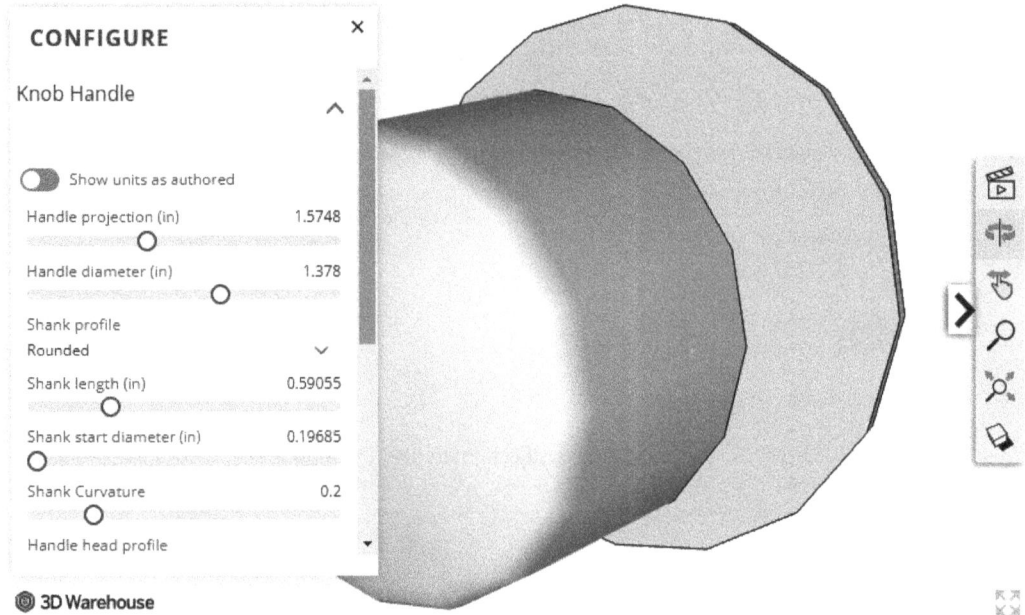

Figure 11.81 – Configure Workflow in 3D Warehouse

Once the Component has been configured, it can be downloaded and saved to a local folder on the user's computer. This will convert the file into a static Component and it can no longer be configured.

> **Note**
> This workflow is only for older versions of SketchUp Pro. Now, let's take a look at how to configure Live Components!

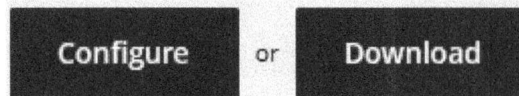

Figure 11.82 – Configure Button (Left), Download Button (Right)

On newer versions of SketchUp Pro, Live Components can be configured directly in the SketchUp model. 3D Warehouse will prompt the user with **Download & Configure**, which is combined into a single workflow.

Figure 11.83 – Download & Configure Button

Once the Live Component has been downloaded and loaded directly into the open SketchUp model, the Component can be configured by right-clicking and choosing **Configure Live Component**:

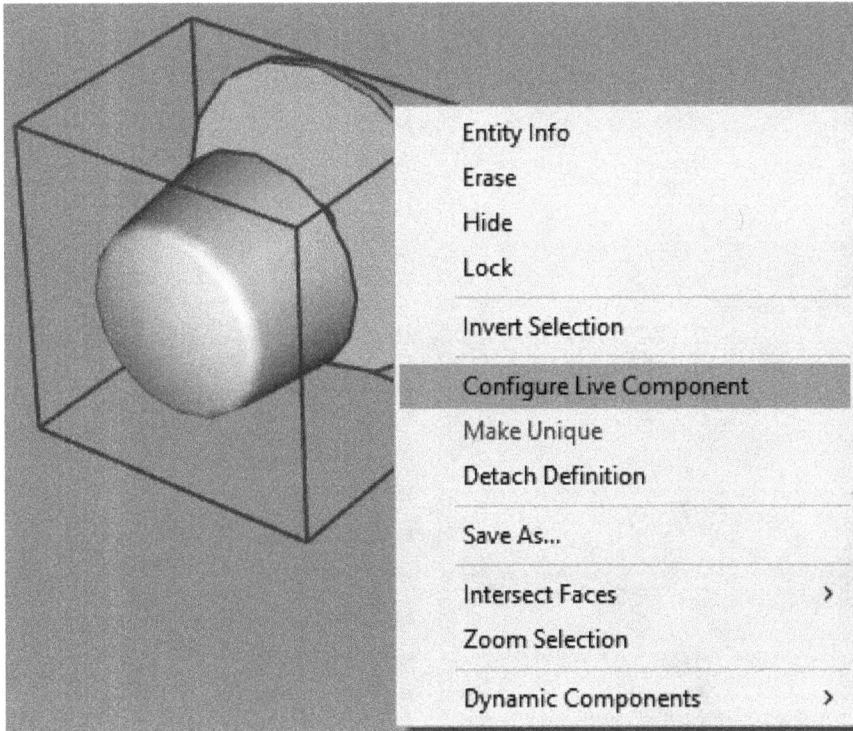

Figure 11.84 – Component Loaded into Model and Right-Clicked | Configure Live Component (Right)

This will open the **Configure Live Component** dialog box, which is linked to 3D Warehouse through an internet connection. This window will allow for all of the Component configuration options to be edited in real time, and the Component will update automatically in real time.

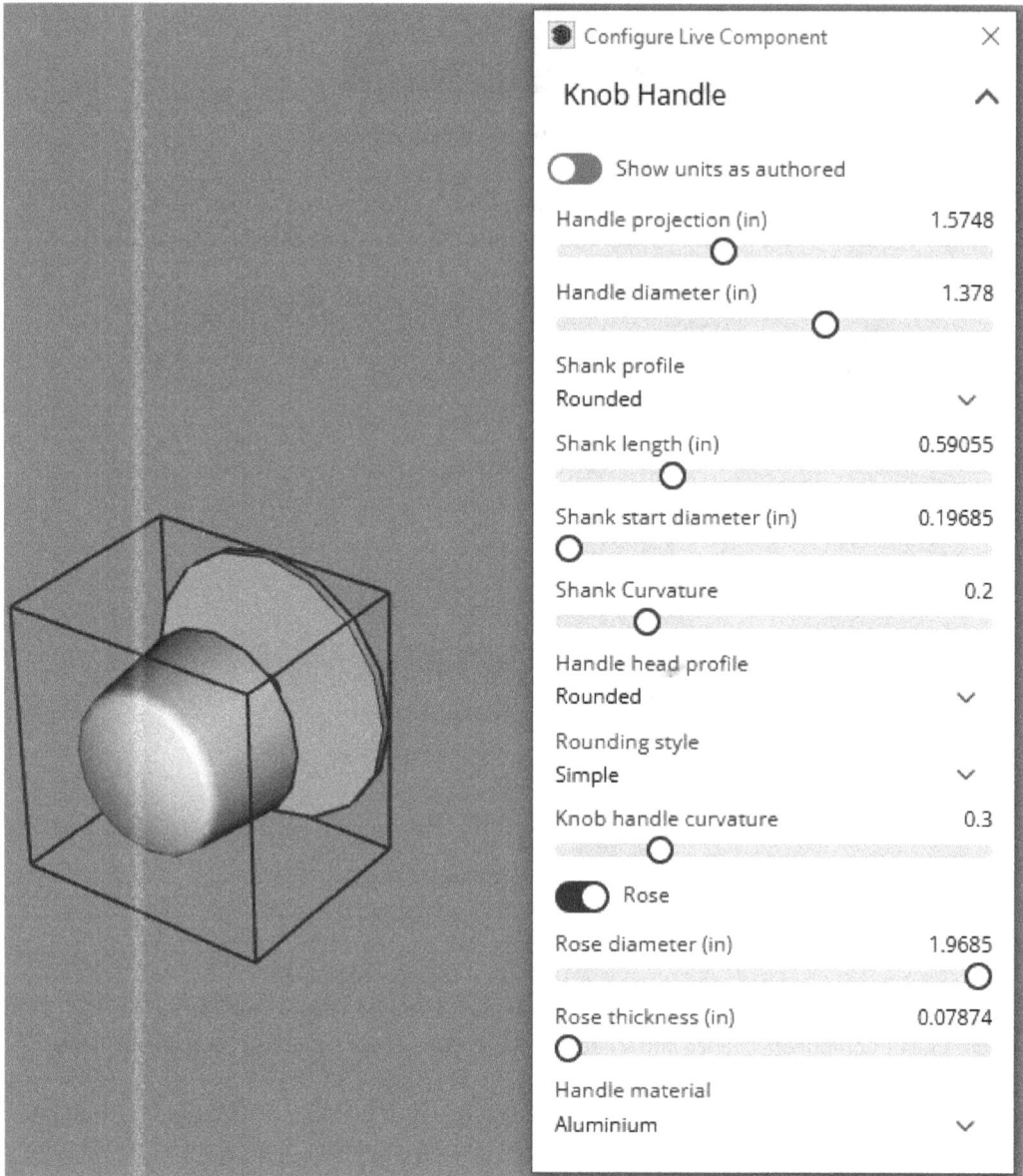

Figure 11.85 – Configuration Options in SketchUp Pro

> **Note**
> Double-clicking a Live Component will also open the Configure dialog box while working in SketchUp!

There is one other specific Live Component option, **Detach Definition**. This option will remove the link to the Live Component from the cloud-based platform, allowing the Component to behave like a normal Component. Once detached, the Component can be activated and edited, or even exploded to use parts of the Component.

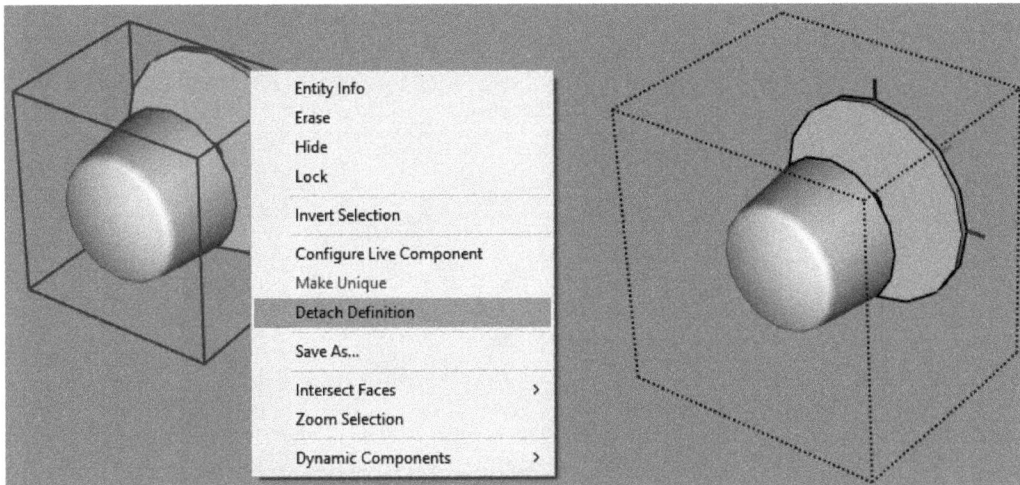

Figure 11.86 – Right-Click | Detach Definition (Left), Component Activated (Right)

Live Components can be a lot of fun, and SketchUp Labs has already released over 150 configurable Live Components in 3D Warehouse. There may be a useful Component for your modeling project – check it out in 3D Warehouse!

Summary

In this chapter, we discussed the Components panel and how to use Components in our models, as well as how to Edit and Update Components. We also covered an overview of Dynamic Components and Live Components. Components are an essential next step to take your SketchUp models to the next level, and Dynamic Components and Live Components can be used to make modeling easier or create interactive and engaging models. The more you work with Components, the more comfortable you will become with them, so I encourage you to play around with Components in your SketchUp models!

In the next chapter, we will look at Importing and Exporting, the Extension Warehouse, and 3D Warehouse.

12
Import, Export, 3D Warehouse, and Extensions

In this chapter, we will talk about the different types of files that SketchUp Pro can import and how these files can be used to create Materials, create Geometry, or simply be placed in the model. Additionally, we will discuss the different options for exporting information from our SketchUp models, including images, 2D and 3D files, and animations.

We will also talk about 3D Warehouse for downloading and uploading files and Sending our SketchUp files to LayOut. We will finish by looking at Extensions in the Extension Warehouse and Extension Manager.

The following topics are covered in this chapter:

- Import Options

  - Importing 3D Files

  - Importing 2D Files

  - Importing Image Files

- Export Options

  - Exporting 3D Files

  - Exporting 2D Files

  - Exporting Image Files

  - Section Slices

  - Animations

  - Send to LayOut

- 3D Warehouse
- Extensions

  - Extension Warehouse
  - Extension Manager
  - Overlays

Import Options

SketchUp Pro is able to import a variety of file types that can help with modeling, visualization, and providing contextual elements to your SketchUp models. In this section, we will look at three primary categories of files that can be imported: 3D files, 2D files, and image files.

> **Note**
>
> Extensions can be added to SketchUp Pro that make it possible to import additional file types. We will discuss extensions later in this chapter. This section will focus on the files that SketchUp Pro can natively import.

SketchUp will allow 3D files, 2D files, and image files to be imported through the same workflow. Files can be imported by selecting the **File** dropdown in the Main Menu and choosing **Import…**:

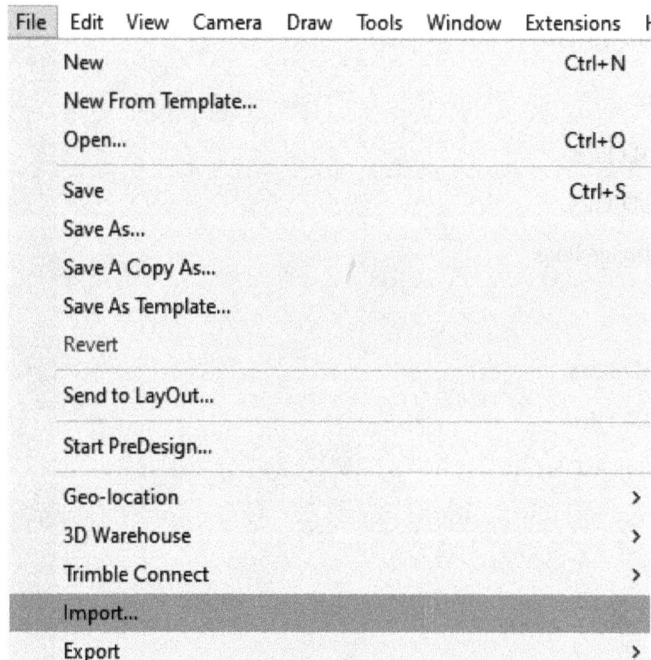

File	Edit	View	Camera	Draw	Tools	Window	Extensions	
New							Ctrl+N	
New From Template…								
Open…							Ctrl+O	
Save							Ctrl+S	
Save As…								
Save A Copy As…								
Save As Template…								
Revert								
Send to LayOut…								
Start PreDesign…								
Geo-location							>	
3D Warehouse							>	
Trimble Connect							>	
Import…								
Export							>	

Figure 12.1 – File –> Import… in the Main Menu

The **Import** Dialog Box will open and prompt the user to select a file for import. By default, the **All Supported Types** option is selected in the file type filter located at the bottom right of the Dialog Box.

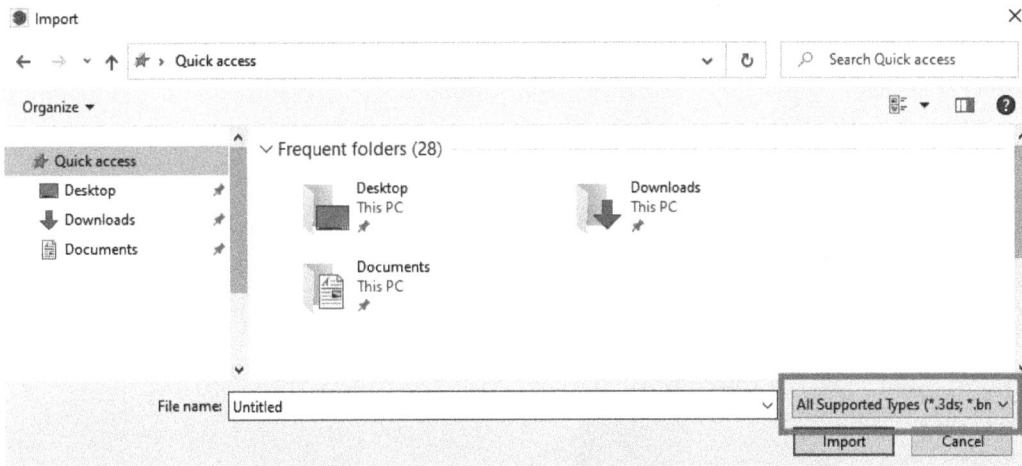

Figure 12.2 – Import Dialog Box with File Type Filter Highlighted

All Supported Types will list all file type extensions that can be imported, including image, 2D, and 3D files. This filter can be changed by clicking on the filter and choosing a different option from the dropdown. Each file type can be selected individually, and **All Supported Image Types** can also be filtered as a group.

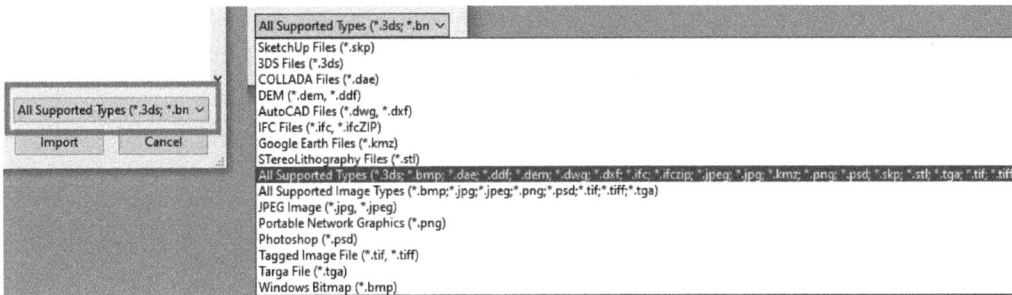

Figure 12.3 – File Type Filter Clicked (Left) and All Supported Image Types Selected (Right)

> **Note**
>
> If you think you are in the right place to find a file but can't find it, make sure to check the filter! The filter will often need to be updated in order to see the file you are trying to import into SketchUp Pro.

Depending on the type of file selected, an **Options** Button will appear, or additional options for the import will be added to the bottom of the Dialog Box. We will discuss these additional settings in the *3D Files*, *2D Files*, and *Image Files* sections of this chapter.

Once the file has been selected, and all options have been set, the file can be imported by clicking on the **Import** Button.

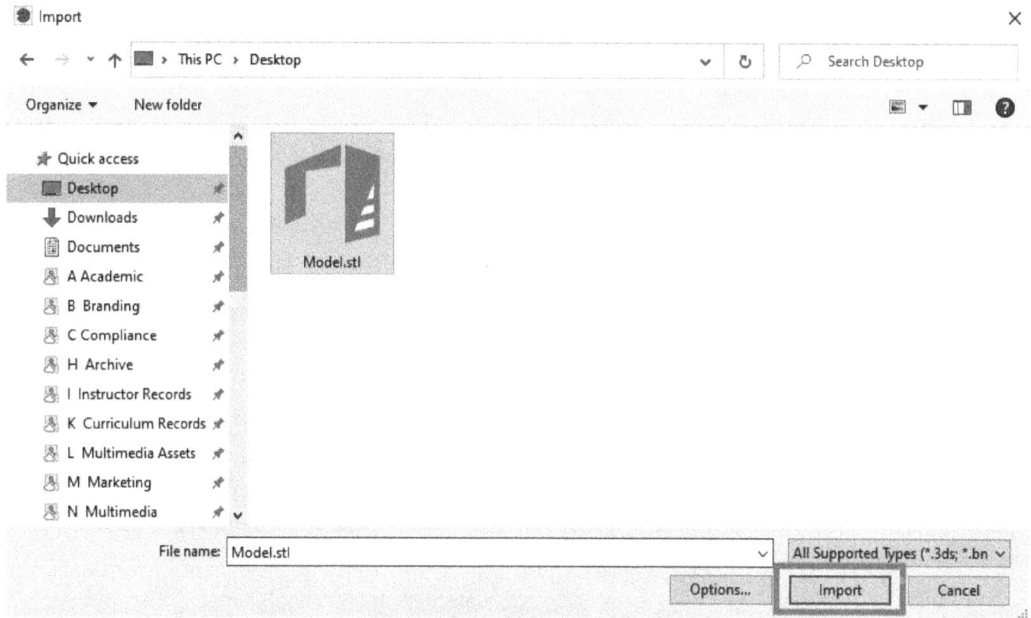

Figure 12.4 – File Selected and Import Button Clicked

In the next three sections, we will explore some more specific details about importing different types of files, and how we can use those files in a SketchUp model.

Importing 3D Files

SketchUp is a 3D modeling software that can import and understand many 3D file types. All 3D files will be added to the SketchUp model as Components when imported. The 3D file types that SketchUp supports are as follows:

File Extension	File Description
SKP	SketchUp Files
3DS	3DS Files
DAE	COLLADA Files
DEM	Digital Elevation Model
DDF	Diamond Directive File
IFC	IFC Files (BIM)
IFCZIP	IFC Zip Files (BIM)
KMZ	Google Earth Files
RVT	Revit (BIM)
STL	STereoLithography Files
TRB	TrimBIM (BIM)

Table 12.1 – Supported 3D Files for Import

SketchUp Files (.SKP)

SketchUp models can be imported into other SketchUp models as Components. All SketchUp model Components saved in Component Collections are saved locally as SKP files. There are no additional import options for SKP files.

SKP files can typically be found in the 3D Warehouse and on many manufacturers' websites promoting the use of their products. SKP files are the easiest files to use in SketchUp, so many organizations understand the importance of providing them directly! We will discuss accessing SKP files from the 3D Warehouse in the *3D Warehouse* section of this chapter.

3D Studio Files (.3DS)

3DS files were originally created for Autodesk 3DS Max and other Autodesk software titles. 3DS files contain 3D vector mesh data, materials, camera settings, and more information. 3DS files are often more complex than SketchUp models, but they can still be useful to import into SketchUp for various workflows.

In the next example, we can see a 3DS model of a building being imported into SketchUp Pro. When a 3DS file is selected, the **Options…** Button appears at the bottom of the **Import** Dialog Box. 3DS files have specific import options, including the **Merge coplanar faces** option and scaling the incoming model. The model can be scaled using **Model Units**, or a standard unit of measure.

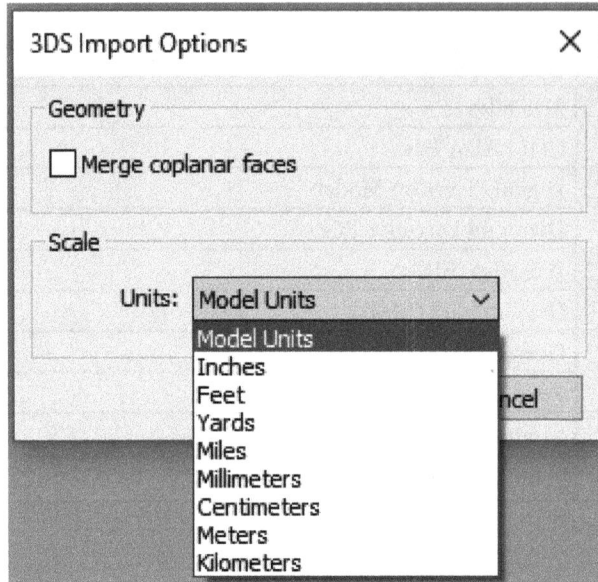

Figure 12.5 – 3DS Import Options Dialog Box with Units Dropdown Open

When the **Import** button is clicked, SketchUp will show the **Import Progress…** Dialog Box while the file is being imported, and the **Import Results** Dialog Box when the import has been completed.

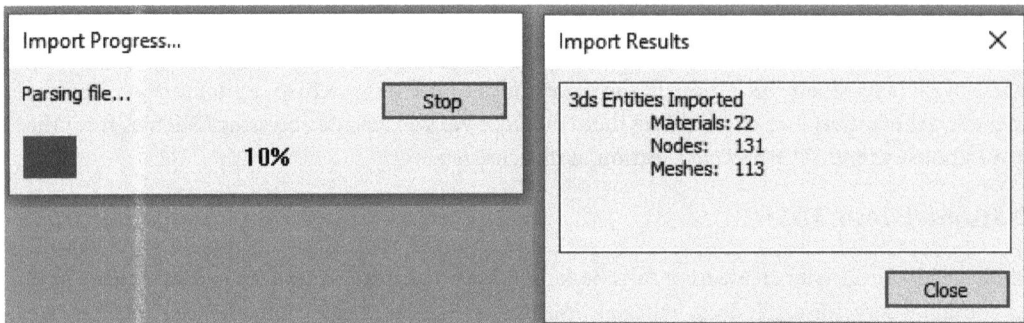

Figure 12.6 – Import Progress Dialog Box (Left) and Import Results Dialog Box (Right)

Once the **Import Results** Dialog Box is closed, the 3DS file can be placed by clicking in the Drawing Area. This will add the file as a new Component to the current SketchUp model. 3DS Max files will often be triangulated and will show all Edges, which may need to be hidden in the SketchUp model.

> **Note**
> The **Soften Edges** Panel can be used to quickly Soften coplanar Faces. This can be seen in the next example, where many triangulated faces are coplanar.

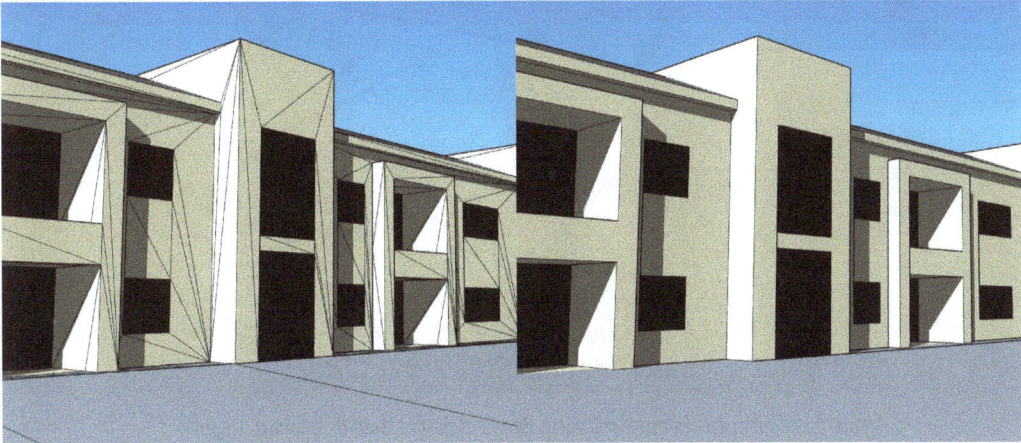

Figure 12.7 – Imported 3DS File with Triangulated Faces (Left) and Coplanar Edges Softened (Right)

Digital Asset Exchange Files (.DAE)

DAE files serve as a general data transfer file type for 3D modeling applications. DAE files are based on the **COLLAborative Design Activity (COLLADA)** XML schema, which allows the file to include geometry, animation, shaders, material, and physics. SketchUp can only read the information that is native to the SketchUp interface, so some data may be ignored when importing DAE files.

In the next example, we can see a 3DS model of a building being imported into SketchUp Pro. When a 3DS file is selected, the **Options…** Button appears at the bottom of the **Import** Dialog Box. DAE files have specific import options, including **Validate Dae file** and **Merge Coplanar Faces**.

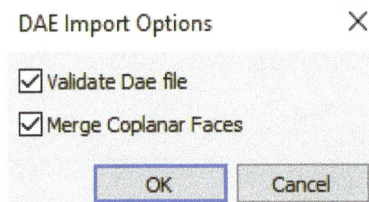

Figure 12.8 – DAE Import Options Dialog Box

When the **Import** button is clicked, SketchUp will show the **Import Progress…** Dialog Box while the file is being imported, but there is no **Import Results** Dialog Box, and the file can be placed by clicking in the Drawing Area. This will add the file as a new Component to the current SketchUp model.

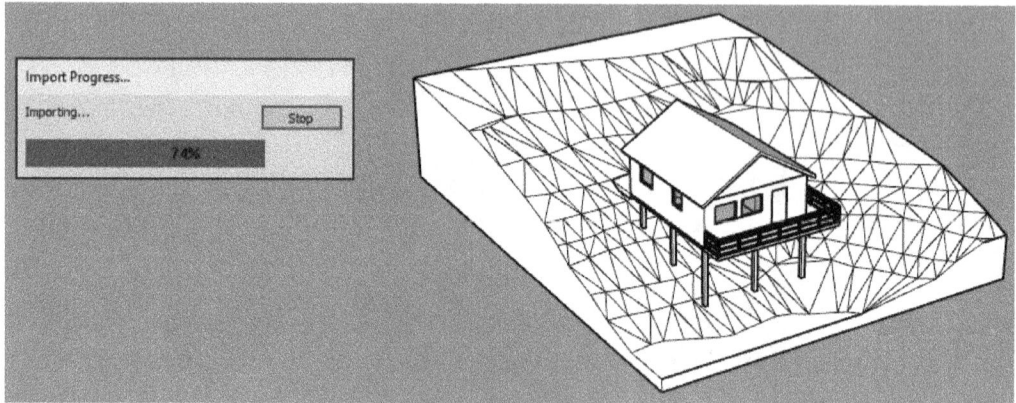

Figure 12.9 – Import Progress Dialog Box (Left) and the File Placed (Right)

DAE files can soften coplanar Faces on import, but non–coplanar Faces may need to be cleaned up when imported.

Digital Elevation (.DEM) and Spatial Data Transfer Standard Format Files (SDTS) (.DDF)

DEM files and SDTS Format files (DDF) are both used to import mesh terrain information. These files are text–based files that contain coordinate information that SketchUp can interpret into 3D mesh points. DEM and DDF files are often created using **Geographic Information System (GIS)** software. The options Dialog Box for DEM and DDF files provides information on the number of points and faces in the model, as well as an option to make a gradient texture color for the mesh.

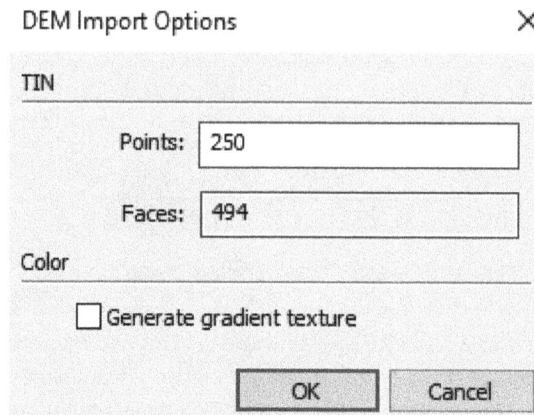

Figure 12.10 – DEM and DDF Import Options Dialog Box

Importing DEM and DDF files into SketchUp is an advanced workflow and will often require additional software platforms to accurately import terrain data. It is recommended that beginner users of SketchUp Pro utilize the **Geo-location** options to bring in site topography into their models.

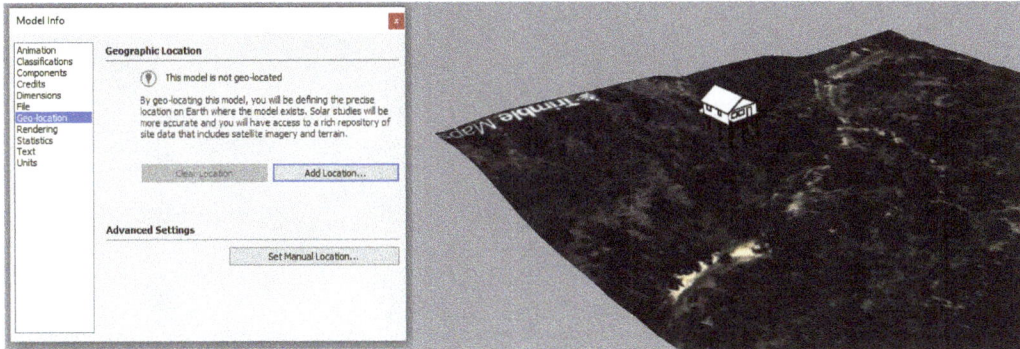

Figure 12.11 – Geolocate Settings (Left) Mesh Data Added to SketchUp (Right)

Autodesk AutoCAD Drawing (.DWG) and Drawing Exchange Files (.DXF)

Autodesk DWG and DXF file types have been used for years for 2D and 3D files, although 2D DWG and DXF files are more commonly in use today. 3D DWG and DXF files were used for many years as the standard for AutoCAD Blocks, which are similar to SketchUp Components.

We will discuss the full Import process for 2D and 3D AutoCAD DWG and DXF files in the *2D Files* section of this chapter.

> **Note**
>
> It is recommended to turn on the **Merge Coplanar Faces** and **Orient Faces Consistently** Import Options when importing 3D DWG and DXF files, the Faces in these models are triangulated and not combined by default. Merging the Faces and Orienting the Faces Consistently will save a lot of time when working with these models!

International Foundation Class Files (.IFC and .IFCZIP)

IFC files are designed to support **Building Information Modeling** (**BIM**). BIM is a general term that can include many file types, including proprietary files from other software. However, IFC files are an internationally agreed-upon classification for how data should be stored in 3D building files. IFC files are platform-neutral and can be read and updated in any BIM software.

In the next example, we can see an IFC model of a chair being imported into SketchUp Pro. IFC files do not have any special Import Options, so the file can just be selected, and the **Import** button can be clicked.

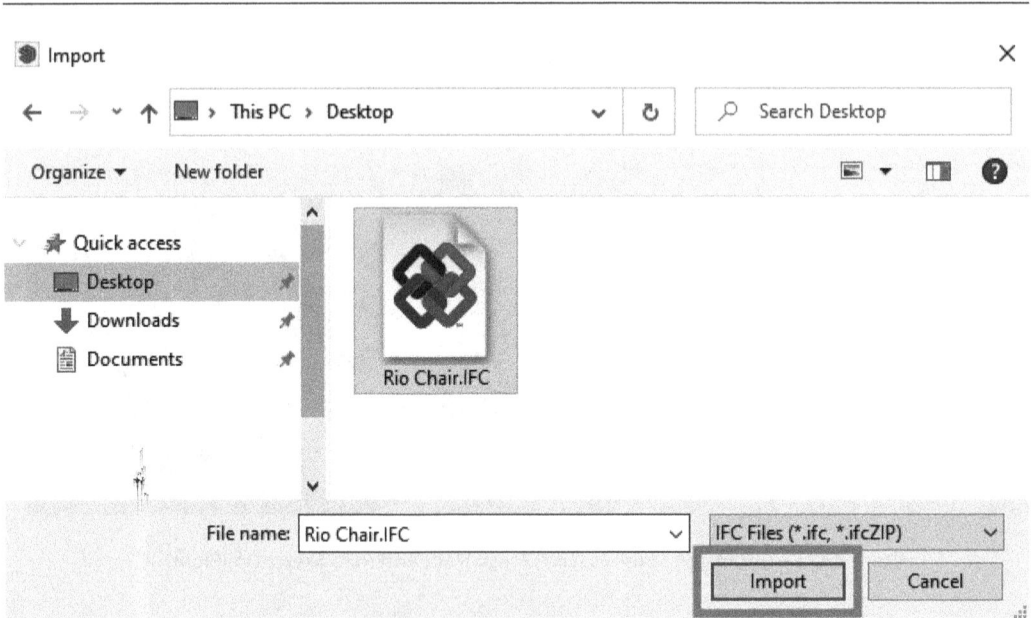

Figure 12.12 – IFC File Selected and Import Clicked

When the **Import** button is clicked, SketchUp will show the **Import Progress...** Dialog Box while the file is being imported, and the **Import Results** Dialog Box when the import has been completed.

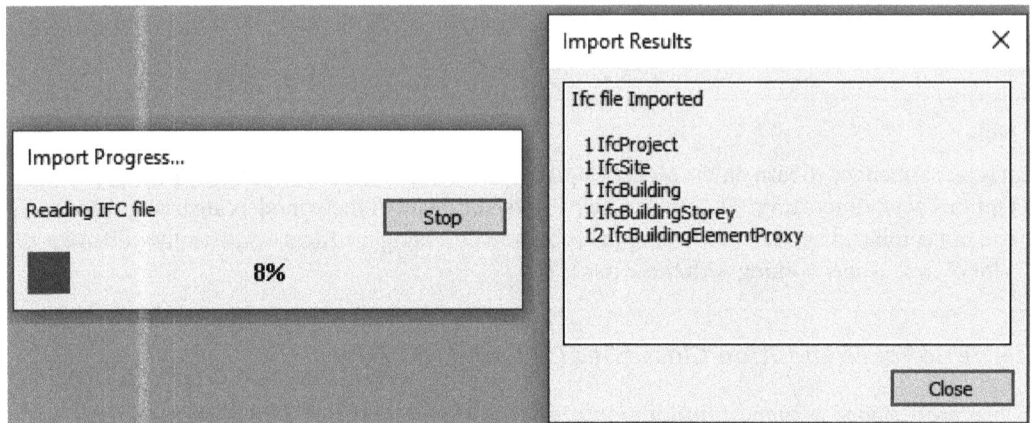

Figure 12.13 – Import Progress... Dialog Box (Left) and Import Results Dialog Box (Right)

Once the **Import Results** Dialog Box is closed, the IFC file will be automatically placed at the Axis Origin. This will add the file as a new Component to the current SketchUp model. Commonly, IFC files seem to be rotated strangely, as different types of modeling software have different settings for which Axis is up.

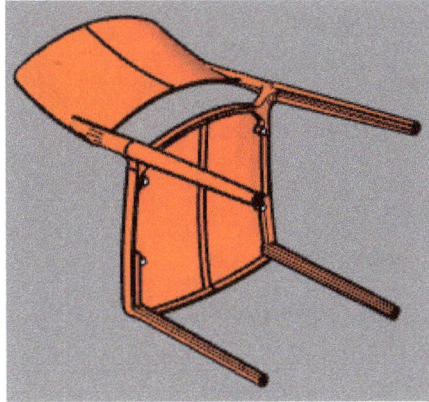

Figure 12.14 – IFC File Imported and Rotated

Google Earth Files (.KMZ)

KMZ files are the compressed version of **Keyhole Markup Language** (**KML**) files. KMZ files are the default file type for spatial data for Google Earth. SketchUp Pro was previously owned by Google, and Google Earth was a major integration into SketchUp Pro during that time. SketchUp Pro now utilizes a different platform for the built–in **Geo–location** and topography import. We discussed this in *Chapter 7, View Options*.

The KMZ files that work in SketchUp have to be created in a very specific way through third–party tools, and it is very unlikely that any SketchUp Pro beginner will ever use KMZ files. There is more information online about KMZ files and SketchUp Pro that describes the specific workflows for importing and using Google Earth data. We will not discuss these workflows in this book.

STereoLithography Files (.STL)

STL is a common 3D file type that was originally created by 3D Systems, a 3D printing and scanning company. STL files may incorrectly be called **Standard Triangle Language** or **Standard Tessellation Language** because STL files are comprised of triangulated Faces and straight Edges. Just like native SketchUp files, STL files do not support curved surfaces or edges, so they work well in SketchUp Pro. STL files are basic 3D files and do not support color, texture, or other attributes.

In the next example, we can see an STL model of a building being imported into SketchUp Pro. When an STL file is selected, the **Options...** Button appears at the bottom of the **Import** Dialog Box. STL files have specific import options, including the **Merge Coplanar Faces** option, changing or preserving the drawing origin and axes, and setting **Scale** options. The model can be scaled using a standard unit of measure.

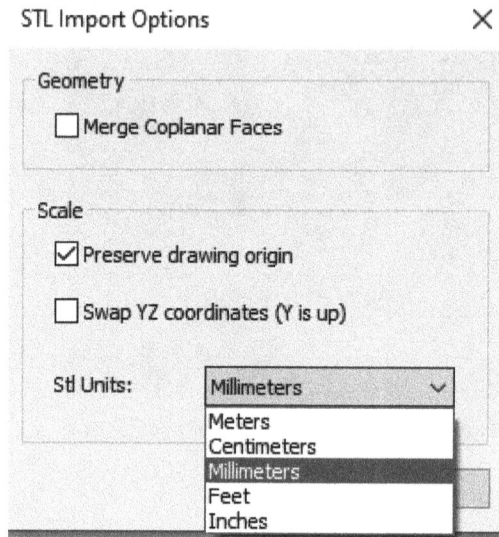

Figure 12.15 – STL Import Options Dialog Box with the Stl Units Dropdown Open

When the **Import** button is clicked, SketchUp will show the **Import Progress** Dialog Box while the file is being imported, which will automatically close when the file is imported. The STL file will be added as a new Component to the current SketchUp model. STL files will be completely triangulated and will show all Edges that may need to be hidden in the SketchUp model. STL files also will default every Face to the Default Material, so the file will appear white or blue.

> **Note**
> Remember to try the **Soften Edges** Panel to quickly Soften coplanar Faces! Or use the **Merge Coplanar Faces** Option in the **Import Options** Dialog Box.

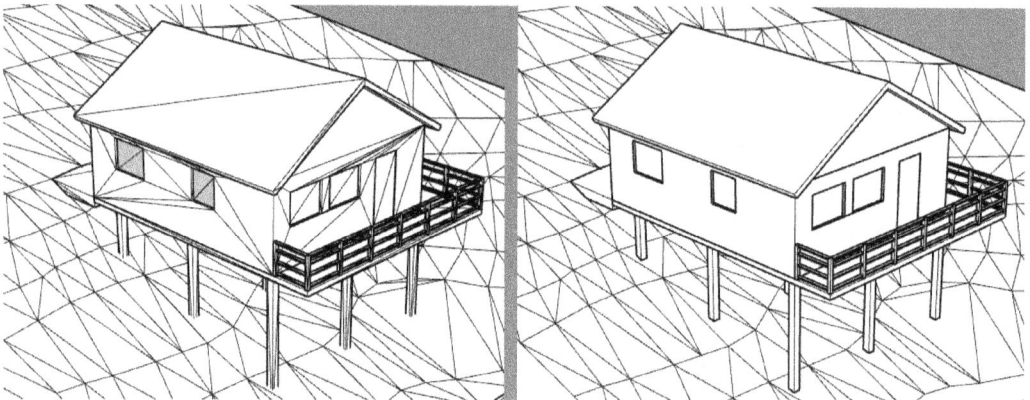

Figure 12.16 – Imported 3DS File with Triangulated Faces (Left) and Coplanar Edges Softened (Right)

STL Objects are often imported as a part of a 3D printing workflow. Because of the popularity of these files, there have been multiple Extensions (plugins) developed to assist in the cleanup and preparation of these files. We will discuss Extensions in more detail later in this chapter.

TrimBIM Files (.TRB)

TRB files (Trimble BIM files) are the proprietary BIM file created by Trimble, the developer of SketchUp Pro. TrimBIM files now have native Import capability since the 2023 release of SketchUp Pro. TrimBim files are **Building Information Modeling (BIM)** files and were designed to act as a connection point for other BIM software, such as Autodesk Revit.

In this next example we can see a TRB model is being imported into SketchUp Pro. TrimBIM files do not have any specific Import options, so the file can be selected, and the **Import** button can be clicked.

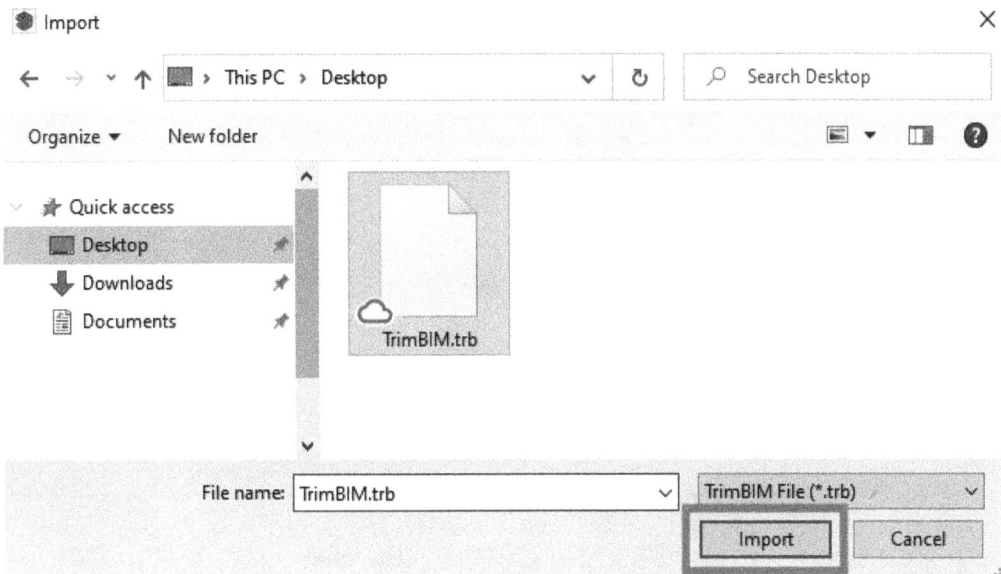

Figure 12.17 – TRB File Selected and Import Clicked

When the **Import** button is clicked SketchUp will import the file and will show the **Import Results** Dialog Box when the import is completed.

Import results   ✕

TrimBIM import succeeded   ⌃

                                                      ⌄

                                              Close

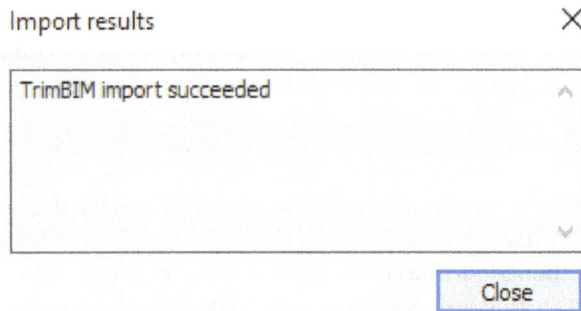

Figure 12.18 – Import Results Dialog Box

Once the Import Results Dialog Box is closed the TRB file will need to be manually placed in the Drawing Area. This will add the file as a new Component in the current SketchUp model.

In the 2023 release of SketchUp Pro TrimBIM files are imported with hidden Edges when the same Material is used on connected Faces. In this next example we can see the SketchUp team member Component. This Component has been exported directly from SketchUp Pro as a TrimBIM file, and can be imported to a new file. When the TrimBIM import is placed, a new Component is generated and can be seen in the Components Panel.

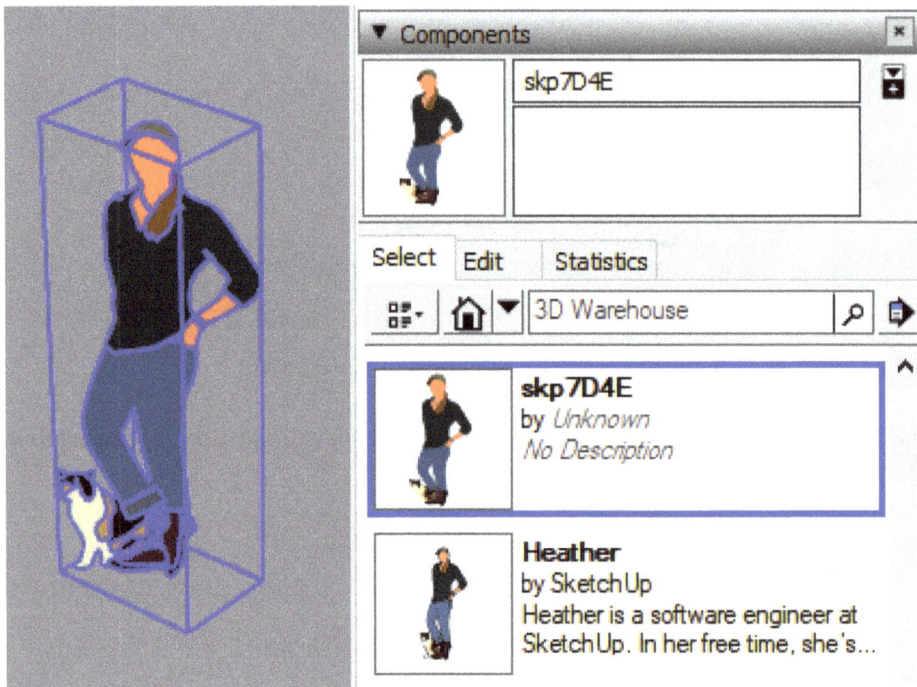

Figure 12.19 – Component Placed (Left); Component Added to Components Panel (Right)

When comparing the original SketchUp Component (Left) and the TrimBIM imported Component (Right) we can see that the Edges have been hidden when the adjoining Faces are painted with the same Material.

Figure 12.20 – Edges Hidden on TrimBIM Import (on Right)

This also happens on 3D Objects that are exported and imported using the same TrimBIM file type. In this next example a simple shape has been created, and the Geometry has been exported as a TrimBIM file.

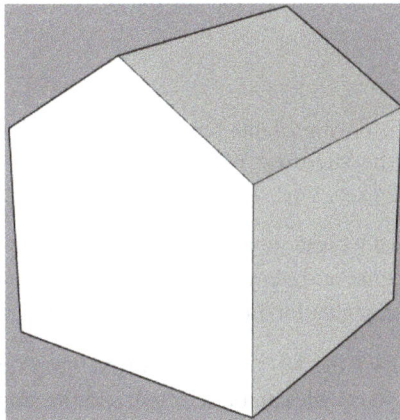

Figure 12.21 – Simple Shape Exported to TrimBIM File

When the Object is imported back into the file, we can see that all the Edges have been hidden, as the entire model used the same Material.

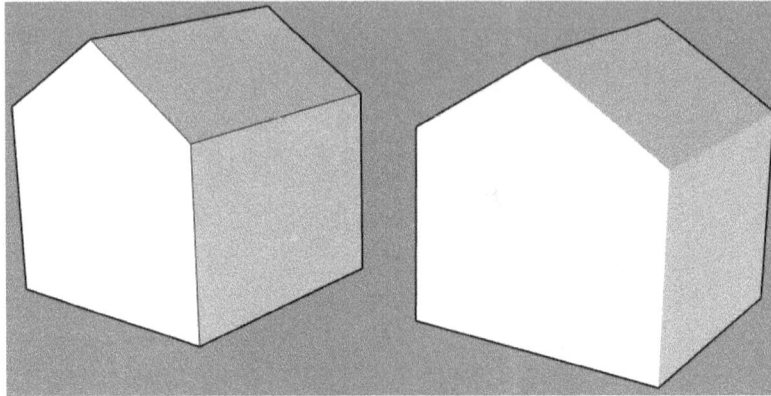

Figure 12.22 – Simple Shape Imported to SketchUp with Edges Hidden (on Right)

> **Note**
> Remember to use the Smooth/Soften Edges Panel in the Default Tray to quickly unhide all hidden Edges for a Component!

Autodesk Revit Files (.RVT)

SketchUp Pro introduced a native Revit Importer with the 2023 release of SketchUp Studio. As of the 2023 release of SketchUp Pro, the Revit Importer is only available with a SketchUp Studio license. SketchUp can import Revit files to SketchUp Pro 2021 and beyond, and Revit files can be imported as long as they are files made in Revit 2011 or newer. The SketchUp Revit Importer does not require a Revit license.

> **Note**
> The SketchUp Revit Importer is only available for Microsoft Windows as of the 2023 release. Autodesk Revit is exclusive to the Microsoft Windows environment as well.

The Revit Importer will automatically create multiple SketchUp objects when importing a Revit file, including Components, Tags, Sections, and Materials. Additionally, SketchUp will optimize the Revit Geometry by reducing redundant triangulation in the file and will reduce the segment count on curves.

In this next example we can see a RVT model is being imported into SketchUp Pro. Revit files do not have any specific Import options, so the file can be selected, and the **Import** button can be clicked.

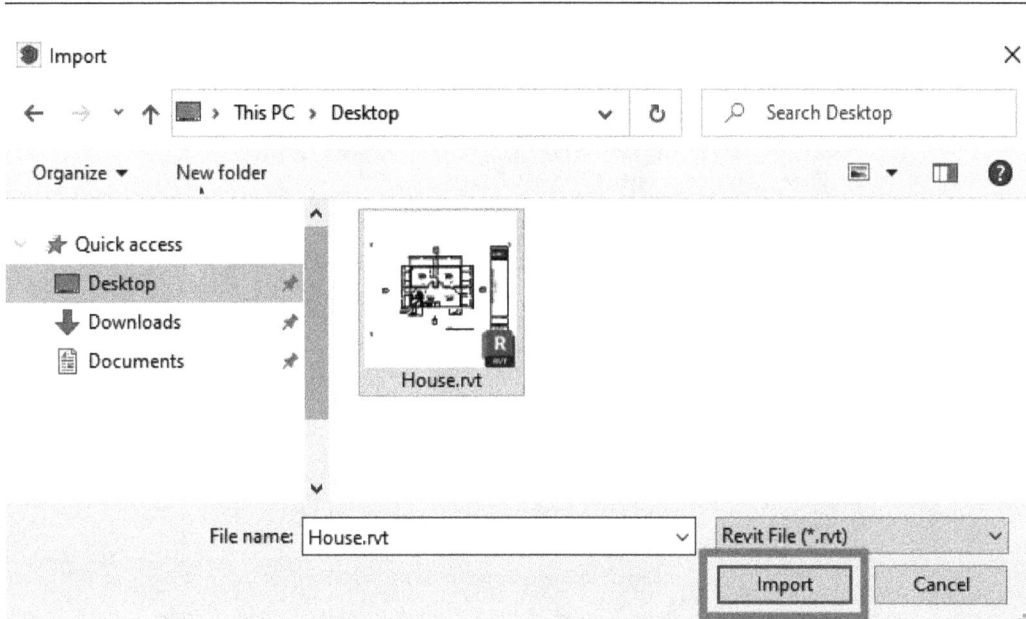

Figure 12.23 – Revit File Selected and Import Clicked

When the **Import** button is clicked SketchUp will show the **Import Progress** Dialog Box while the file is being imported.

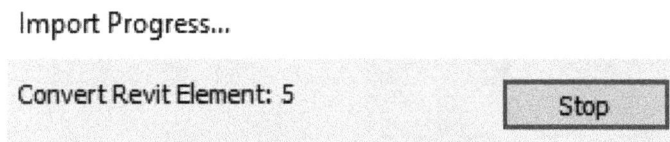

Figure 12.24 – Import Progress Dialog Box

If a SketchUp Studio license is not associated with the SketchUp Pro license SketchUp will show a Licensing dialog box which will ask the user to sign in to SketchUp Studio. The Revit file will continue to be imported in the background.

Licensing

Importing Revit files is a feature only available in SketchUp Studio. Please purchase a license or activate a Trial version.

| Upgrade |
| Activate a Studio trial |
| Cancel |

Figure 12.25 – Licensing Dialog Box for SketchUp Studio

With a current version of SketchUp Studio licensed, the **Import Results** Dialog Box will show when the import is completed.

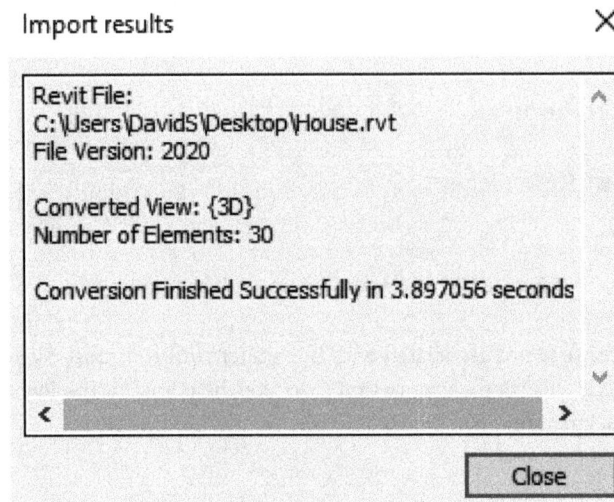

Import results   ✕

Revit File:
C:\Users\DavidS\Desktop\House.rvt
File Version: 2020

Converted View: {3D}
Number of Elements: 30

Conversion Finished Successfully in 3.897056 seconds

Close

Figure 12.26 – Import Results Dialog Box

Once the Import Results Dialog Box is closed the Revit file will automatically be placed in the SketchUp model and this will add the file as a new Component. Revit files are comprised of individual BIM components called Revit Families. These Revit Families will also be added to the model as Components, but only after the main Revit file Component is exploded. In this example we can see the imported Revit file Component in the Components Panel, and after the Revit file Component has been exploded the additional Components are added to the Components Panel.

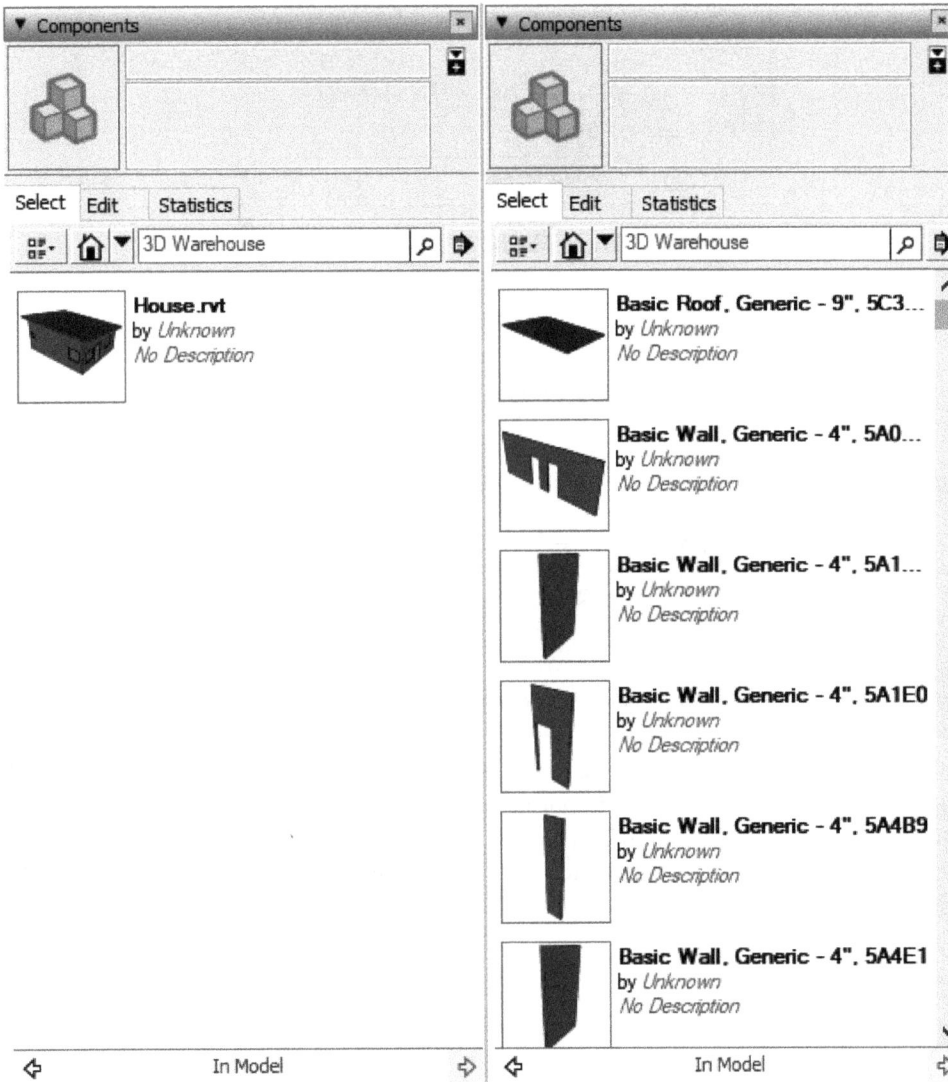

Figure 12.27 – One Component in the Components Panel (Left)
Additional Revit Family Components Added (Right)

Even if the Revit Families are the same Object in Revit, once they are imported into SketchUp they will be created as Unique Components. In this example three Single–Flush doors were the same Revit Family, but SketchUp has added them as three Unique Components with the Revit door ID included in the name.

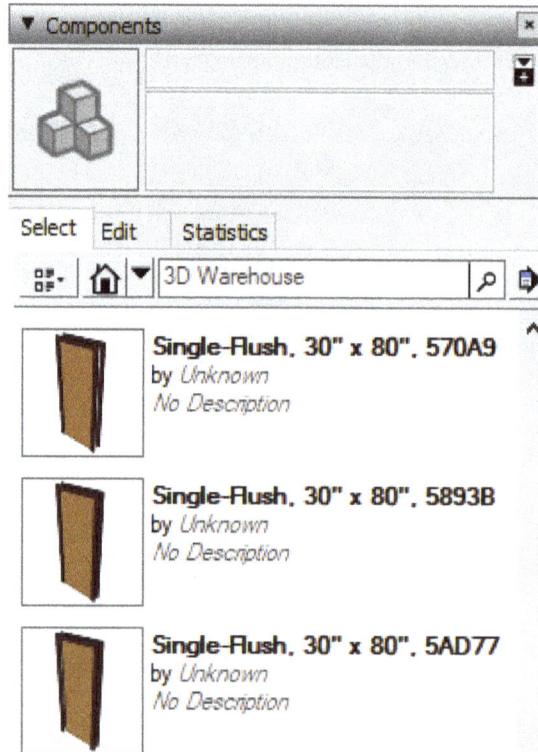

Figure 12.28 – Revit Family Components Added as Unique Components

In addition to the Components being added from Revit, SketchUp will also import the Tags, Materials, and Views from Revit. The Tags will be imported from the Revit Layers, with SketchUp Edges and Faces being imported to the Untagged Tag, and the individual Family Components being assigned to the appropriate Tag. The Tags can be edited without exploding the top–level Revit Component.

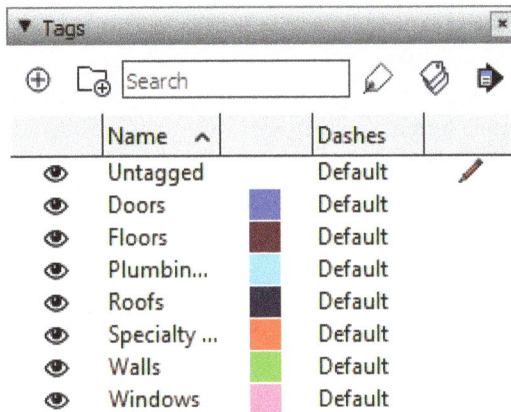

Figure 12.29 – Revit Layers Imported to SketchUp Tags

The Revit Materials will also be imported as Materials and can be found in the Materials Panel in the Default Tray. SketchUp will not attempt to combine the Revit Materials with SketchUp Materials. For instance, in this example the Revit "Glass" Material has been imported, and will not be combined with the SketchUp "Translucent Glass Blue" Material.

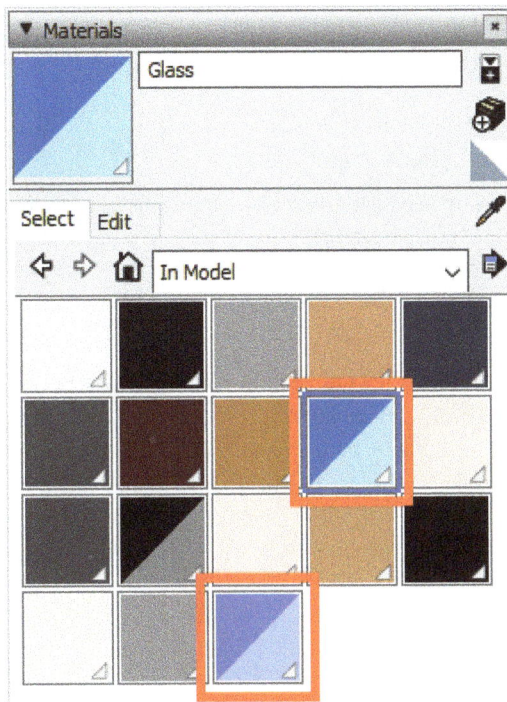

Figure 12.30 – Revit Glass Material and SketchUp Translucent Glass Blue Material

The individual Family Components are not imported with nested Groups, so the individual Faces are assigned the Materials. This can be frustrating, as Painting the Component will not update any Materials. The Component must be activated, and individual Faces need to be Painted using the Paint Bucket Tool. We can see this in the next example, where a wall Component should be updated to a brown color. When the component is clicked no Material change occurs, and only when the Component is activated and the individual Face is clicked with the Paint Bucket Tool does the wall update.

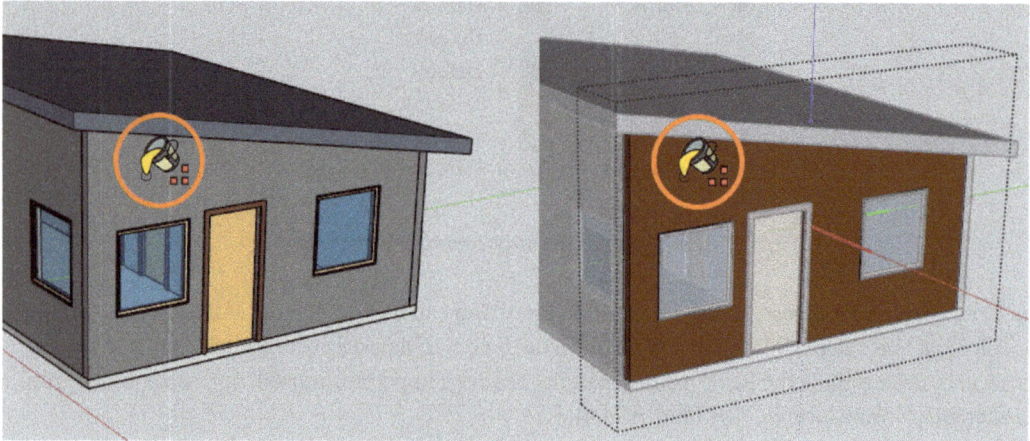

Figure 12.31 – Component Clicked with No Change (Left); Component
Activated and Individual Face is Painted (Right)

Note

It is recommended to clean up the individual Revit Family Components by creating nested Groups and removing the materials from individual Faces. Once nested groups are established, or if the entire Component can use the same Material, then the Group or Component can be Painted as a whole.

SketchUp will also import some Revit Views as Section Planes. SketchUp does not save camera views natively, and it does not automatically create SketchUp Scenes for Revit views. However, the view can be simulated by activating a Section Plane, then choosing the Align View option from the right-click contextual menu. In this example we can see the Revit Import has brought in two Section Planes, representing the Revit Views for the roof plan and Level 1 floor plan. The floor plan Section Plane can be right-clicked and made the Active Cut, then right-clicked again to Align View.

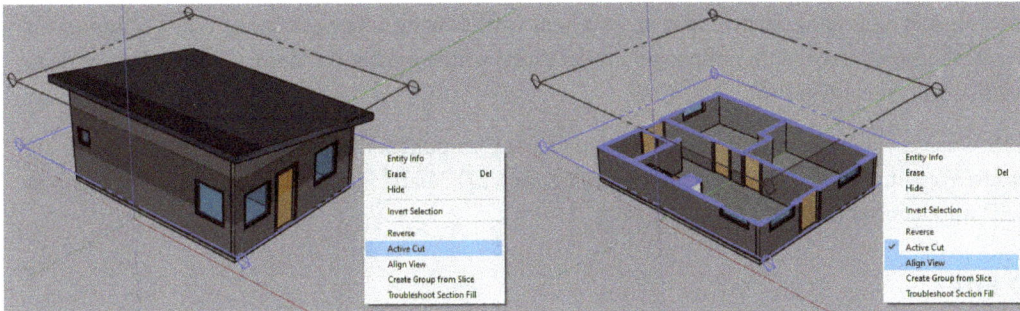

Figure 12.32 – Right–Click | Active Cut (Left) Right–Click Align View (Right)

In this case, the Align View has activated the same option as using the Top view from the Views Toolbar. In order to make this view appear as it would in Revit, the Parallel Projection Camera Setting should be activated from the Camera Menu dropdown, which will remove the perspective warping.

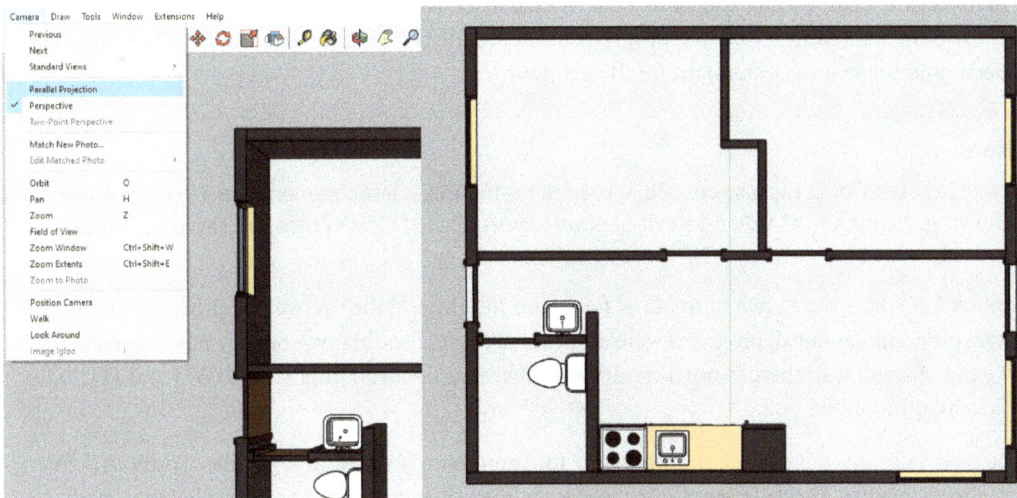

Figure 12.33 – Camera | Parallel Projection (Left) Updated SketchUp Floorplan View (Right)

This view can now be saved as a Scene. Or, if you are using Layout for drawing creation, this functionality is built into the default Layout workflow.

Importing Revit files into SketchUp is an excellent workflow for interoperability, but it is important to remember to ensure that the imported model meets your working standard. Good luck importing Revit files into SketchUp Pro!

All 3D Objects will be imported into SketchUp Pro as Components, but that is not the case for all 2D Objects. We will discuss 2D Objects in two categories, 2D Model Files and Image Files. Let's look at the 2D import options now!

2D Files

2D model files are Imported in the same way that 3D files are imported. Two types of 2D files will be imported with a similar result to the 3D file types: Autodesk DWG and DXF.

.DWG and .DXF

Autodesk is a software company that creates complimentary 2D and 3D drafting software programs in the Architecture, Engineering, and Construction industries. AutoCAD was one of the first Computer–Aided Drafting and Design software programs used broadly in the industry, and the DWG file type has been used for years as a standard for 2D geometry.

> **Note**
> AutoCAD DWG files can also contain 3D geometry, but this is less common than a 2D AutoCAD Drawing. Many AutoCAD and Revit files are exported as 3D DWG files and saved as Blocks.

Autodesk DXF files were created to allow for easier interoperability between AutoCAD and other software programs, as not all programs were able to read DWG files. Many programs now can recognize DWG and DXF files, so there is not a significant difference between importing DWG and DXF files for SketchUp Pro users.

In the next example, we can see the workflow for importing DWG and DXF files. AutoCAD Files are grouped together in the **Import** selection dropdown. When a DWG or DXF file is selected, the **Options…** Button appears at the bottom of the **Import** Dialog Box. DWG and DXF files have specific import options including **Merge coplanar faces, Orient faces consistently, Import Materials**, changing or preserving the drawing origin, and setting the **Scale** options. The model can be scaled using a standard unit of measure.

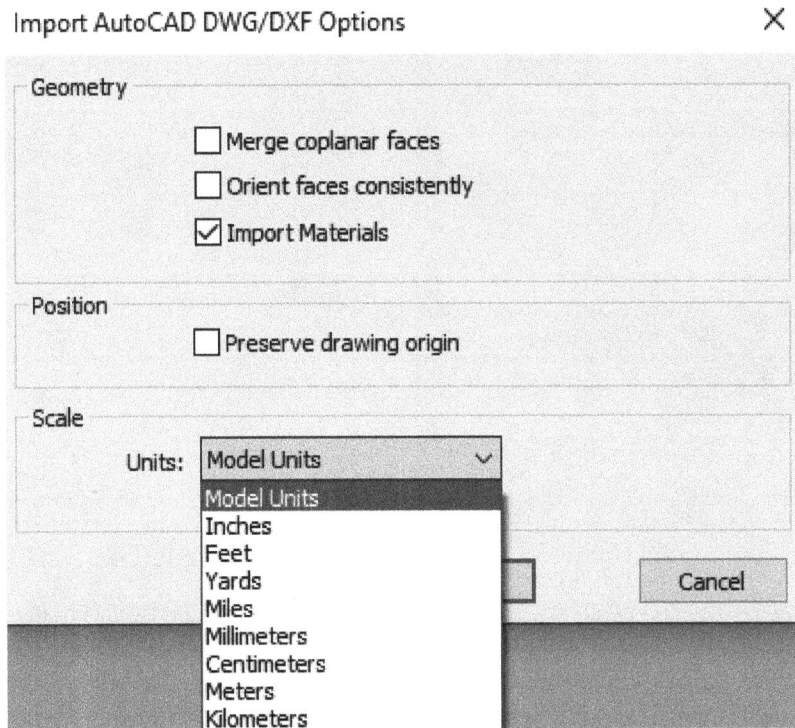

Figure 12.34 – AutoCAD DWG/DXF Import Options Dialog Box with the Units Dropdown Open

Merge coplanar faces and **Orient faces consistently** are primarily used in 3D DWG or DXF imports, but **Import Materials** and **Preserve drawing origin** are important for 2D and 3D files. Imported materials from Autodesk software may have different appearance outcomes in SketchUp Pro, and it can be recommended to turn this Option off if you would like to work with SketchUp Materials instead. **Preserve drawing origin** will keep the DWG or DXF model relative to the (0,0,0) coordinate in the original file, which can move it dramatically relative to the SketchUp Origin.

When the **Import** button is clicked, SketchUp will show the **Import Progress...** Dialog Box while the file is being imported, and the **Import Results** Dialog Box when the import is completed.

550   Import, Export, 3D Warehouse, and Extensions

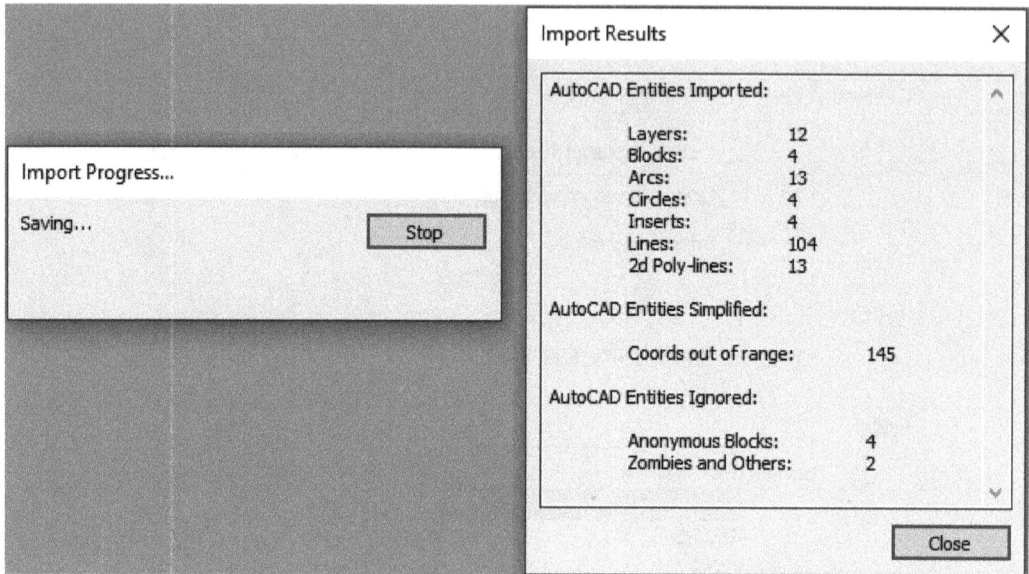

Figure 12.35 – The Import Progress... Dialog Box (Left) and Import Results Dialog Box (Right)

SketchUp may show a warning in the **Import Results** Dialog Box that the Geometry is very far from the Origin. This is common with Civil 3D files, or files that use real–world coordinates. If this warning appears, it is common that the import will fail, and SketchUp will show an **Import Failed** Dialog Box. It is recommended that the DWG or DXF file is opened in AutoCAD and the Geometry should be moved closer to the AutoCAD Origin before importing.

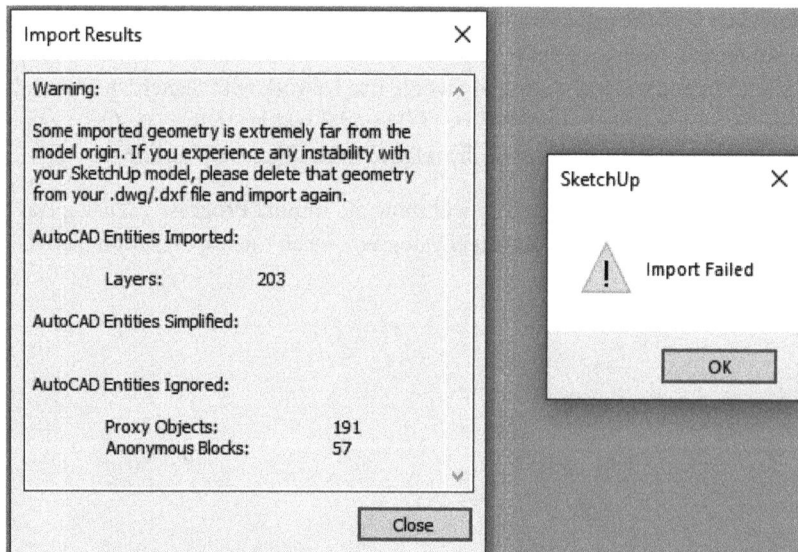

Figure 12.36 – Warning in the Import Results Dialog Box (Left) and the Import Failed Dialog Box (Right)

Once the **Import Results** Dialog Box is closed, the DWG or DXF file will be automatically placed at Axis Origin. This will add the file as a new Component to the current SketchUp model. DWG/DXF files are typically oriented in the normal direction, with the 2D drawing appearing to be flat on the ground plane.

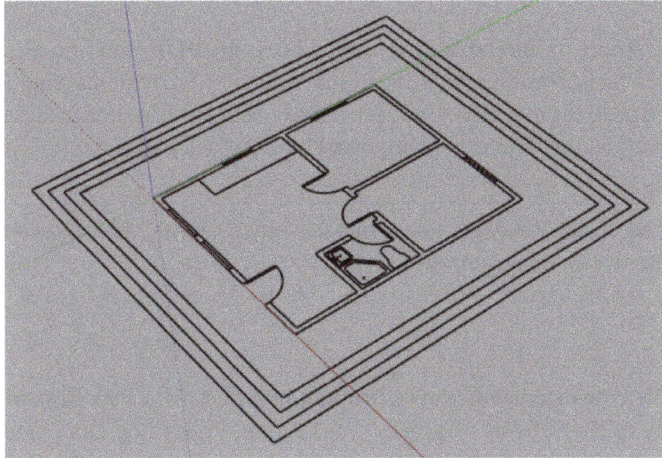

Figure 12.37 – DWG/DXF File Imported

DWG/DXF files can also bring in AutoCAD layers, which are added to the **Tags** Panel in the **Default** tray. These Tags can be used while modeling or cleaned up/removed, but it is important to remember that these AutoCAD layers do not retain the same AutoCAD options. Once imported into SketchUp Pro, the layers have been converted into Tags, and the Geometry is now "sticky". Hiding Tags may stick hidden and visible Geometry together and editing the Geometry may cause unintended results.

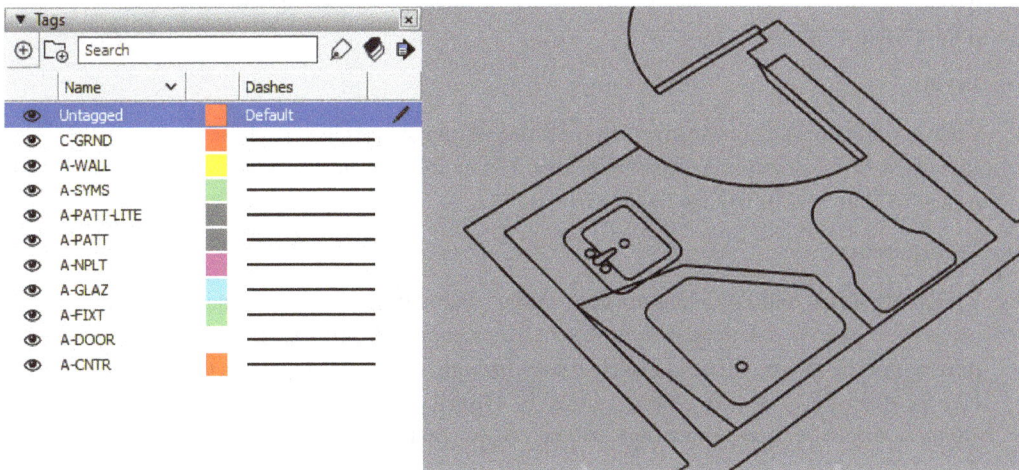

Figure 12.38 – AutoCAD Layers Converted into Tags (Left) and the Hidden Geometry Moved (Right)

Importing AutoCAD drawings is a common workflow for many professional disciplines, including bringing in site plans for landscape design, floor plans for architectural design, or 2D drawings for industrial or mechanical design.

> **Note**
>
> Are you trying to import a **Portable Document Format** (**PDF**) file into SketchUp Pro? Unfortunately, PDFs cannot be imported natively into SketchUp Pro without using an Extension. Try converting vector PDFs into DXFs or DWGs, and raster PDFs to image files using third-party software, and then that files can be imported into SketchUp Pro!

AutoCAD drawings can be 2D or 3D, but image files can only be brought in as 2D assets. In the next section, we will discuss the six image import file types.

Image Files

Six different image file type extensions can be brought into SketchUp Pro: Bitmap (`.BMP`), JPEG (`.JPG` or `.JPEG`), Portable Network Graphics (`.PNG`) files, Photoshop Documents (`.PSD`), Tag Image File Format (`.TIF` or `.TIFF`), and Targa (`.TGA`) files. Images can be brought in in three different ways, and we will discuss those options later in the chapter. First, let's look at a few differences between these files:

- **Similarities**:

  SketchUp Pro will treat all of these image file types the same way once they are imported into SketchUp. Specific programs might be used to create and edit these images, but SketchUp Pro will use them all in the same ways. Using one file type over another does not matter for most uses in SketchUp Pro, so if one of those files is available, there is no need to convert it into a different file extension.

- **Layers**:

  TIFF and PSD files will have layers in their native image formats, but these layers are not the same type of information as SketchUp Tags. When an image is imported into SketchUp Pro, it will be "flattened" into a single image.

- **Transparency**:

  TIFF, PSD, TGA, and PNG files are commonly used with transparency, meaning that not all pixels are colored in the image. When SketchUp Pro imports a transparent image, the transparent portions of the image will be completely see-through, and it will appear as though there is no face for those parts. This is different from the **Opacity** slider for SketchUp Materials, which applies a value of transparency (opacity) on opaque materials.

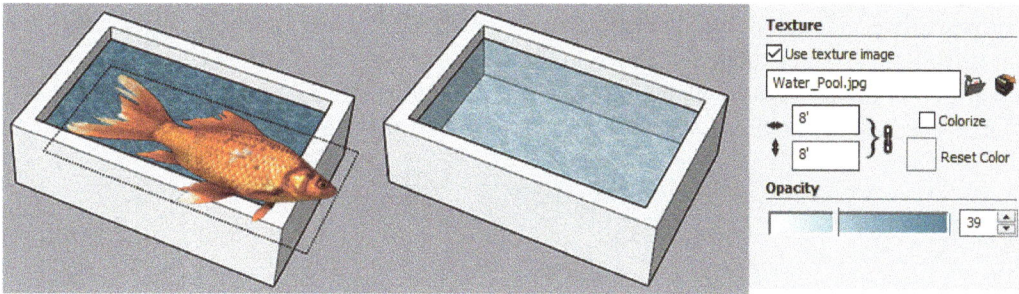

Figure 12.39 – Transparent Image (Left) versus Transparent
SketchUp Material and the Opacity Slider (Right)

There are a few included SketchUp Materials that use the image transparency feature, however. In the **Landscaping, Fencing, and Vegetation** Collection, there is a collection of fencing Materials, all of which appear to be an image pattern with a white background, but these images are all transparent and will have a transparent appearance when applied to a Face.

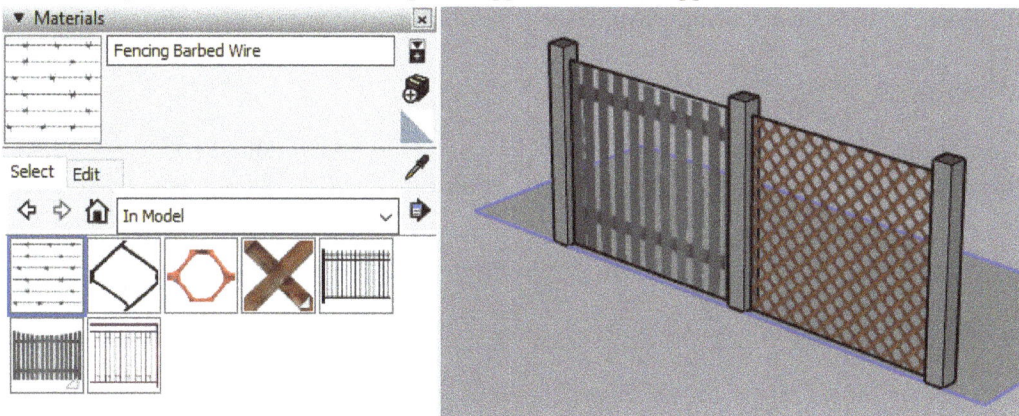

Figure 12.40 – Fencing Materials (Left) and the Transparent Image Material Applied (Right)

> **Note**
> Even though JPEG and BMP images are not known for transparency, some versions of these files can support transparency. Even when working with other file types, it is always a good idea to review the image file before it is imported to make sure that it is designed appropriately for use in SketchUp Pro.

- **File Size**:

    Certain image file types are larger than other file types because of additional information that is specific to the file editor for which they were originally created. Additionally, some files are more compressed than others. If you are working with many imported images, it can be important to use "lighter" image files to manage the size of your SketchUp model.

Now that we have understood some of the similarities and differences between the file types, let's look at the three ways in which they can be used in SketchUp Pro. With the **Import** Dialog Box open, when any of the image types are highlighted in the dropdown, or the **All Supported Image Types** options is highlighted, the three **Use Image As** options will be shown below the selection window:

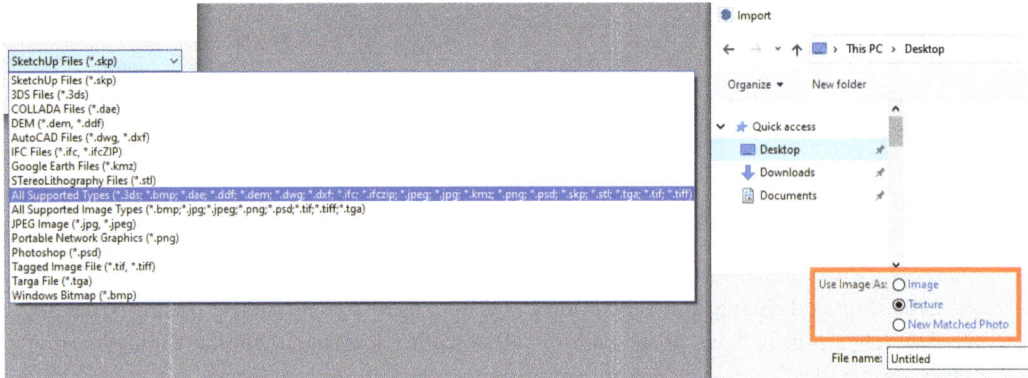

Figure 12.41 – All Supported Image Types Highlighted (Left) and Use Image As Options (Right)

Use Image As Texture

We have already discussed creating a Texture Material from an image import in *Chapter 8, Materials*. Please review the *Create Material From Image Import* section of that chapter for a deep dive into creating and editing Texture Materials.

Creating an Image or creating a New Matched Photo will be discussed in detail in the next two sections.

Use Image As Image

Images in a SketchUp Pro model are a unique type of Object. They are referenced as Images in the **Entity Info** Panel, but cannot be found in the **Components** or **Materials** Panels. When they are not selected, they appear to have no Edges!

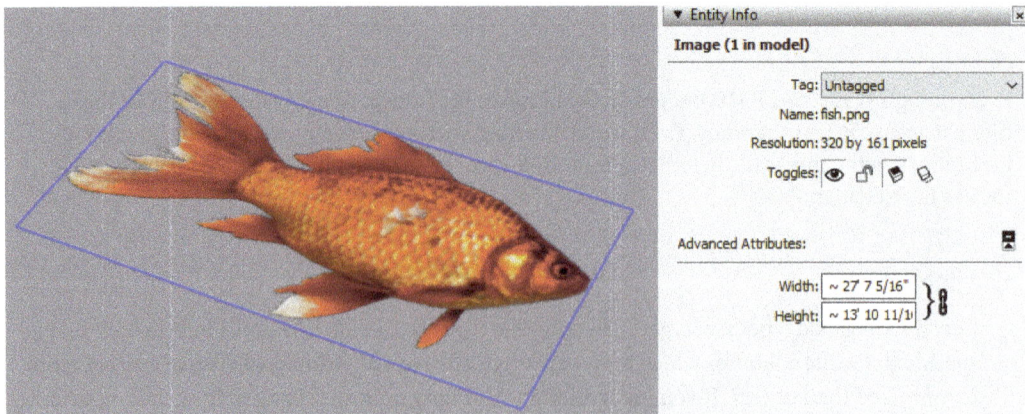

Figure 12.42 – Image Selected (Left) and the Entity Info Panel (Right)

> **Note**
> We can manipulate Images to show their Edges and Face and use them as Texture Materials. We will cover these workflows later in this section.

Images can be Imported by choosing the **Image** option in the **Import** Dialog Box and hitting the **Import** Button.

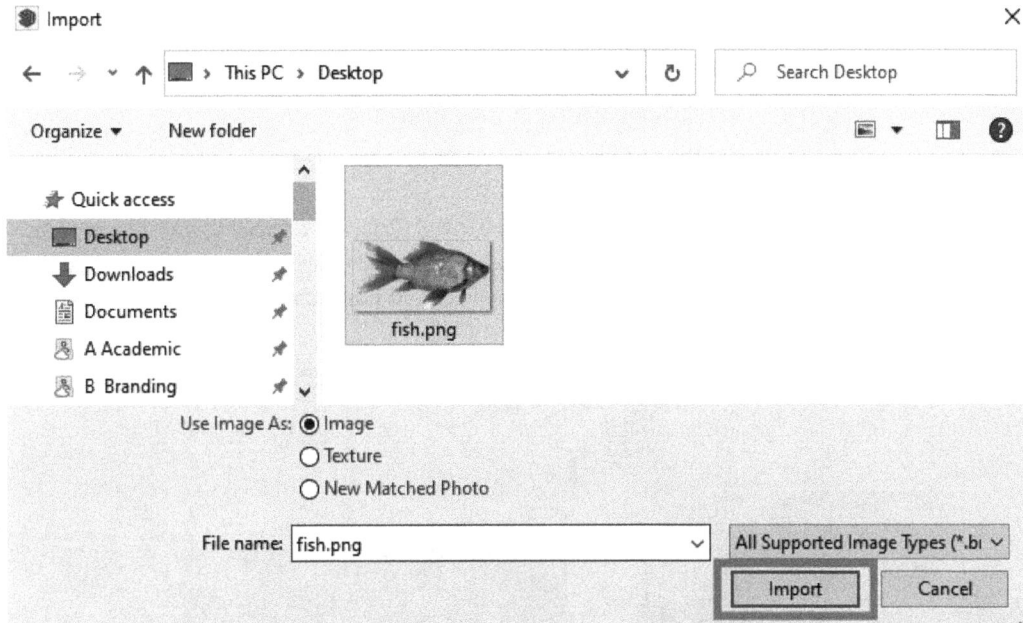

Figure 12.43 – Image Selected and Import Clicked On (Right)

Once the **Import** Dialog Box closes, the Image can be placed in the model. The Image can be placed with two clicks on the screen, with the first click being at the bottom left of the Image and the second click at the top right. An Image does not need to be snapped to any existing Geometry or placed onto a Face, which is different from Texture Materials. In this example, the Imported Image has been placed vertically perpendicular to the Green Axis with two clicks. This is a great way to add elements to a SketchUp model without needing to model complex shapes!

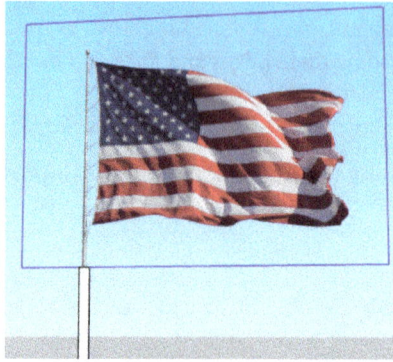

Figure 12.44 – Image Placed Vertically in Model

Once an Image is imported into the model, it can be Moved, Rotated, or Scaled, but it cannot be edited using **Push/Pull**, **Follow Me**, or **Offset**. An Image is a special Object, and the Face and Edges are not able to be used in **Edit** workflows. However, the Face and Edges can still be referenced when using the **Drawing** tools.

Figure 12.45 – Image File Selected (Left) and Image File Referenced with the Pencil Tool (Right)

The Face and Edges of an Image cannot even be seen when using **Show Hidden Geometry**. The Image Object must be exploded by right–clicking on the Image and choosing **Explode**. This will convert the Image Object into a rectangular Face bounded by four Edges and will add the Image to the **Materials Panel** as a Texture Material.

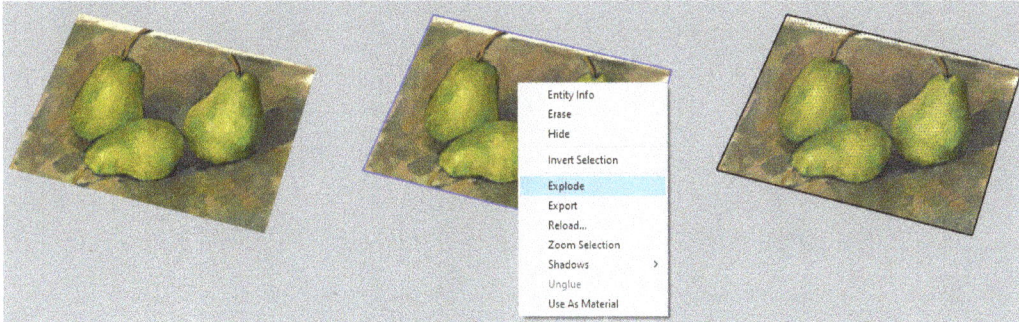

Figure 12.46 – Right–Clicking Explode for the Image (Left), its Face (Middle), and Texture Material (Right)

Image Objects have other unique right–click menu options, including **Export**, **Reload…**, **Shadows**, and **Use As Material**. The **Export** option will allow the Image to be saved to your local computer, which is helpful if the Image was originally imported on a different computer. The **Reload…** option allows an updated version of the image to be located to replace the current Image. This is the same workflow as Importing the Image, although the Image will remain in the same location. If the computer recognizes the original file path for the Image file, SketchUp will ask the user whether they would like to choose a new Image.

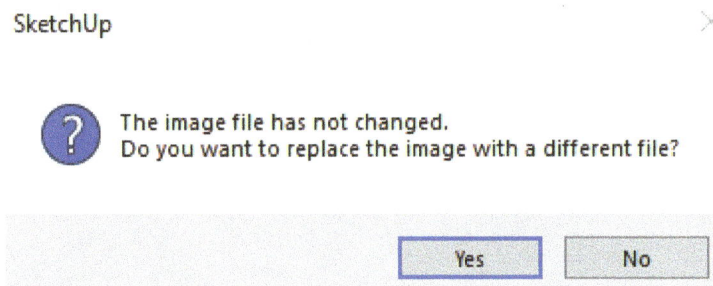

Figure 12.47 – Reload Image Dialog Box

The **Shadows** option allows the Image to individually override the model settings for Casting and Receiving Shadows. However, SketchUp Pro does not recognize transparent or non–transparent pixels in an Image, so when Casting Shadows, the entire rectangular Face will Cast the shadow.

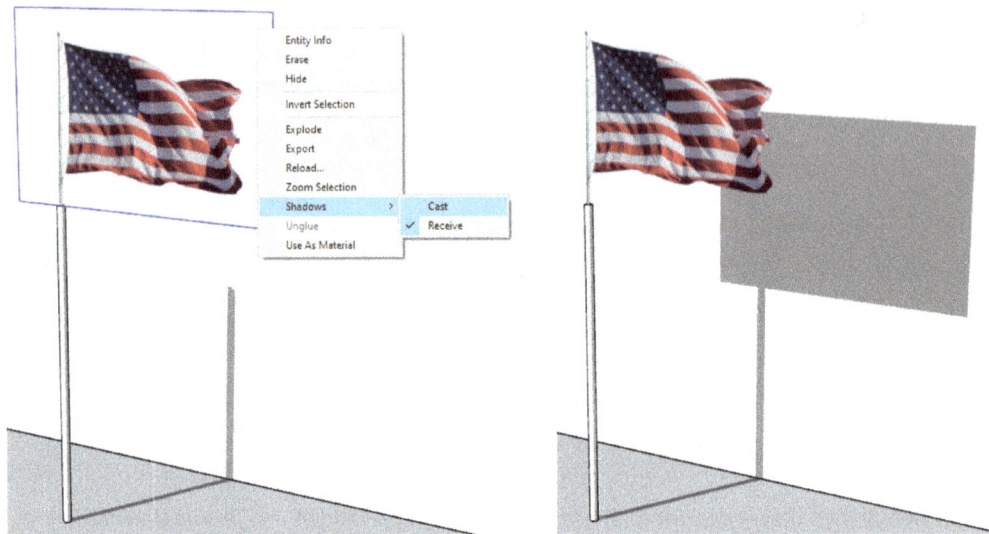

Figure 12.48 – Right–Clicking on Shadows –> Cast (Left) and the Rectangular Shadow Cast (Right)

Image Objects are a quick and easy way to add 2D elements to a SketchUp model with very small modeling requirements. Matched Photos are different from standard SketchUp Images, as they allow Images to be used to create 3D models.

Use Image As New Matched Photo

Matched Photos are images that are set as a Scene background that make it appear like you are drawing on the object in the image. SketchUp Pro can set the camera position, field of view, and Axes to align the perspective of the view on top of what is in the photo, allowing you to draw on top of a photo. This can be very helpful to quickly sketch a model for an existing building or object, although the dimensions will not be extremely precise.

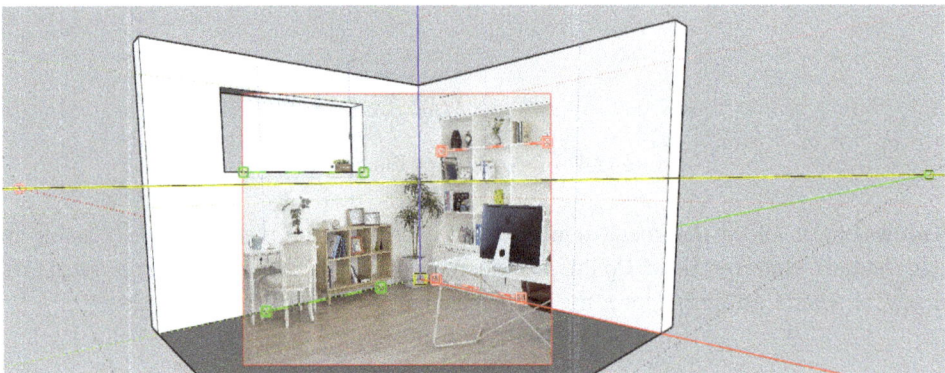

Figure 12.49 – Matched Photo Scene and Geometry

Matched Photos can also be used as Projected Texture Materials, projecting on all Geometry in the same direction as the original photo. This will not color a model with perfect Materials, but it will allow an Object to have quick and dirty Texture Materials that can be viewed from additional viewing angles.

Figure 12.50 – Matched Photo Scene (Left) and Projected Texture Materials on all Geometry (Right)

Matched Photos are imported using the same workflow in the **Import** Dialog Box, by going to **File** -> **Import**, selecting an Image, and choosing **New Matched Photo** in the **Import** options.

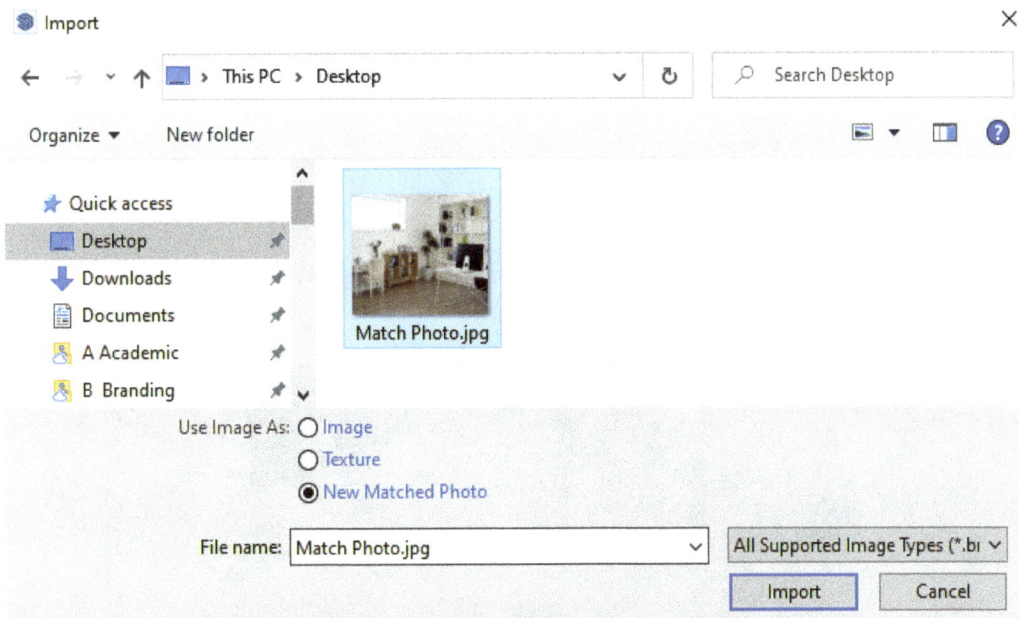

Figure 12.51 – Import Dialog Box with New Matched Photo Selected

When the **Import** Button is clicked, the **Match Photo** Panel will open on the Default Tray. The **Match Photo** Panel also has a **New Matched Photo** button, which looks like a plus sign.

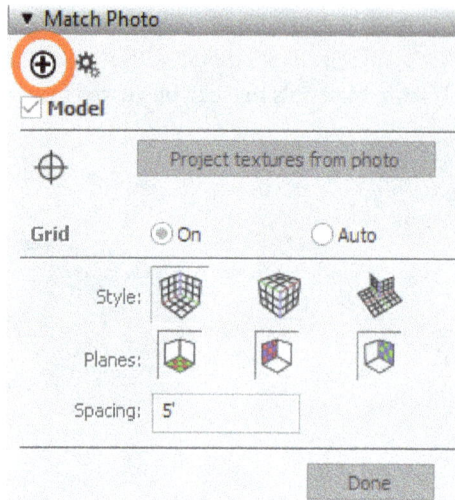

Figure 12.52 – The Match Photo Panel and the New Matched Photo Button

The **Match Photo** interface can be confusing when it first opens, but we can recognize the colors of these elements from other areas of our SketchUp model. The green and red lines represent the setup points for the Red and Green Axes, and the yellow line represents the new horizon line. The red and green grid lines help to see more alignment points throughout the photo, and the Axes will show where the new Origin in the model will be.

Figure 12.53 – The Match Photo Interface

> **Note**
>
> All of the Match Photo settings will be saved in a new Scene named after the photo, and when the Scene is activated, all of the settings can be edited in the **Match Photo** Panel. If you Pan or Orbit the camera, activate the Scene to return to the Matched Photo. **Undo** and **Redo** do not work with Matched Photo edits, so be careful when making big changes!

The red and green lines can be aligned with straight lines in the photo – the two red lines should point to the left–hand vanishing point and the two green lines should point to the right–hand vanishing point.

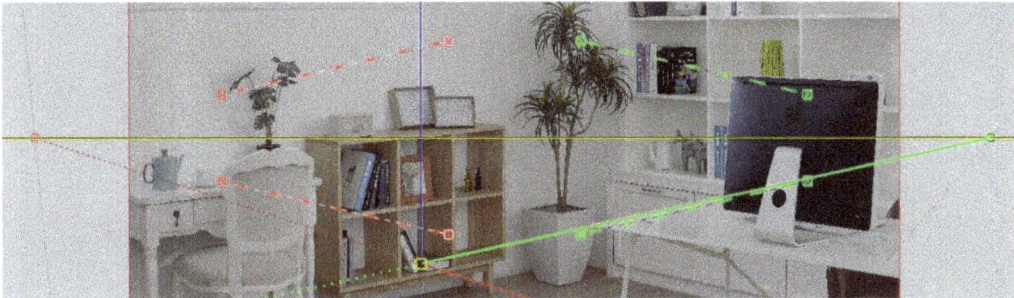

Figure 12.54 – Red Lines Point to the Left–Hand Vanishing Point and
Green Lines Point to the Right–Hand Vanishing Point

We typically cannot see the vanishing points in a photo itself, so the vanishing points will often be outside the photo's bounds. If the photo has relatively little camera warping/distortion, the Blue Axis should point up along the image if the red and green lines are set up properly. In this next example, we can see that one of the green lines is not aligned with the photo, and the Blue Axis is angled and does not align with the vertical lines in the photo.

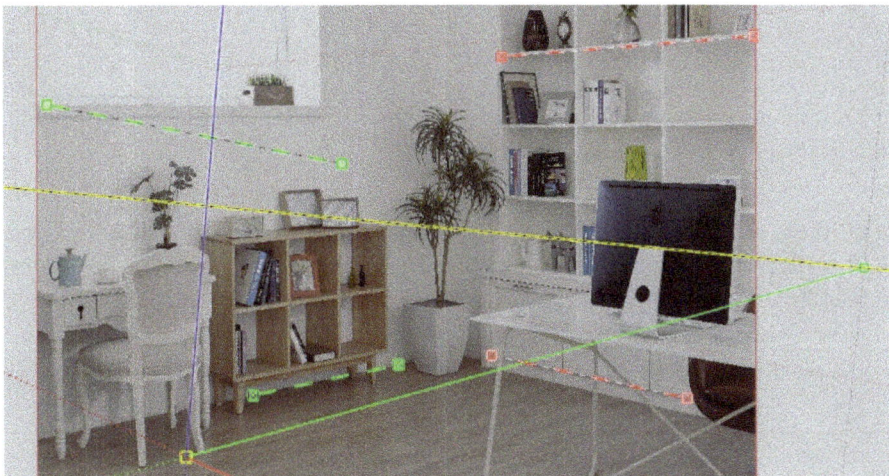

Figure 12.55 – Green Line Not Aligned and the Axis at an Angle

When the green line is aligned with the photo geometry, we can see the Blue Axis appears to be vertical. When the Axes appear to be aligned to the photo, we can click and drag on the yellow square that represents the new origin to drag it to a specific part of the image, typically a corner of the geometry.

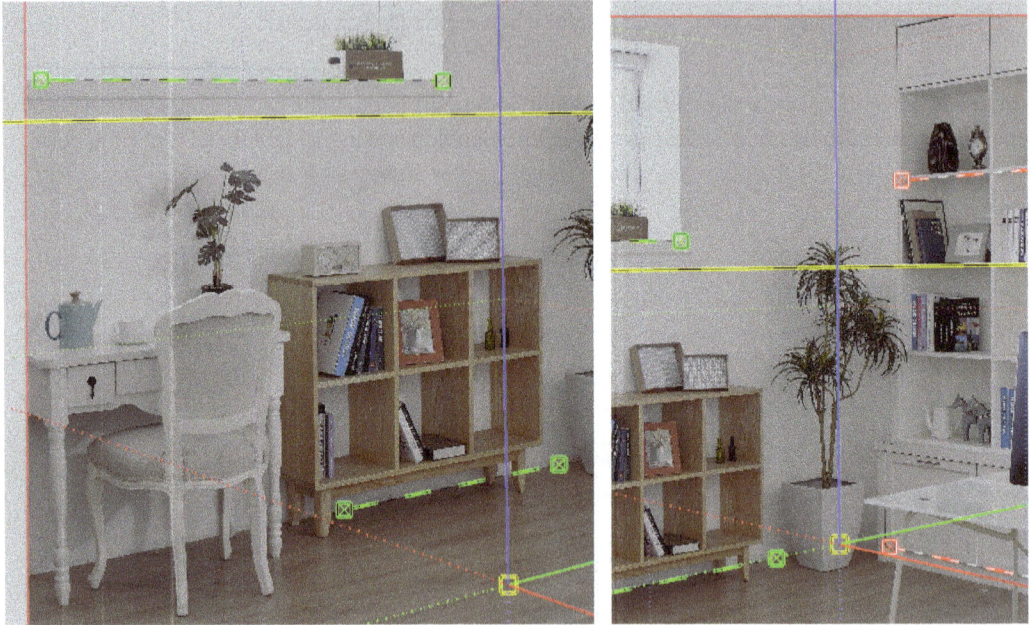

Figure 12.56 – Green Line Adjusted and Vertical Axes (Left) versus
the Axes Dragged to New Position (Right)

> **Note**
> The Match Photo workflow works best with large photos and when you zoom very close into a photo! You should be able to see individual pixels in order to be sure that you are aligning the green and red lines as best as you can!

The size of the photo can be adjusted relative to the existing geometry in the model. In this example, we have existing Geometry in the model, and we will use the Match Photo workflow to project textures onto the model. This building will be in the background of the final model and can be represented with simple photo textures.

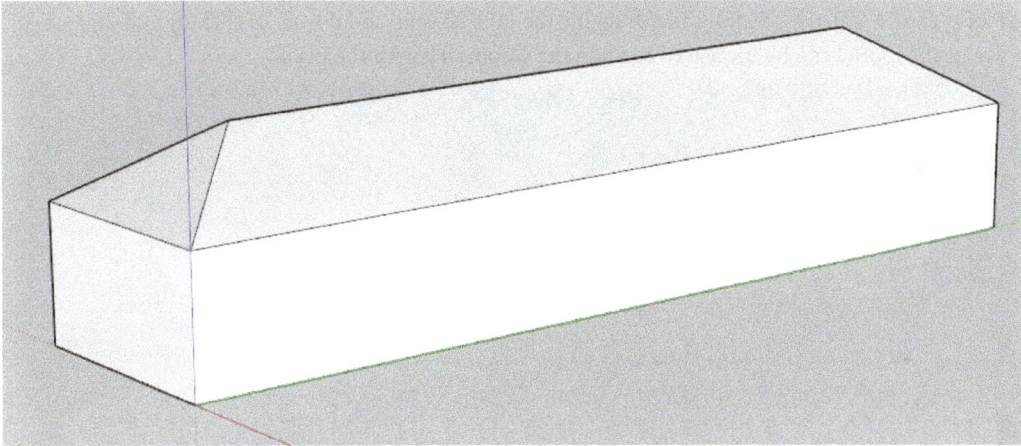

Figure 12.57 – The Existing Geometry in the Model

After choosing **Import -> New Matched Photo**, we can see that the photo is much larger than the existing Geometry. Even with the red and green lines adjusted onto the Geometry, the building appears to be much smaller than the image.

Figure 12.58 – Matched Photo Adjusted to a Small Building

The model can be scaled by clicking and dragging the Blue Axis up and down. By clicking and dragging, the size of the model can be increased to align the Geometry with the photo.

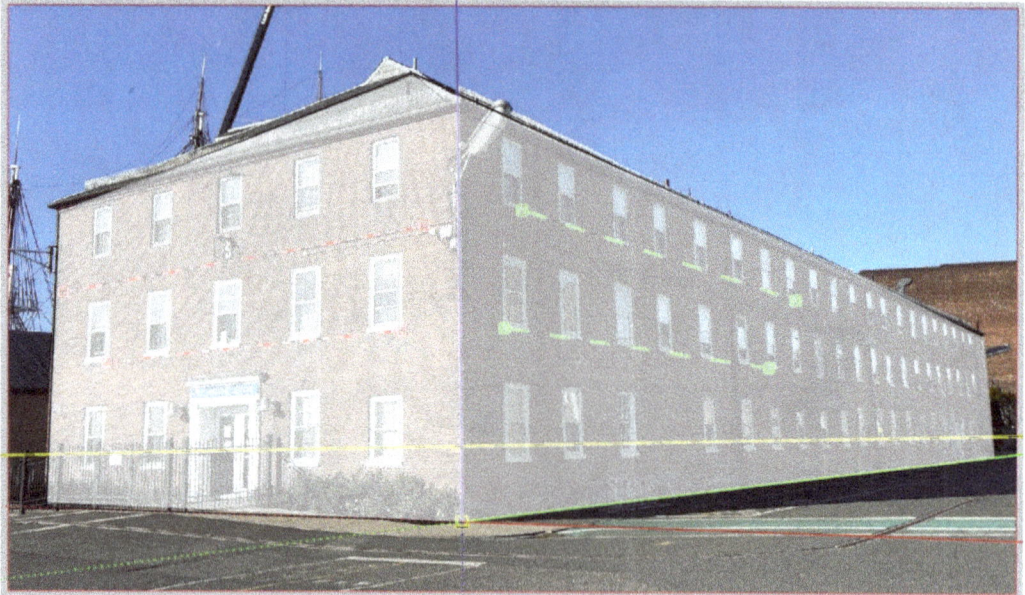

Figure 12.59 – The Model is Scaled and the Geometry Aligns with the Matched Photo

Now that the Matched Photo is complete, it can be Projected onto the Geometry. This can be done by Right–Clicking and choosing **Project Photo**, or by clicking the Project Textures From the **Photo** Button in the **Match Photo** Panel. If no Geometry is selected, SketchUp will attempt to Project the photo onto all Faces. This is not always desirable. In this example, we can see that the front and side wall received appropriate Projected Texture Materials, but Materials were also Projected onto the roof, which is not at the right angle for receiving a Projected Material. There are also trees in the SketchUp model that received the building Texture Material, which we would like to avoid.

Figure 12.60 – Roof and Trees Showing an Undesirable Projected Texture

Faces and Objects can be preselected in the model before the Matched Photo is right–clicked, and these will be the only Faces and Objects that will receive the Projected Materials. In this next example, we have preselected the two wall Faces in the model, then Right–Clicked –> **Project Photo**. When the camera is orbited around the model, we can see the roof and trees have not received the Projected Texture Material.

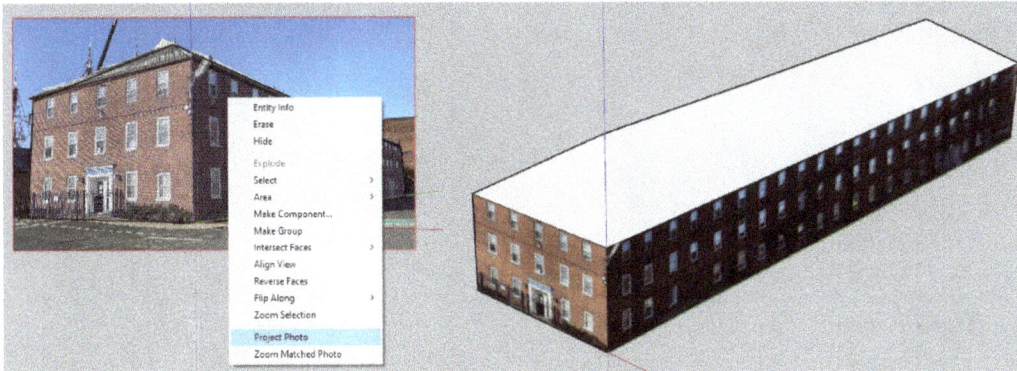

Figure 12.61 – Wall Faces Preselected (Left) with Right–Click –>
Project Photo (Middle) to Orbit the Camera (Right)

It is very difficult to align models with photos, even if the model has exactly the right dimensions. This can be because of camera warping/distortion, which will happen with any photo. Additionally, panoramic, stitched, and cropped photos should be avoided because of the associated distortion.

Try playing around with Matched Photos for yourself. Try turning on and off the different **Grid Styles**, **Planes**, and **Grid Spacing** in the **Match Photo** Panel to see how they impact aligning different types of photos! When you are done with a Matched Photo, it can be deleted by deleting the associated Scene from the Scene Bar or the **Scenes** Panel.

Export Options

Importing files into SketchUp Pro is only one of the ways SketchUp works with outside files – SketchUp Pro can also export files! SketchUp can export even more files than it can import. We will discuss exporting files in six categories – 3D Files, 2D Files, Image Files, Animations, 2D Section Slices, and Send to LayOut. Let's begin by looking at 3D Files.

Exporting 3D Files

SketchUp models can be exported by going to **File** –> **Export** –> **3D Model…**. This workflow will export the entire 3D Model.

| File | Edit | View | Camera | Draw | Tools | Window | Extensions | Help |

New   Ctrl+N
New From Template...
Open...   Ctrl+O

Save   Ctrl+S
Save As...
Save A Copy As...
Save As Template...
Revert

Send to LayOut...

Start PreDesign...

Geo-location   >
3D Warehouse   >
Import...
Export   >   3D Model...
                                                              2D Graphic...
Print Setup...   Section Slice...
Print Preview...   Animation...
Print...   Ctrl+P

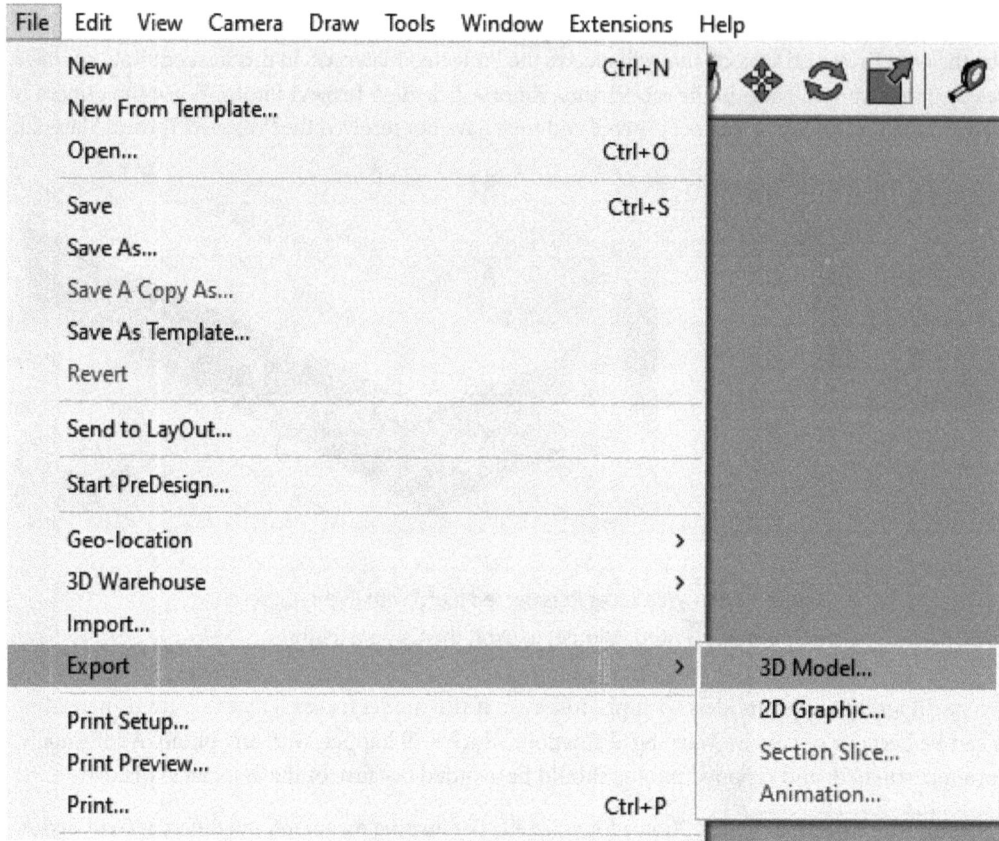

Figure 12.62 – File –> Export –> 3D Model…

This will open the **Export Model** Dialog Box. The file can be saved in many file types and these can be chosen from the **Save as type:** dropdown. Many of these file types have unique Export Options, and these can be accessed by clicking on the **Options…** Button.

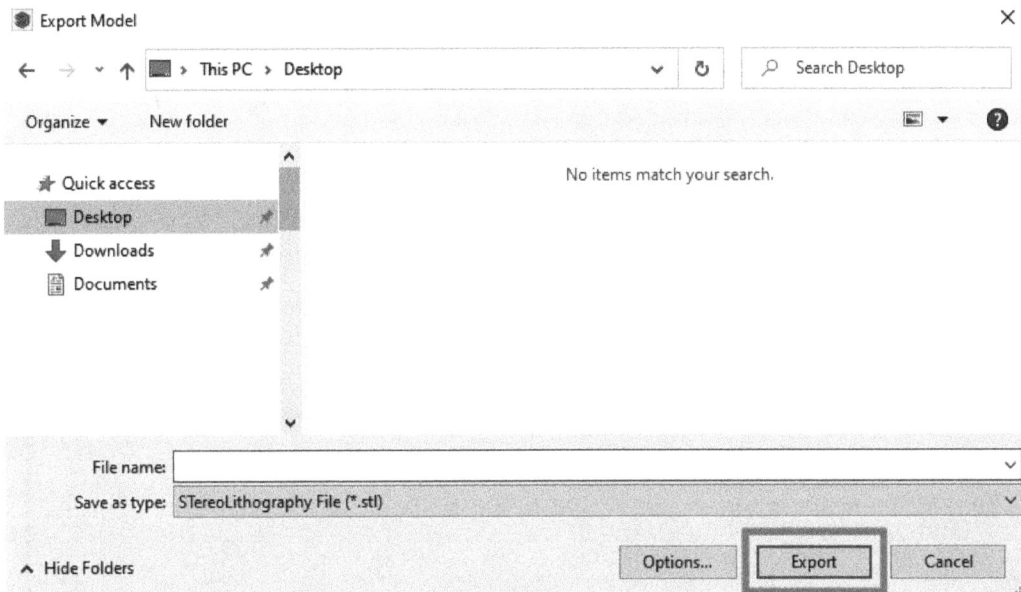

Figure 12.63 – The Export Model Dialog Box and Options Button

We have already discussed the files that SketchUp can import that contain 3D Geometry, including other SketchUp files (SKP), 3DS, DAE, DEM, DDF, DWG, DXF, IFC, KMZ, and STL files. With the exception of DAE and DEM files, SketchUp Pro can also natively export to all those file types as well! As we discussed in previous sections, each of those file types has limitations to what it supports, including materials, complex planar faces, and other model data. It is important to understand what the complete workflow will be when choosing the export file type. If at all possible, keeping the file as a SketchUp file is preferred!

We discussed these file types in the *Importing 3D Files* section of this chapter. In this section, we will quickly discuss the different Export Options for these files.

.SKP

SKP Files cannot be Exported through the **File** -> **Export** -> 3D Model workflow because the Export workflow is viewed as similar to a **Save As** workflow.

SketchUp models can be Saved by going to **File** -> **Save As**, which will open the **Save As** Dialog Box. This workflow is commonly used when trying to save a SketchUp model to an earlier version. In the **Save As** Dialog Box, the **Save as type:** selector will show all available SketchUp versions.

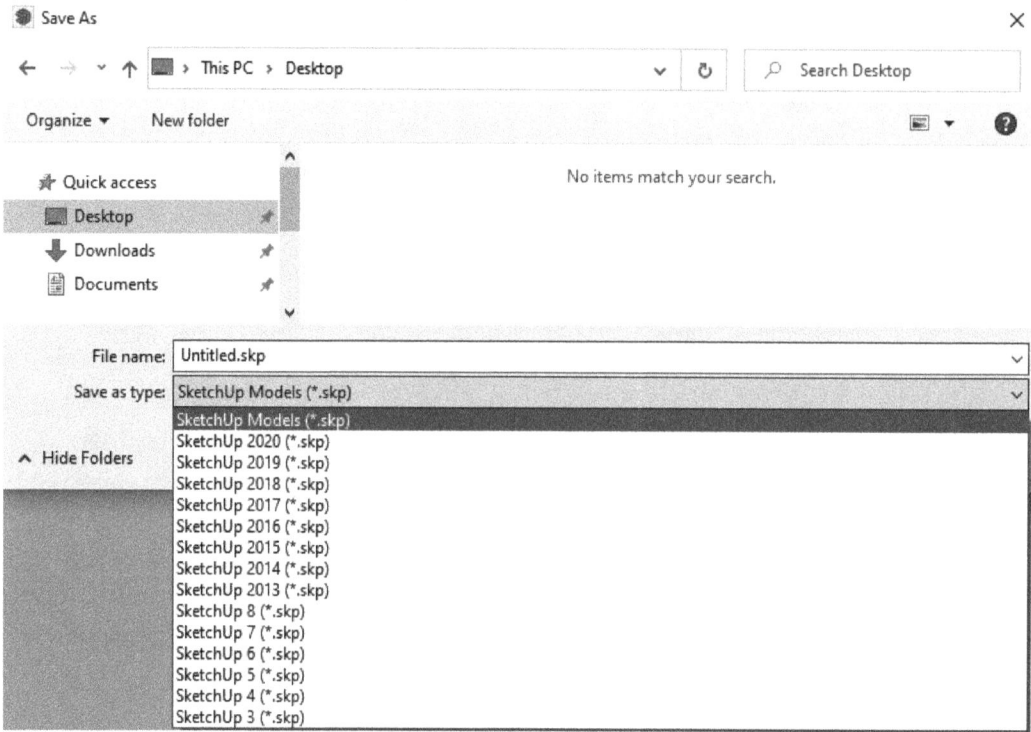

Figure 12.64 – The Save As Dialog Box with the Save As Type Dropdown Open

SketchUp models can also be saved by using the Save As workflow for specific Components. Components can be selected in the Drawing Area or the **Components** Panel and Right–Clicking –> **Save As**. This will open a similar **Save As** Dialog Box where the individual Components can be saved. The Components can also be saved to previous versions of SketchUp Pro.

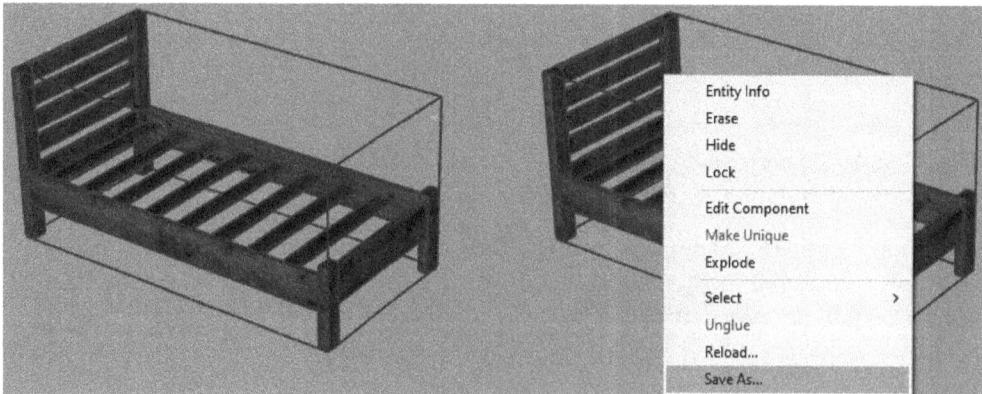

Figure 12.65 – Component Selected (Left) with Right–Click –> Save As (Right)

3DS

3DS Files have many Export Options, which are broken into four sections: **Export**, **Materials**, **Camera**, and **Scale**. The Export options allow for assigning hierarchy based on **Full hierarchy,** by Tag, by Material, or by a single object. Additionally, 3DS files can be exported for just the current selection, which cannot be done for all file types. Two–sided Faces can be exported in two ways – either by material or geometry. Additionally, stand–alone Edges can be exported independently of connected Faces.

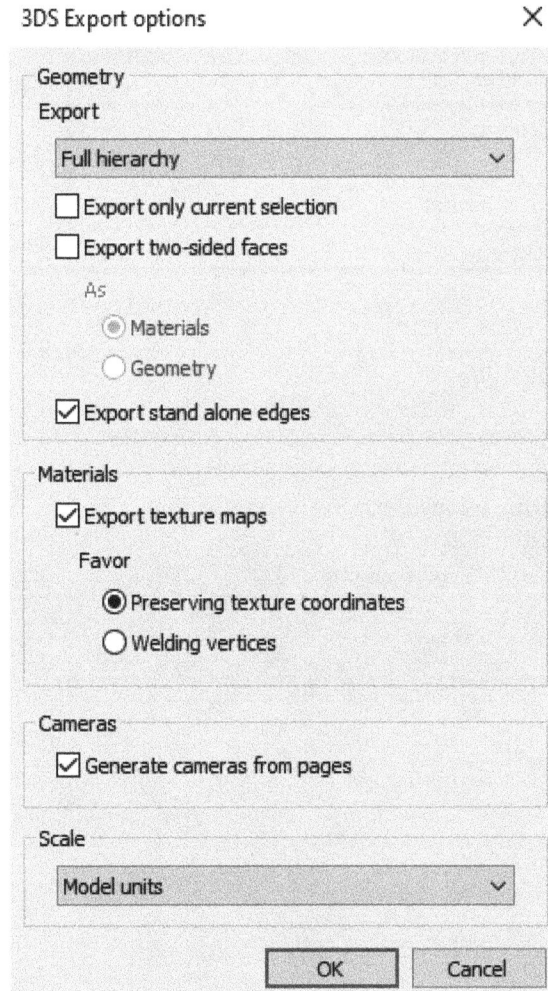

Figure 12.66 – The Export Section of the 3DS Export options Dialog Box

The **Materials** section has options for exporting texture maps, which can include texture coordinates and can weld vertices. Welding vertices can connect endpoints of Edges that are very close to one another but are not perfectly aligned.

The **Camera** section has one option to **Generate cameras from pages**, and the **Scale** section allows for the export scale to be set to **Model units** or a standard unit of measure.

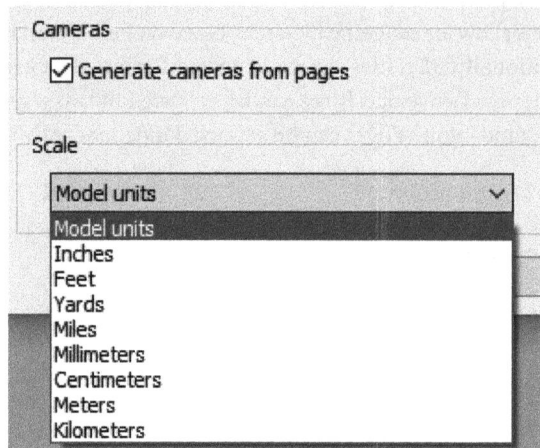

Figure 12.67 – The 3DS Export options Dialog Box with the Scale Dropdown Open

Autodesk DWG and DXF files

AutoCAD DWG and DXF files have the same Export Options. The **DWG/DXF Export options** Dialog Box allows for the version of AutoCAD to be selected, as well as options to include **Faces**, **Edges**, **Construction geometry**, **Dimensions**, **Text**, and **Materials**.

Figure 12.68 – The DWG/DXF Export options Dialog Box

> **Note**
> AutoCAD is commonly backward-compatible for working in AutoCAD, but some **Computer-Aided Manufacturing (CAM)** workflows will prefer older versions of AutoCAD. Most elements of a SketchUp model can be preserved in later versions of AutoCAD.

DAE Files

COLLADA DAE files have unique export options, including sections for **Geometry**, **Materials**, and **Credits**. The **Geometry** section includes options to **Export two-sided faces**, **Export edges**, **Triangulate all faces**, **Export only selection set**, **Export hidden geometry**, and **Preserve component hierarchies**. The **Materials** section has the option to **Export texture maps** and the **Credits** section has the option to **Preserve credits**.

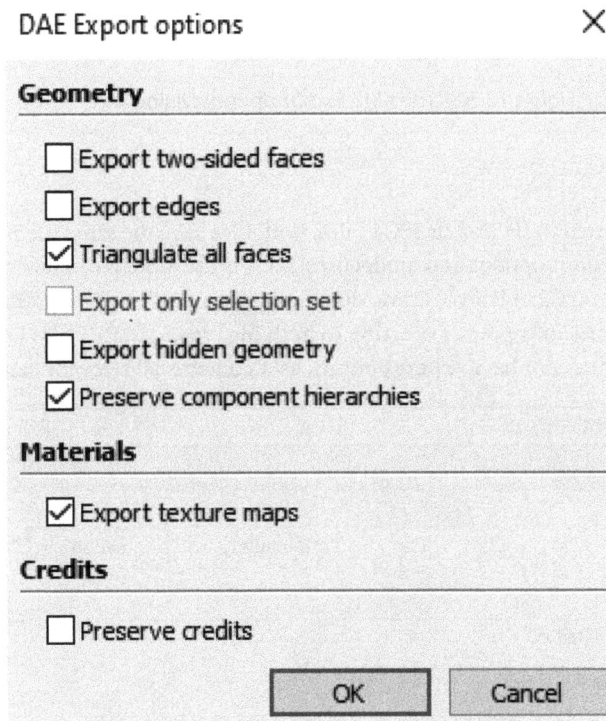

Figure 12.69 – The DAE Export options Dialog Box

> **Note**
> The **Export only selection set option** will be grayed out if no selection was made before choosing **Export -> 3D Model**.

.KMZ

Google Earth Files only have two additional export options – **Export hidden geometry** and **Preserve credits**.

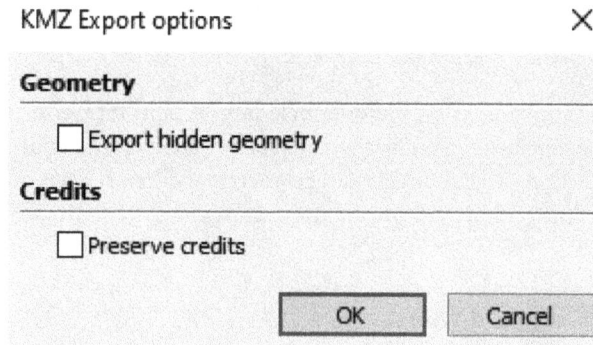

Figure 12.70 – The KMZ Export options Dialog Box

.IFC Files

IFC models can be exported to IFC2x3 or IFC4 Files. Both files have the same file extension (`*.ifc`), but they have different export options and model uses. IFC4 is the more recent file version released for use in BIM models, while IFC2x3 is an older version. Both files are still used in professional workflows, so SketchUp Pro continues to support exporting to both file types. IFC2x3 files have export options in SketchUp, but IFC4 does not have export options, as it contains all relevant data in the model.

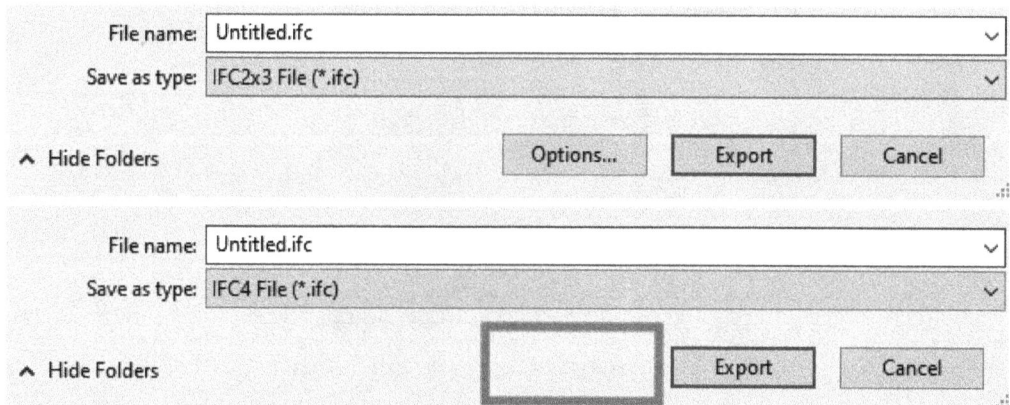

Figure 12.71 – The IFC2x3 Options Button (Top) and the IFC4 equivalent with No Options Button (Bottom)

IFC2x3 files have a unique **IFC Export options** Dialog Box, as specific IFC elements can be selected for export. By default, all IFC elements will be selected, which will make the selection panel appear blue.

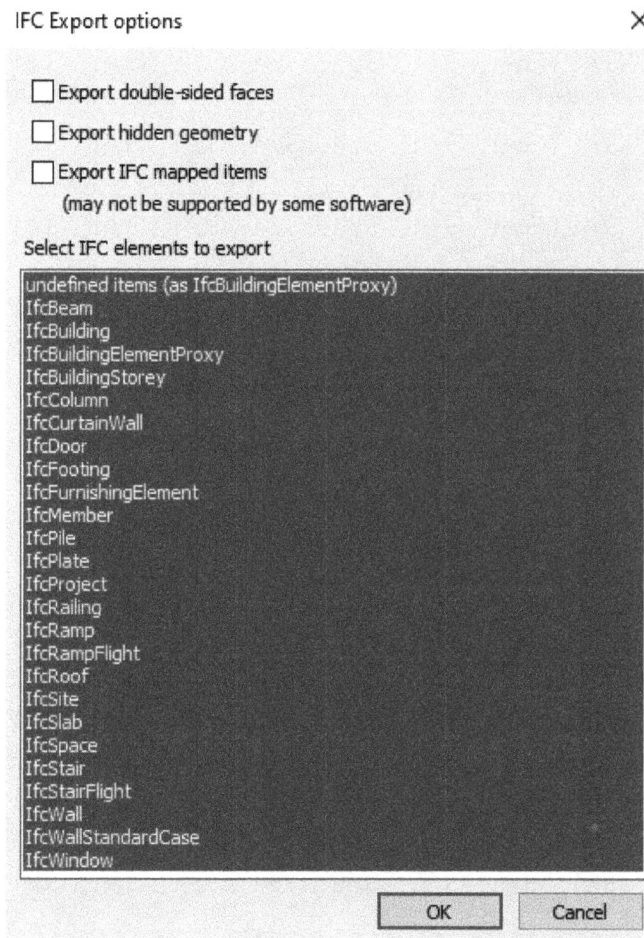

Figure 12.72 – The IFC2x3 Export options Dialog Box

Individual IFC elements can be selected with a single click, and the *Ctrl* and *Shift* modifier keys can be used to add and remove elements to the selection.

The **IFC Export options** Dialog Box also has options to **Export double–sided faces**, **Export hidden geometry**, and **Export IFC mapped items**.

STL Files

STL Files have a few important Export Options, which are broken into three sections: **Geometry**, **File Format**, and **Scale**. The export options allow for the current selection to be exported if the Geometry is preselected. Additionally, the file can be exported to **ASCII** or **Binary** format. The file can be set to swap the YZ Coordinates, which can be important when switching between a design workflow to a manufacturing or 3D printing workflow. The model can be scaled using **Model units** or a standard unit of measure.

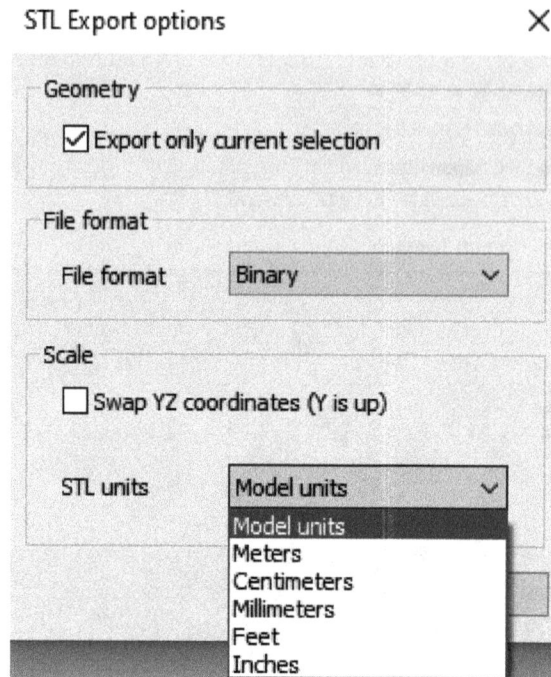

Figure 12.73 – The Export Section of the STL Export options Dialog Box with the STL units Dropdown Open

There are five additional file types that SketchUp Pro can export. Let's take a look at each of these file types in more detail.

TrimBIM Files (.TRB)

TRB files have some of the fewest Export Options, only having Export Selection Set Only and Export Hidden Geometry. It may be expected that there are few Options as this is the native BIM file type for Trimble, the parent company that publishes SketchUp Pro. These two Options are self–explanatory, but they should be considered when creating a TrimBIM export.

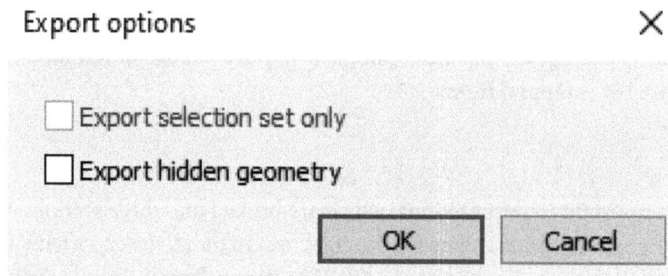

Figure 12.74 – TRB Export Options Dialog Box

TrimBIM exports may be required when creating a backup file, or when trying to collaborate with a different Trimble software. TrimBIM files are not recommended when trying to work collaboratively with Autodesk software, or any software not created by Trimble.

Autodesk Filmbox (FBX)

The FBX file type was originally developed by Kaydara for Kaydara MotionBuilder, which was acquired by Autodesk. FBX files are a common file format, typically for solid–body modeling and rendering/animation.

FBX files are often thought of as the best file type to use when moving information between software platforms because they contain a large variety of model elements, even more than SketchUp can natively create! FBX files can contain 3D models (both triangulated mesh and NURBS surfaces and curves), scene hierarchy, animations, lighting, materials, and more detailed animation modeling elements.

FBX files have three export options sections: **Geometry**, **Materials**, and **Scale**. The **Geometry** section includes options to **Export only current selection**, **Triangulate all faces**, **Export two–sided faces**, and **Separate disconnected faces**. The **Materials** section has the option to **Export texture maps** and the **Scale** section has the option to Swap the YZ Coordinates and scale using **Model units** or a standard unit of measure.

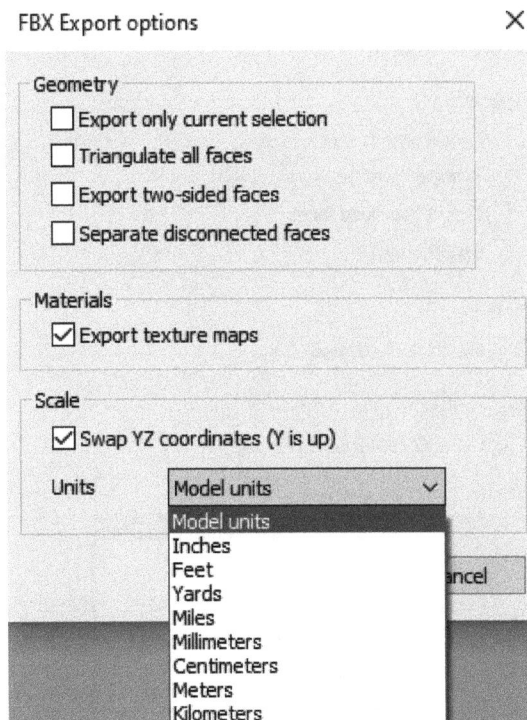

Figure 12.75 – The FBX Export options Dialog Box with the Units Dropdown Open

> **Note**
> The **Export only current selection** option will be grayed out if no selection was made before choosing **Export -> 3D Model**.

Object Files (.OBJ)

The OBJ file type was originally developed by Wavefront Technologies. OBJ files are the standard for transferring models between software platforms. All OBJ files are text-based files that list the 3D points for each face of a 3D model, and that is all! OBJ files do not contain any other information except for the faces/surfaces for a 3D model. The file type is very simple but it can lead to large files if the 3D geometry is very complex.

OBJ files are commonly used in 3D printing workflows because there is no unnecessary data for the 3D printer to recognize – only the 3D object.

OBJ files have very similar Export Options to FBX Files, including **Geometry**, **Materials**, and **Scale** sections. The **Geometry** section includes options to **Export only current selection**, **Triangulate all faces**, **Export two-sided faces**, and **Export edges**. The **Materials** section has the option to **Export texture maps**, and the **Scale** section has the option to Swap the YZ Coordinates and scale using **Model units** or a standard unit of measure.

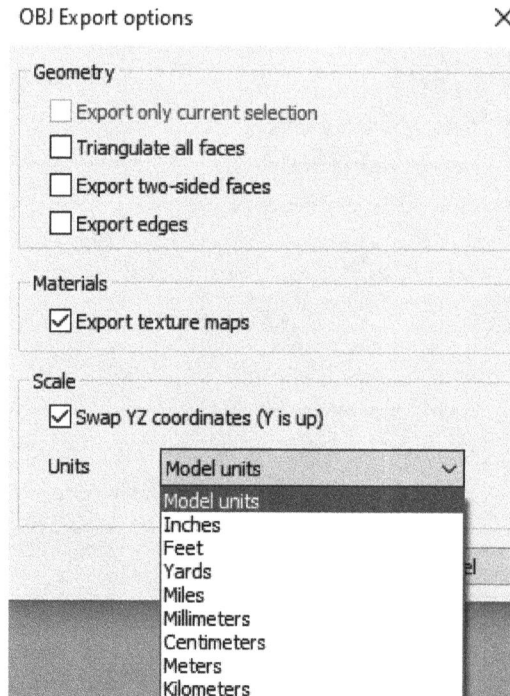

Figure 12.76 – The OBJ Export options Dialog Box with the Units Dropdown Open

> **Note**
>
> The **Export only current selection** option will be grayed out if no selection was made before choosing **Export -> 3D Model**.

GL Transmission Files (.GLTF and .GLB)

GL Transmission Files (**GLTF**) and **GL Transmission Binary** (**GLB**) Files are files that are commonly used in **Virtual Reality** (**VR**), **Augmented Reality** (**AR**), gaming, and web–based applications. GLTF files are coded in JSON format, and GLB files are coded in Binary format. Because of these encoding structures, information including 3D geometry, textures/materials, shaders, camera locations, and animation can be stored in these files.

These files do not have any specific export options from SketchUp Pro, so there is no **Options** button next to the **Export** button in the **Export Model** Dialog Box.

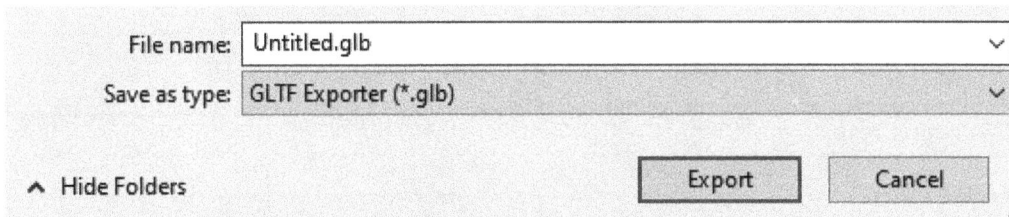

Figure 12.77 – No Options Button in the Export Model Dialog Box

Virtual World Files (.WRL)

WRL files are programmed in **Virtual Reality Modeling Language** (**VRML**) and are typically used in Blender to store 3D models. WRL files contain 3D information in plain text, similar to OBJ files, but WRL files can contain more data about the geometry.

VRML files have unique export options compared to the other file types, including **Output texture maps**, **Ignore back face of material**, and **Output edges**. Additionally, files can be exported using VRML standard orientation, generate cameras, allow mirrored components, and check for material overrides.

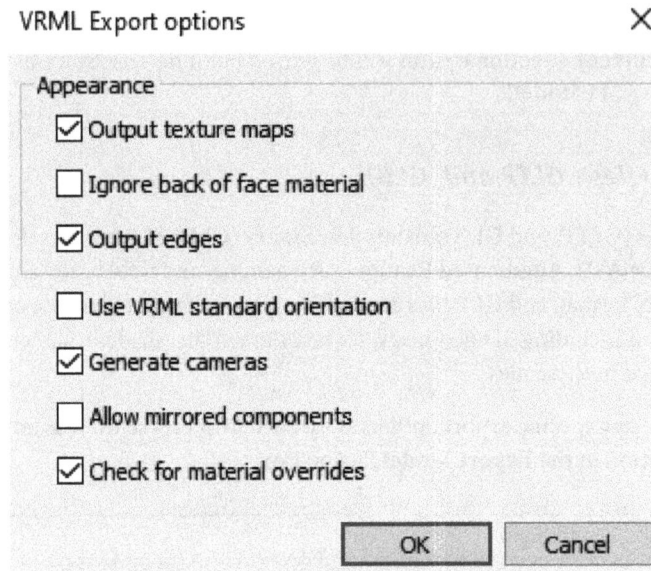

Figure 12.78 – The VRML Export options Dialog Box

XSI Files (.XSI)

XSI files were originally created by Softimage to save 3D character information. XSI files are complex files and contain triangulated mesh surfaces/faces, materials, shadows, and more detailed animation modeling elements. XSI files are now commonly associated with Microsoft InfoPath.

All of the 3D file export workflows follow the same process, although there are some differences in the export options.

XSI Files have identical export options to OBJ Files, including **Geometry**, **Materials**, and **Scale** sections. The **Geometry** section includes options to **Export only current selection**, **Triangulate all faces**, **Export two-sided faces**, and **Export edges**. The **Materials** section has the option to **Export texture maps** and the **Scale** section has the option to Swap the YZ Coordinates, and scale using **Model units** or a standard unit of measure.

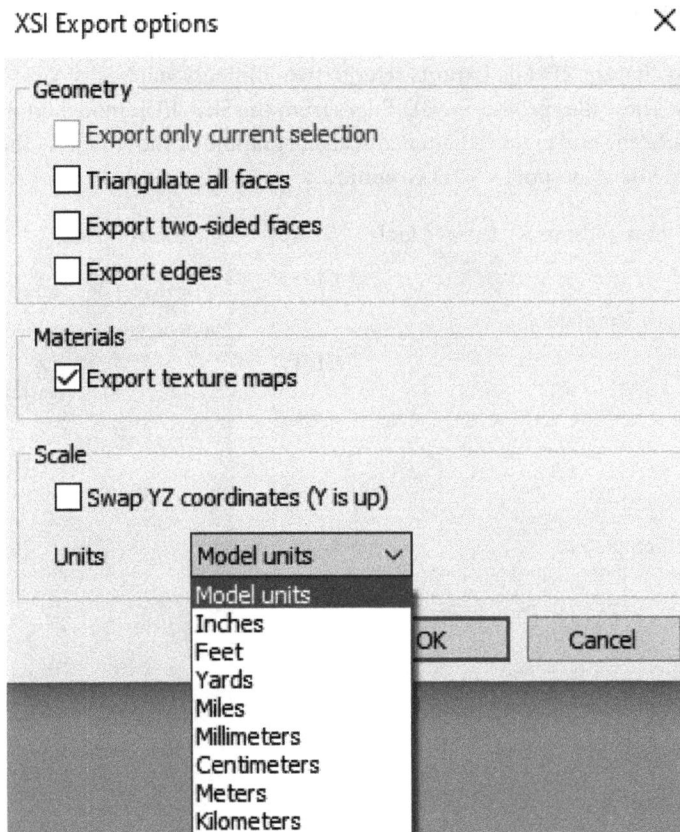

Figure 12.79 – The XSI Export options Dialog Box with the Units Dropdown Open

> **Note**
> The **Export only current selection** option will be grayed out if no selection was made before choosing **Export –> 3D Model**.

It is very unlikely that you will ever need to export more than one or two of these file types! In the next section, we will discuss how we can convert our 3D models into exported 2D Files.

Exporting 2D Files

In the context of this chapter, 2D File Exports refer to two-dimensional Vector Graphics, including DWG and DXF files. These files project the 3D Edges from the SketchUp model onto the projection plane of the Camera/Scene and create a flattened scalable graphic of those Edges. These files can be Exported by going to **File** -> **Export** -> **2D Graphic...**.

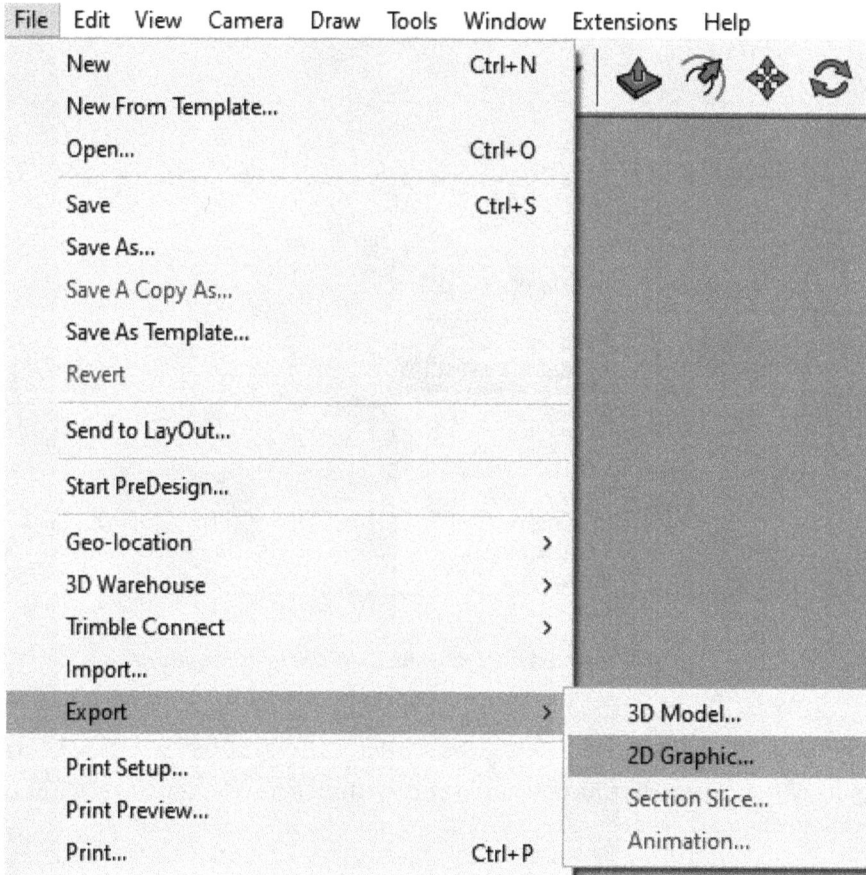

Figure 12.80 – File -> Export -> 2D Graphic...

> **Note**
> These DWG and DXF Files are different than the fully 3D files discussed in the *Exporting 3D Files* section. Be sure that you are choosing the right workflow for your project!

When exported as 2D graphics, DWG and DXF files have the same Export Options. The files can be selected independently from the **Save as type:** dropdown, but the **Options** Button will open the same **DWG/DXF Export options** Dialog Box.

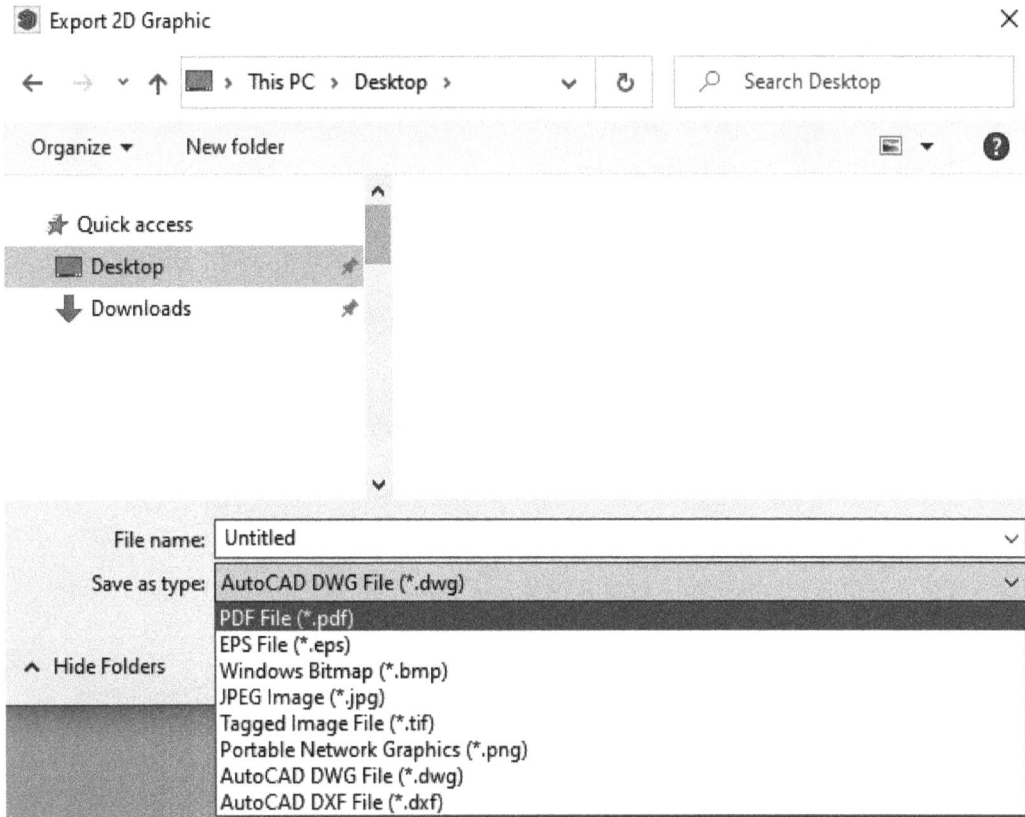

Figure 12.81 – The Export 2D Graphic Dialog Box with the Dropdown Selector Open

This **DWG/DXF Export options** Dialog Box has options for the **AutoCAD version**, **Drawing scale & size**, **Profile lines**, **Section lines**, and **Extension lines** settings.

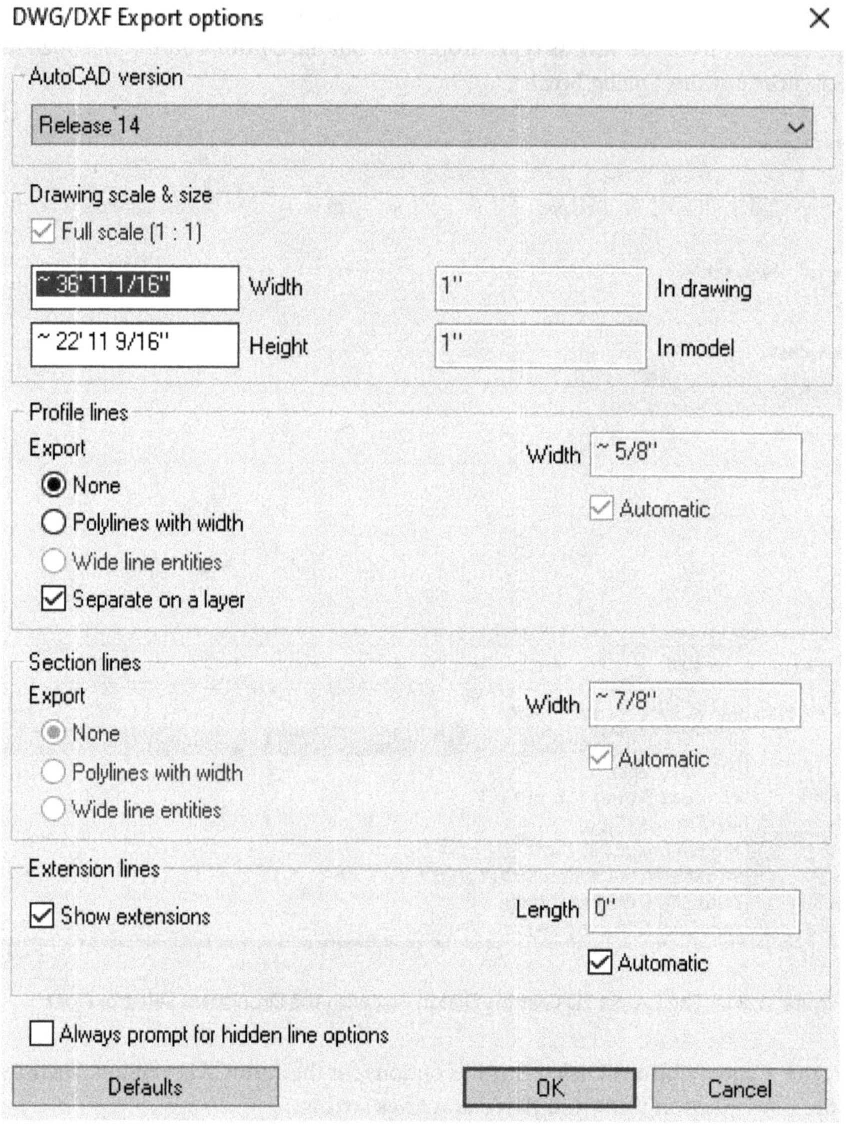

Figure 12.82 – The DWG/DXF Export options Dialog Box

Note

By default, DWG and DXF 2D files do not Export information such as the line weight, Material colors or Textures, or Faces; only the vector Edge Geometry is projected. EPS and PDF files do have the ability to Export those options, and we will discuss EPS and PDF files in the *Image Files* section of this chapter.

The AutoCAD version can be set from the dropdown, although the Edges exported from SketchUp Pro will be identical in the different AutoCAD versions. The **Drawing scale & size** section has options that convert the size of the visible drawing area into the real dimensions for an exported 2D file. This can be very difficult to set correctly, especially because the file is projecting lines to create a 2D image. In this case, it is suggested to approximate the desired size of the final image and set the width appropriately.

The **Profile lines** section has a default setting of **None**, but this can be toggled to **Polylines with width** or **Wide line entities**. It is recommended to choose **None** or **Polylines with width**. If **Polylines with width** is selected, they can be placed on a separate layer, and the width can be overridden by unchecking the **Automatic** checkbox and specifying a **Width** value.

> Note
> Placing Profile Edges on a separate layer does not duplicate the Edges in the resulting DWG. The Edges will simply be set on two layers of the DWG, which can then be set to different appearance settings in a different software program.

Section Lines are similar to Profile Lines and can be exported with the same set of options. Section Lines are always added to their own layer.

Extension Lines can be included or excluded from the DWG export, as they are typically only included for visual interest, and may not be desired in CAD software. The length of the Extension Lines can also be set specifically if desired.

Always prompt for hidden line options is at the bottom of the Dialog Box and can be turned on to show a prompt for hidden lines. Additionally, there is a **Defaults** button, which will restore defaults across the Dialog Box.

When exported, the resulting DWG will only have Edge Geometry, and will not differentiate between Groups or Components. All Lines in the DWG/DXF file will be set to black/white, although the Lines may be on different layers as specified in the **Export Options** Dialog Box.

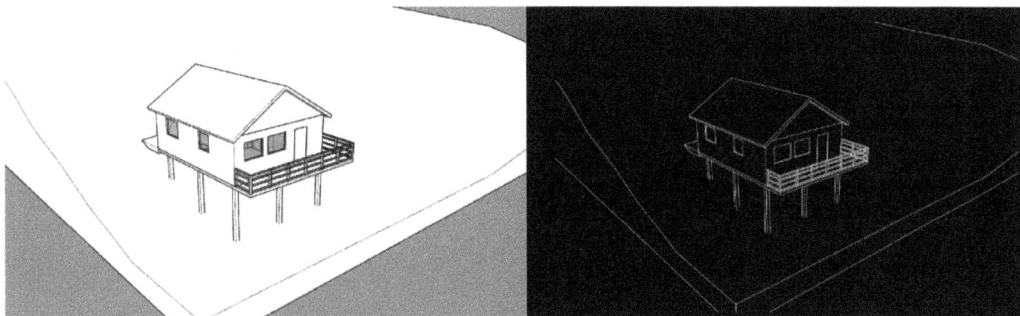

Figure 12.83 – SketchUp Drawing Area (Left) and the Exported 2D AutoCAD Drawing (Right)

These exported 2D Files look very strange when viewed in a perspective view, as they are a projected view of a perspective, to begin with! In this next example, a 2D Graphic DWG has been re–imported into SketchUp Pro, which, when viewed in 3D, appears to be drawn incorrectly. It is recommended that 2D Graphic DWGs be used only in image processing software for visual presentations and not brought back into SketchUp for additional design.

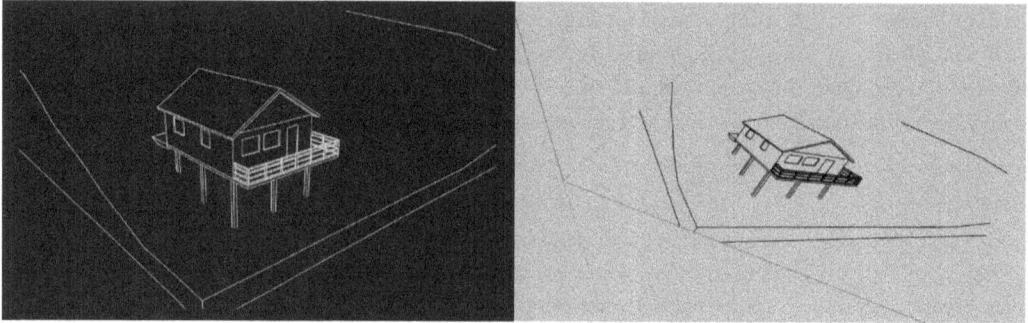

Figure 12.84 – Exported 2D AutoCAD Drawing (Left) and an
AutoCAD Drawing Imported to SketchUp Pro (Right)

As mentioned earlier, these files are Vector files but only contain Edge Geometry. In the next section, we will discuss Vector Image Files (EPS and PDF), and Raster Image Files (BMP, JPG, TIF, and PNG).

Image Files

In the context of this chapter, Image Files mean 2D Image files that contain an accurate visual representation of the Drawing Area, including Colors/Textures, Shadows, and Geometry. These files are Vector Image Files, which are infinitely scalable and made from 2D Profiles, and Raster Image Files, which are not infinitely scalable and are made of Pixels. In this example, we can see the same Scene has been exported as an EPS and a PNG.

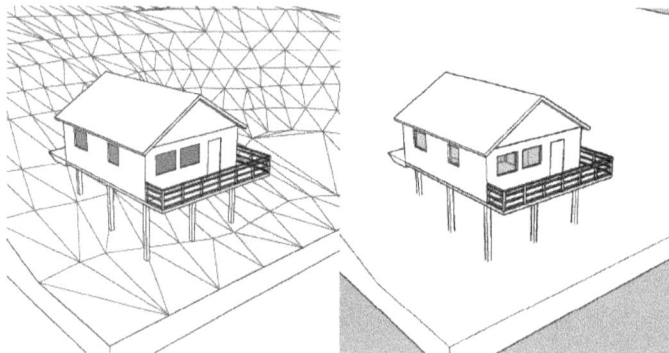

Figure 12.85 – An EPS Vector Export (Left) and a PNG Raster Export (Right)

When zoomed in, we can see that the EPS file is scalable and does not pixelate, while the PNG is limited to the number of pixels it was originally exported at.

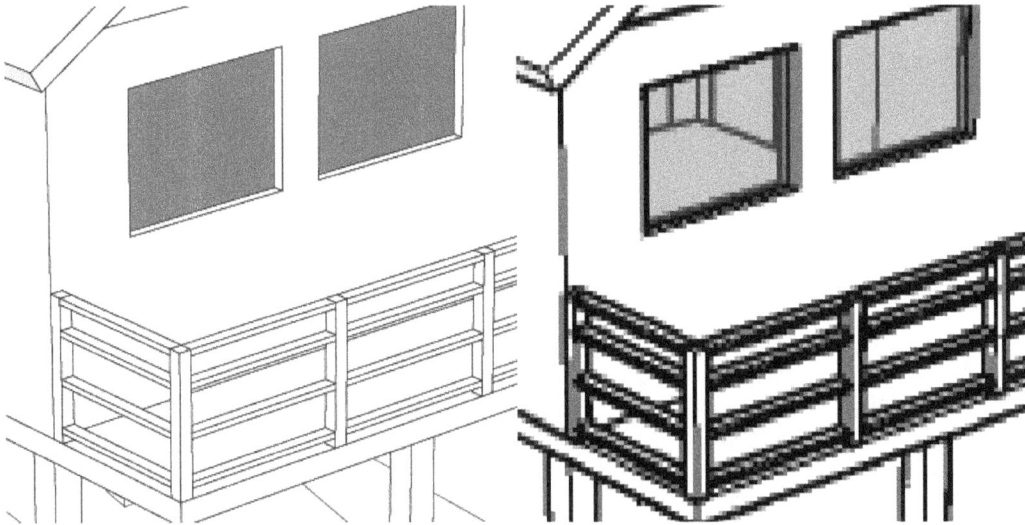

Figure 12.86 – Zoomed EPS Vector Export (Left) Zoomed PNG Raster Export (Right)

The advantage of Raster over Vector is that Raster can be edited using more common software applications. Many post–production workflows are performed in Adobe Photoshop, which specializes in Raster files.

In the following sections, we will review the Export Options for these file types.

Vector Image Files (.PDF and .EPS)

PDF files were created by Adobe to originally create a universal file format for documents that were not required to be opened by one specific software. PDF files are commonly considered to be converted word–processing documents, but that is not the only use case. PDF files can now contain text, raster and vector graphics, and even interactive 3D elements. PDFs exported from SketchUp Pro contain a 2D representation of the 3D Drawing Area view, where the Edges and Faces are converted into paths and filled profiles.

Encapsulated Postscript (**EPS**) files were also originally created by Adobe, but these files were primarily designed to be image files. EPS files are vector files created using paths and filled profiles. EPS files have fewer functions than PDF files, and most modern workflows prefer to use PDF files instead of EPS files.

PDF and EPS files have similar export options to DWG/DXF, where the image will be converted from real-world dimensions to represent the final size of an image. SketchUp has the option to set **Drawing size** to **Full scale (1 : 1)**, or specific **Width** and **Height** values can be set. The remaining options are similar to DWG as well and include **Match screen display (auto width)**, **Show profiles**, **Extend edges**, **Always prompt for hidden line options**, and **Map Windows fonts to PDF base fonts**.

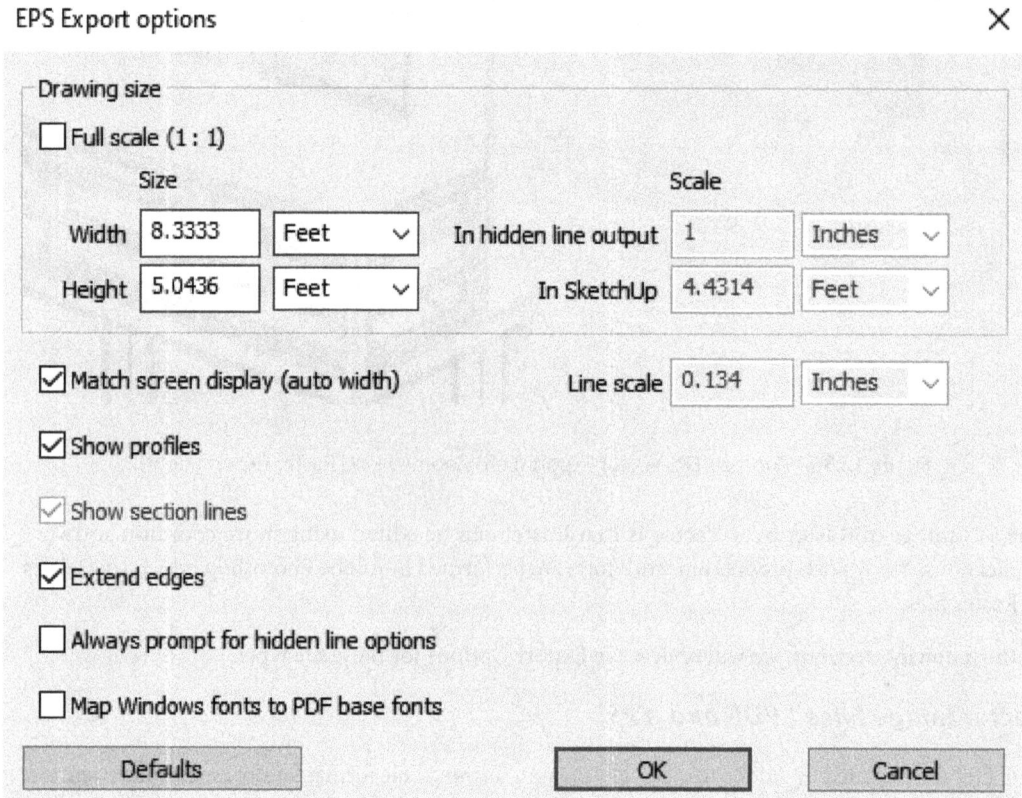

Figure 12.87 – The PDF and EPS Export options Dialog Box

Exported PDF and EPS files will have transparent backgrounds and will not export the ground plane, horizon line, or sky gradient. PDF and EPS files will also ignore Shadows, Fog, Guides, Axes, and Sections. Exported PDF and EPS files will focus on modeled Geometry and Objects.

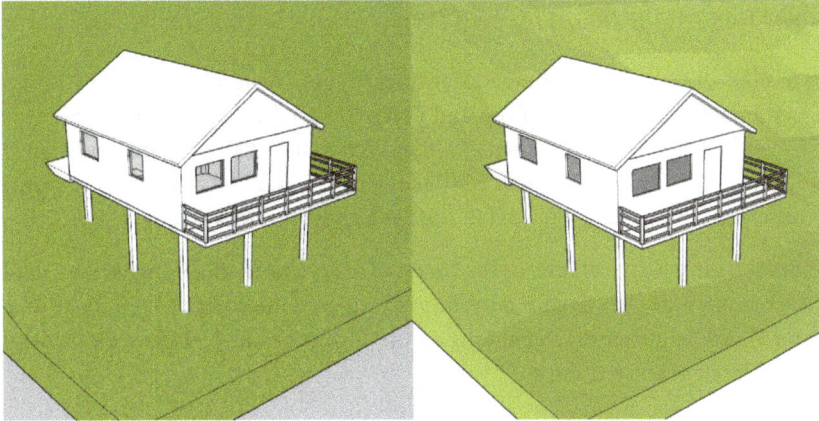

Figure 12.88 – SketchUp Model Drawing Area (Left) and the Exported PDF File (Right)

> **Note**
> SketchUp Tags (formerly Layers) will not be exported to PDF layers using this workflow. PDF documents exported through this workflow will not have PDF layers.

EPS and PDF files exported as 2D Graphics will be exported with the Shaded visual style, not Shaded with Textures. The Shaded visual style averages the color of Texture Materials, which are Raster files. EPS and PDF files exported as 2D Graphics are vector files, and only support paths and filled profiles, which are filled with a single color. In this example, we first see a SketchUp model with the Shaded with Textures face style, then with the Shaded face style, and the resulting exported EPS file using the Shaded appearance.

Figure 12.89 – Shaded With Textures (Left) and the Shaded (Middle) Exported EPS File (Right)

Raster Image Files (BMP, JPG, TIF, and PNG)

SketchUp Pro has the ability to export four Raster Image types, Windows BMP, JPEG Image (JPEG and JPG), Tagged Image File (TIFF and TIF), and PNG. We briefly discussed these file types in the *Importing Image Files* section of this chapter. The main differences between these four file types are that JPEG files have better compression than BMP files and TIF and PNG files support transparency. These unique file characteristics are reflected in the file **Export options** Dialog Boxes.

All four files have the same Export Options for **Image size** and **Anti-alias**. **Image size** can be set to match the pixels that the Drawing Area is currently showing on the computer monitor, or a specific size can be set for **Width** and **Height**. Depending on the size of the image, **Line scale multiplier** may be used to create thinner or thicker lines.

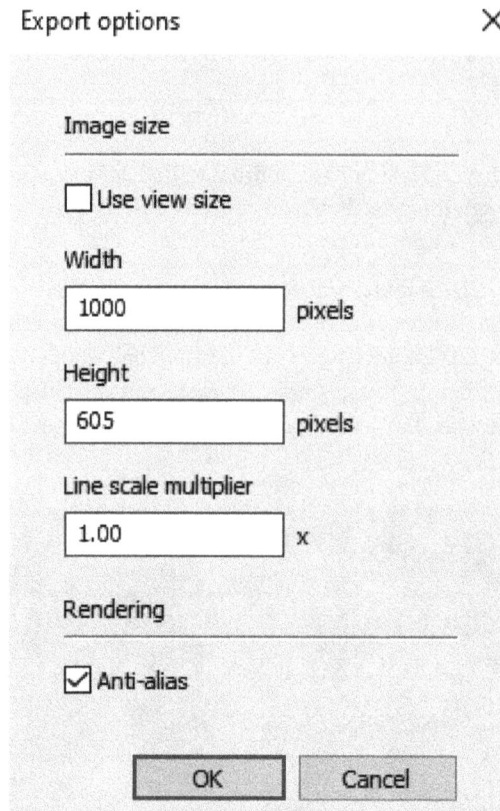

Export options   ✕

Image size
_____

☐ Use view size

Width
┌─────────────────────────┐
│ 1000   │   pixels
└─────────────────────────┘

Height
┌─────────────────────────┐
│ 605   │   pixels
└─────────────────────────┘

Line scale multiplier
┌─────────────────────────┐
│ 1.00   │   x
└─────────────────────────┘

Rendering
_____

☑ Anti-alias

┌──────────┐   ┌──────────┐
│   OK   │   │ Cancel │
└──────────┘   └──────────┘

Figure 12.90 – The BMP Export options Dialog Box

The JPEG **Export options** Dialog Box includes a slider to set the level of compression for the image. The TIF and PNG Dialog Boxes include a **Transparent background** option, which ignores the ground plane, horizon line, and sky gradient.

Export options ✕

Image size

☐ Use view size

Width

| 1000 | pixels

Height

| 605 | pixels

Line scale multiplier

| 1.00 | x

Rendering

☑ Anti-alias

JPEG compression

Smaller file ———————▮ Better quality

[OK]   [Cancel]

Export options ✕

Image size

☐ Use view size

Width

| 1000 | pixels

Height

| 605 | pixels

Line scale multiplier

| 1.00 | x

Rendering

☑ Anti-alias

☐ Transparent background

[OK]   [Cancel]

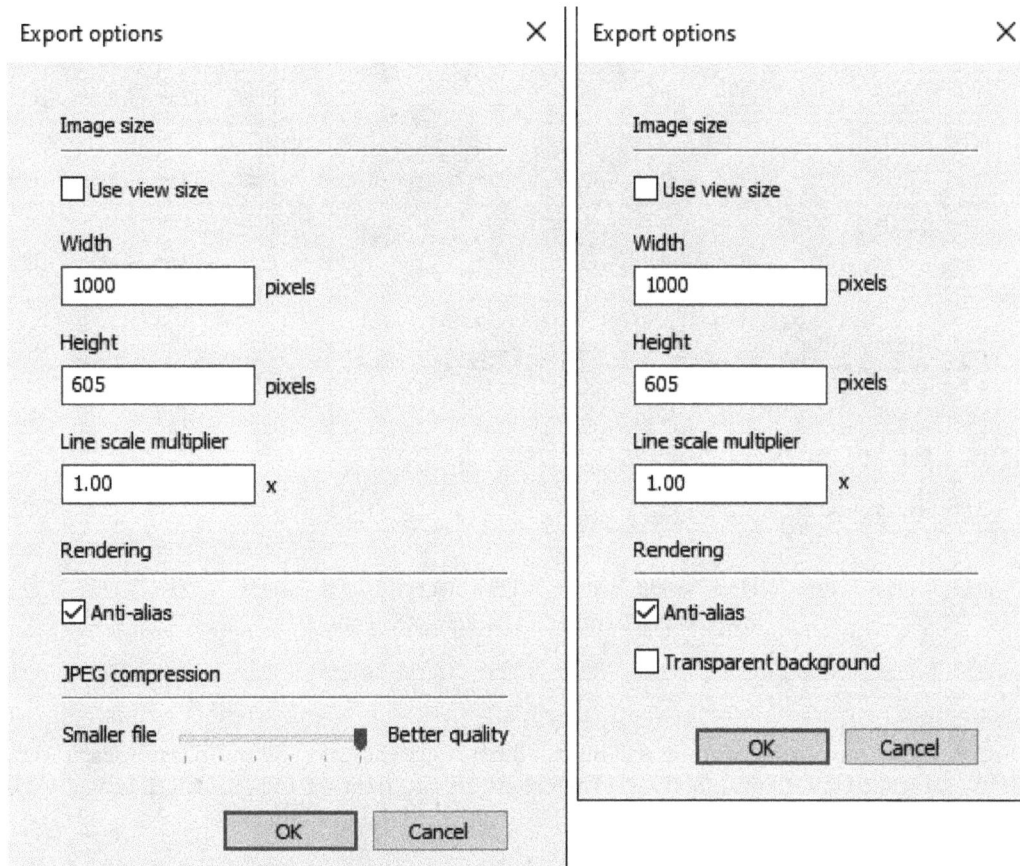

Figure 12.91 – The JPEG Export options Dialog Box (Left) and the
TIF and PNG Export options Dialog Box (Right)

Exported Raster Image files will match the current Style exactly, including exporting Texture Materials, Fog, Shadows, Sections, and Guides. Axes will not be included in the final image export. The Sky will also be included in any transparent Raster Image files.

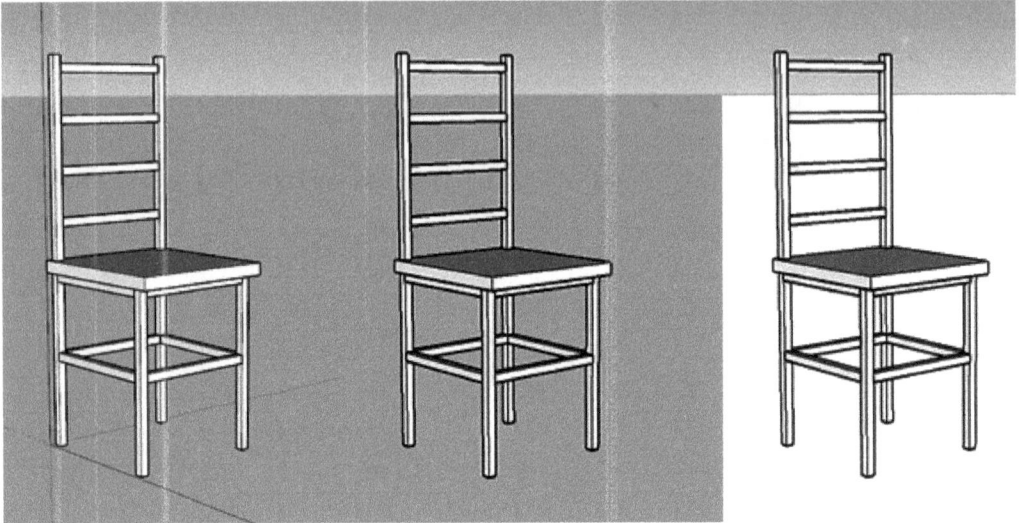

Figure 12.92 – Drawing Area (Left), JPEG Image (Middle), and PNG
Image with a Transparent Background (Right)

> **Note**
> Fog, Shadows, and visible Sections will impact the transparency of PNG and TIF images. It is recommended to turn these options off before exporting a transparent PNG or TIF image.

Some image post–processing workflows include a combination of multiple exported images with both raster and vector file types! Explore these image types in your own projects!

Section Slices

Section Slices are 2D DWG/DXF files that are created using the Active Section Cut. Section Slices cannot be exported if there are no Section Cuts in the SketchUp model, or if there is not an Active Section Cut. Section Slices can be Exported by clicking **File** –> **Export** –> **Section Slice…**

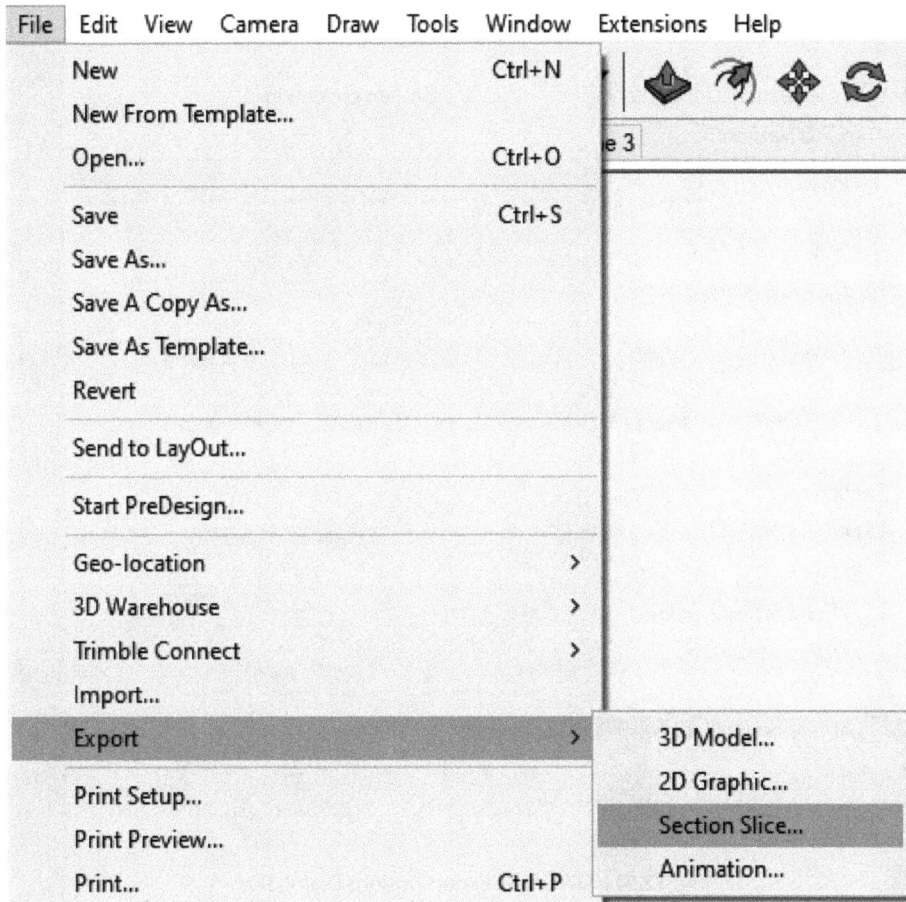

Figure 12.93 – File –> Export –> Section Slice

Section Slices exported as DWG or DXF files have identical Export Options. Section Slices can be exported as **True section (Orthographic)** or **Screen projection**.

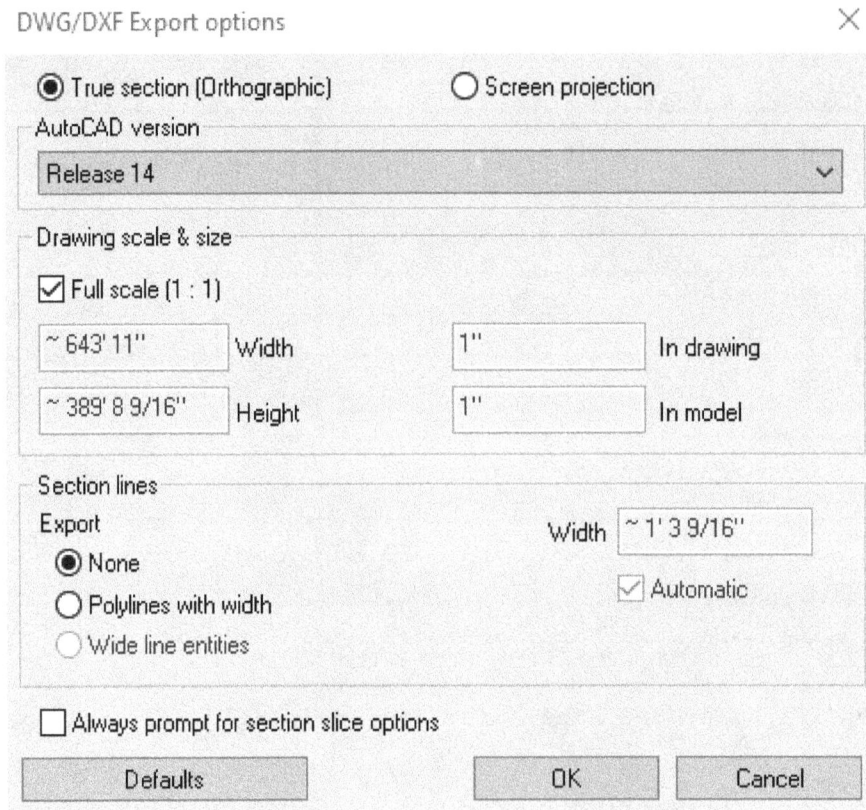

Figure 12.94 – DWG/DXF Export options Dialog Box

The next example shows the SketchUp model cut with an active section, and an example of a True Section and a Screen Projection Section.

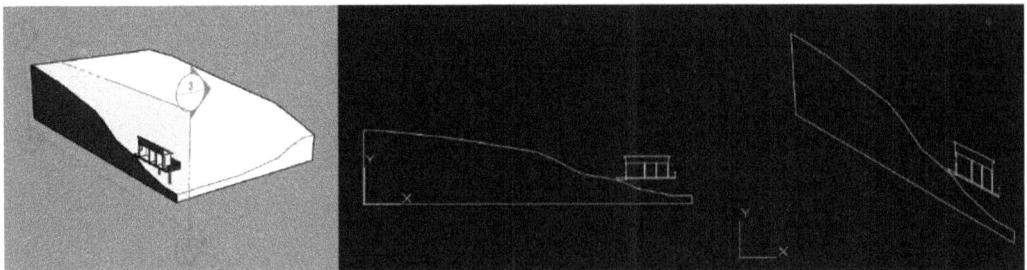

Figure 12.95 – SketchUp Model (Left), True Section (Middle), and Screen Projection Section (Right)

Screen Projection Section Slices are identical to 2D Graphics exported as DWG/DXF files that isolate the SectionCutEdges layer. Any Geometry not cut by the Section is ignored in the Section Slice export.

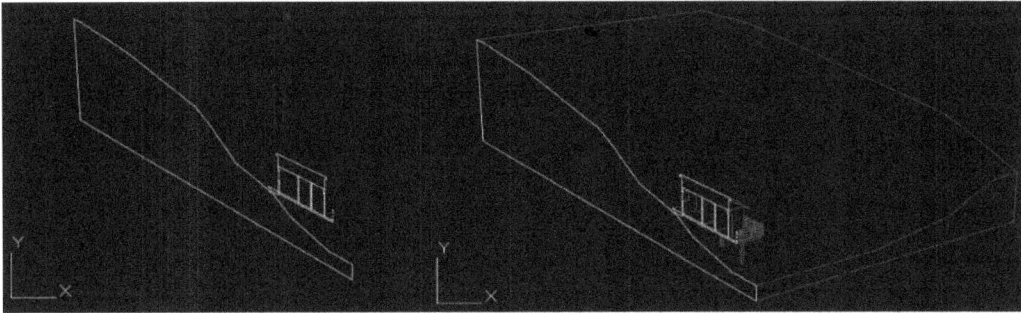

Figure 12.96 – 2D Graphic DWG Export (Left) and Screen Projection Section Slice (Right)

True Section (Orthographic) is created perpendicular to the Section Cut with an Orthographic camera. True Sections can be created regardless of the camera view. Similar to Screen Projection Section Slices, True Sections will also isolate the SectionCutEdges layer in the DWG export.

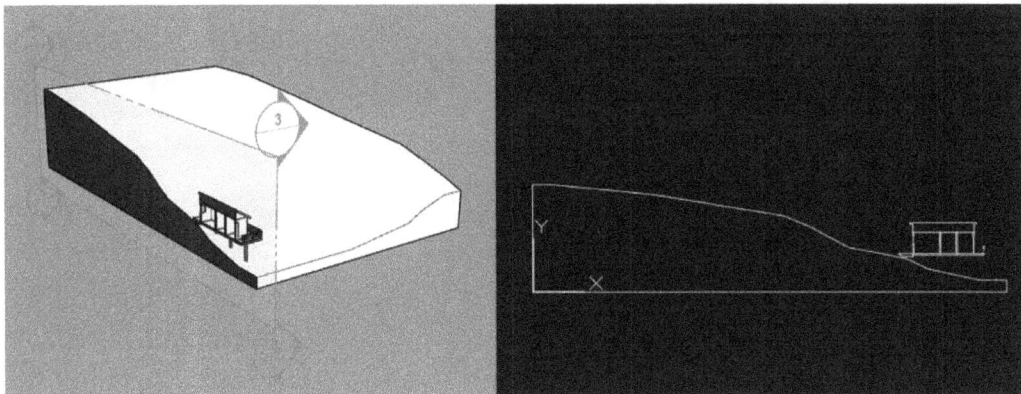

Figure 12.97 – SketchUp Model Drawing Area (Left) and True Section Slice (Right)

True Sections can be visualized in SketchUp using the Parallel Projection camera, right–clicking on the Section Cut, and choosing **Align View**.

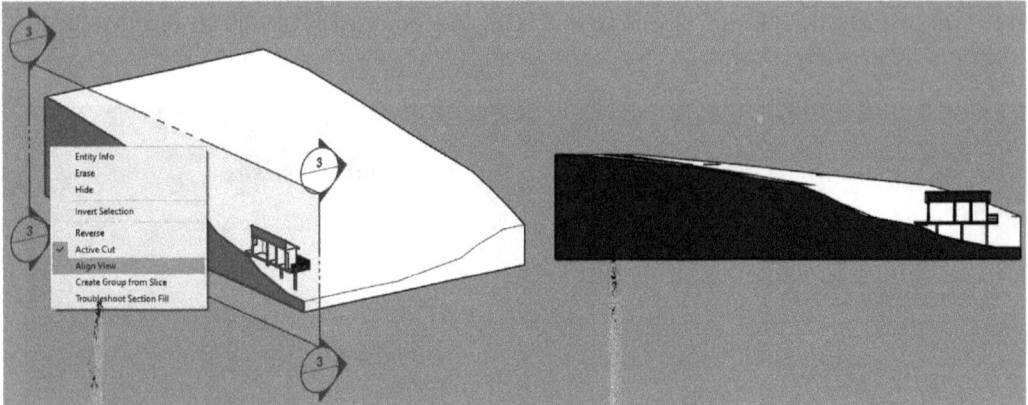

Figure 12.98 – Parallel Projection Camera and Right–Click –> Align View (Left) Section Cut View (Right)

This workflow can be used to get the same view as a Section Slice, but a 2D Graphic DWG can be exported to include the Geometry beyond the section cut.

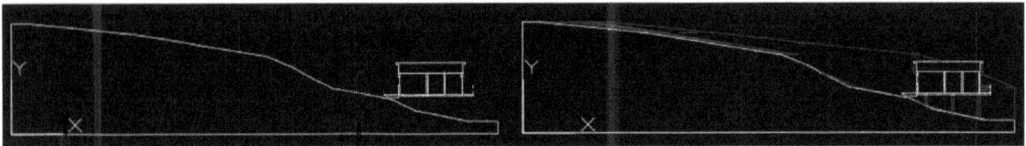

Figure 12.99 – True Section Slice (Left) and 2D Graphic Exported (Right)

The Export Options for Section Slices are very similar to 2D Graphics DWG/DXF exports, with **AutoCAD version**, **Drawing scale & size**, and **Section lines** sections. The **Full scale (1 : 1)** option is appropriate for True Sections, as Orthographic views will retain true dimensions without any perspective distortion.

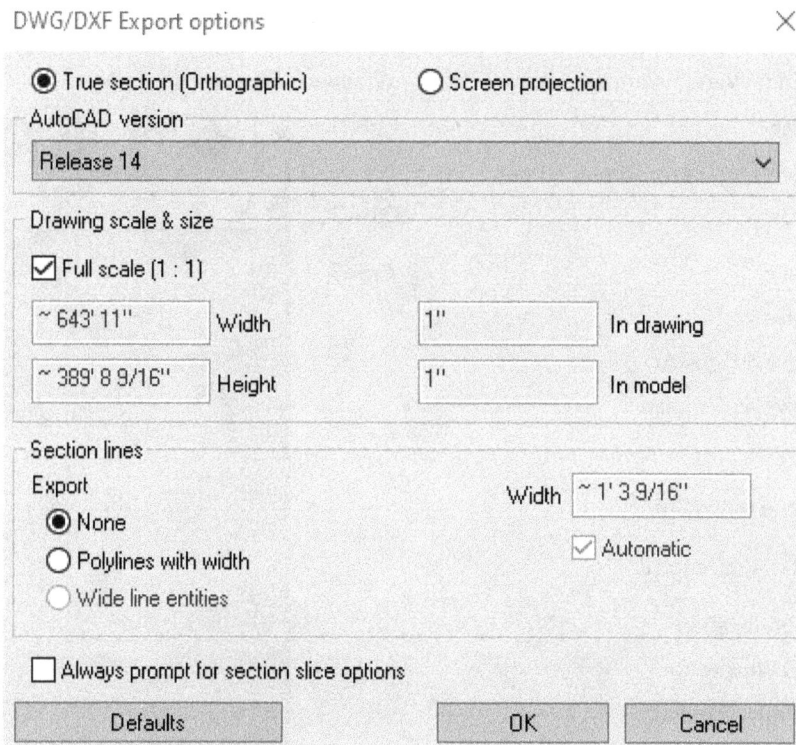

Figure 12.100 – Section Slice Export options Dialog Box

> **Note**
> The **Section lines** Export option will create 2D DWG Geometry even when set to **None**. The **None** option will produce AutoCAD Lines, while the **Polylines with width** option allows the AutoCAD polylines to be specified with a width. All lines and polylines will be separate segments, but only polylines can support a width parameter in AutoCAD DWG files.

Section Slices can be used to create extremely helpful drawings for AutoCAD or another CAD drafting software. It can be helpful to save scenes of different active section cuts if many Section Slices need to be exported on a project. Try integrating Section Slices into your next project!

Animations

We have discussed how to create Animations in SketchUp Pro using Scenes in *Chapter 7, View Options*. This section will focus on the Export Options for Animations.

In order to create an Animation, there must be more than one scene in the model.

Animations can be exported by clicking **File –> Export –> Animation....**

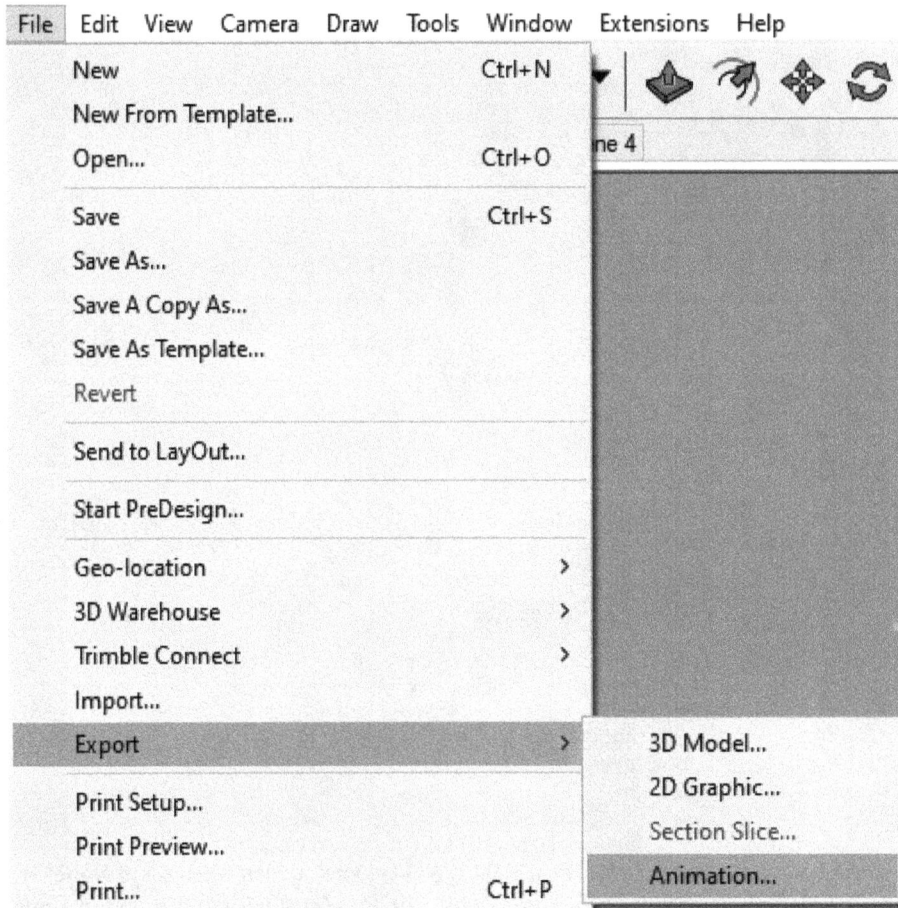

Figure 12.101 – File –> Export –> Animation…

Animations can be exported as a video file (MP4) or as a series of image files (BMP, JPG, TIF, or PNG). MP4 video files will be exported as a single file and image files will have a different image exported for every frame of the animation. Image files will be named sequentially but will not be placed in a specific folder or ZIP file. It is recommended that image Animations are exported to a folder that is specifically for storing image files for that Animation.

All Animation file types have the same primary Export Options in the **Export options** Dialog Box. The Export Options include **Resolution, Aspect ratio, Width, Height, Line scale multiplier, Frame rate**, and **Loop to starting scene**. If **Resolution** is set, it will automatically set the **Aspect Ratio, Width**, and **Height** options. Using the Custom Resolution Option will allow the other options to be set to a custom value.

Export options   ✕

Resolution
[720p HD   ⌄]

Aspect ratio
[16:9 Wide   ⌄]

Width
[1280]   pixels

Height
[720]   pixels

Line scale multiplier
[1.00]   x

[Preview frame size   ]

Frame rate
[24   ⌄]   frames/second

☑ Loop to starting scene

☐ Always prompt for animation options

[Restore defaults   ]

              [OK   ]   [Cancel]

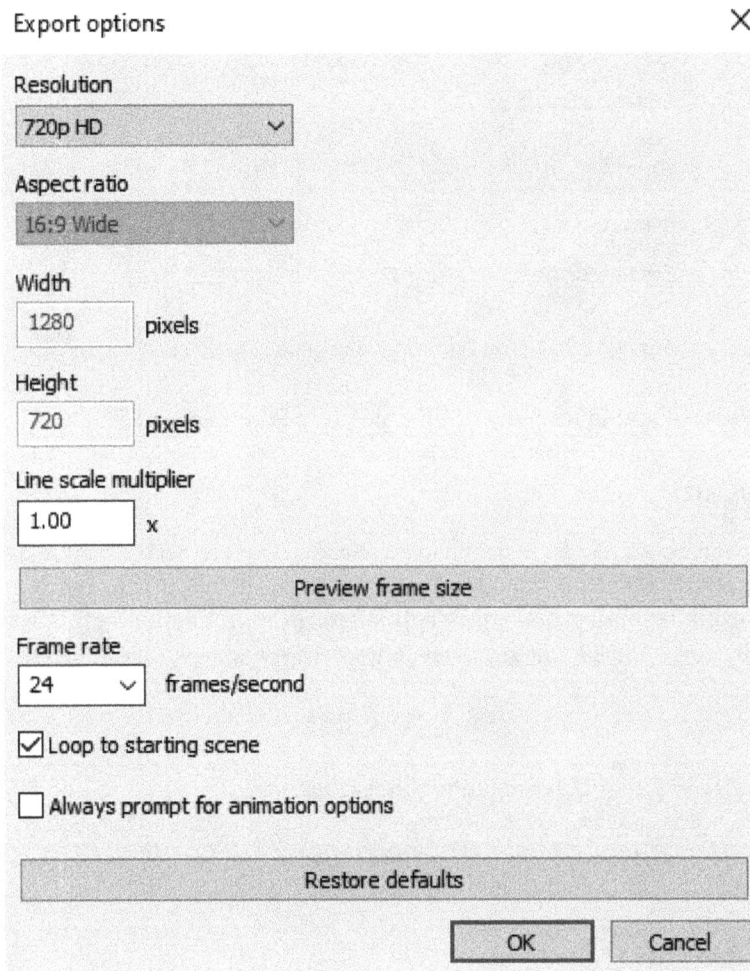

Figure 12.102 – Animation Export options Dialog Box

> **Note**
> MP4 Animations have an additional option to turn on **Anti-alias** Rendering.

Once the **Export** Button is clicked, an **Exporting Animation…** Dialog Box will appear, showing the progress of the Animation being created. The Animation can be canceled if necessary.

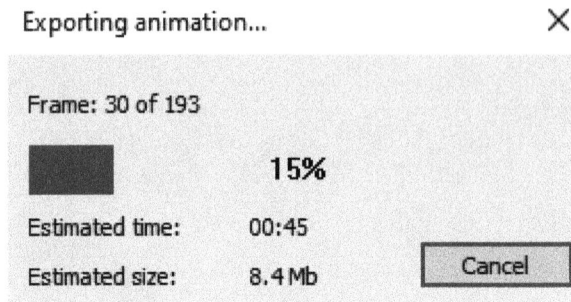

Figure 12.103 – The Exporting Animation… Dialog Box

The Animation will be shown in the order that the SketchUp Scenes appeared.

Send to LayOut

The last export option we will discuss in this chapter is **Send to LayOut**. Send to LayOut is technically not an export workflow, but rather a linking of the file to SketchUp Pro's sister software, LayOut. LayOut is a tool to create construction documents for architecture, interior design, woodworking, or other design drawings. The LayOut logo is an abstract shape that looks like three stacked rings.

Figure 12.104 – The LayOut Logo

SketchUp models can be sent to LayOut by clicking **File | Send to LayOut…**. Files must be saved before they can be sent to LayOut, and if the file is not saved, SketchUp will prompt asking you to save the file.

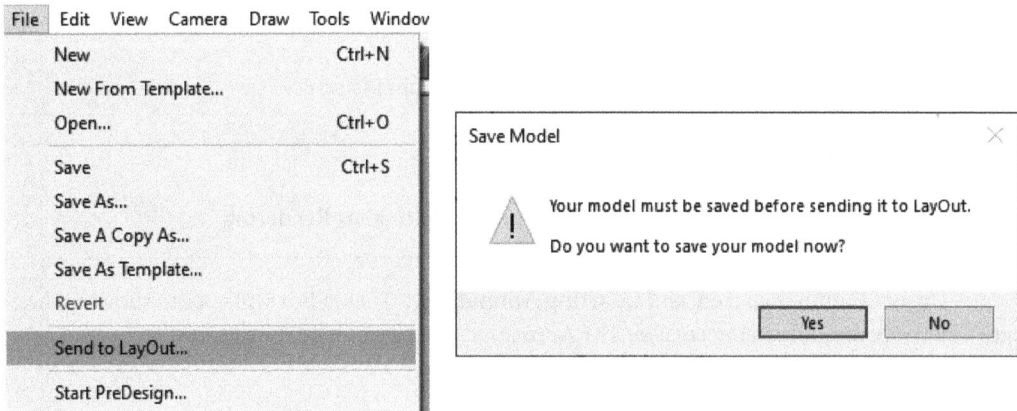

Figure 12.105 – File –> Send to LayOut… (Left) and the Save Model Prompt (Right)

This will open the LayOut software so that the SketchUp model can be used to create drawings and construction documents. When LayOut opens, the **Choose a Template** Dialog Box will open. A **Paper**, **Storyboard**, or **Titleblock** page can be selected.

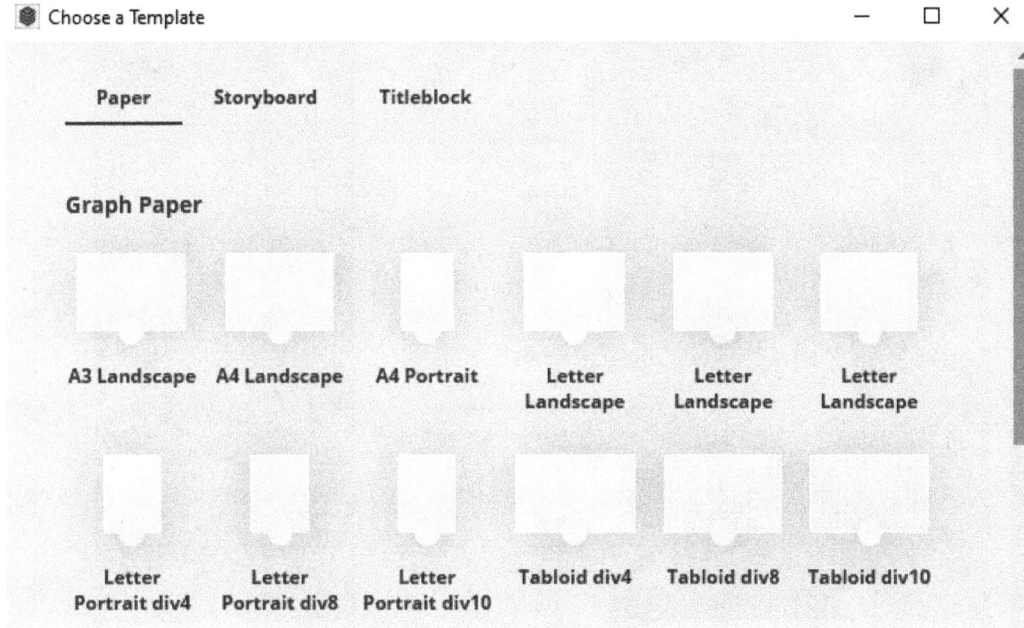

Figure 12.106 – The Choose a Template Dialog Box

The SketchUp model will automatically be placed on the page showing the last saved view. This is a link to the SketchUp model – not an image – and can be changed in a number of ways! The best way to edit the view is to Right–Click on the view, which will show the options to **Arrange**, **Center**, **Flip**, **Edit 3D View**, **Open with SketchUp**, **Update Model Reference**, **Relink Model Reference**, **Perspective**, **Standard Views**, **Styles**, **Scenes**, **Render Mode**, **Shadows**, and **Scale**.

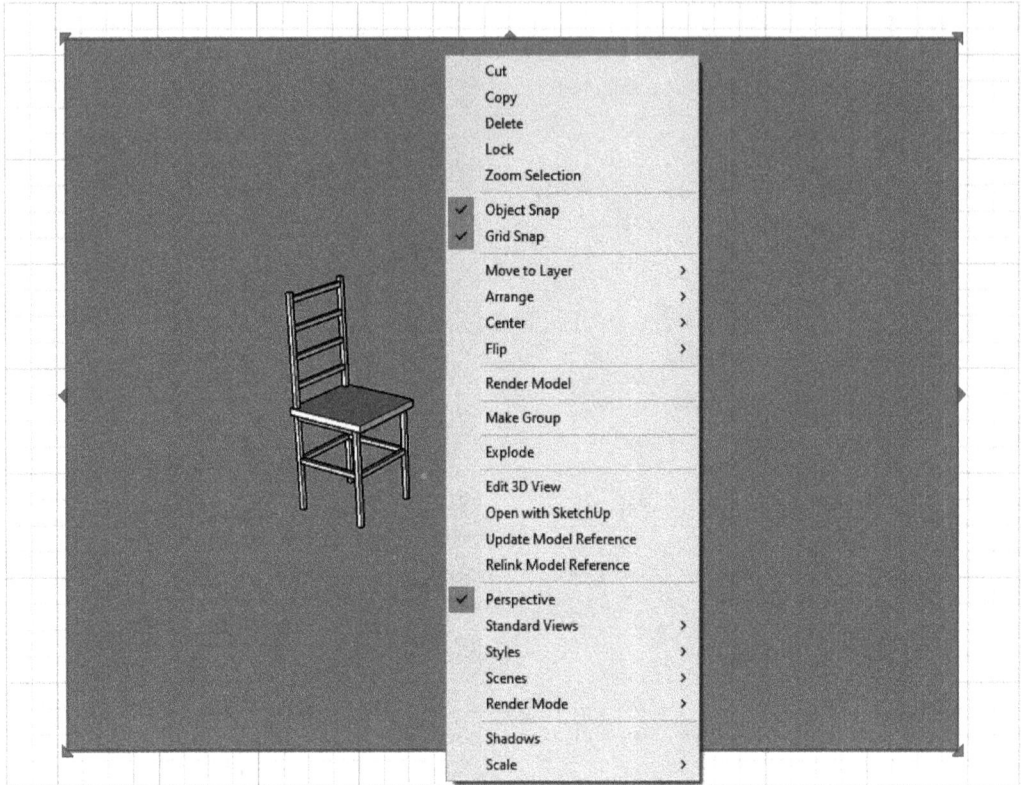

Figure 12.107 – The View Right–Clicked to Show the Right–Click Menu

The edit options for the view reference the settings from the SketchUp model, so only **Styles**, **Scenes**, and **Shadows** that have been set in the SketchUp model will be available. The Scene can also be Edited in a 3D View directly in LayOut, where **Orbit**, **Pan**, and **Zoom** are available to change the view.

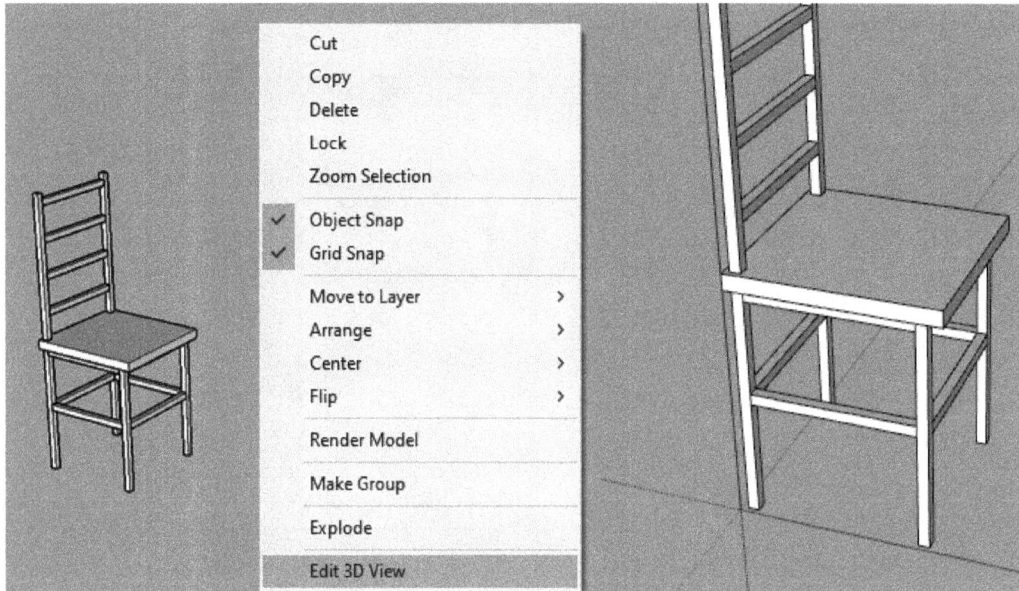

Figure 12.108 – Right–Click –> Edit 3D View (Left) and the View Orbited and Zoomed (Right)

Views can be arranged on the sheet, copied to create more views, and LayOut Geometry can be added to create notes and diagrams. Additionally, dimensions and tables can be added in more advanced workflows. These tools can be found on the Main Toolbar at the top of the program.

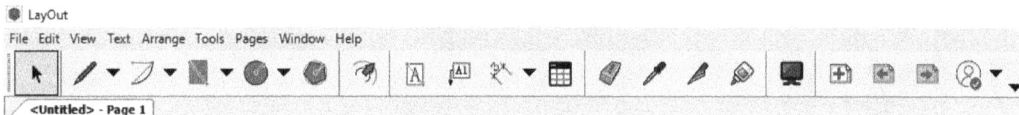

Figure 12.109 – The Main Toolbar with the Drawing and Edit Tools

Like SketchUp Pro, LayOut has a **Default Tray** window that can be edited by going to the Window dropdown in the Menu Bar. The **Default Tray** window in LayOut includes **Colors, Shape Style, Pattern Fill, SketchUp Model, Scaled Drawing, Dimension Style, Text Style, Pages, Layers, Scrapbooks,** and **Instructor**.

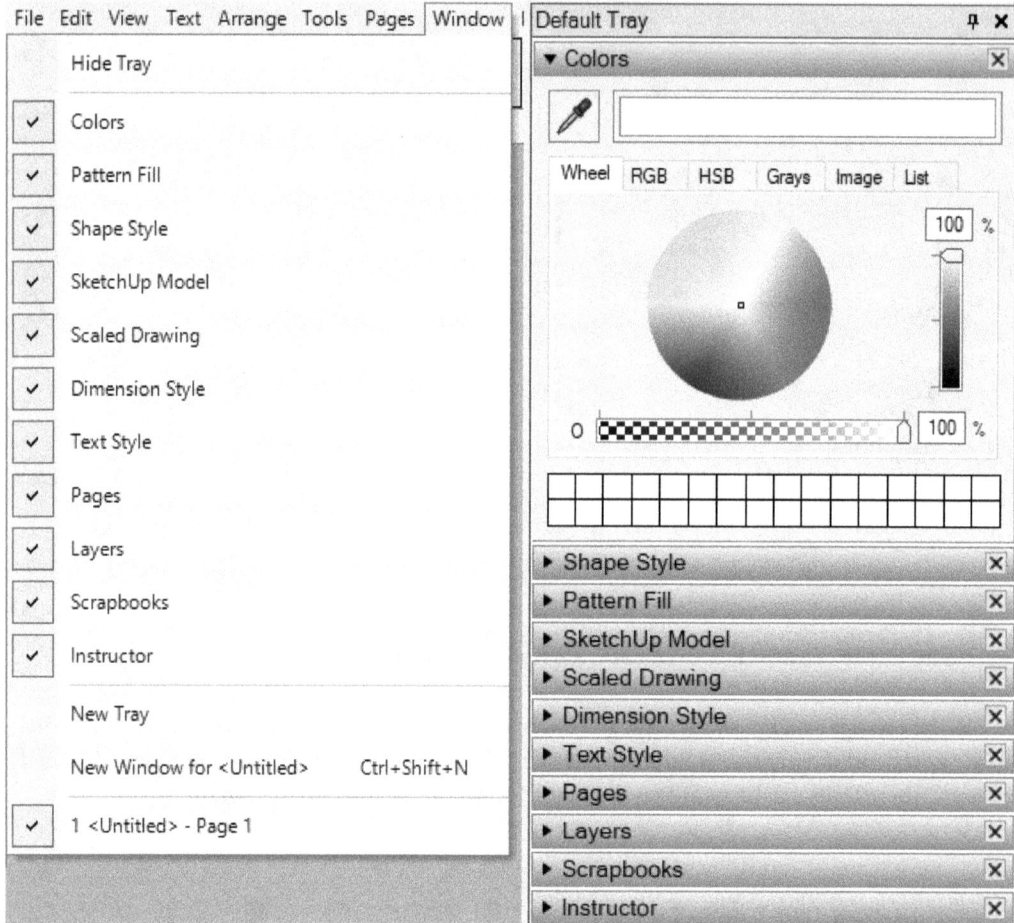

Figure 12.110 – The Window Dropdown (Left) and the Default Tray Panels (Right)

Most **Default Tray** Panels will activate once the correct Object is selected on the page. The **Pages** Panel is the best way to add, remove, or organize pages in multi–page documents. It is also a quick way to switch between different pages for editing the document. The **Layers** Panel allows for Geometry to be placed on stacked Layers, and the **Scrapbooks** Panel acts as a repository for 2D Graphics that can be added to the drawings.

In this next example, we can see a LayOut Page that shows a 3D **Perspective View** section, a **Floorplan** section using a Scene and Active Section Cut, **Site Section**, and a **Sun Study** section using multiple copies of the same 3D view. Additional Objects have been added, such as dimensions, titles/text, and some Scrapbook Graphics.

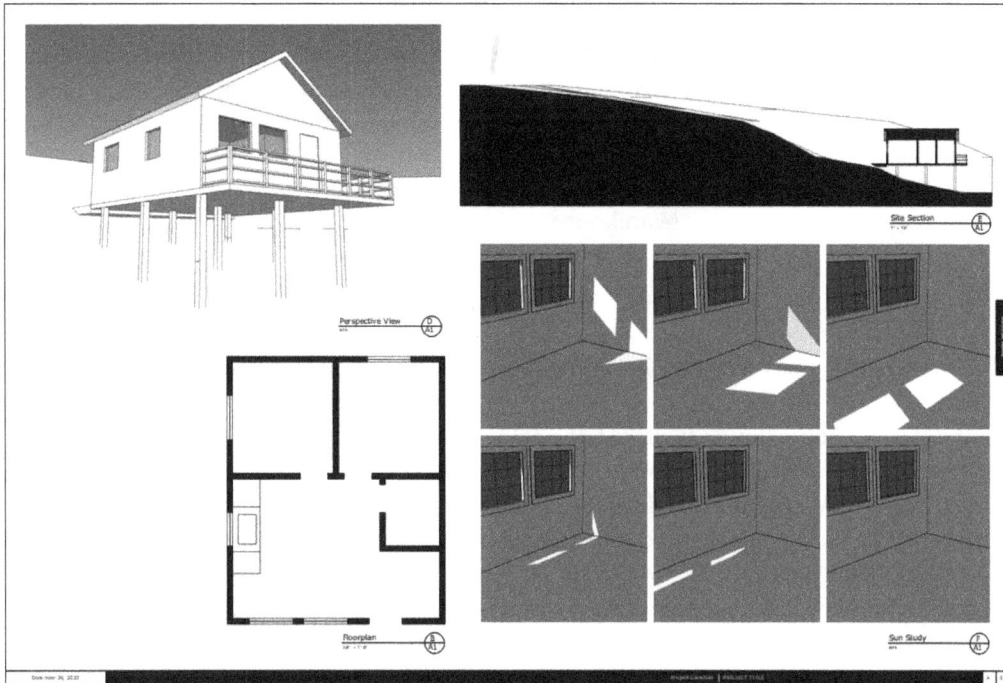

Figure 12.111 – An Example LayOut Page with Multiple Views and Graphics

LayOut is a great program that links directly to your SketchUp models! Give LayOut a try for your next project!

3D Warehouse

We have introduced 3D Warehouse in multiple chapters in this book, including *Chapter 11, Working with Components*, when we introduced Dynamic and Live Components, and *Chapter 8, Materials*, and *Chapter 9, Entity Info, Outliner, and Tags Dynamically Organize Your Models*, when discussing how to clean up Components downloaded from 3D Warehouse. 3D Warehouse is the largest collection of downloadable, searchable 3D models that work seamlessly with SketchUp Pro. Models in 3D Warehouse can be downloaded directly into any SketchUp model, saving time and effort while working on projects. The 3D Warehouse logo is an abstract representation of the old SketchUp Pro logo, which is a cube within a curved shell. In SketchUp, the 3D Warehouse logo is often shown with a downward–facing arrow representing downloading SketchUp models.

Figure 12.112 – The 3D Warehouse Logo and the 3D Warehouse Logo with a Download Arrow

The 3D Warehouse Button can be found on the **Getting Started** Toolbar and the **Warehouse** Toolbar in SketchUp. The Warehouse Toolbar also includes the **Share Model** Tool, which is represented by the 3D Warehouse logo and an upward–facing arrow representing uploading your own SketchUp model to 3D Warehouse. 3D Warehouse is a combination of user–submitted models, company models, and SketchUp–designed and curated models.

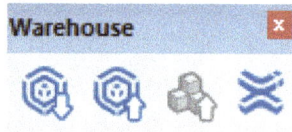

Figure 12.113 – The Warehouse Toolbar Showing 3D Warehouse and Share Model Buttons

When the 3D Warehouse Button is clicked, the **3D Warehouse** dialog box will be opened. The 3D Warehouse Dialog Box is a version of the 3D Warehouse website, and an internet connection is required to view 3D Warehouse and download models. The 3D Warehouse home page will show a mix of recently viewed Components, featured Components, and Catalogs from fixture and furniture manufacturers. Many manufacturers will create 3D models of their products and include them in 3D Warehouse in the hope that their products are specified for construction projects!

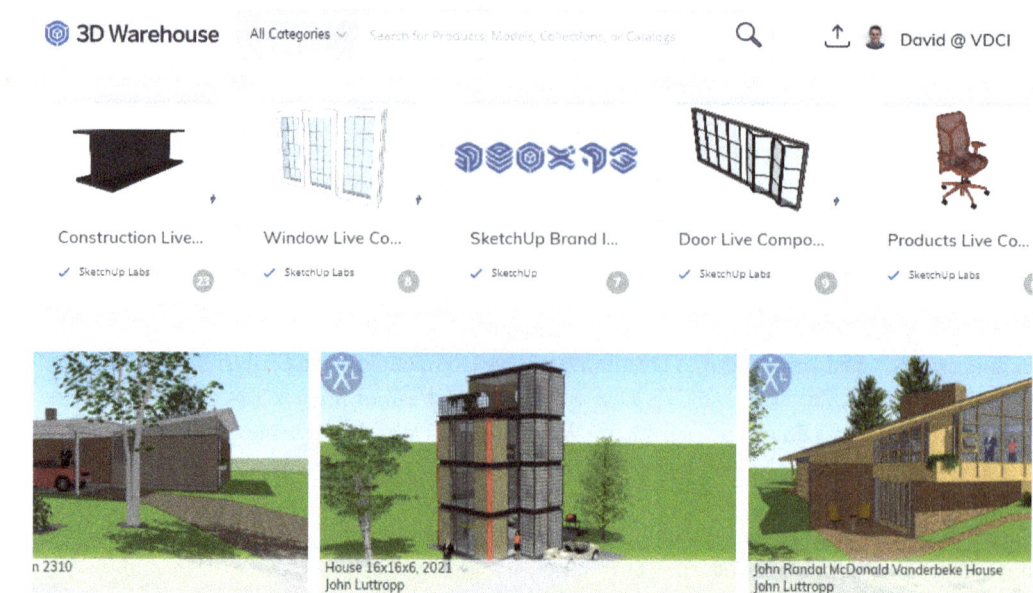

Figure 12.114 – The 3D Warehouse Dialog Box

> **Note**
> You can also access the 3D Warehouse by going to `https://3dwarehouse.sketchup.com/`.

3D Warehouse is a great way to search for any model to use in your SketchUp projects. The search bar at the top of the Dialog Box can be used to enter a search term and the magnifying glass can be clicked to search.

In this next example, the phrase `stained glass window` has been typed, and 3D Warehouse has suggested more specific searches that might be helpful. When the magnifying glass is clicked, 3D Warehouse will show results for that search.

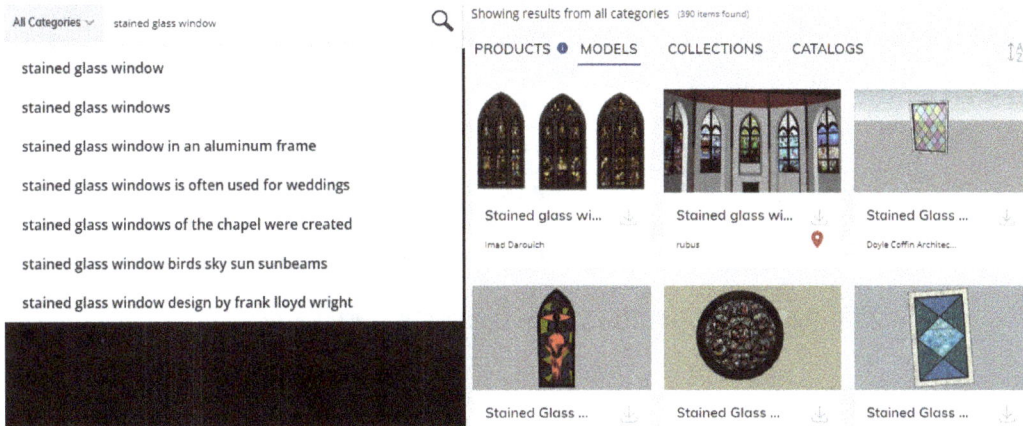

Figure 12.115 – 3D Warehouse Search Suggestions (Left) and Search Results (Right)

3D Warehouse searches can be filtered by using the Categories dropdown to the left of the search bar, or on any results page in the **CATEGORY** and **PROPERTIES** filter panels on the left side of the Dialog Box.

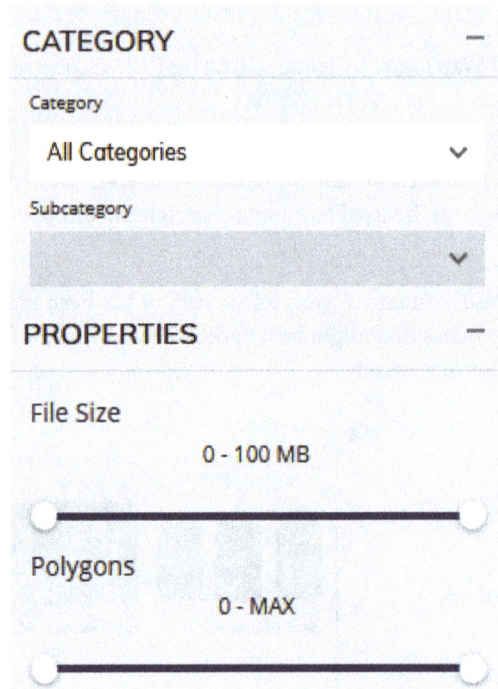

Figure 12.116 – CATEGORIES and PROPERTIES Filter Panels

3D Warehouse will return four categories for each search: **PRODUCTS, MODELS, COLLECTIONS,** and **CATALOGS**. Products represent models made by companies and Models represent user–submitted SketchUp models. Collections are groups that can contain either Products or Models, while Catalogs are curated catalogs of Products. If a search returns no results for any of these Categories, SketchUp will show an **IT'S EMPTY** message.

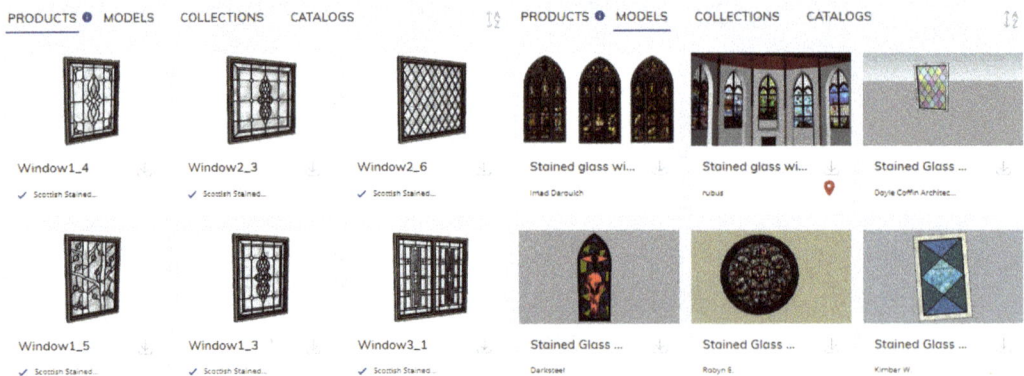

Figure 12.117 – 3D Warehouse Search Result: Products (Left) and Models (Right)

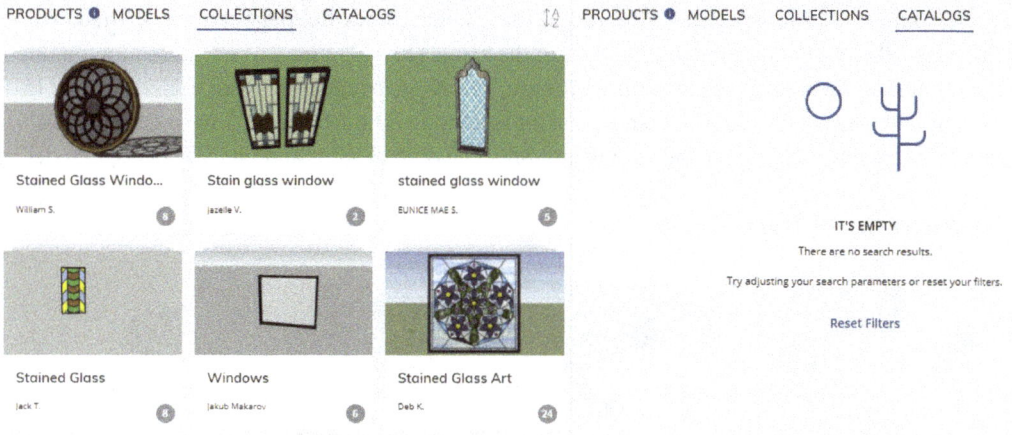

Figure 12.118 – 3D Warehouse Search Results: Collections (Left) and Catalogs reading "IT'S EMPTY" (Right)

Any search result can be clicked to open the page or popup for that Product, Model, Collection, or Catalog. Collections and Catalogs will open pages that will further show the individual Models and Products.

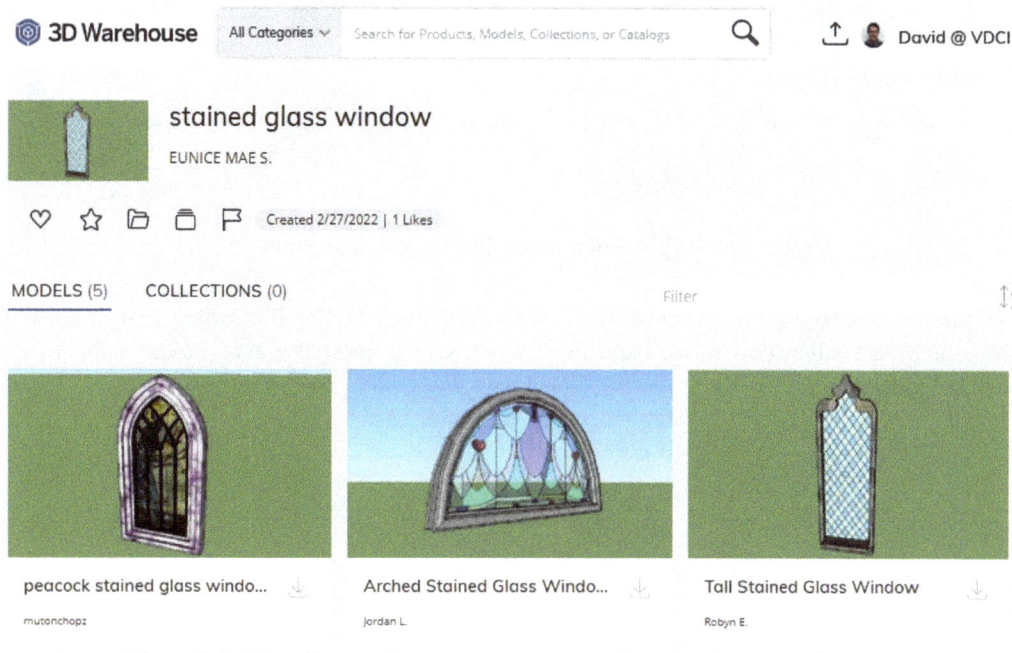

Figure 12.119 – The COLLECTIONS Page

Once a Model or Product is clicked, the individual SketchUp model page will be opened. The SketchUp model can be Downloaded directly into the active SketchUp model or to a location on the local Computer by clicking on the **Download** Button. Additionally, the model can be liked, added to favorites, or added to a Collections folder. Model information is at the bottom of the page, which can be helpful when understanding the total number of **Materials** or **Polygons** and **File Size** of each model.

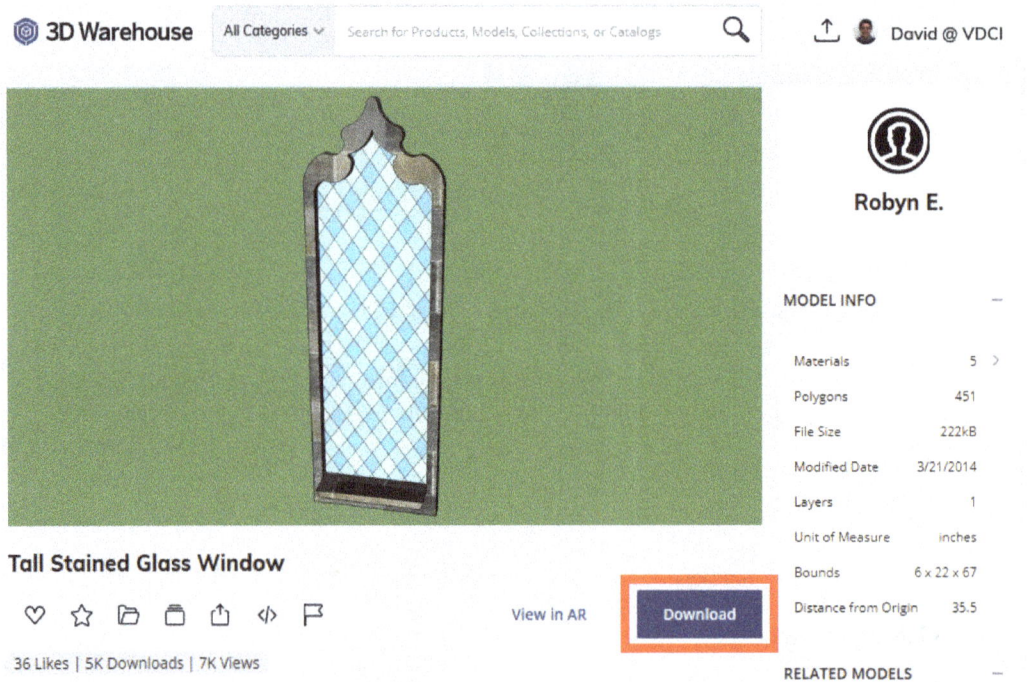

Figure 12.120 – The Model Page with the Download Button

When the **Download** Button is clicked, the **Load Into Model?** Dialog Box will appear. If **Yes** is clicked, the Model will be downloaded and added to the **Components** Panel of the currently open SketchUp Model.

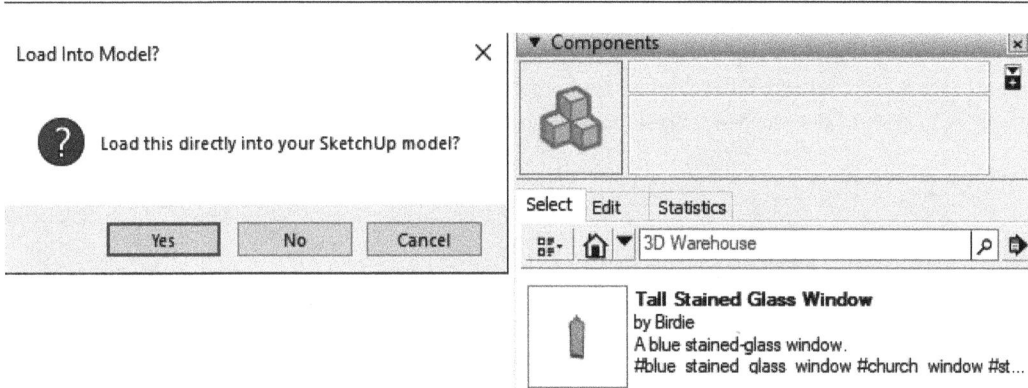

Figure 12.121 – The Load Into Model? Dialog Box (Left) with Yes Clicked
and the Model Added to the Components Panel (Right)

If **No** is clicked, a **Save As** Dialog Box will open, and the SketchUp Model can be saved to the local computer. These SketchUp models can be opened or Imported as Components into other SketchUp models.

Models can be uploaded to 3D Warehouse by clicking the **Share Model** Button or the **Share Component** Button. Sharing models and Components is what makes 3D Warehouse a powerful modeling tool! When the **Share Model** Button is clicked, SketchUp will prompt you to click on the **Save** Button for the model and then **Purge Unused** Materials, Components, Tags, and Styles. Purging the model of unused items lowers the total file size and is helpful for not giving unnecessary information to others who may use your model.

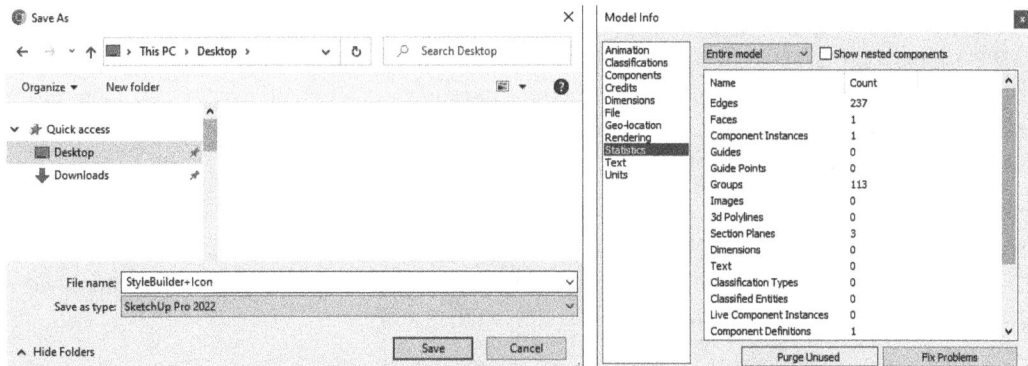

Figure 12.122 – The Save Model Dialog Box and the Purge Model Dialog Box

The **Share Component** Button will appear to be grayed out until a Component is selected in the Drawing Area. When a Component is selected, the **Share Component** Button will return to full color, which is represented by a dark gray stack of cubes and an upward–facing arrow. When the **Share Component** Button is clicked, no Save Model or Purge Dialog Boxes are prompted.

Both of these workflows will open the 3D Warehouse Upload Model dialog box. This page will walk through each step of setting **Title**, **Description**, and **Category** information for the Model, as well as the option to set it to **Private Model** or **Disable Comments**.

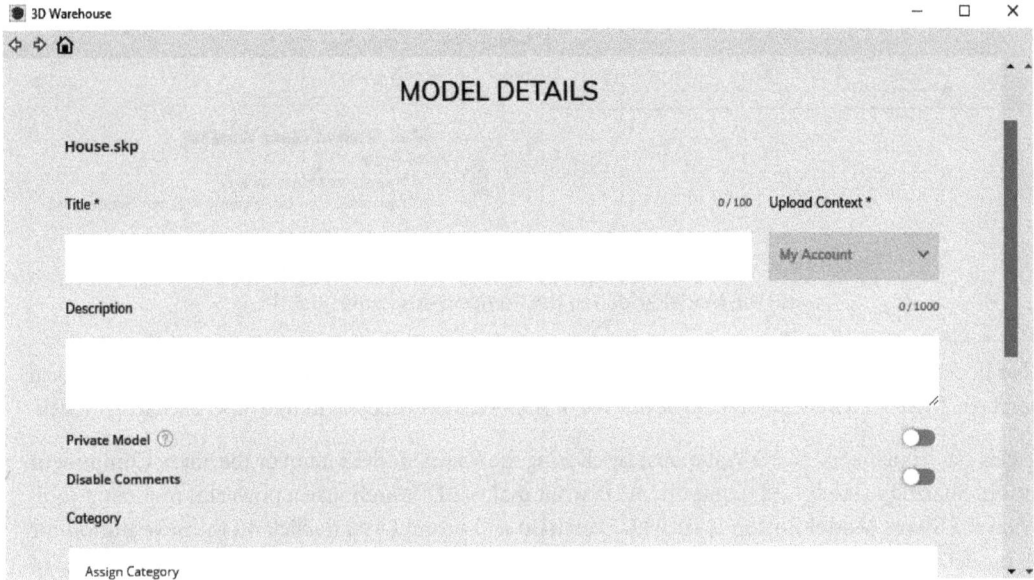

Figure 12.123 – The 3D Warehouse Upload Model Dialog Box

Once the model settings are correct, the **Upload** Button can be clicked, and the model will be added to 3D Warehouse! Your 3D Warehouse Models can be viewed by clicking on your profile picture or name in the upper-right corner of the 3D Warehouse and choosing **My Public Page** or **My Content**. **My Public Page** will be how others see your profile, and 3D Warehouse models can be sorted into folders and collections, made private, or deleted.

3D Warehouse is one of the most powerful tools that SketchUp users use to save time and create amazing models! Explore 3D Warehouse to find downloadable models for your next SketchUp project! One other powerful tool that SketchUp users use to save time and effort in SketchUp Pro is Extensions and the Extension Warehouse, which we will look at in the next section.

Extensions

Extensions are packages of tools that can be plugged into SketchUp Pro. Extensions include tool packages for Materials, Renderings, Editing Tools, Tools for curves, splines, and complex 3D shapes, architecture, engineering, and construction objects, tools for 3D printing, and many more! SketchUp Pro was created by Trimble and is not an open source software program. However, the SketchUp team has recognized how powerful user improvements can be, so SketchUp is very supportive of allowing Extensions to be added to SketchUp Pro.

Extension Warehouse

The Extension Warehouse is similar to 3D Warehouse, but the Extension Warehouse allows you to search for Extensions and download them to the local installation of SketchUp Pro. The Extension Warehouse button can be found on the Getting Started Toolbar and Warehouse Toolbar. The Extension Warehouse button is an abstract flat X shape.

Figure 12.124 – The Extension Warehouse Button (Left) and the Warehouse Toolbar (Right)

When the Extension Warehouse Button is clicked, the **Extension Warehouse** Dialog Box will be opened. The **Extension Warehouse** Dialog Box is a version of the Extension Warehouse website, and an internet connection is required to view the Extension Warehouse and download Extensions. The Extension Warehouse homepage will show **Featured Extensions**, **Top Extensions**, and **Top Developers**.

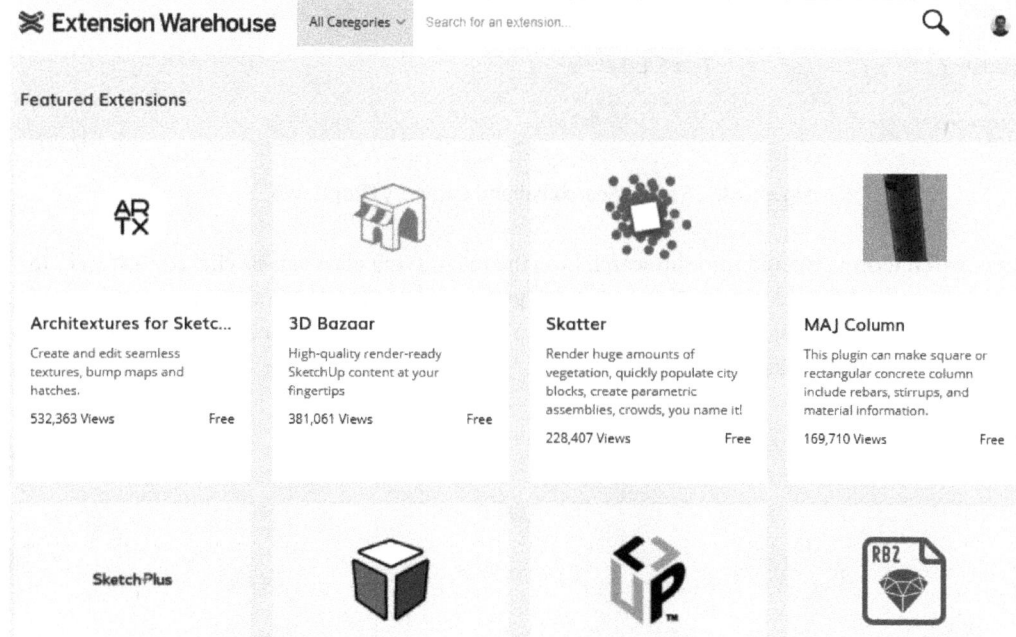

Figure 12.125 – The Extension Warehouse Dialog Box

> **Note**
> You can also access the Extension Warehouse by going to `https://extensions.sketchup.com/`.

Like the 3D Warehouse, the Extension Warehouse also has a search bar at the top of the Dialog Box. There is also an **All Categories** sorter to refine your search, with categories related to tools, industries, and specific workflows.

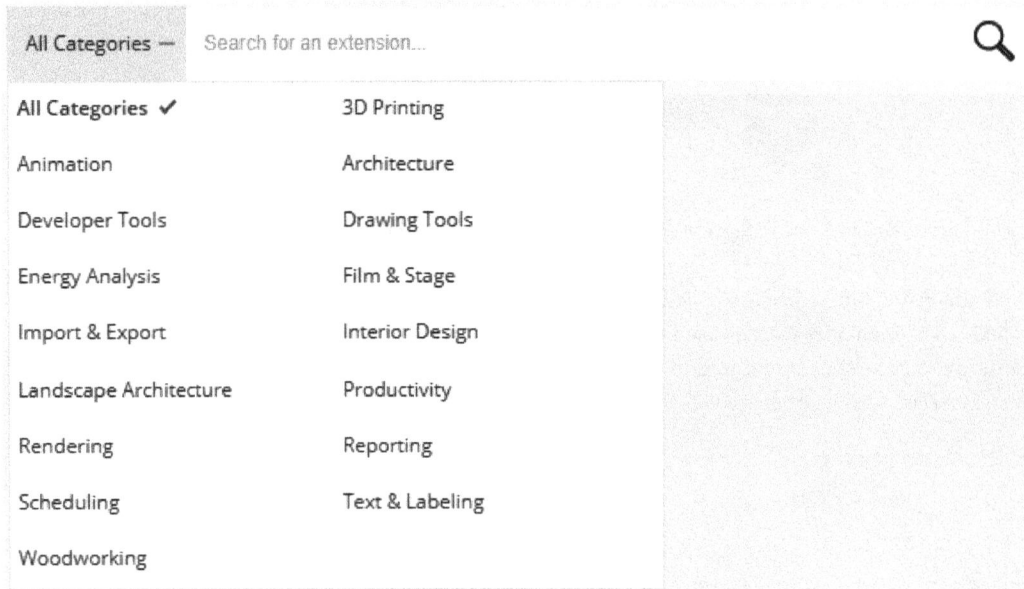

All Categories — Search for an extension...	Q
All Categories ✔	3D Printing
Animation	Architecture
Developer Tools	Drawing Tools
Energy Analysis	Film & Stage
Import & Export	Interior Design
Landscape Architecture	Productivity
Rendering	Reporting
Scheduling	Text & Labeling
Woodworking	

Figure 12.126 – The Search Bar and Category Dropdown

Once a search term is entered into the search box, the magnifying glass can be clicked to search. In this next example, the word `textures` has been typed, and the Extension Warehouse has suggested more specific searches that might be helpful. When the magnifying glass is clicked, the Extension Warehouse will show results for that search.

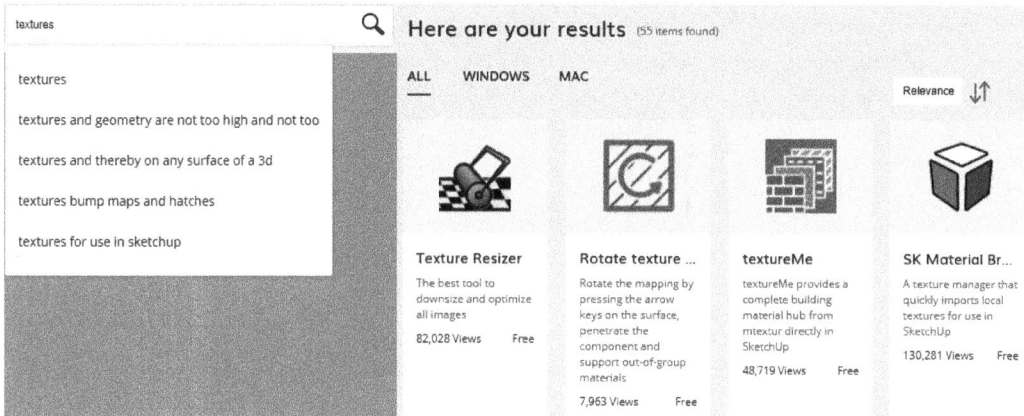

Figure 12.127 – Extension Warehouse Search Suggestions (Left); Search Results (Right)

In addition to the Category dropdown filter, Extension Warehouse searches can also be filtered in the **SKETCHUP VERSION** and **EXTENSION TYPE** filter panels to the left side of the Dialog Box.

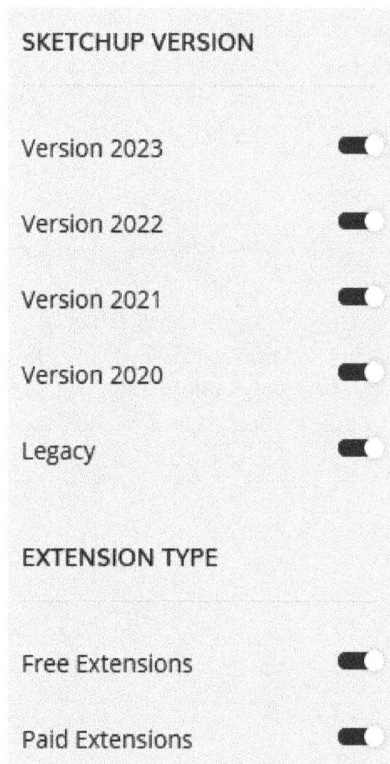

Figure 12.128 – SKETCHUP VERSION and EXTENSION TYPE Filter Panels

The Extension Warehouse will allow the search results to be sorted into Windows, Mac, or All. It is recommended to select the operating system that you are currently working on in order to find Extensions that will work with your version of SketchUp Pro. The search results can also be sorted by **Relevance**, **Price**, or the date the Extension was created.

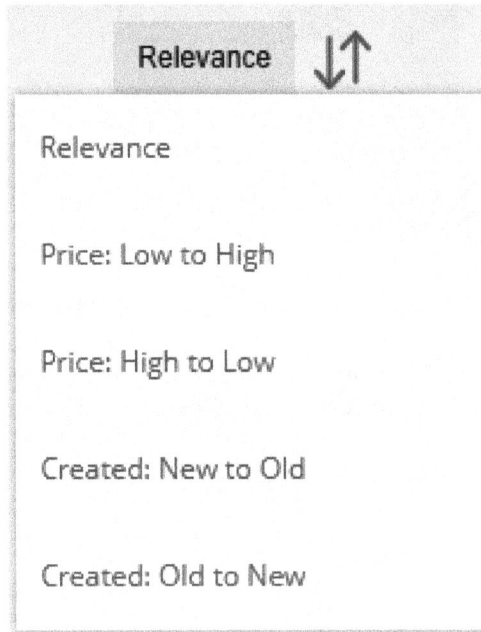

Figure 12.129 – Extension Warehouse Search Result Sorting Options

Any search result can be clicked to open the page for that Extension. The individual page for each Extension will show some featured images of how the Extension works, the **Price**, **Version**, **SketchUp Compatibility**, **OS Compatibility**, and **Languages** information. The Extension Creator is also featured on the right side of the screen, and a link to contact the developer is available.

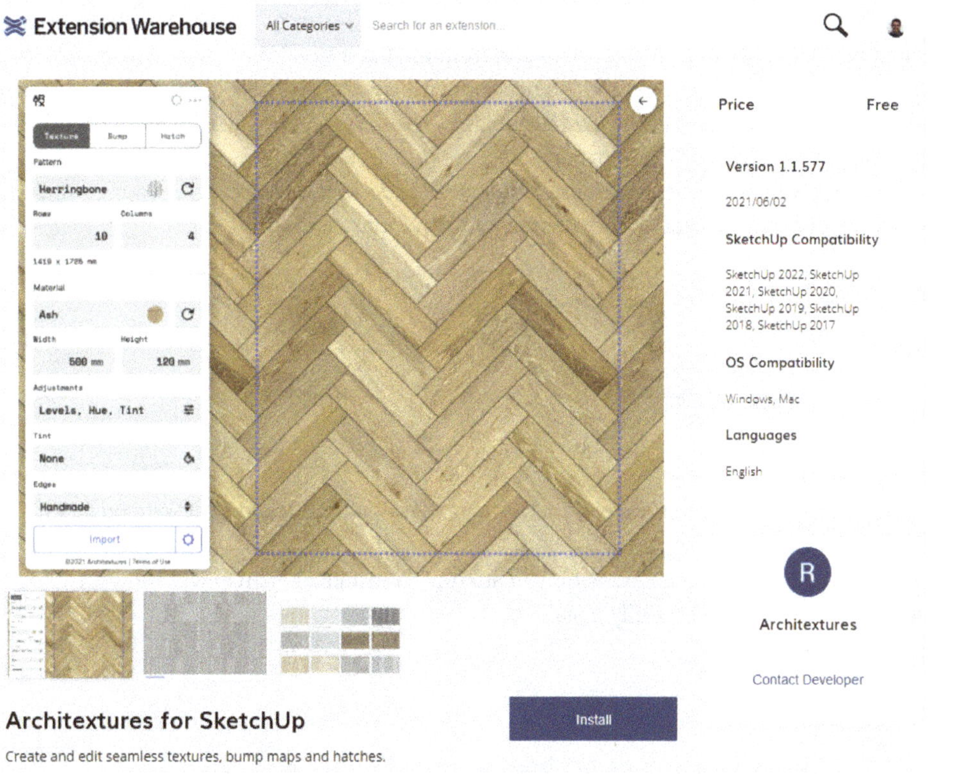

Figure 12.130 – The Extension Page in the Extension Warehouse Dialog Box

The Extension Title, tags, and more information are below the images. The **Install** button can also be found below the images. If the developer has created other Extensions, they will be listed in the **Also By This Developer** section in the bottom right of the Dialog Box.

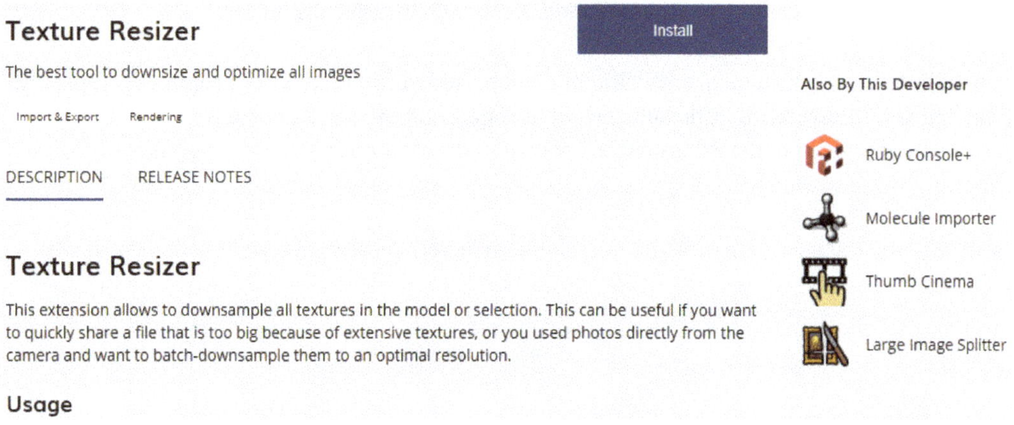

Figure 12.131 – Description, the Install Button, and Also By This Developer

Extensions can be Installed by Clicking the **Install** Button. A Terms of Service agreement will need to be signed before installing your first Extension from the Extension Warehouse. Any Extension can be downloaded, and if the Version or OS is not compatible with your current version of SketchUp, Pro the Extension Warehouse will show a warning. In some cases, especially with SketchUp Versions, the Extension will work, but it is important to ensure that an Extension works before it is installed. A second warning will appear if an Extension requires access to the filesystem of your computer. Be sure to only download and install trusted Extension Warehouse Extensions.

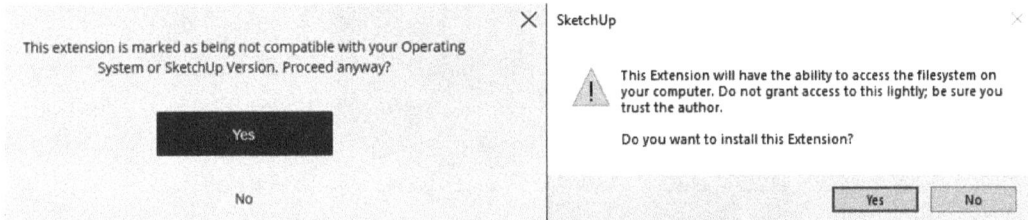

Figure 12.132 – An Extension Warehouse Compatible Warning (Left) and Trust Warning (Right)

If all warnings are agreed to, the Extension will attempt to install. SketchUp will show a Dialog Box if the Extension has been successfully installed.

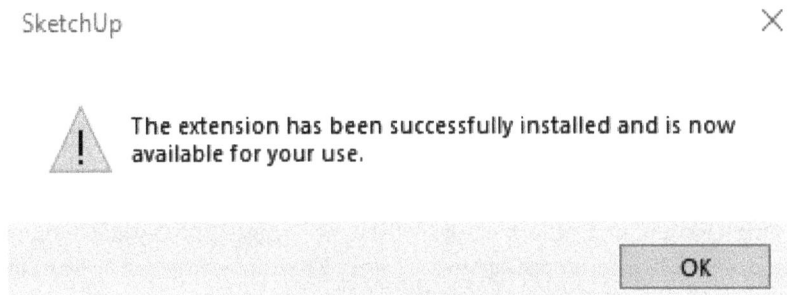

Figure 12.133 – The Extension Successfully Installed Dialog Box

When an Extension is installed, a new Toolbar will be added to the toolbars dropdown. Typically, the Toolbar will be turned on by default and will be floating above the Drawing Area. The Extension Tools will also be added to the extensions dropdown in the Menu Bar. The Extension is ready to use!

Figure 12.134 – The Toolbar Floating Above the Drawing Area (Left) and the Extensions Dropdown (Right)

Compared to creating and uploading Components to 3D Warehouse, developing and uploading Extensions is a complex process that requires knowledge of the Ruby API. Documentation on Ruby and the Ruby Console can be accessed in the Extensions Dropdown in the Menu Bar, but creating Extensions is outside of the scope of this book.

Extensions can be installed into SketchUp to be used right away! In addition to installing Extensions, it can be helpful to keep an eye on updates for Extensions, and sometimes remove Extensions that we no longer use. We use the Extension Manager to help with these processes.

Extension Manager

The **Extension Manager** Button is the same as the Extension Warehouse image with a gear in the lower–right corner. The **Extension Manager** Button can be found on the Getting Started and Large Tool Set Toolbars. Clicking the **Extension Manager** Button will open the **Extension Manager** Dialog Box.

Figure 12.135 – The Extension Manager Button

Extension Manager is where Extensions can be Enabled or Disabled, Updated, and Uninstalled. Extensions can also be installed from the local computer by clicking the **Install Extension** Button at the bottom of the Dialog Box. The **Home** Tab of Extension Manager allows individual Extensions to be Enabled or Disabled, or an entire set of Extensions can be toggled by clicking on the column headings. Individual Extension information can also be viewed by clicking the arrow on the right of each Extension.

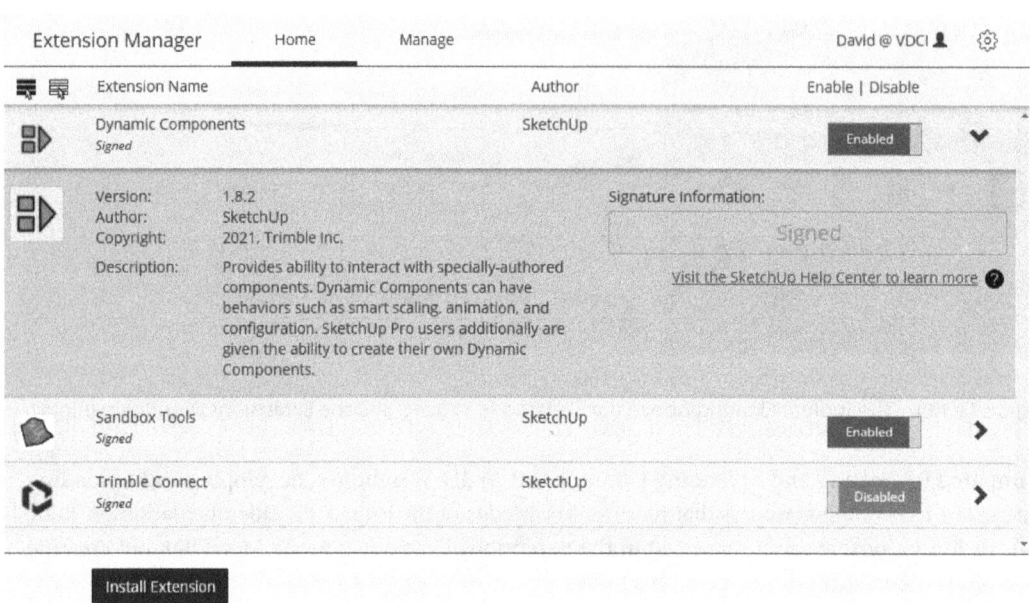

Figure 12.136 – The Extension Warehouse Dialog Box's Home Tab with Extension Details Unrolled

Note

SketchUp Pro will come pre-loaded with some Extensions, usually Dynamic Components, Sandbox Tools, and Trible Connect.

Extensions can be Updated and Uninstalled on the **Manage** Tab. The Extension details are the same on both tabs.

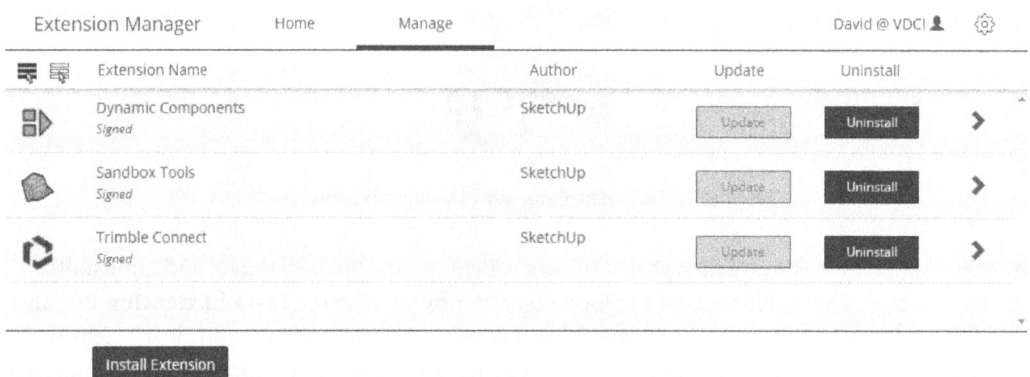

Figure 12.137 – The Extension Warehouse Dialog Box's Manage Tab

Try to find some Extensions in the Extension Warehouse that can improve your modeling speed, create photorealistic renderings, or even add camera and animation tools. Be careful to make sure that each Extension is trusted and use Extension Manager to Uninstall Extensions that you no longer need!

Overlays

Overlays are a feature that were added in the 2023 release of SketchUp Pro. Overlays can be found in their own Default Tray panel. This panel can be activated by going to **Window | Default Tray | Overlays**.

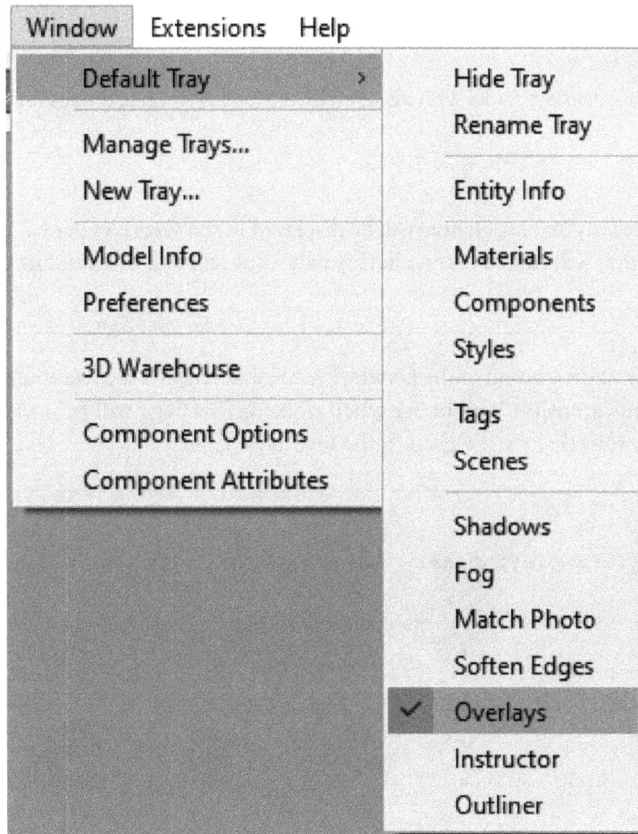

Figure 12.138 – Window | Default Tray | Overlays

The **Overlays** panel shows a select set of Extensions that can be overlayed onto the Drawing Area in real time. This is different than how most Extensions work in SketchUp, where some feature or tool of the Extension must be activated outside of a modeling workflow. The Overlays panel shows the eligible installed Extensions in a list, with a check box to the left of each Extension, and six dots to the right of the Extension that can be used to rearrange the Extension list in the panel.

Figure 12.139 – Overlays Panel with Installed Extensions with Checkboxes on the Left

> **Note**
>
> Only installed Extensions that are eligible will be displayed in the **Overlays** panel. If an Extension cannot be overlayed, it will not be shown in this panel but will still be found in the Extension Manager.

There is also a hyperlink at the bottom of the Overlays panel that reads "Discover more", and will launch the **Overlay** page of the Extension Warehouse when clicked. This page will be updated by SketchUp with the eligible Extensions that can be used in the Overlays panel.

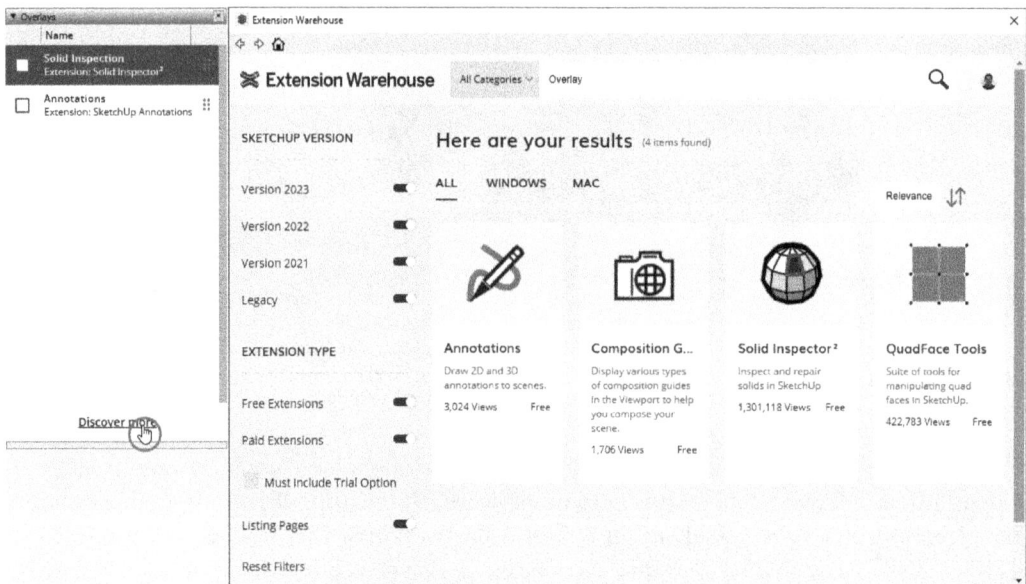

Figure 12.140 – Discover More Hyperlink Clicked (Left); Overlay Page of the Extension Warehouse (Right)

When eligible Extensions are installed they will automatically appear in the **Overlays** panel. Some Extensions may have Overlays–specific features, and some features may not work with the Overlays interface. SketchUp will prompt the user if any action must be taken to correctly use the Extensions features. In this next example the Annotations Toolbar is turned on, and when the Annotate Scene in 2D Space Tool is clicked, SketchUp will first provide a warning dialog box prompting the user to enable the Overlay.

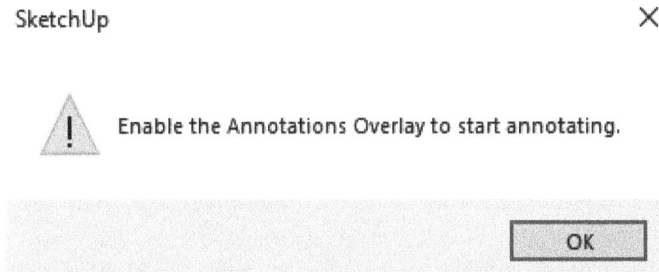

Figure 12.141 – SketchUp Warning to Enable the Annotations Overlay

The Annotations Overlay can be enabled by clicking the checkbox on the left of the panel, and then the Annotations Tools can be used as intended.

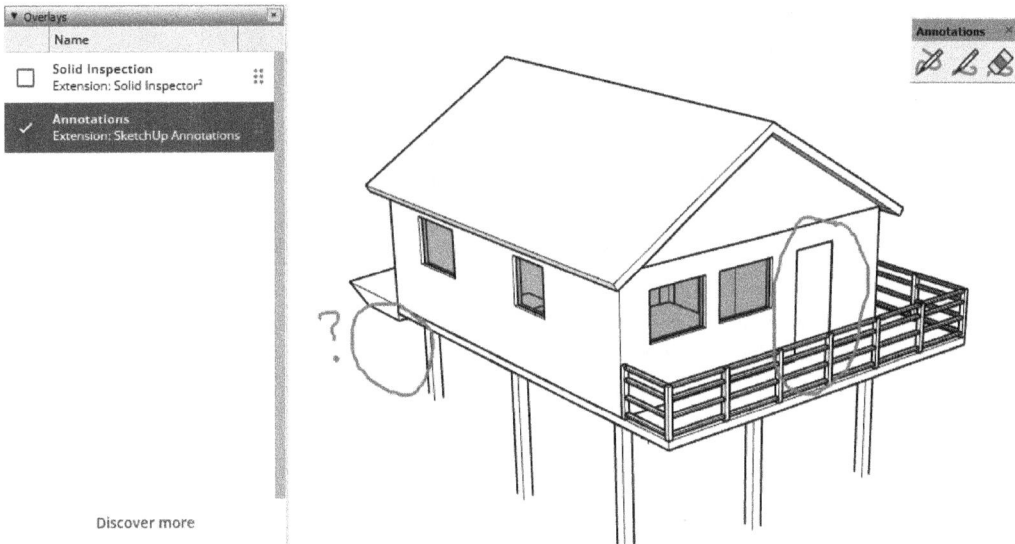

Figure 12.142 – Annotations Overlay Enabled (Left); Annotations Created in SketchUp Scene (Right)

Some eligible Extensions will have more functionality when used outside of the Overlays panel. In this next example we can see the Solid Inspection Toolbar activated on the left, and the Solid Inspection Overlay is activated on the right.

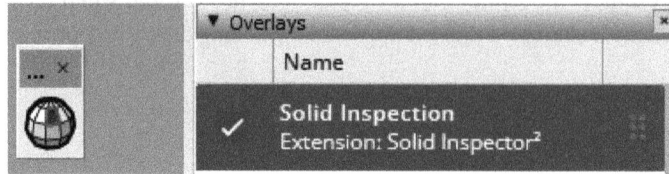

Figure 12.143 – Solid Inspection Toolbar (Left); Solid Inspection Overlay (Right)

The Solid Inspection Overlay Tool will run a one–time check of the model and will display a full description of the errors discovered during the check. Additionally, the Solid Inspector dialog box will prompt the user to automatically **Fix** some errors if possible. In this example we can see a model with 4 Face Holes has been checked, and the **Fix** button has been clicked to **Fix** all the errors.

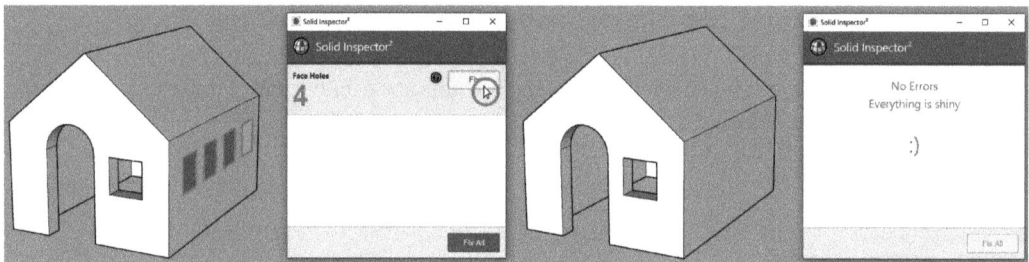

Figure 12.144 – Solid Inspector Dialog Box & Fix All Button Clicked
(Left) Model Fixed with No Additional Errors (Right)

The Solid Inspection Overlay will work similarly in that it will also highlight errors such as Surface Borders, Face Holes, External Faces, and Nested Instances, but it will not provide any way to learn more about the errors, nor will it provide any way to automatically fix the errors. We can see the red outlines shown on the model when the Solid Inspection Overlay is turned on, but the Solid Inspector dialog box is not activated.

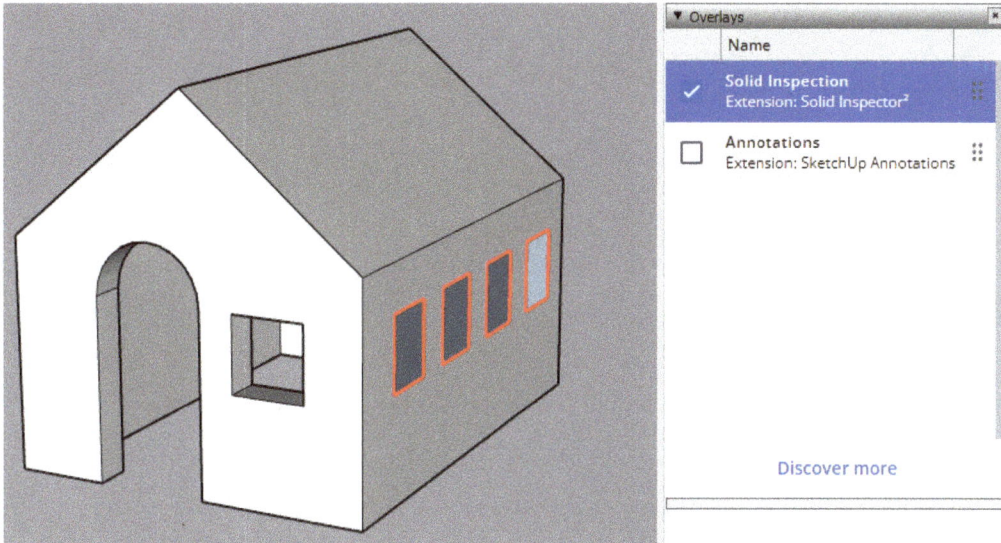

Figure 12.145 – Solid Inspection Overlay Enabled and Red Outlines Shown

I encourage you to check the **Discover More** link for yourself, and try the Extensions out for yourself. More Extensions are expected to be added to the Overlay panel in the future, so there may be one that fits your modeling needs!

Summary

In this chapter, we discussed Importing 3D, 2D, and Image files, and Exporting 3D, 2D, Image, Section, and Animation files. We also looked at an overview of Send to LayOut and the LayOut interface, 3D Warehouse, and Extensions. Being able to work with other files quickly and effectively will allow your SketchUp models to go from simple to stunning!

I encourage you to use 3D Warehouse to find models that you can use in your own SketchUp projects, but also to share the models that you create! SketchUp models will always be created with varied levels of detail, materials, and special features. No matter the complexity of your model, if it was helpful to you, there is a good chance that it would be helpful to someone else using SketchUp! Extensions are not necessary for modeling in SketchUp, but they can make some workflows much easier, and can also take your SketchUp models to the next level!

Good luck getting started with SketchUp Pro. Have fun working in SketchUp, and I hope to see some of your SketchUp models in the future!

Index

S

‹packt›

Packtpub.com

Subscribe to our online digital library for full access to over 7,000 books and videos, as well as industry leading tools to help you plan your personal development and advance your career. For more information, please visit our website.

Why subscribe?

- Spend less time learning and more time coding with practical eBooks and Videos from over 4,000 industry professionals

- Improve your learning with Skill Plans built especially for you

- Get a free eBook or video every month

- Fully searchable for easy access to vital information

- Copy and paste, print, and bookmark content

Did you know that Packt offers eBook versions of every book published, with PDF and ePub files available? You can upgrade to the eBook version at packtpub.com and as a print book customer, you are entitled to a discount on the eBook copy. Get in touch with us at customercare@packtpub.com for more details.

At www.packtpub.com, you can also read a collection of free technical articles, sign up for a range of free newsletters, and receive exclusive discounts and offers on Packt books and eBooks.

Other Books You May Enjoy

If you enjoyed this book, you may be interested in these other books by Packt:

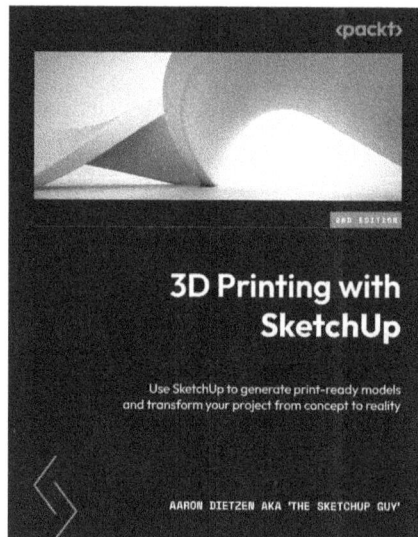

3D Printing with SketchUp

Aaron Dietzen

ISBN: 978-1-80323-735-0

- Understand SketchUp's role in the 3D printing workflow
- Generate print-ready geometry using SketchUp
- Import existing files for editing in SketchUp
- Verify whether a model is ready to be printed or not
- Model from a reference object and use native editing tools
- Explore the options available for adding onto SketchUp for the purpose of 3D printing (extensions)
- Understand the steps to export a file from SketchUp

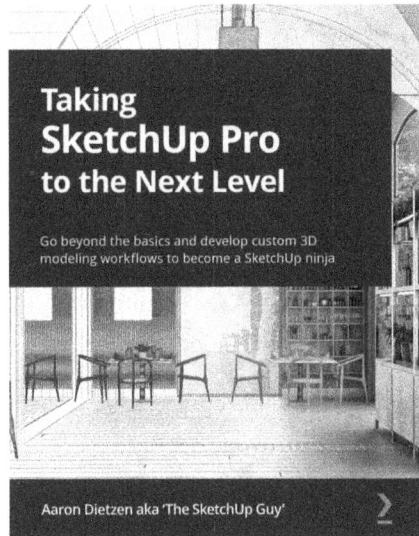

Taking SketchUp Pro to the Next Level

Aaron Dietzen

ISBN: 978-1-80324-269-9

- Recap the basics of navigation and SketchUp's native modeling tools
- Modify commands, toolbars, and shortcuts to improve your modeling efficiency
- Use default templates, as well as create custom templates
- Organize your models with groups, components, tags, and scenes
- Analyze your own modeling workflow and understand how to improve it
- Discover extensions and online repositories that unlock the advanced capabilities of SketchUp
- Leverage your existing SketchUp Pro subscription for even better results

Packt is searching for authors like you

If you're interested in becoming an author for Packt, please visit `authors.packtpub.com` and apply today. We have worked with thousands of developers and tech professionals, just like you, to help them share their insight with the global tech community. You can make a general application, apply for a specific hot topic that we are recruiting an author for, or submit your own idea.

Share Your Thoughts

Now you've finished *Getting Started with SketchUp Pro*, we'd love to hear your thoughts! Scan the QR code below to go straight to the Amazon review page for this book and share your feedback or leave a review on the site that you purchased it from.

`https://packt.link/r/1789800188`

Your review is important to us and the tech community and will help us make sure we're delivering excellent quality content.

Download a free PDF copy of this book

Thanks for purchasing this book!

Do you like to read on the go but are unable to carry your print books everywhere?

Is your eBook purchase not compatible with the device of your choice?

Don't worry, now with every Packt book you get a DRM-free PDF version of that book at no cost.

Read anywhere, any place, on any device. Search, copy, and paste code from your favorite technical books directly into your application.

The perks don't stop there, you can get exclusive access to discounts, newsletters, and great free content in your inbox daily

Follow these simple steps to get the benefits:

1. Scan the QR code or visit the link below

https://packt.link/free-ebook/9781789800180

2. Submit your proof of purchase
3. That's it! We'll send your free PDF and other benefits to your email directly